Chemical Process Simulations using Aspen Hysys

Chemical Process Simulations using Aspen Hysys

Khalid W. Hameed
University of Baghdad, Iraq

WILEY

Published by John Wiley & Sons, Inc., Hoboken, New Jersey. Published simultaneously in Canada.

Library of Congress Cataloging-in-Publication Data Applied for:

Hardback ISBN: 9781394293278

Cover Design: Wiley
Cover Image: © Rudchenko Liliia/Shutterstock

Set in 9.5/12.5pt STIXTwoText by Straive, Chennai, India

To my creator who taught humans things they did not know before;

To the last prophet Mohammad and to the preceding prophets, peace be upon all of them;

To my first lovely teachers in this life: mom and dad;

To my biochemical engineering students;

To AspenTech, Inc., who sponsored this work;

To the rest, I say: my sincere and warm greetings to all of you.

Contents

Preface

Aspen HYSYS is a software program used for the simulation and plant design of chemical processes. It is a market-leading process modeling tool for conceptual design, optimization, business planning, asset management, and performance monitoring. It is designed by the Canadian company Hyprotech, founded by researchers from the University of Calgary in 1996. In May 2002, AspenTech acquired and modified Hyprotech, including HYSYS. So, the HYSYS software was renamed by AspenTech to Aspen HYSYS. In this text, for brevity, the software is mostly named HYSYS.

Due to the accuracy of this software, it is widely used in industries such as oil and gas production, gas processing, petroleum refining, chemical processing, and air separation. The study shows the advantages of HYSYS and why it is preferred over other software. This book gives an introduction to the workings of HYSYS in various chemical fields. Aspen HYSYS is a core element of AspenTech's AspenONE Engineering applications. Aspen HYSYS has established itself as a very intuitive and easy-to-use process simulator in the oil and gas refining industry. Users with little prior knowledge of Aspen HYSYS can pick up and train themselves in its modeling capabilities. Some of the very intuitive capabilities include a highly interactive process flow diagram for building and navigating through large simulations. The software also provides a very flexible and easy-to-use distillation column modeling environment. Additionally, the interactive nature of HYSYS enables users to build and use their models quickly and effectively. Aspen HYSYS offers a comprehensive thermodynamics foundation for accurate calculation of physical properties, transport properties, and phase behavior for the oil and gas and refining industries. This software uses mathematical models to simulate the behavior of chemical and physical systems, allowing engineers to predict how a process will behave under different conditions and to test different process scenarios before implementing them in the real world. A comprehensive library of unit operation models, including distillation columns, reactors, heat transfer operations, rotating equipment, controllers, and logical operations in both steady state and dynamic environments, execute the economic evaluation.

A crucial tool for many design, control, and optimization activities is chemical process simulation. Chemical and process engineers are capable of producing intricate computations quickly, as well as assessing various options and operational scenarios. As a result, chemical process simulators are now a crucial teaching tool for chemical engineering, and an appreciation of them will open up new avenues for development in the conceptual and detailed engineering domains.

Simulation is a situation in which a particular set of conditions is created artificially in order to study or experience something that could really exist in reality. It is the act of pretending that something is real when it is not. A computer simulation is an attempt to model a real-life or hypothetical situation on a computer so that it can be studied to see how the system works.

In this textbook, the principle of calculation will be presented basically to learn and understand the basic calculations in chemical engineering as a review before using the software and if he/she likes in hand calculation and compares with the results obtained by software.

The advantages of the Aspen HYSYS software can be summarized as follows:

- Contains the physical and thermodynamic properties of hundreds of chemicals.
- Performs the material and energy balance as well as unit conversion.
- Performs the design calculations for the equipment.
- Perfect for analytical/numerical minds (how the process would behave under different conditions), e.g. Case steady research.
- Executes the dynamics and control process as well as steady state.
- Performs the economic evaluation.
- Obtains a report of the plant details.
- Eliminates routine and save of time, i.e. make easier/faster work as well as accuracy of results.
- The software can present a virtual lab
- It can be applied to the operating conditions by software first, then in the practical field to avoid errors and obtain the optimum conditions.
- Excellent for your curriculum vitae as an engineer.

There are two modes of simulation: steady-state mode and dynamic mode.

Steady-state mode: Initially, process simulation was used to simulate steady state processes. Steady-state models perform the mass and energy balance of a stationary process (a process in an equilibrium state), but any changes over time have to be ignored.

Dynamics mode: Dynamic simulations require increased calculation time and are mathematically more complex than a steady-state simulation. It can be seen as a repeatedly multiply steady-state simulation (based on a fixed time step) with constantly changing parameters.

In this text, the Aspen HYSYS software will be explained and understood simply by solving examples dealing with unit operation in the chemical engineering process.

2024
Iraq

Khalid W. Hameed

Acknowledgments

The author acknowledges with thanks, appreciation, and gratitude for the assistance provided by Aspen Technology Incorporation for the great efforts this incorporation has made in providing help, guides, tutorial manuals, videos, etc., so that the learner becomes a professional in the field of the company's software. The author thanks Natarajan Priyadarshini, Michael Leventhal, and Sabeen Aziz for their help in organizing and arranging this book and their quick response to emails.

1

Starting with Aspen HYSYS

This chapter begins by explaining how to start Aspen HYSYS and how to select the right components and fluid package for simulation purposes. Knowing how to start Aspen HYSYS and get familiar with its desktop is very important in this chapter. The second part is about how to enter and reenter the simulation environment, and get familiar with the simulation flowsheet. In this part, users will be informed of some important features of Aspen HYSYS. The last part deals with how to add and specify material streams for simulation. Variable specification is one of the important steps that users need to understand when dealing with Aspen HYSYS.

Upon completion of this chapter, the reader will be able to:

- Start the Aspen HYSYS
- Select the Components
- Determine and pick an appropriate fluid package
- Switch to the Simulation Environment
- Insert and define the material streams.

Mostly in this book and for making it concise, Aspen HYSYS will be called HYSYS.

Chemical Process Simulations using Aspen Hysys, First Edition. Khalid W. Hameed.

1.1 HYSYS Menu Command to Ribbon Map

1.1.1 HYSYS Basis Environment

Command Type	Arrive to the Command
New Case	File ǀ New ǀ Case
Open	File ǀ Open ǀ Open Case
Example Cases	Properties Environment ǀ Resources ǀ Examples
Save	File ǀ Save
Save as	File ǀ Save as
Close Case	File ǀ Close Case ǀ Close Case
Close All	File ǀ Close Case ǀ Close All Cases
Print	File ǀ Print
Print Windows Snapshot	File ǀ Print ǀ Print Snapshot
Print Setup	File ǀ Print Setup
Exit	File ǀ Exit

1.1.2 Simulation Menu

Command Type	Arrive to the Command
Input Summary	Home ǀ Summaries ǀ Input
Optimizer	Home ǀ Analysis ǀ Optimizer
Event Scheduler	Dynamics ǀ Modeling Options ǀ Event Scheduler
Integrator	Dynamics ǀ Run Integrator
Dynamics	Dynamics ǀ Dynamic Simulation ǀ Dynamics Assistant
Solver Active/Holding	Home ǀ Run ǀ Solver Active ǀon Hold
Enter Basis Environment	Navigation Pane ǀ Properties
Oil Output Settings	Properties Environment ǀ Oil Manager ǀ Oil Output Settings
Costing Options	Economics ǀ Prepare ǀ Cost Options

1.1.3 Flow Sheet Menu

Command Type	Arrive to the Command
Add Stream	Flowsheet/Modify I Flowsheet I Additional Command I Add Stream
Add Operation	Flowsheet/Modify I Flowsheet I Additional Command I Add Operation
Find Object	Flowsheet/Modify I Tool I Find Object
Flowsheet Summary	Home I Summaries I Flowsheet Summary
Notes Manager	View I Show Notes Manager
Palette	View I Model Palette or Flowsheet/Modify I Palette I Models and Streams
Extensions	A tab on the Palette
Reaction Package	Properties Environment I Home tab I Navigate I Reactions or Navigation Pane I Reactions
Fluid Package/Dynamics Model	Properties Environment I Home tab I Navigate I Fluid Package or Navigation Pane I Fluid Package
Dynamic Initialization	Dynamics I Modeling Options I Dynamic Initialization

1.1.4 PFD Menu

Command Type	Arrive to the Command
Select Objects	Navigation Pane I UnitOps
Show Hidden Objects	Flowsheet/Modify I Display Options Hide Object I Reveal
Break a Connection	Flowsheet/Modify I Flowsheet I Break Connection
Swap Connection	Flowsheet/Modify I Flowsheet I Swap Connection
Auto Position All	Flowsheet/Modify I Flowsheet I Additional Commands I Auto Position All
Auto Position Selected	Flowsheet/Modify I Flowsheet I Additional Commands I Auto Position Selected
Select Mode	Flowsheet tab (windows) shortcut menu I Mode
Drag Zoom	Flowsheet/Modify I Flowsheet I Drag

1.1.5 Tools Menu

Command Type	Arrive to the Command
Workbooks	Home I Summaries I Workbook
PFDs	View I Show Flowsheet
Report Manager	Home I Summaries I Reports
Excel Report	Home I Summaries I Reports
Text Report	Home I Summaries I Reports
Data-Fit	Navigation Pane I Data Fits or Home I Analysis I Data Fit
Face Plates	Dynamics I Tool I Face Plates
Dynamics Assistant	Dynamics I Dynamic Simulation I Dynamics Assistant
Control Manager	Dynamics Tools I Control Manager
Dynamic Profiling Tool	Dynamics I Tool I Profile
Snapshot Manager	Dynamics I Modeling Options I Snapshot Manager
Script Manager	File I Script Manager or Customize I Tool Script Manager
Case Security	Home I Case I Case Security
Preferences	File I Options

1.1.6 Edit Menu

Command Type	Arrive to the Command
Cut	Home I Clipboard I Cut
Copy	Home I Clipboard I Copy
Copy Special	Home I Clipboard I Copy
Paste	Home I Clipboard I Paste

1.1.7 Hot Keys

Hot Key Commands in Flowsheet Menu	
Access Notes Manager	**CTRL G**
Access Object Navigator	F3
Add Material Stream	F11
Add Operation	F12
Show/Hide Object Palette	F4
Show/Hide RefSYS Object Palette	F6

Hot Key Commands in Tool Menu	
Access Controller Face Plates	CTRL F
Access Databook	CTRL D
Access Dynamics Assistant	CTRL Y
Access PFDs	CTRL P
Access Reports	CTRL R
Access Utilities	CTRL U
Access Workbook	CTRL W

Hot Key Commands in Tool Menu	
Go to the Column Runner in the Column Sub-flowsheet	CTRL T
Reset the Column	SHIFT F6
Run the Column	F6
Stop Column Solver	CTRL BREAK

Hot Key Commands in Help Menu	
Access Help	F1

Hot Key Commands in for Property Views	
Access Editing Cell Function	F2
Access Stream Composition Property View (from Workbook Property View)	CTRL K
Close Active Property View	CTRL F4

Hot Key Commands in for Property Views	
Go to Next Item Within Active Property View	TAB
Go to Next Tab Within in the Active Property View	CTRL SHIFT N
Go to Next Property View	* CTRL F6 * CTRL TAB
Go to Previous Item Within Active Property View	SHIFT TAB
Go to Previous Tab in the Active Property View	CTRL SHIFT P
Undo	CTRL Z

General Hot Key Commands	
Access Pull-Down Commands from the Menu Bar	* F10 * ALT
Start/Stop Integrator	F9
Stop Calculations	CTRL BREAK

PFD Hot Key Commands	
Center PFD Objects with Cursor	* PERIOD key * C
Break Stream Connection	B
Default Orientation	N
Delete Selected Object	DELETE
Displays Inlet Nozzle Elevation	SHIFT I
Displays Stream Mass Flow Rates	SHIFT M
Displays Stream Molar Flow Rates	SHIFT F
Displays Stream Name (Default)	SHIFT N
Display Object Description	SHIFT R
Display Outlet Nozzle Elevation	SHIFT O
Display Stream Pressure	SHIFT P
Display Stream Temperature	SHIFT T
Mirror about X-axis	X
Mirror about Y-axis	Y
Open Selected Object	* V * E
Pan the Object	* ARROW keys * SHIFT ARROW keys

Hot Key Commands in for Property Views	
Rotate 90°	1
Rotate 180°	2
Rotate 270°	3

PFD Hot Key Commands	
Select Object Label	L
Toggle Between Move/Attach Options	CTRL
Toggle Between previous Zoom Option	Z
Undo	CTRL Z
Zoom All Object	Home
Zoom In Object	PAGE UP
Zoom Out Object	PAGE DOWN

1.2 Starting Aspen HYSYS

In Search, type "Aspen HYSYS V9" and click on its icon Aspen HYSYS V9. The HYSYS window will appear as in Figure 1.1.

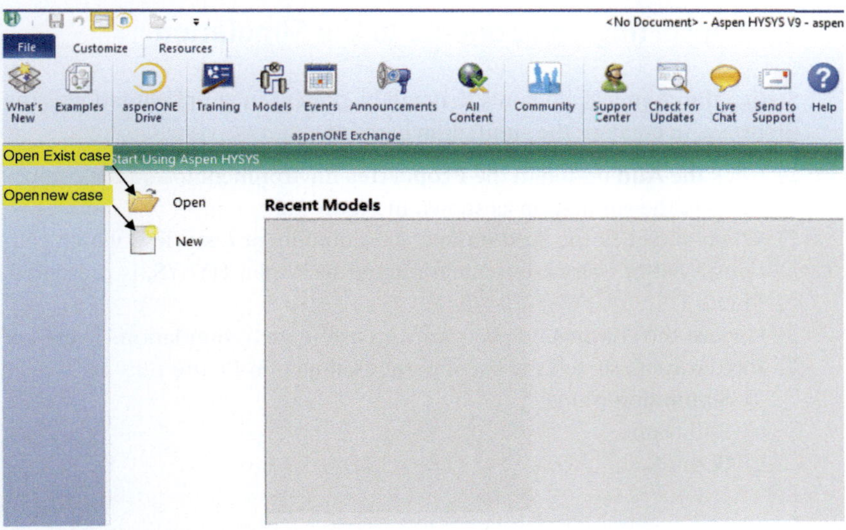

Figure 1.1 Window of HYSYS when opened first.

1.3 Properties Environment

The fluid package idea was utilized by HYSYS to organize all the data required to execute flash and physical property calculations. With this method, you may describe any piece of information (such as a property package, hypothetical components, interaction parameters, responses, and tabular data) inside a single entity [1, 2].

This strategy offers the following four main benefits:

- Because all related information is specified in one place, it is simple to create and modify the information.
- Fluid packages are fully described entities that may be saved and used in any simulation.
- Component lists are fully specified entities that may be saved apart from Fluid Packages and used in any simulation.
- More than one fluid package may be utilized in a single simulation. They are, nevertheless, described in the common basis manager.

1.4 Creation of a New Simulation

To open a new case, press **Ctrl + N**, or click **New** Case icon  New as in Figure 1.1. Your simulation is referred to as a "case" in HYSYS. The Simulation Basis Manager, where all of the components and their parameters may be defined, will then be opened as a result [3, 4].

1.5 Inserting Components to the Simulation

Setting the chemical elements that will be present in your simulation is the first stage in creating the simulation base [5].

1) Click the **Add** button in the **Properties** Environment to add the components to the simulation as shown in Figure 1.2.
2) When you click the **Add** button, the Component List View, which contains a list of every component offered by Aspen HYSYS, is presented, Figure 1.3.
3) Choose the components you want to use in your simulation. There are three ways to search the list of parts, as shown in Figure 1.3:
   a) Simulation Name
   b) Full Name
   c) Formula

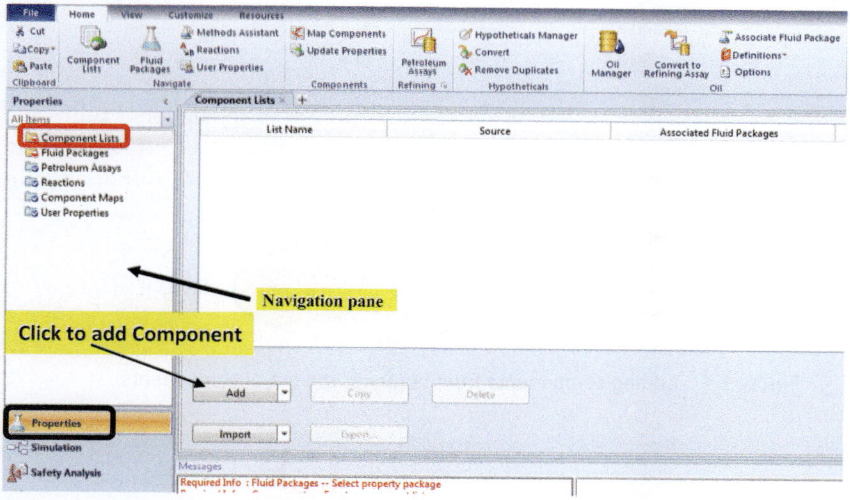

Figure 1.2 Window of New case in HYSYS, Properties environment.

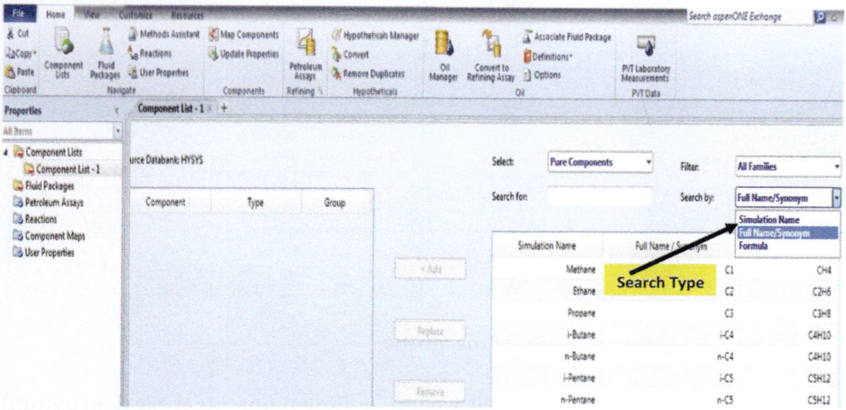

Figure 1.3 How you can select the components from library in HYSYS.

4) After you've found the appropriate component, either double-click on it or select it with the mouse and click the **Add** button to add it to the simulation's list of components.

For instance, type "Ethane" in the Search field, highlight it in the list of components, and then click the **Add** button, Figure 1.4.

Double-click on any component shown in the component table of Figure 1.4 to view its properties as shown in Figure 1.5; the window of ethane properties will appear.

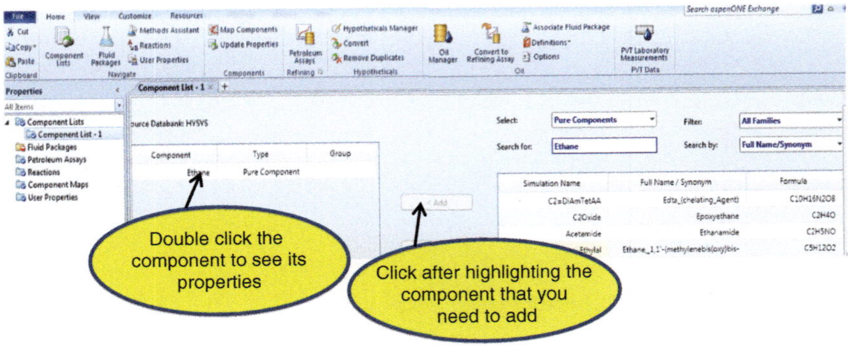

Figure 1.4 Adding components from library to the list of components.

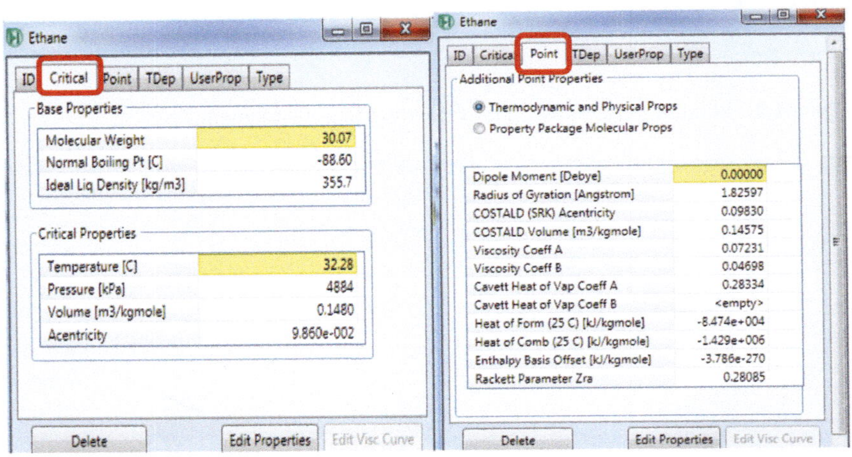

Figure 1.5 Some properties of the ethane.

Properties such as molecular weight, boiling point, ideal gas density, and critical temperature and pressure are found in the **Critical** tab; the heat of formation and of combustion are found in the **Point** tab, Figure 1.5. Antoine vapor pressure and Gibbs free energy are found in the **TDep** tab.

1.6 Adding the Fluid Package

One of the most important steps in process simulation is choosing the right property approach. The adequacy of the property approach utilized determines the accuracy and believability of the simulation results [6].

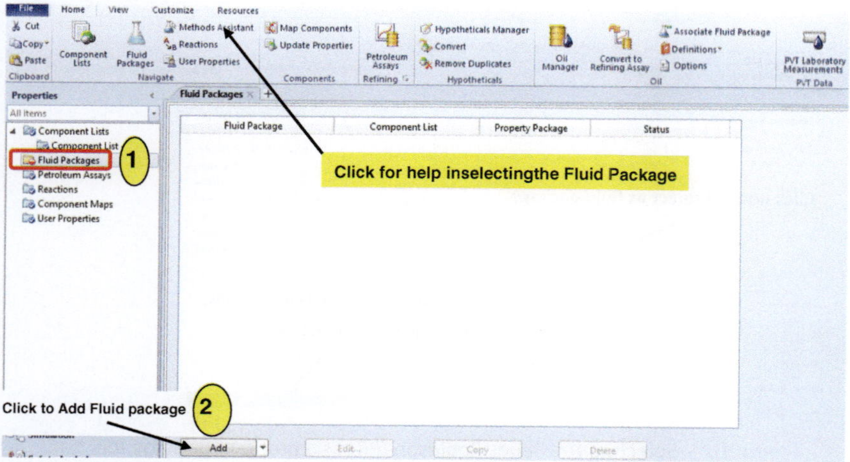

Figure 1.6 Adding property package.

The fluid package is used to calculate the fluid/thermodynamic properties of the components and mixtures in your simulation (such as enthalpy, entropy, density, and vapor–liquid equilibrium).

Click Fluid Packages under the Component List-1, OR click the icon

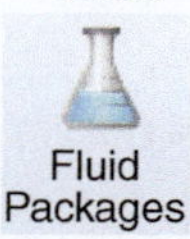

in the **Home** tab of the Ribbon to select proper fluid package, Figure 1.6.

1) Select the Fluid Package by clicking **Add** button as shown in Figure 1.6.
2) From the list of **Property Package Selection**, select the desired Fluid package.
3) Once the desired model has been located, select it by click once (*no need to double-click*). For example, select **Peng-Robinson** (**PR**) property package for your simulation, Figure 1.7.

1.7 Methods for Choosing Thermodynamics Model (Fluid Package)

When selecting a thermodynamic model, it might be good to at least follow a rational process to choose which model to test out first.

You may forecast the characteristics of mixtures ranging from clearly defined light hydrocarbon systems to complicated oil mixes and extremely

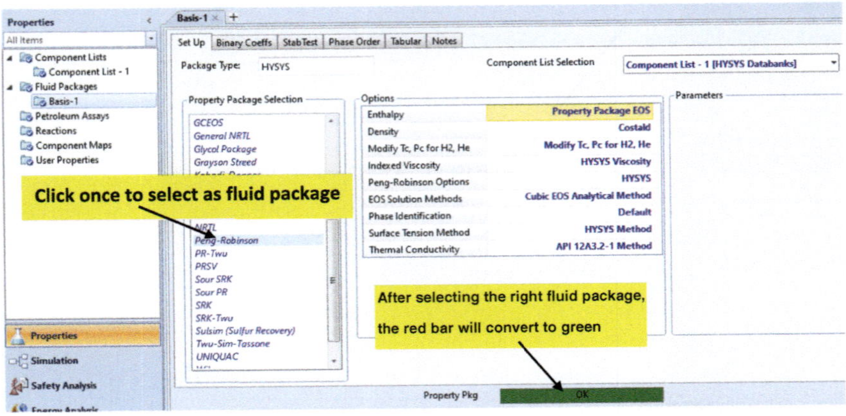

Figure 1.7 Selecting the Peng–Robinson (PR) as a property package for simulation.

Table 1.1 Typical systems and their recommendation property package.

Type of System	Recommended Property Method
TEG Dehydration	Peng–Robinson (PR)
Sour Water	PR, Sour PR
Cryogenic Gas Processing	PR, Peng–Robinson Stryjek Vera (PRSV)
Air Separation	PR, PRSV
Atm. Crude Towers	PR, PRSV Options, Grayson Streed (GS)
Vacuum Towers	PR, PRSV Options, GS (<10 mmHg), Braun K10, Esso Tabular
Ethylene Towers	Lee Kesler Plocker
High H_2 System	PR, Zudkevitch–Joffee (ZJ)
Reservoir Systems	Steam Package, Chao Seader (CS) or GS
Hydrate Inhibition	PR
Chemical Systems	Activity Models (e.g. NRTL, UNIQAUC, and Wilson), PRSV
HF Alkylation	PRSV, NRTL (Non-Random Two Liquid)
TEG Dehydration with Aromatics	PR
Hydrocarbon Systems Where H_2O Solubility in HC is important	Kabadi Danner
System with select gases and light HC	MBWR

PR = Peng–Robinson; PRSV = Peng–Robinson Stryjek–Vera; GS = Grayson–Streed; ZJ = Zudkevitch Joffee; CS = Chao–Seader; NRTL = Non-Random-Two-Liquid.

non-ideal (non-electrolyte) chemical systems using the property packages offered by HYSYS. For the more complex hydrocarbon systems, HYSYS offers improved equations of state Peng–Robinson and Stryjek–Vera Modification of the Peng–Robinson (PR and PRSV), semi-empirical and vapor pressure models, steam correlations for precise steam property predictions, and activity coefficient models for chemical systems. You are urged to learn more about how to use each equation because each one has its own unique inherent limits.

The Peng–Robinson (PR) property package is often suggested for use in oil, gas, and petrochemical industries. Please refer to the Aspen HYSYS Simulation Basis Manual for further information [5–7].

Table 1.1 lists some typical systems and their recommended correlations.

If you *can't select* the proper fluid package, **Methods Assistant** Methods Assistant in the **Home** tab of the Ribbon can be used; click next, then select **Specify component type** or select **Specify process type**, Figure 1.8. If you choose Specify component type, according to our example, select **Hydrocarbon system**, Figure 1.9. From the message that appears, click **No**; the software will tell you to Use Peng–Robinson,

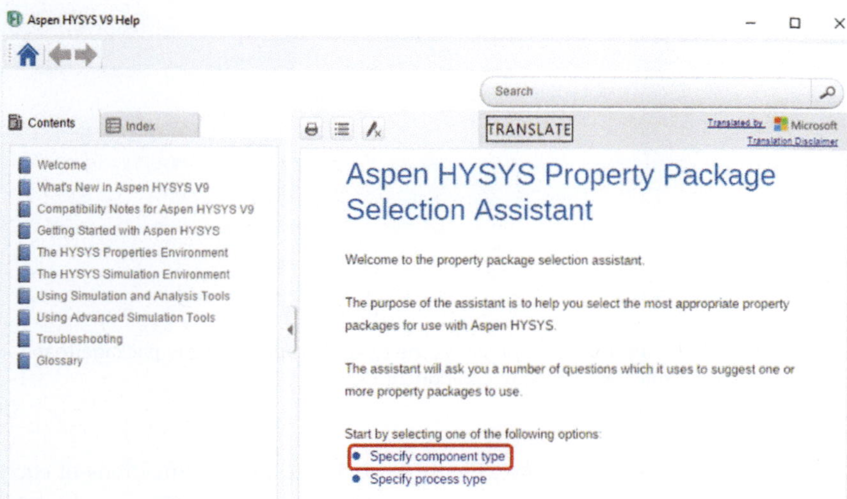

Figure 1.8 Using method assistant to select the proper property package according to component type.

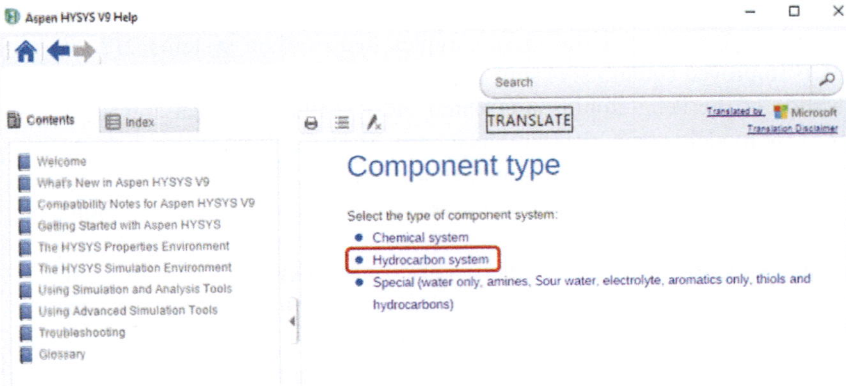

Figure 1.9 Choose the hydrocarbon system as the component type.

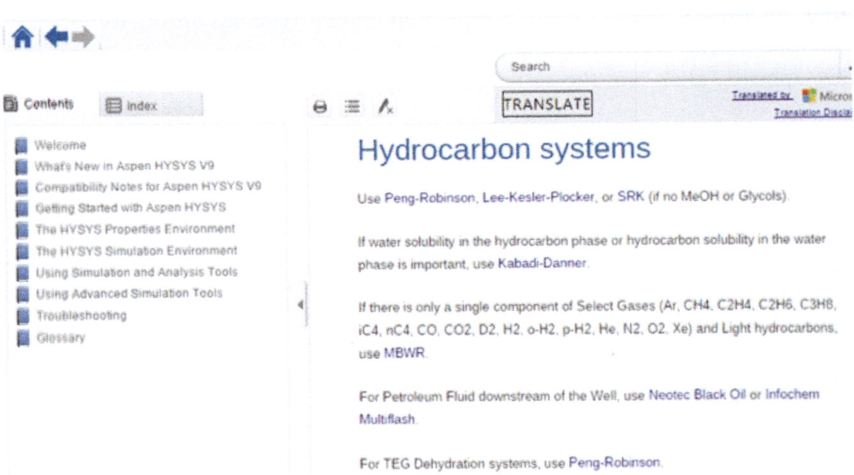

Figure 1.10 Method assistant shows the type of proper property package that you have to select according to component type.

Lee–Kesler–Plocker or SRK, Figure 1.10. To see the specifications of each model, you can click or choose this method, e.g. choose Peng-Robinson, Figure 1.11.

It is difficult to use the first method during the practical application of the software unless the table (1.1) is at hand, but the second method is possible and is easier by relying on the software itself.

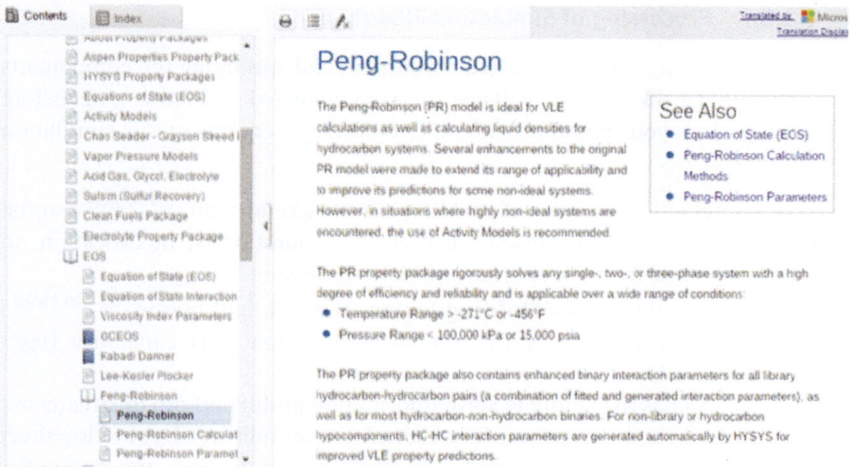

Figure 1.11 Specification of Peng–Robinson model.

1.8 Simulation Environment

After you've finished all the required input in the properties environment, you can start your simulation. Click **Simulation** button in the left bottom 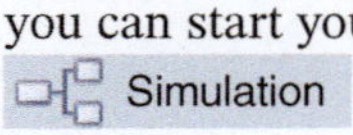

of Figure 1.12.

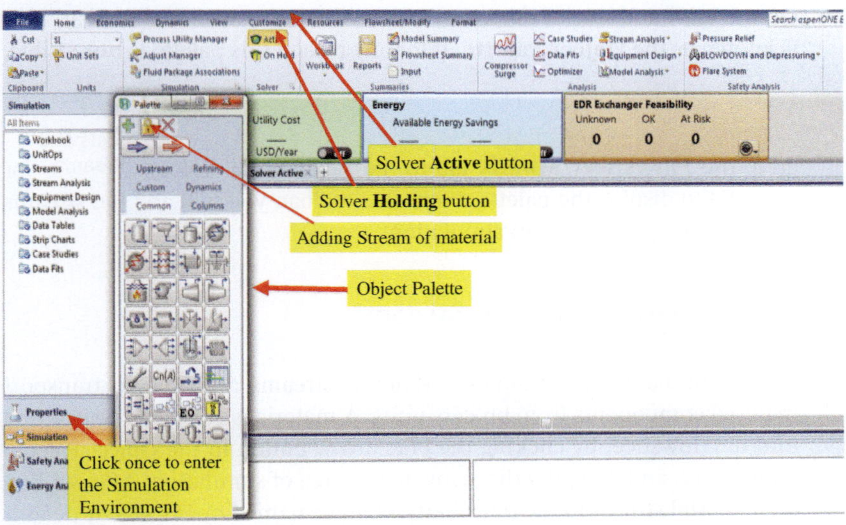

Figure 1.12 Simulation environment of the Aspen HYSYS.

1.8.1 Processing of Simulation Flowsheet

After entering the simulation environment and specifying the components and fluid package, you will see the view depicted in Figure 1.12. Before moving on, you should attend to the following elements of this simulation window:

1) As default, the solver of HYSYS is in the **green** light 

, so the HYSYS solves the flowsheet after every addition or modification, in contrast to the bulk of other simulation software.
   Clicking the **Solver Holding** button (the **red** light button 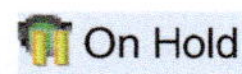

) in the Home tab of the ribbon will disable this capability (see Figure 1.12).
2) HYSYS has the ability to solve information problems both upstream and downstream. Therefore, it is crucial to carefully read your flowsheet specification to make sure you aren't giving HYSYS any contradictory information. If you don't, you'll get an error and the simulation won't work.
3) When the basis of the simulation has to be changed, the **Properties** Environment needs to be re-entered. Simply click 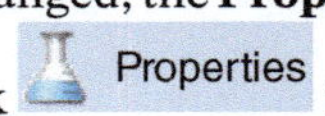

icon located in the left bottom.

1.8.2 Object Palette

On the left-hand side of Figure 1.12, you will notice a vertical toolbar. This is known as the **Object Palette**. If for any reason this palette is **not** visible, go to the **Flowsheet/Modify** tab in the Ribbon and select 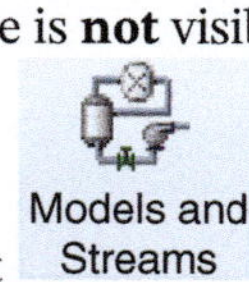

or press **F4** to display the palette. From this toolbar, you will add the streams and unit operations to your simulation.

1.9 Creating Material Streams

In the simulation environment, Material Streams are used to transport material components from process units. A material stream can be inserted into the flowsheet by clicking the blue arrow button in the Object Palette, Figure 1.12, and dragging the arrow to the area of simulation. After adding the material stream to the flowchart, it will be named by HYSYS by default in ascending numerical order, with "1" being assigned to the first stream that is produced. This name is subject to change at any time, Figure 1.13 [5].

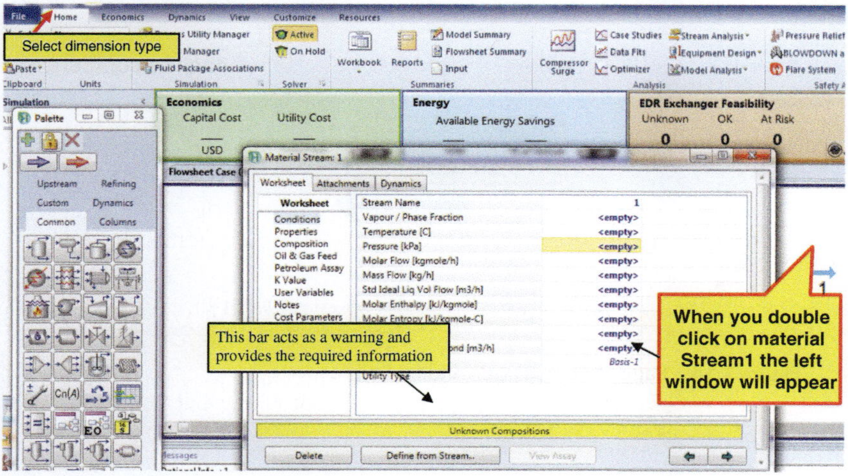

Figure 1.13 Windows of material streams.

1.9.1 Conditions of Material Streams

To enter the conditions of the stream's material, double-click the **arrow** button and enter 4 necessary variables. These variables are:

Composition, flowrate, and two of each of temperature, pressure, or vapor/phase fraction are the four necessary input stream variables.

The yellow message bar at the bottom of the window acts as a warning and provides the required information (unknown compositions), as seen in Figure 1.13. Just do as the letter asks; compositions, for instance, are the first thing you need to add. To adjust the composition of the stream, select the "Composition" option from the list under Worksheet to open the window depicted in Figure 1.14. Within this box, the user specifies the composition of the stream. Keep in mind that only the components you defined in the Properties Environment will be listed here. You can make several changes to the composition by choosing the "Edit" option. The user has the option to select each component's mass fractions, liquid volume fractions, or fluxes in addition to the HYSYS default of mole fractions. If the user specifies fractions, they must all add up to one. To indicate 1 mol fraction of ethane, enter a mole fraction of 1 in the Ethane section (Figure 1.14), then click OK.

The yellow caution message bar indicates that an input temperature must be provided for this stream. The window in Figure 1.15 will open once you choose the "Conditions" option from the list under Worksheet to determine the stream's temperature. The user can change the temperature

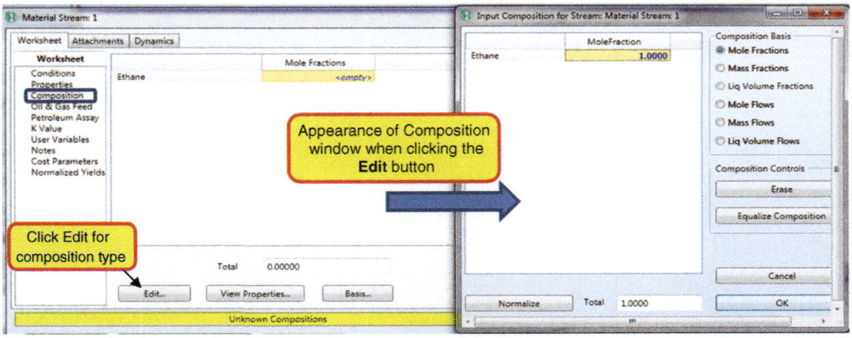

Figure 1.14 Page of Composition and its Edit.

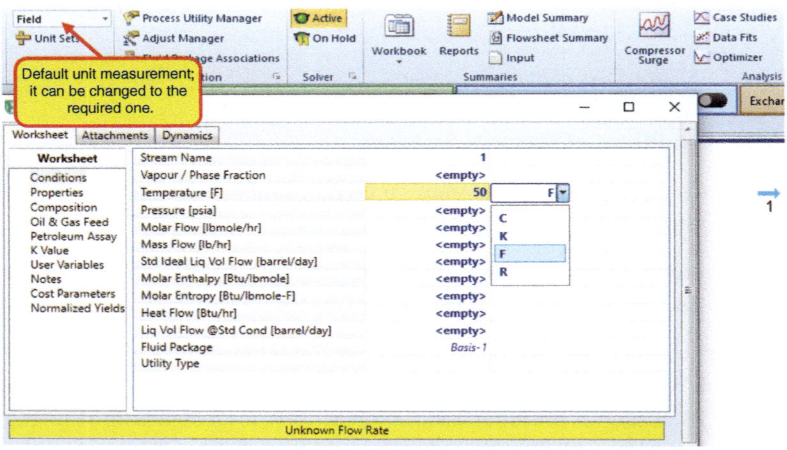

Figure 1.15 How you can enter the temperature in the stream of material and its dimensions.

of the stream within this box. It is not required to supply values for a stream's criteria in the specified default units. A drop-down arrow displays in the unit's box next to the cell when the user starts typing a value in one of the cells. The user can select any dimension for the required value by selecting this drop-down arrow, and HYSYS will automatically convert the number to the set's default unit. Enter 50 and select "F" from drop list as the temperature. The yellow caution message bar that appears next indicates that you need to set the input pressure for this stream. Enter 1000 in the pressure field and choose kPa from the drop-down list (Figure 1.16); the software will convert it to 145.0 psia.

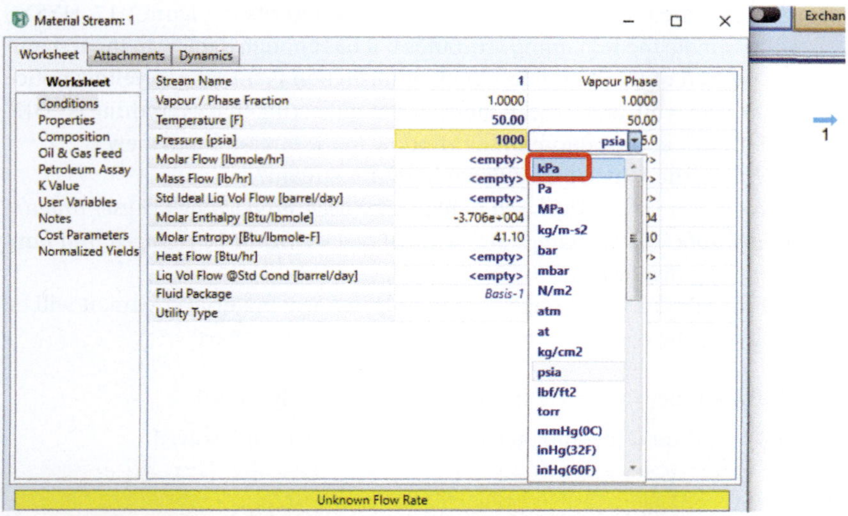

Figure 1.16 Entering the pressure in the stream of material and select kPa.

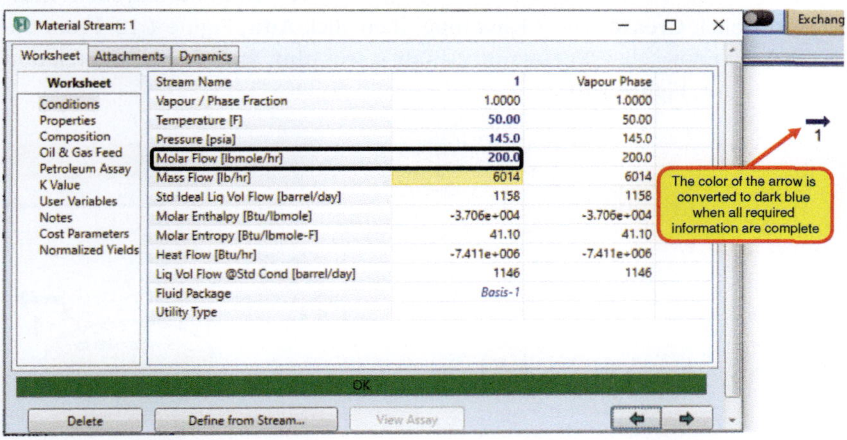

Figure 1.17 After entering the molar flow rate of the stream (all required information is complete), the color of the arrow is converted to dark blue.

The final variable that needs to be supplied is the flowrate. You have the option of providing a molar or mass flowrate. To represent the flowrate as shown in Figure 1.17, enter 200 lbmol/h in the molar flowrate part of the identical window of Figure 1.16.

After a stream has enough data to be fully described, Figure 1.17, HYSYS will estimate the remaining attributes if it has enough data from the remaining portion of the flowsheet. If any information is missing, a yellow notice bar will tell you, and a green message bar indicating that everything is "OK" will display at the bottom of the window in the stream input view.

What is the Vapor/Phase Fraction of this stream?

Note: *The values displayed in blue have been supplied by the user and are* ***modifiable****, but the values displayed in black* **cannot** *be changed and have been computed by HYSYS.*

If the HYSYS has enough data to fully define a material stream, it will be shown by the color code for that stream on the flowchart:

1 Dark Blue = properly specified and completely solved

1 Light Blue = incompletely specified, properties not solved.

1.9.2 Phase Diagram

To get the **phase diagram** and find the relationships between **P-T**, **P-V**, **V-S**, etc., double-click the material stream, press **attachment** button, click **Analysis**, click **Create**, select **Envelope**, then click **Add,** Figure 1.18. From the new window, click **Performance** tab, select **plot**, and choose any relation you want, Figure 1.19.

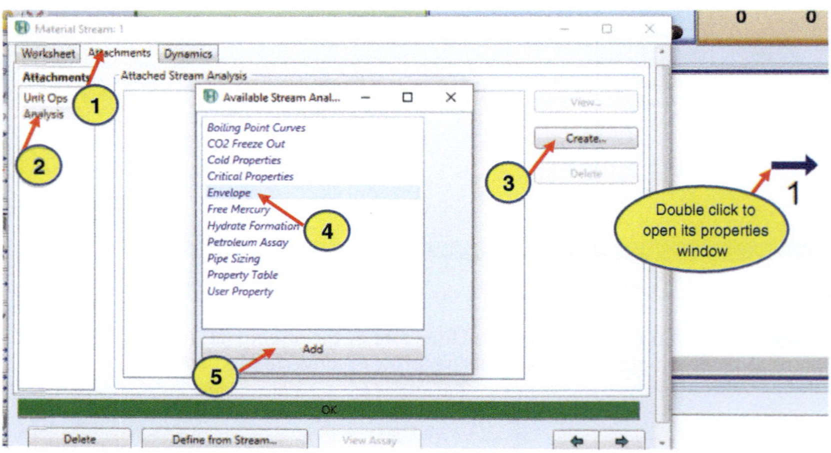

Figure 1.18 Arriving at the Envelope of the material stream.

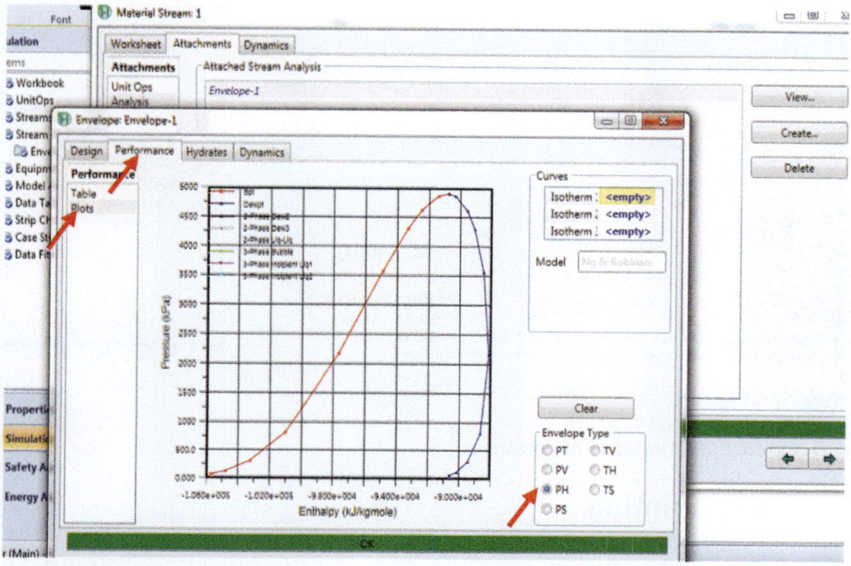

Figure 1.19 Plotting the pressure–enthalpy (PH) of the ethane material stream.

1.10 Summary

The first section of this chapter started with instructions on how to open Aspen HYSYS and get acquainted with its desktop environment. We also argued about the process of selecting the components connected to the simulation. It is important to select the right fluid/thermodynamic package; hence, we have included a flowchart to assist users in selecting the suitable thermodynamics models.

The second section discussed familiarizing yourself with the simulation flowsheet and how to enter and exit the simulation environment. In this section, users are also informed about various important HYSYS features.

The third section of this chapter deals with how to add and specify material streams for simulation as well as how to get the **phase diagram** and the relationships between **P-T**, **P-V, V-S**, etc. One of the crucial procedures that users of Aspen HYSYS must comprehend is the defining of variables. Users must give at least four variables for HYSYS to determine the remaining attributes when they want to define streams, particularly materials.

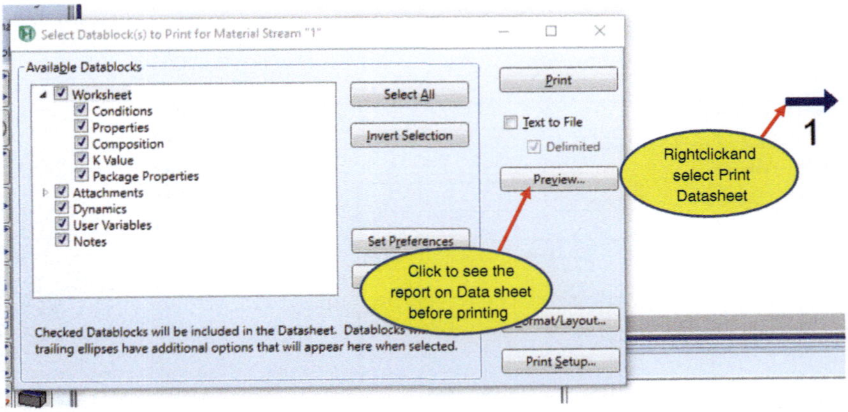

Figure 1.20 Window of datasheet.

1.11 Printing

(1) The process flow diagram can be printed in the PDF file by right clicking on any space and select Print PFD.

(2) The report of the process flow diagram with all conditions and properties, etc. can be got by selecting all unit operations in the PFD, then right clicking on any unit operations and selecting Print Datasheet..., then choosing **preview** to view the report before printing, Figure 1.20.

Problems

P.1.1 Create one materials stream of methane using the following conditions:
Fluid Package: Peng–Robinson, Temperature: 35°F, Mass Flowrate: 100 lb/h, Pressure: 90 psia.
Determine the vapor fraction of this stream. Replace the temperature by vapor fraction of 0.8. Now, how does one calculate the temperature?

P.1.2 Repeat the problem (P.1.1) by deleting the pressure, enter a temperature of 160°F, and specify the vapor fraction equal to 1.
What is the value of pressure of this stream?

P.1.3 By the same condition in (P.1.2), change the temperature to 70°F. Now, what is the value of enthalpy of this stream?

P.1.4 Construct a material stream of ethanol that has the following conditions:
Fluid Package: NRTL, Mass Flowrate: 300 kg/h, Pressure: 3 bar, Temperature: 100°C
The vapor fraction will be .....

P.1.5 By the same condition in (P.1.4), change the pressure to 2 bar. The vapor fraction will be ......

P.1.6 For the same condition in (P.1.4), replace the temperature with vapor fraction equal to 0.5.
The value of new temperature will be ....

References

1 (2017). *Aspen HYSYS User Guide Version 10.0*. Aspen Technology.
2 Tremblay, D. and Mantrala, V. (2014). *Jump Start: Aspen Simulation Workbook in Aspen HYSYS® V8*. Aspen Technology, Inc.
3 (2012). *Getting Started in Aspen HYSYS V8.0*. Aspen Technology, Inc.
4 (2002). *HYSYS 3.2 Operation Guide, © Hyprotech, a subsidiary of* Aspen Technology, Inc.
5 Mohd. Kamaruddin Abd Hamid, *HYSYS®: An Introduction to Chemical Engineering Simulation For UTM Degree++ Program*, 2007, University Teknology Malaysia.
6 Ghasem, N. (2022). *Computer Methods in Chemical Engineering*. Taylor and Francis Group, 2e published by CRC Press ©.
7 Smith, J.M., Van Ness, H.C., and Abbott, M.M. (2001). *Introduction to Chemical Engineering, Thermodynamics*, 6e. McGraw-Hill Companies, Inc.

2
Physical and Thermodynamics Properties

Most of the chemical calculations (fluid flow, heat transfer, mass transfer, reaction design, etc.) are needed for physical and thermodynamic properties. They can be obtained from the properties from the references or from networks, but the available properties are found in the standard conditions (e.g. 20°C and 1 atm), and perhaps more than one reference is needed to obtain the required properties. In our calculations, we must convert the properties to the conditions of our case, and if there is more than one component (a mixture of components), the problem will be more difficult and time consuming. In the HYSYS, most physical and thermodynamic properties are available for many components and can be estimated for any conditions required and for any mixture of components.

2.1 Fluid Package

In this section, effect of type of fluid package on the estimated properties of a component will be seen. The fluid package that offers a sufficiently accurate depiction of the system must be chosen by the user. It takes expertise and a solid understanding of the system thermodynamics to choose an appropriate fluid package [1, 2]. To choose the most accurate model, it is frequently necessary to compare several models to the measured data. The softwares for process simulation allow for model parameter adjustment. In certain instances, modifying certain model parameters could be imperative to attain a more accurate depiction of the observed data [3, 4].

Chemical Process Simulations using Aspen Hysys, First Edition. Khalid W. Hameed.

2.1.1 Problem Statement

In this example, you will select two types of fluid package simultaneously. Find the mass density of *n*-pentane at 550 K and 16 atm using the following equations of state:

- Soave–Redlich–Kwong (SRK)
- Peng–Robinson (PR)

The procedure of adding a component and selecting the fluid package are the same as mentioned in Chapter 1.

2.1.2 Simulation

1) Open a new case and enter the following values:
   Components: *n*-pentane
   After selecting (SRK) as a fluid package, the software will name it (Basis-1). To add another type of fluid package, click the **Fluid Package** folder, click the **Add** button, and select (Peng–Robinson), the software will name it to (Basis-2), Figure 2.1. Rename the Basis-1 to SRK by right clicking the (Basis-1), then choose **Rename.** By the same way, rename the Basis-2 to PR.
   Before switching to the simulation environment, change the Units of Measure by clicking **File** and select **Options**. Under the **Available Units sets,** select **SI**, then click **Copy** to create special units with new name (NewUser), Figure 2.2.
   Change the dimension of Pressure to (atm) and Temperature to (K), click OK.
2) Switch to the Simulation environment.

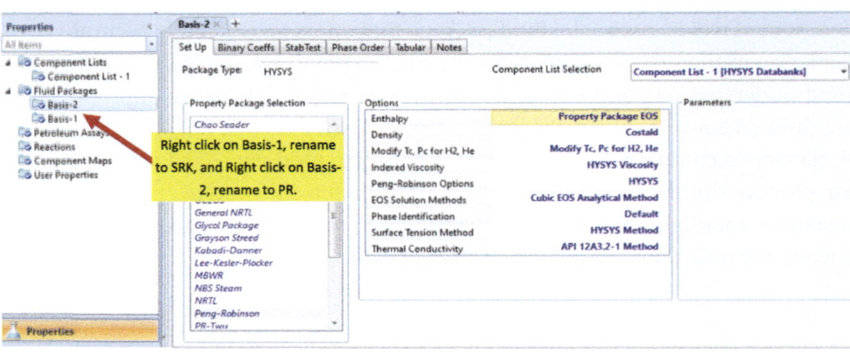

Figure 2.1 Adding two types of fluid package, Basis-1 and Basis-2, then rename them to SRK and PR.

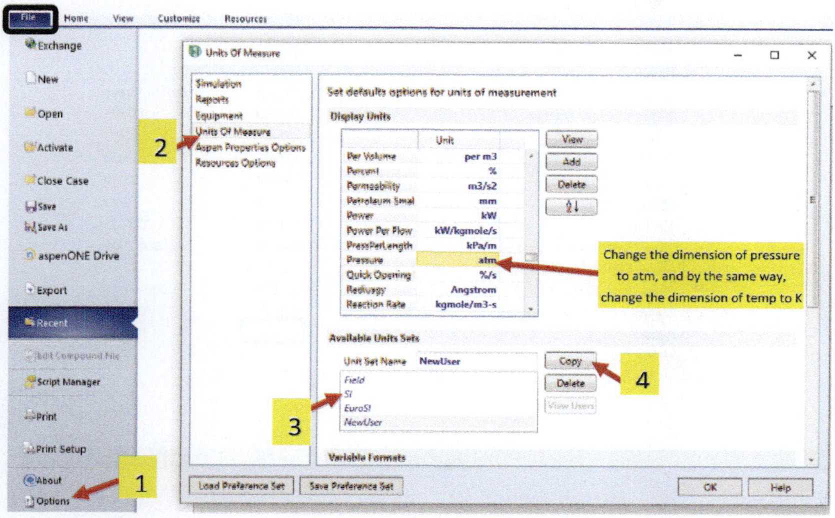

Figure 2.2 Changing the Units of Measure to a special one in the Properties environment, File | Options.

3) In the Object Palette, double-click the Stream icon. The software automatically adds a material and names it (1).
4) Enter the conditions of the stream-1 as in Figure 2.3:
   As default, the software will select the Basis-1 (SRK), since it is the first fluid package.
5) Add another material stream, the software will name it (2). Enter the same conditions of stream-1, but for the fluid package, select (PR), Figure 2.4.
6) Show the short results of two streams by right clicking each, and selecting Show Table, Figure 2.5.
7) Double-click the table of stream-1, PFD Table-1 pops up, click Add Variable, from Variable list, select Mass Density, click OK, Figure 2.6, then close the PFD Table-1. Repeat the same procedure for table-2. The show results of streams 1 and 2 will look like as in Figure 2.7.

It is clear that the mass densities of stream-1 and stream-2 are different. This means that the fluid package must be chosen carefully and cautiously.

2.1.3 Analysis of the Property Using Case Study Tool

This section will analyze the mass density of *n*-pentane as a function of pressure and temperature changes. To achieve this analysis, do the following [2, 5, 6]:

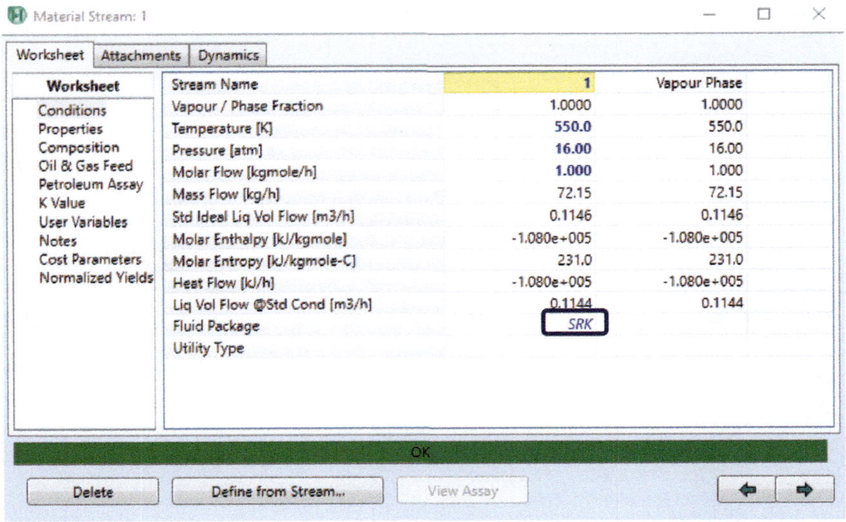

Figure 2.3 On entering the conditions of the stream-1, the software automatically selects SRK as a fluid package since it is the first fluid package.

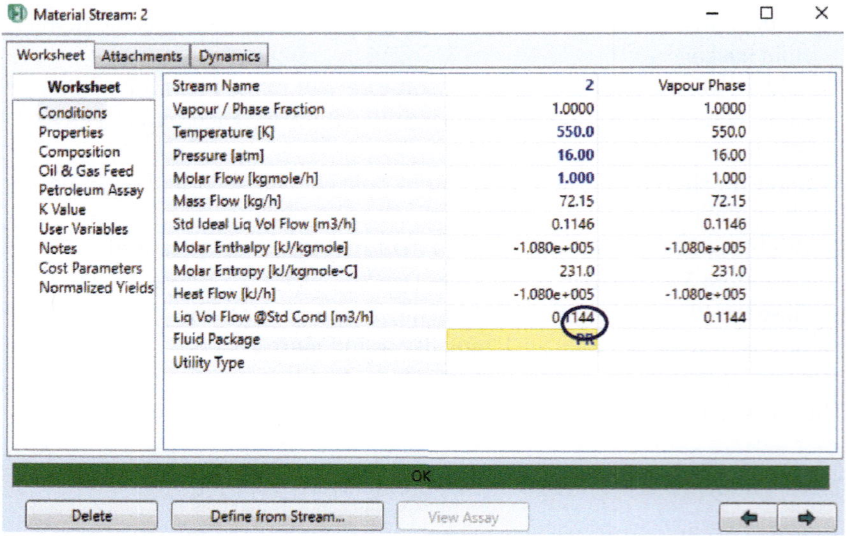

Figure 2.4 On entering the conditions of the stream-2, same as conditions of stream-1 but select (PR) as a fluid package.

1		
Temperature	550.0	K
Pressure	16.00	atm
Molar Flow	1.000	kgmole/h

2		
Temperature	550.0	K
Pressure	16.00	atm
Molar Flow	1.000	kgmole/h

Figure 2.5 Streams 1 and 2 with their results.

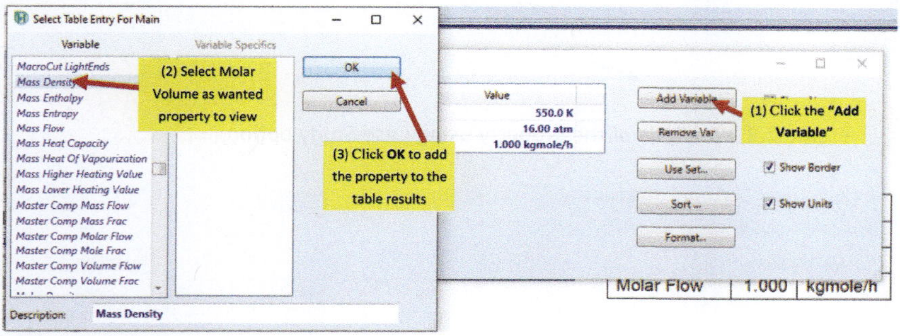

Figure 2.6 How to add any property to the Results of table.

1		
Temperature	550.0	K
Pressure	16.00	atm
Molar Flow	1.000	kgmole/h
Mass Density	28.31	kg/m3

2		
Temperature	550.0	K
Pressure	16.00	atm
Molar Flow	1.000	kgmole/h
Mass Density	28.77	kg/m3

Figure 2.7 : Compare the mass densities of stream 1 with property package SRK and stream 2 with property package PR.

1) In the Navigation pane, click Case Studies or in the **Home** tab of the Ribbon click the Case Studies icon as shown in Figure 2.8.
2) Next, click the **Add** button, then **Set up** view displays as shown in Figure 2.9.
3) In the **Independent Variables** group click **Find Variables**; the **Variable Navigator** window will show. In the **Object** column select stream 1,

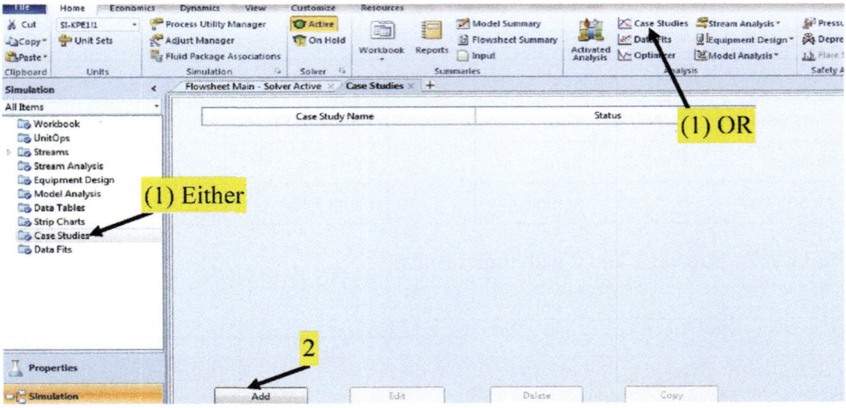

Figure 2.8 Analysis of the property using Case Study option.

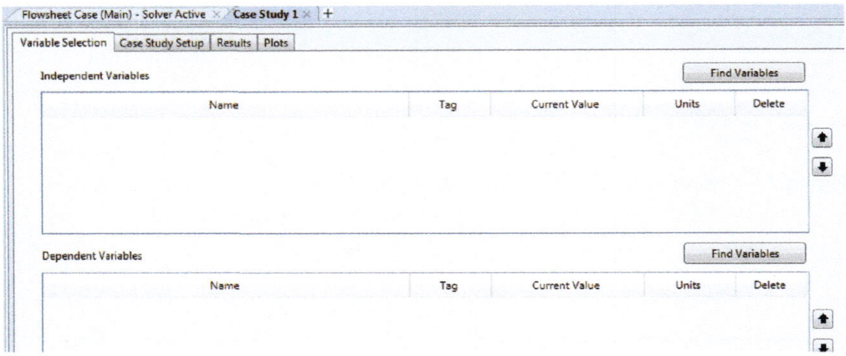

Figure 2.9 Window of case study when first open.

and in the **Variable** column, select Pressure, then click the arrow ➡, Figure 2.10. By the same way, select Temperature as another independent variable, click the arrow, then click the **Done** button, Figure 2.10.

4) In the **dependent Variables** group, click **Find Variables**; the **Variable Navigator** window will pop up. In the **Object** column select stream 1, and in the **Variable** column, select **Mass Density**, click the arrow, then click the **Done** button. The final window will appear as in Figure 2.11.
5) Click **Case Study Setup**; Give the Start, End, and Step Size of the Pressure (atm) and Temperature (K) as shown in Figure 2.12. The number of points on the graph will be estimated (#Steps).
6) Click the **Run** button to Run the analysis. Once the analysis finished, click **Results** to view the results as shown in Figure 2.13.

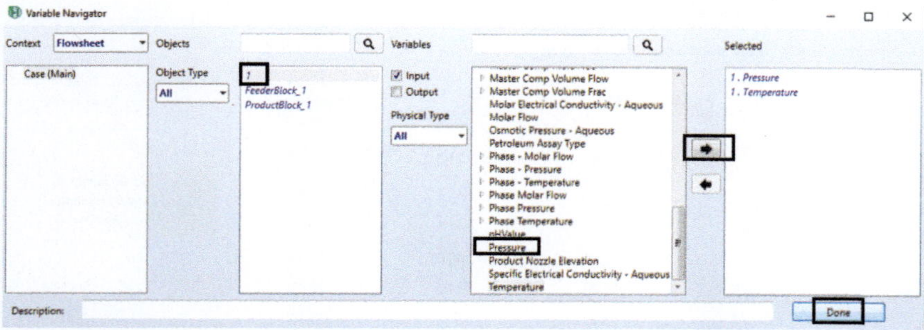

Figure 2.10 Adding the independent variables to the case study.

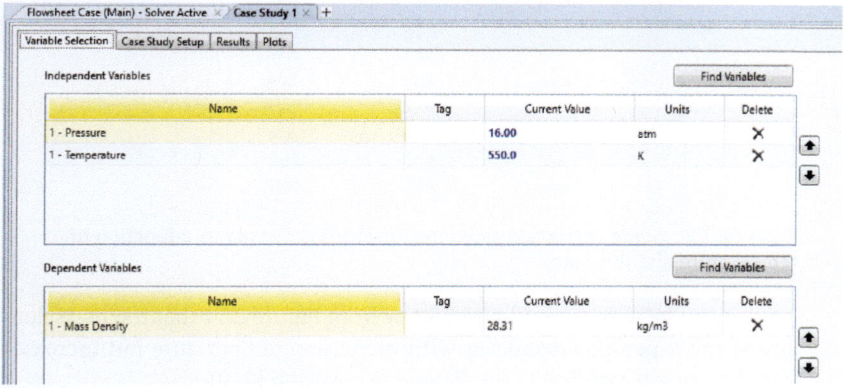

Figure 2.11 Final window of case study after adding the dependent and independent variables.

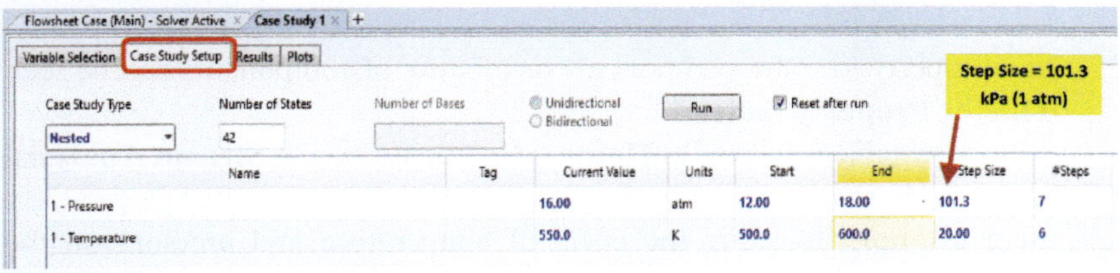

Figure 2.12 How you can enter the range and steps of independent variable.

7) It can be send the results to excel and draw the mass density as a function of two parameters (temperature and pressure) as shown in Figure 2.14.

What can you conclude from this graph?

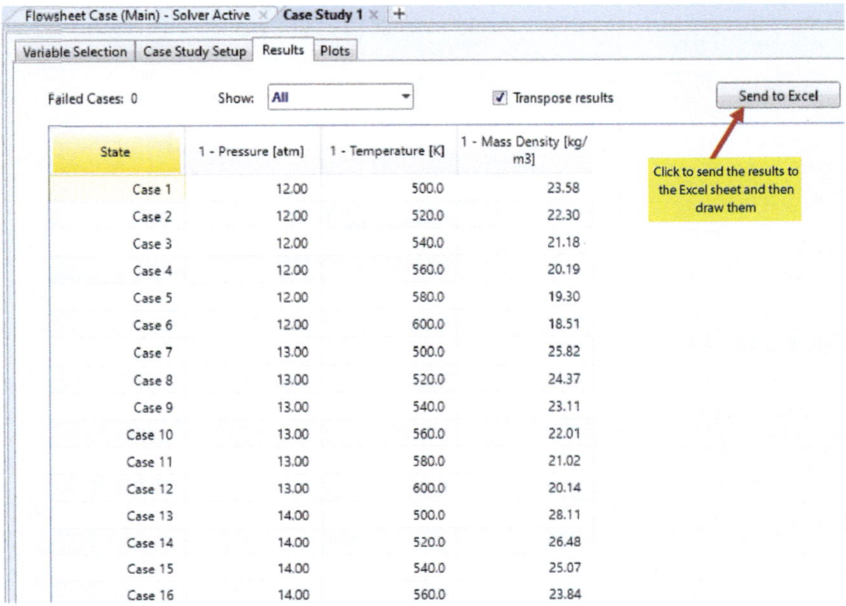

State	1 - Pressure [atm]	1 - Temperature [K]	1 - Mass Density [kg/m3]
Case 1	12.00	500.0	23.58
Case 2	12.00	520.0	22.30
Case 3	12.00	540.0	21.18
Case 4	12.00	560.0	20.19
Case 5	12.00	580.0	19.30
Case 6	12.00	600.0	18.51
Case 7	13.00	500.0	25.82
Case 8	13.00	520.0	24.37
Case 9	13.00	540.0	23.11
Case 10	13.00	560.0	22.01
Case 11	13.00	580.0	21.02
Case 12	13.00	600.0	20.14
Case 13	14.00	500.0	28.11
Case 14	14.00	520.0	26.48
Case 15	14.00	540.0	25.07
Case 16	14.00	560.0	23.84

Figure 2.13 Some of the tabulated results of mass density as a function of temperature and pressure.

So, it can be seen from Figure 2.13 or from Figure 2.14 that the mass density of the *n*-pentane decreases with increasing temperature but increases with pressure according to the following equation [7–9]:

$$\text{Mass density} = \frac{M}{V} = \frac{P \cdot Mw}{RT} \tag{2.1}$$

2.1.4 Property Table

Any property for pure components or mixture of components can be seen from the **Property table.**

Property table is located at **Home** tab of the Ribbon → **Stream Analysis**, Figure 2.15.

For example, to study the effect of temperature and pressure on the mass density of the *n*-pentane, click **Property Table** as mentioned above; Figure 2.16 will appear.

In the **Object** Column, select stream 1, and click **OK**, then click **Property Table-1** from Navigation Pane as in Figure 2.17.

Figure 2.18 will appear; set the Independent variables (Temperature and Pressure) as in Figure 2.18, where the range of temperature is (450–600 K) and the range of pressure is (12–18 atm).

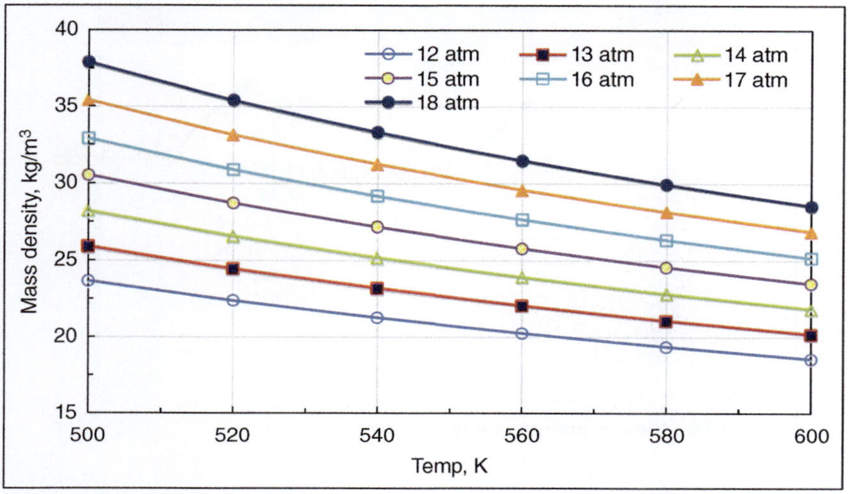

Figure 2.14 Plotting the mass density versus temperature and pressure using Excel.

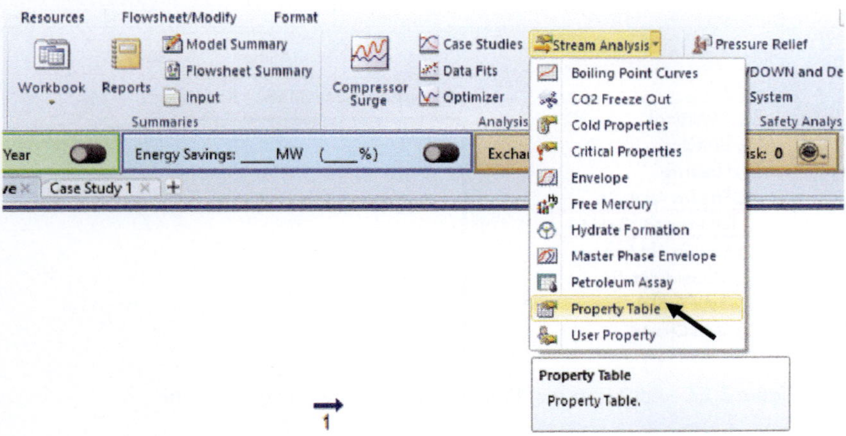

1		
Temperature	550.0	K
Pressure	16.00	atm
Molar Flow	1.000	kgmole/h
Mass Density	28.31	kg/m3

Figure 2.15 Arriving at the property table.

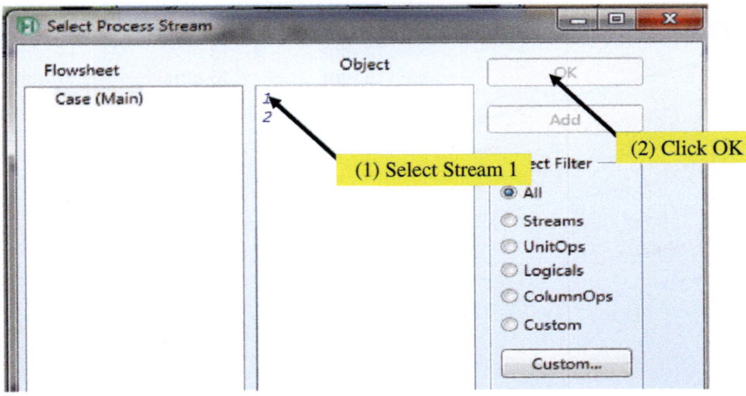

Figure 2.16 View of property table.

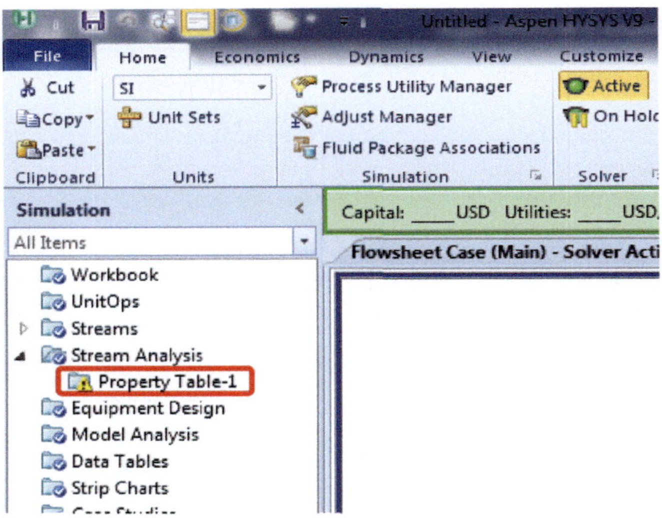

Figure 2.17 Appearance of Property Table in the Navigation Pane.

Then select the **Dep. Prop** (dependent properties).

From the new window, Click **Add.**

From the **Variable** group, Select Mass Density and click **OK,** Figure 2.19.

Click the **Calculate** button, then click the **Performance** tab. To see the results as graph, click **Plot,** then click View Plot button, Figure 2.20 will

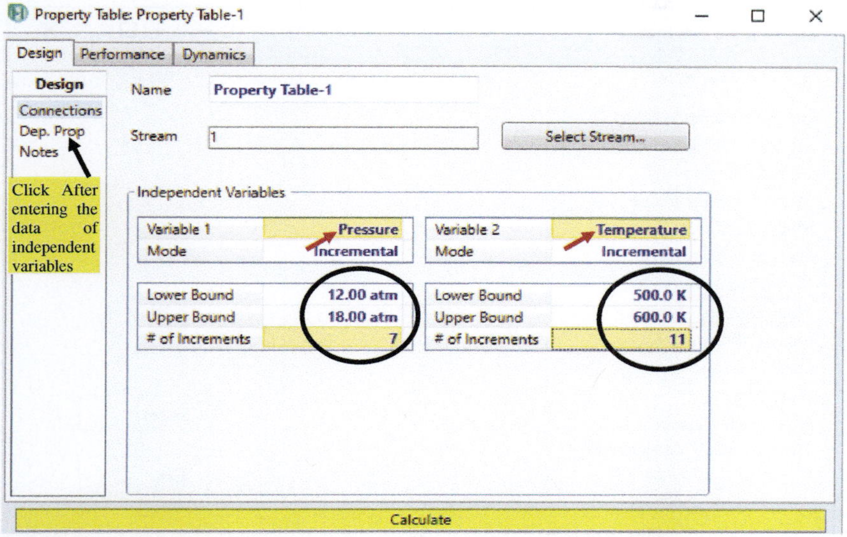

Figure 2.18 Specifying the range of independent variables in the property table.

appear. So according to Eq. (2.1), the mass density increases with increasing pressure and decreases with increasing temperature.

Figure 2.20 is similar to Figure 2.14.

2.1.5 Summary

Using Aspen HYSYS, you have a fairly straightforward issue to find the precise mass density of a pure component. When using Aspen HYSYS, the computations are pre-programmed and the parameters are saved in a database. In this section, you can use Show Table to present the results of streams/Unit Operation. The Show Table is set up to display information about any object type in addition to stream and general unit operation information (streams, pipes, controllers, separators, etc.). The results of two alternative equations of state, Soave–Redlich–Kwong (SRK) and Peng–Robinson (PR), can also be compared.

You should also use case studies to examine the process feature. The Case Study is designed to track how important process variables respond to one or more parameters in your steady-state process. You may see the outcomes on a table or plot once the Case Study has been solved. Also the property table can be used to study the effect of any parameter on any variable within the same stream.

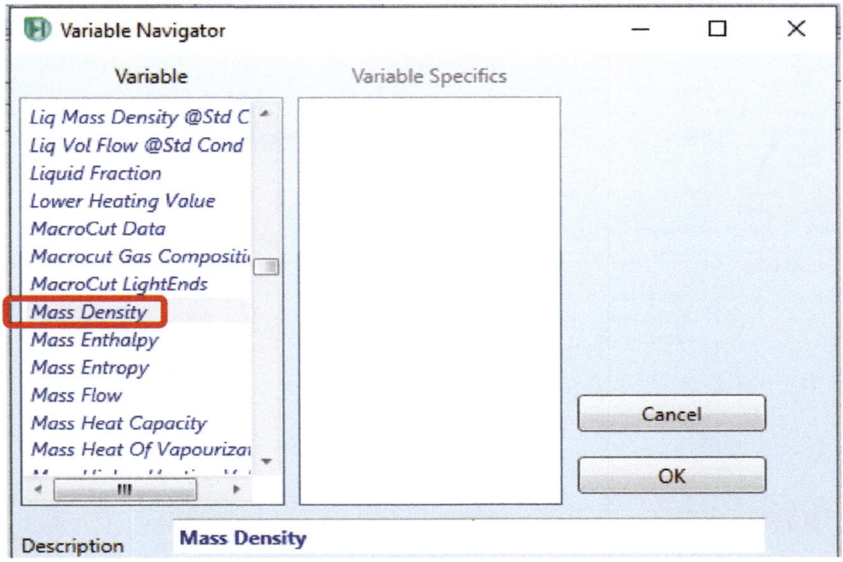

Figure 2.19 Adding the Mass density as dependent variable.

2.2 Steam Table

Steam tables are defined as the thermodynamic data that contain the properties of water or steam [10]. These data are commonly used by engineers. It is normally used to obtain the following properties using steam pressure for saturated steam temperature and saturated temperature for saturated steam pressure.

2.2.1 Problem Statement

Steam table of saturated water vapor in the range (40–400°C), step 4 °C, can be created to calculate some thermodynamics properties such as saturated pressure, mass enthalpy, mass entropy, and mass heat of vaporization.

2.2.2 Simulation

1) Open a new case; from the components list, add water and add **NBS Steam** as Property package.
2) Enter the Simulation environment and add material stream, named "Steam" and enter 1 for composition.
3) As in Figure 2.15, From the Ribbon, Home tab, select the **Stream Analysis** → Property Table, then click **Property Table-1** from Navigation Pane.

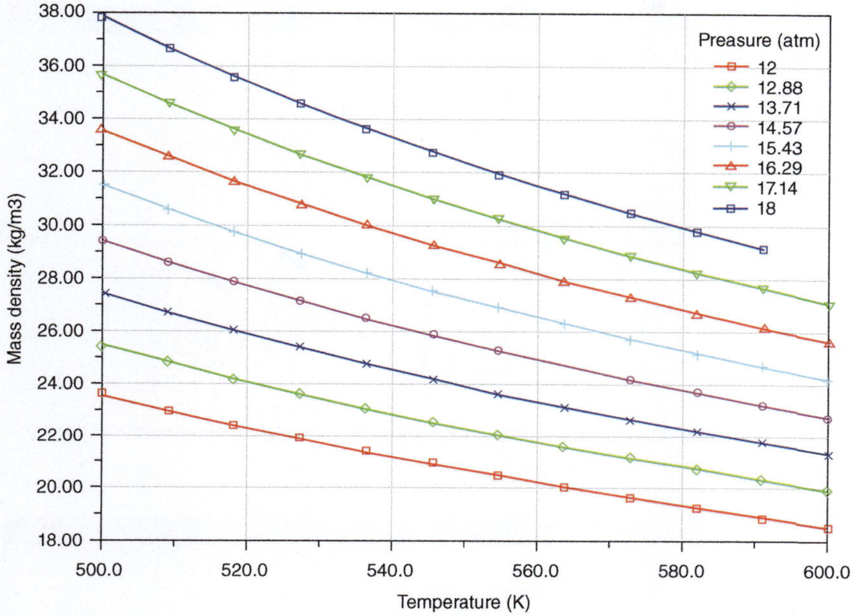

Figure 2.20 Plotting the mass density versus pressure and temperature.

4) Range of Independent Variables:
   Variable 1 = temperature: Enter the range; Lower Bound = 40°C, Upper Bound = 400°C, and
   Number of increments = 90 (*this shows the properties, each 4°C; i.e. (400 – 40)/4 = 90 points, if you want the properties, each 2°C; enter the #increments = 180, and so on*), Figure 2.21.
   Variable 2 = Vapor fraction, Mode = State, and enter the state value = 1 as in Figure 2.21.
5) Select the **Dep. Prop** (Dependent properties) and add the following variables:
   Pressure, Mass Density, Mass Enthalpy, Mass Entropy, and Mass Heat of Vaporization as in Figure 2.22. More variables can be added as required.
6) Click the **Calculate** button. To see the results, click the **Performance** tab, Figure 2.23.

2.2.3 Summary

The steam table is important through which the properties of vapor and liquid can be found. In the references the steam table available within specific conditions, from the results obtained, the HYSYS has the ability to estimate the required properties and create the steam table at any conditions.

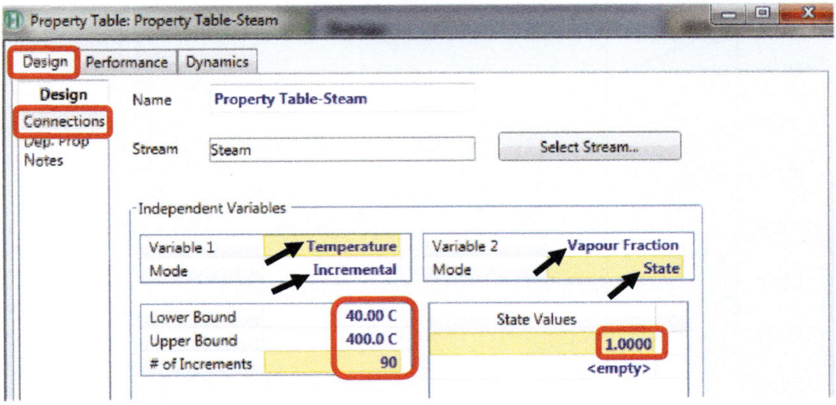

Figure 2.21 Entering the range of temperature and state values in the Property Table.

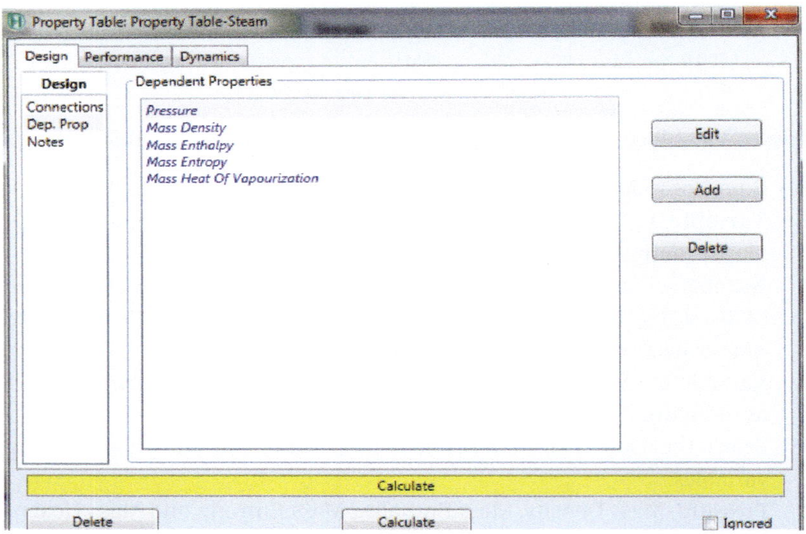

Figure 2.22 Adding the required dependent variables of saturated steam.

2.3 Bubble-Point and Dew-Point Calculations

Bubble point: It is the point where the first bubble of vapor is formed [7, 11]. In the Aspen HYSYS, the bubble-point calculation is performed by inserting the **Vapor/phase fraction** = 0 in the conditions of the **Material**

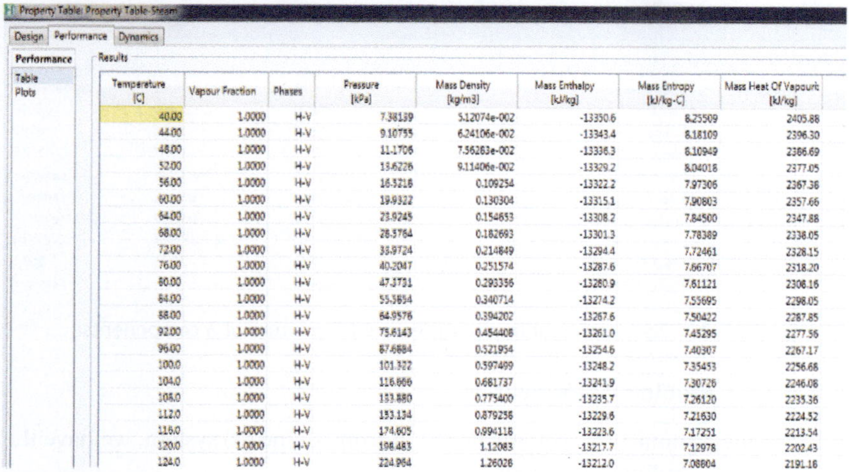

Temperature [C]	Vapour Fraction	Phases	Pressure [kPa]	Mass Density [kg/m3]	Mass Enthalpy [kJ/kg]	Mass Entropy [kJ/kg-C]	Mass Heat Of Vapourk [kJ/kg]
40.00	1.0000	H-V	7.38139	5.12074e-002	-13350.6	8.25509	2405.88
44.00	1.0000	H-V	9.10755	6.24106e-002	-13343.4	8.18109	2396.30
48.00	1.0000	H-V	11.1706	7.56263e-002	-13336.3	8.10949	2386.69
52.00	1.0000	H-V	13.6226	9.11406e-002	-13329.2	8.04018	2377.05
56.00	1.0000	H-V	16.5216	0.109254	-13322.2	7.97306	2367.36
60.00	1.0000	H-V	19.9322	0.130304	-13315.1	7.90803	2357.66
64.00	1.0000	H-V	23.9245	0.154653	-13308.2	7.84500	2347.88
68.00	1.0000	H-V	28.5764	0.182693	-13301.3	7.78389	2338.05
72.00	1.0000	H-V	33.9724	0.214849	-13294.4	7.72461	2328.15
76.00	1.0000	H-V	40.2047	0.251574	-13287.6	7.66707	2318.20
80.00	1.0000	H-V	47.3731	0.293356	-13280.9	7.61121	2308.16
84.00	1.0000	H-V	55.5854	0.340714	-13274.2	7.55695	2298.05
88.00	1.0000	H-V	64.9576	0.394202	-13267.6	7.50422	2287.85
92.00	1.0000	H-V	75.6143	0.454408	-13261.0	7.45295	2277.56
96.00	1.0000	H-V	87.6884	0.521954	-13254.6	7.40307	2267.17
100.0	1.0000	H-V	101.322	0.597499	-13248.2	7.35453	2256.68
104.0	1.0000	H-V	116.666	0.681737	-13241.9	7.30726	2246.08
108.0	1.0000	H-V	133.880	0.775400	-13235.7	7.26120	2235.36
112.0	1.0000	H-V	153.134	0.879256	-13229.6	7.21630	2224.52
116.0	1.0000	H-V	174.605	0.994118	-13223.6	7.17251	2213.54
120.0	1.0000	H-V	198.483	1.12083	-13217.7	7.12978	2202.43
124.0	1.0000	H-V	224.964	1.26028	-13212.0	7.08804	2191.16

Figure 2.23 Results of steam table of saturated steam in the range of (40 – 400°C).

Stream Name	1	Liquid Phase	Vapour Phase
Vapour / Phase Fraction	0.0000	1.0000	0.0000
Temperature [C]	10.00	10.00	10.00
Pressure [kPa]	146.8	146.8	146.8
Molar Flow [kgmole/h]	100.0	100.0	0.0000
Mass Flow [kg/h]	6233	6233	0.0000
Std Ideal Liq Vol Flow [m3/h]	10.60	10.60	0.0000
Molar Enthalpy [kJ/kgmole]	-1.622e+005	-1.622e+005	-1.361e+005
Molar Entropy [kJ/kgmole-C]	61.25	61.25	145.2
Heat Flow [kJ/h]	-1.622e+007	-1.622e+007	0.0000

Figure 2.24 Result of bubble-point pressure when the temperature is given.

Stream, where if the temperature is given, the bubble pressure will have been calculated (= vapor pressure), Figure 2.24, and if the pressure is given, the bubble temperature will have been calculated.

Dew point: It is the **point** where the first drop of liquid is formed [7, 11]. In the Aspen HYSYS, the dew-point calculations are performed by inserting the **Vapor/phase fraction** = 1 in the conditions of the **Material Stream**, where if the temperature is given, the dew pressure will have been calculated, and if the pressure is given, the dew temperature will have been calculated.

Flash calculation: If the temperature and pressure are given and the **Vapor/Phase fraction** result between (0–1), that means the pressure is the flash pressure.

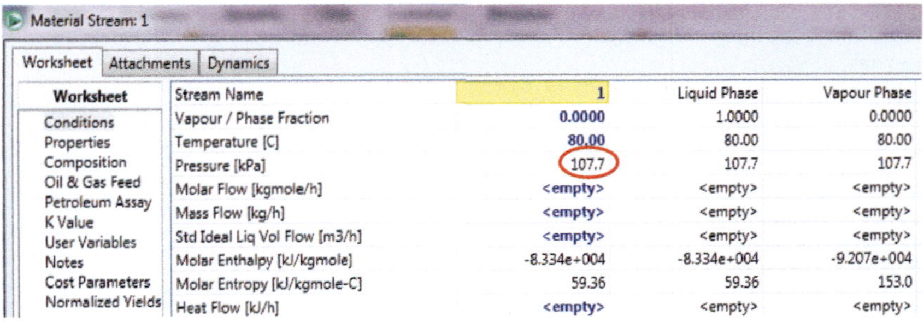

Figure 2.25 Result of bubble-point pressure for mixture of 3 components.

2.3.1 Problem Statement

For the acetone (1)/ acetonitrile (2)/ nitromethane (3) system, we have the following Antoine equations [10]:

$$\ln P_1^{sat} = 14.5463 - \frac{2940.46}{T + 237.22} \tag{2.2}$$

$$\ln P_2^{sat} = 14.2724 - \frac{2945.47}{T + 224} \tag{2.3}$$

$$\ln P_3^{sat} = 14.2043 - \frac{2972.64}{T + 209} \tag{2.4}$$

"*The equations above will not be used by the software; for view only.*"

Here T is in °C and the vapor pressure is in kPa. Assuming that Raoult's law is approximate to this system, calculate:

(a) P and y_k, given that $T = 80°\text{C}$, $x_1 = 0.25$, $x_2 = 0.35$, and $x_3 = 0.4$.
(b) P and x_k, given that $T = 70°\text{C}$, $y_1 = 0.5$, $y_2 = 0.3$, and $y_3 = 0.2$.
(c) T and y_k, given that $P = 80\,\text{kPa}$, $x_1 = 0.3$, $x_2 = 0.45$, and $x_3 = 0.25$.
(d) T and x_k, given that $P = 90\,\text{kPa}$, $y_1 = 0.6$, $y_2 = 0.2$, and $y_3 = 0.2$.

2.3.2 Simulation

1) Open a new case and add the following components: acetone, acetonitrile, and nitromethane (CH_3NO_2)
   Fluid Package: Antoine,
2) Enter the Simulation Environment.
   a) It is a bubble pressure.
3) Add a stream with Vapor/Phase Fraction = 0, $T = 80°\text{C}$ as in Figure 2.25. Set the molar fractions: $x_1 = 0.25$, $x_2 = 0.35$, $x_3 = 0.4$.

The estimated pressure is 107.7 kPa, and the estimated composition will be as in Figure 2.26.

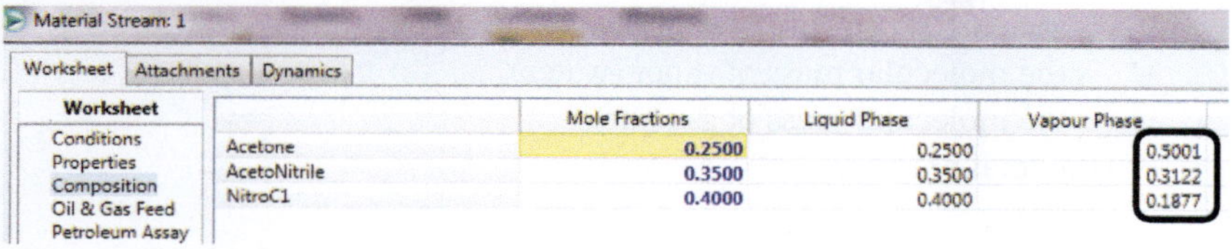

Figure 2.26 Composition of the vapor phase for 3 components after calculating the dew-point pressure.

Note: $\sum y_i = \sum K_i x_i$

And for a *single* component, bubble point = dew point.

The solution of remaining branches is similar to the branch (a).

2.3.3 Summary

The software provides an example calculation to get the bubble point/dew point and vapor/liquid composition of the acetone (1)/ acetonitrile (2)/ nitromethane (3) system. It also explains the calculation of the bubble point and dew point of mixtures using Raoult's law. It also describes how to determine the mixture's dew-point temperature and liquid composition, pointing out that the three equations in this system have three unknowns and are nonlinear, making manual solution difficult and requiring the use of software.

2.4 Humidity Calculations

Humidification is the increase in the amount of vapor present in a gas stream by passing the gas over a liquid which then evaporates into the gas [12].

Dehumidification is the reduction of the vapor present in the gas.

2.4.1 Definitions

Humidity ($\mathscr{H}$) is mass of vapor associated with unit mass of dry gas, kg liquid/kg dry gas; it can be estimated by the following equation [12]:

$$\mathscr{H} = \frac{P_W}{P - P_W}\left(\frac{M_W}{M_A}\right) \tag{2.5}$$

$$\begin{aligned} \textit{Relative humidity (RH)} &= \frac{\textit{Partial pressure of vapor in gas}}{\textit{Partial pressure of vapor in saturated gas}} \times 100 \\ &= \frac{P_w}{P_{wo}} \times 100 \end{aligned} \tag{2.6}$$

where P_w is the partial pressure of vapor (water) in the gas (air).

P_{wo} is the partial pressure of vapor (water) in the saturated gas (air).
M_w is the molecular mass of vapor (water).
M_A is the molecular mass of gas (air).
P is the total pressure.

2.4.2 Problem Statement

With air at a temperature of 30°C, atmospheric pressure, and a relative humidity 60%, calculate the humidity of this air.

2.4.3 Simulation

1) Open a new case and add the components: Air and water.
2) Select the Peng–Robinson as fluid package.
3) Switch to the simulation environment.
4) Add a stream material with the following conditions:
   Name: Air, $T = 30°\text{C}$, $P = 1\ \text{atm}$, composition: 1 for air, and give a basis of mass flow rate = 100 kg/h.
5) Add Saturator icon 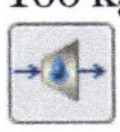 from object palatte.
6) Double-click the saturator icon and connect the input and output streams as in Figure 2.27.

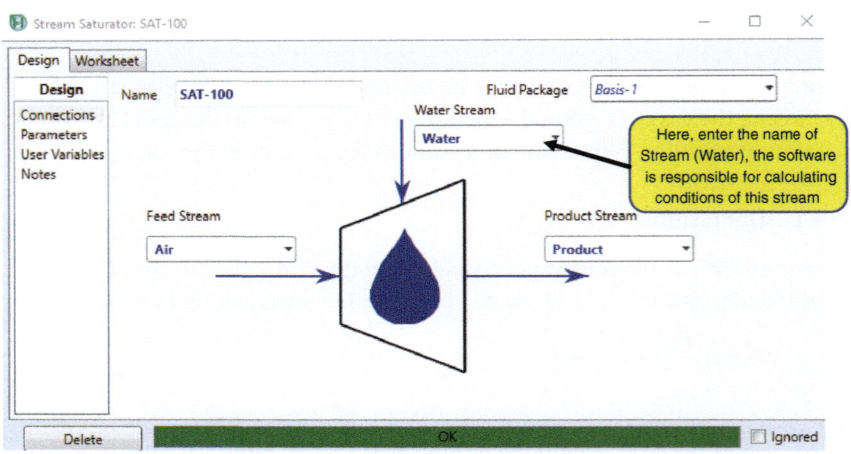

Figure 2.27 Window of Saturator operation with its streams.

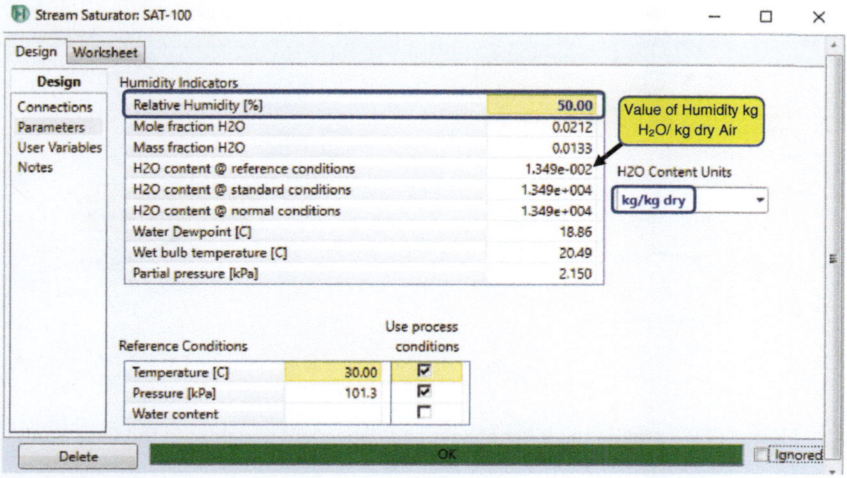

Figure 2.28 Estimating the humidity at $RH = 50\%$.

For the Water-stream, Only type (Water), and don't enter any data, the software calculates all the conditions and properties of the Water-stream.

7) In the Parameters page of Design tab, enter the Relative Humidity % = 50, and select the H2O Content units (kg/kg Dry), Figure 2.28.
The humidity (H_2O content @ reference conditions) is estimated to be 1.349×10^{-2} kg water/kg dry air.

8) Study the effect of temperature and relative humidity of air on the Humidity using case studies.
   (8-1) From the Home Ribbon, Open the **Case Studies**  , then click the **Add** button.
   Add the Independent Variables (Air's temperature, and Relative Humidity of the Saturator), and add the Dependent Variable (H_2O content @ Reference Conditions) as in Figure 2.29.
   (8-2) Click the **Case Study Setup** to give the range of Independent variables: Temperature (20 – 50°C, step = 5°C) and Relative Humidity (20–100%, step = 10%) as in Figure 2.30.
   (8-3) Click the **Run** button and wait a few seconds. The results will present inside the **Results** tab. These results can be plotted by clicking the **Send to Excel**, then drawn as in Figure 2.31.

It can be seen from Figure 2.31 that the humidity increased with increasing relative humidity and temperature of the air.

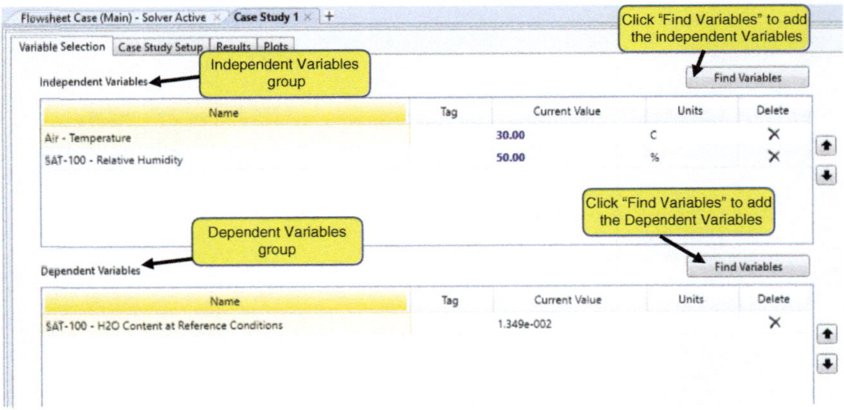

Figure 2.29 Adding the Independent and dependent variables in the Case Study.

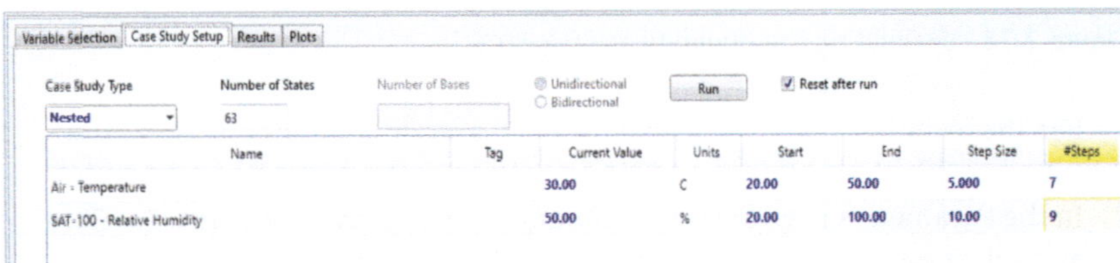

Figure 2.30 Adding the range of independent variables (Temp: 20 – 50°C, step 5 °C, and Relative Humidity: 20–100%, step 10%).

2.4.4 Summary

The humidity of the air is a measure of the amount of water or vapor per unit mass of dry air. It is affected by three parameters: temperature, pressure, and relative humidity. It can be studied to determine the effect of one parameter or more on the value of humidity at constant compared to the others. Humidity can be calculated by equations or using humidity chart that is available in the references, but the software has the potential to calculate the humidity with any parameter or more at constant compared to the others.

Problems

P.2.1 Find the molar density of ethylene (Ethene) gas at 25 bar and 500 K using Peng–Robinson (PR) as a fluid package.

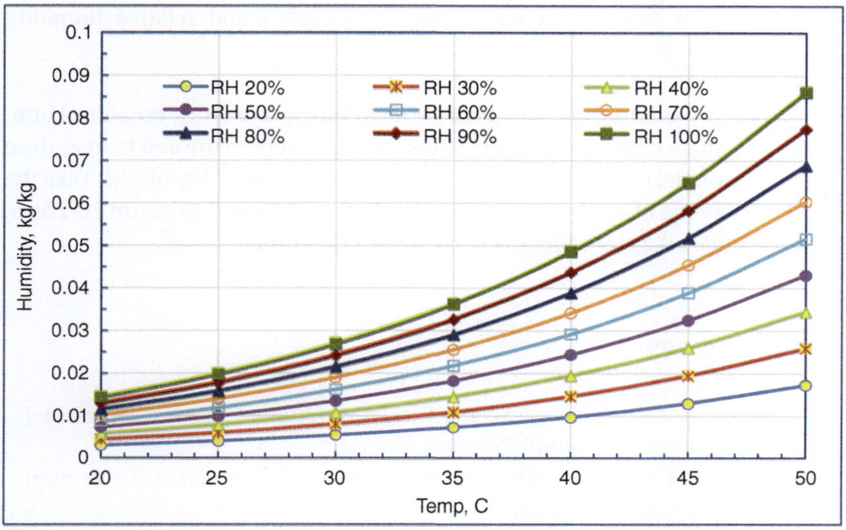

Figure 2.31 Plotting the Humidity versus temperature at different Relative humidity (*RH*).

P.2.2 Using Peng–Robinson and *PRSV*, find the mass density of ethyleneglycol at 10 bar and 573 K. Is there a difference between the two equations of state?

P.2.3 Create a steam table of saturated liquid water for the properties: Pressure, Mass Density, Mass Enthalpy, Mass Entropy, and Mass Heat of Vaporization between 40 and 400°C.

P.2.4 For the acetone(1)/methanol(2)/water(3) system, and Wilson equation as fluid package, make the following calculations:

(a) **Bubble pressure:** where $T = 338.15$ K, $x_1 = 0.3$, $x_2 = 0.4$.

(b) **Dew pressure:** where $T = 338.15$ K, $y_1 = 0.3$, $y_2 = 0.4$.

(c) **Pressure and T-flash:** where $T = 338.15$ K, $P = 0.5 \times (p_{bubble} + P_{dew})$, $z_1 = 0.3$, $z_2 = 0.4$.

P.2.5 Draw the vapor pressure of water as a function of temperature (50 – 200°C).

P.2.6 With air at a temperature of 25°C, 120 kPa pressure, and a relative humidity 40%, calculate the humidity of this air. Plot the humidity as a function of temperature (20 – 50°C) and relative humidity (30–90%).

P.2.7 If the initial conditions of air are temperature 20°C, pressure 1 atm, mass flow rate 200 kg/h, study the effect of pressure and temperature on the humidity of air at a constant relative humidity of 60%. Take the range of temperature (25 – 45°C, step 5 °C) and pressure (1–2 atm, step 0.2 atm). Plot the results in an Excel sheet.

References

1. Dimian, A.C., Bildea, C.S., and Kiss, A.A. (2014). *Integrated Design and Simulation of Chemical Processes*. Elsevier.
2. Chaves, I.'n.D.'o.G. (2016). *Process Analysis and Simulation in Chemical Engineering*, ©. Springer.
3. (2017). *Aspen HYSYS User Guide, Version 10.0*. Aspen Technology.
4. (2002). *HYSYS 3.2 Operation Guide, © Hyprotech, a subsidiary of* Aspen Technology, Inc.
5. Mohd. Kamaruddin Abd Hamid, *HYSYS®: An Introduction to Chemical Engineering Simulation For UTM Degree++ Program*, 2007, University Teknology Malaysia.
6. Ghasem, N. (2022). *Computer Methods in Chemical Engineering*. Taylor and Francis Group, 2e published by CRC Press ©.
7. Poling, B.E., Prausnitz, J.M., and O'Connell, J.P. (2001). *The Properties of Gases and Liquids*, 5e. McGraw-Hill Education.
8. Riggs, J.B. and Himmelblau, D.M. (2012). *Basic Principles and Calculations in Chemical Engineering*, 8e. Englewood Cliffs, NJ: Prentice-Hall.
9. McCabe, W., Smith, J., and Harriott, P. (2004). *Unit Operations of Chemical Engineering*, 7e. McGraw Hill.
10. Smith, J.M., Van Ness, H.C., and Abbott, M.M. (2001). *Introduction to Chemical Engineering, Thermodynamics*, 6e. McGraw-Hill Companies, Inc.
11. Elliott, J.R. and Lira, C.T. (2012). *Introductory Chemical Engineering Thermodynamics*, 3e. Prentice Hall.
12. Coulson, J.M. and Richardson, J.F. (1999). *Chemical Engineering, Volume 1, Fluid flow, Heat Transfer, and Mass Transfer*, 6e. Butterworth–Heinemann.

3
Pressure Change Equipment

3.1 Pump

An incoming liquid stream's pressure is raised by the pump action. The Pump determines an unknown pressure, temperature, or pump efficiency based on the information provided. The incompressibility of the inlet fluid is assumed by the pump operation [1]. In order to simulate the pumping process, the user will run a pump operation in HYSYS. The user will gain knowledge on how to attach streams to unit activities like pumps. Aspen HYSYS uses the following assumptions and equations in calculating the unknown Pump unit operation variables:

- Calculating the ideal power of the pump required to raise the pressure of the liquid:
  The calculations are based on the standard pump equation for power, which uses the pressure rise, the liquid flow rate, and the density [2, 3]:

$$Power\ Required_{ideal} = \frac{(P_{out} - P_{in}) \times Flow\ \ Rate}{Liquid\quad Density} \times 100\% \tag{3.1}$$

  where:
  P_{out} is the pump outlet pressure,
  P_{in} is the pump inlet pressure.
- Calculating the actual power of the pump:
  The actual power requirement of the Pump is defined in terms of the Pump Efficiency:

$$Efficiency(\%) = \frac{Power\ \ Required_{ideal}}{Power\ \ Required_{actual}} \times 100\% \tag{3.2}$$

  When the efficiency is less than 100%, the excess energy goes into raising the temperature of the outlet stream. Combining the above equations leads to the following expression for the actual power requirement of the Pump:

Chemical Process Simulations using Aspen Hysys, First Edition. Khalid W. Hameed.

$$Power\ Required_{actual} = \frac{(P_{out} - P_{in}) \times mass \quad Flowrate}{Liquid \quad Density \times Efficiency \quad (\%)} \times 100\% \tag{3.3}$$

The actual power is also equal to the difference in heat flow between the outlet and inlet streams [3, 4]:

$$Power\ Required_{actual} = (Heat\ Flow_{outlet} - Heat\ Flow_{inlet}) \tag{3.4}$$

The pump calculations that HYSYS performs assume that the liquid is incompressible. The density is constant, and the liquid volume is independent of pressure.

3.1.1 Problem Statement

A liquid water at 25°C, 1 atm, and a mass flow rate of 0.5 kg/s is pumped to 4.5 atm by a pump having an efficiency of 60%. Calculate the power of the pump and the outlet water temperature. Fluid package: NBS Steam.

3.1.2 Simulation

1) Open a new case and enter the following values:
   Components: Water,
   Fluid Package: NBS Steam.
2) Switch to **Simulation** environment.
3) Add a stream of material with the following values.

Name	Feed
Temperature	25°C
Pressure	1 bar
Mass Flow rate	0.5 kg/s
Composition	1

4) From the Object Palette, add a pump icon , Figure 3.1.
5) Double-click the pump icon and connect the streams as in Figure 3.2. From the dropdown list of the **Inlet** cell, choose Feed and, in the **Outlet** cell, type Product (since the product stream is unknown in the simulation field).
   The arrow of blue color is for material stream, while the red color stream is for energy.

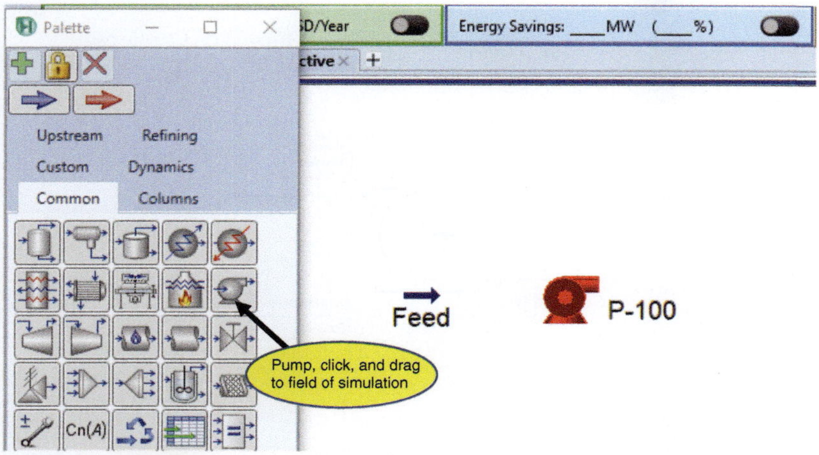

Figure 3.1 Adding a Pump and feed stream in the HYSYS.

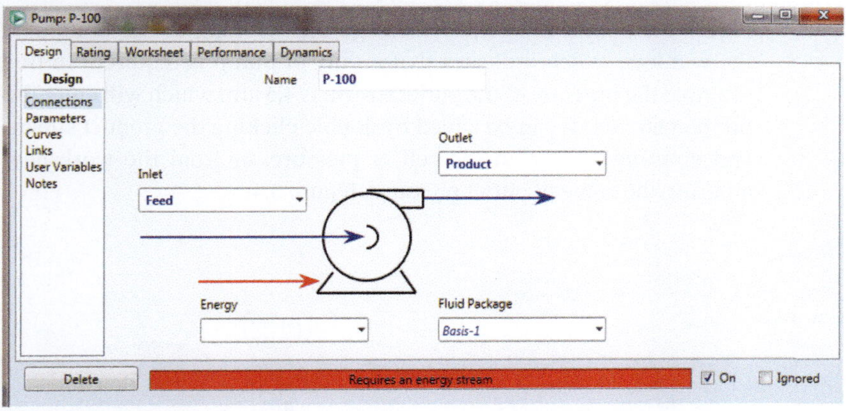

Figure 3.2 Window of Pump with material streams, red color of bottom bar indicates defect of one of streams or more is required; for this case, the energy stream is required.

The color of bottom bar and the displayed message is an indication of what required to add.

When the color of bottom bar is red, that means one or more streams are required to be added; when the color of the bottom bar is yellow, this indicates that the duty of the pump is unknown. The duty of the pump indicates what is pressure out, pressure drop, pressure ratio, or outlet temperature [5].

6) In the energy cell, type (*E*) as an energy stream as shown in Figure 3.3.

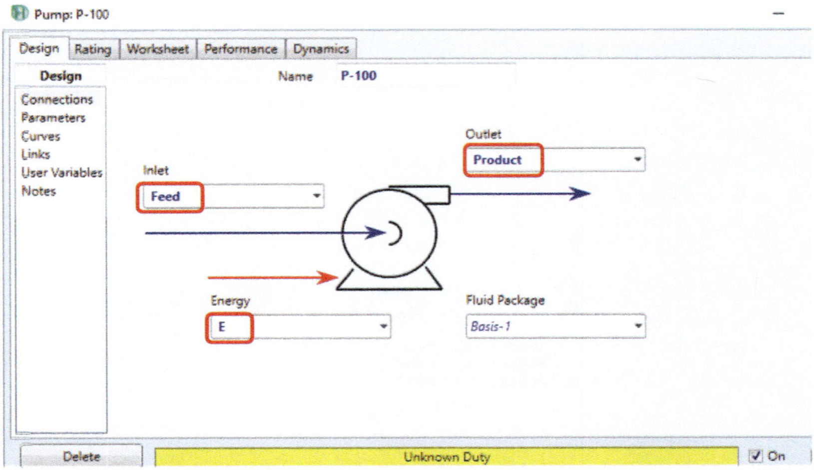

Figure 3.3 Connecting the Pump with its Streams.

7) Figure 3.3 shows the yellow message bar warning at the bottom of the window, which indicates that a duty of pump is required. In this example the pressure of the outlet stream is 4.5 atm which will enter the product stream. It can be added by double-clicking the product stream and entering 4.5 atm in the cell of pressure, or from the worksheet entering the value of outlet pressure, Figure 3.4.

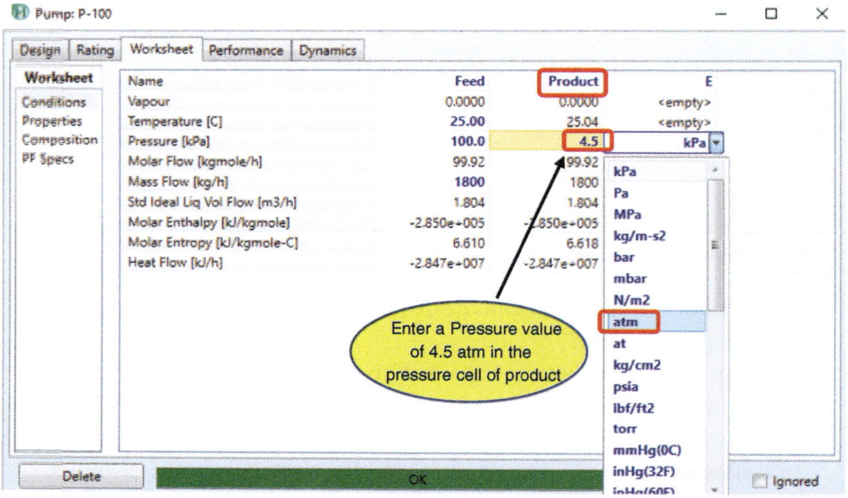

Figure 3.4 The pump is converged after specifying the pressure of product.

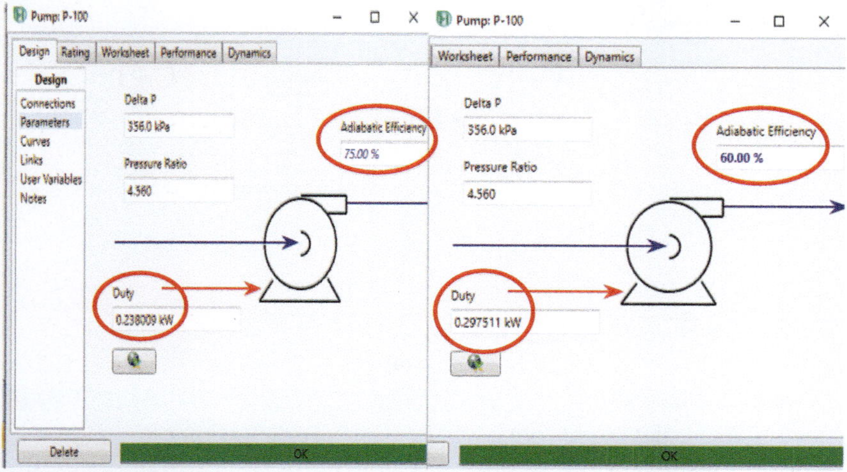

Figure 3.5 When the Pump efficiency is reduced, the duty will increase.

8) Once a pump has enough information, a green bar appears at the bottom of the window, indicating that everything is "OK" (See Figure 3.4).
9) Specifying the Pump Efficiency
   The default efficiency for the pump is 75%. To change the efficiency, do the following:
   - Click the **Design** tab of the pump window, and select **Parameters** Page.
   - In the Adiabatic Efficiency box of the parameters page, enter 60. The units should be in percent, and the power required will increase as shown in Figure 3.5.
10) After the efficiency is entered, the streams of the pumps should be solved. Click **Worksheet** tab to view the results.

What is the water's temperature in the two cases?

If the outlet temperature is 26°C, what is the efficiency of the pump?

Figure 3.5 shows that, when the efficiency decreases, the duty of the pump (power required) increases and the outlet stream temperature increases.

3.1.3 Other Results

Many results can be obtained for the Pump operation such as:

- Net positive suction head (NPSH), Inertia, Electric motor, and Speed versus Torque curve are obtained from **Rating** tab, Figure 3.6a and b.
- The pressure head, Velocity head, and Total Power are obtained from **Performance** tab, Results page, Figure 3.6c.

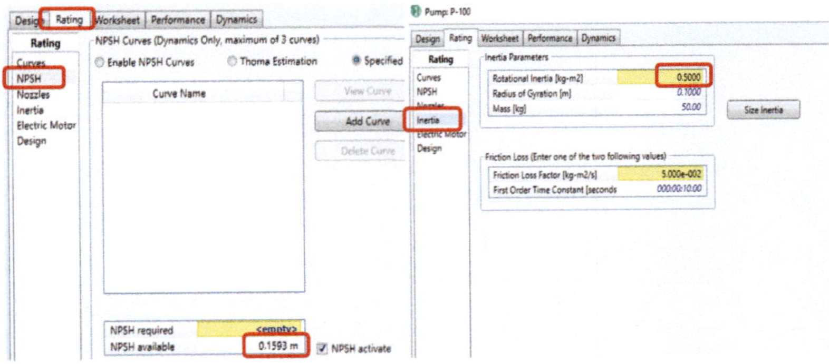

(a)

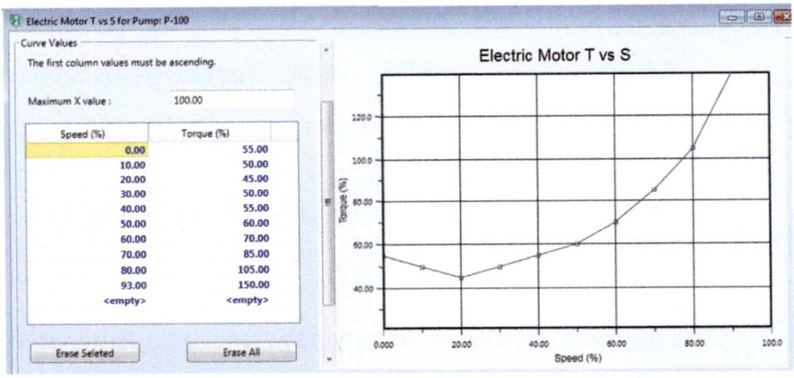

(b)

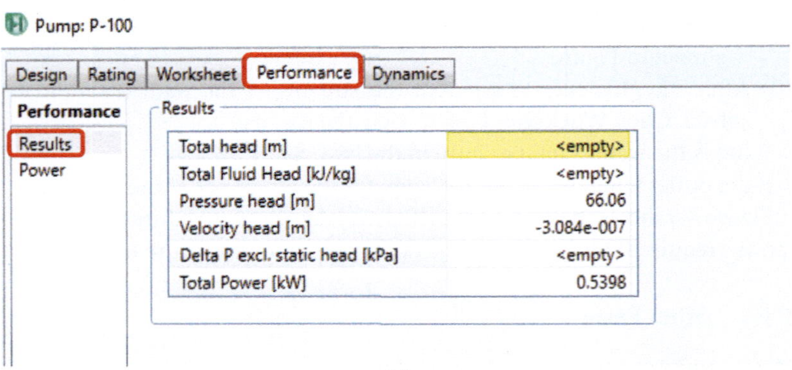

(c)

Figure 3.6 (a) Results of NPSH and Inertia. (b) Torque versus speed curve (Rating tab, Electric Motor page, click Speed versus Torque Curve). (c) Results of pressure head, velocity head, and total power.

3.1.4 Summary

The goal of the issue in the section of pump was to determine the pump's output temperature and pump's power given the pump efficiency. The user has also received instruction on how to attach the streams to unit functions like pumps. The user was instructed to calculate the needed power and pump output temperature at the conclusion of this section when the pump efficiency was provided. On the other hand, HYSYS can calculate the pump efficiency when the output temperature is known. In this example, the pump's efficiency was just 60%, raising the water's temperature by 0.02°C. This is brought on by the fact that a low-efficiency pump requires more energy to push the liquid in order to achieve the same output pressure as a more efficient pump.

3.2 Expander

The expander, sometimes referred to as an expansion engine or turbine, is used to decrease the pressure of a high-pressure inlet gas stream to produce an outlet stream with low pressure and high velocity. An expansion process involves converting the internal energy of the gas to kinetic energy and finally to shaft work. Either a stream characteristic or an expansion efficiency is calculated by the expander. Depending on the information provided, the Expander can solve problems using a variety of techniques. The flow, pressure change, applied energy, and efficiency all play a role in the solution [2, 3]. Regarding what you can specify and what it calculates, the Expander offers a great deal of versatility. To avoid inconsistencies, you must be careful not to enable too many of the solution options [5].

For an Expander, the efficiency is given as the ratio of the actual power produced in the expansion process to the power produced for an isentropic expansion [2, 3, 6]:

$$Efficiency\,(\%) = \frac{Fluid\ Power\ Produced_{actual}}{Fluid\ Power\ Produced_{isentropic}} \times 100\% \tag{3.5}$$

For an adiabatic Centrifugal Expander, Aspen HYSYS calculates the centrifugal expansion rigorously by following the isentropic line from the inlet to outlet pressure. Using the enthalpy at that point, as well as the specified efficiency, Aspen HYSYS then determines the actual outlet enthalpy. From this value and the outlet pressure, the outlet temperature is determined. For a polytropic Centrifugal Expander, the path of the fluid is neither adiabatic nor isothermal. For a 100% efficient process, the only condition is mechanical reversibility. For an irreversible process, the polytropic efficiency is less than 100%. The work determined for the mechanically reversible process is multiplied by an efficiency to give the actual work.

The work for a mechanically reversible process can be determined from [2, 4]:

$$W = \int Vdp \tag{3.6}$$

where:

W is the work
V is the volume
dP is the pressure difference.

The actual power is equivalent to the heat flow (enthalpy) difference between the inlet and outlet streams.

$$Power\ Produced_{actual} = Heat\ Flow_{inlet} - Heat\ Flow_{outlet} \tag{3.7}$$

3.2.1 Problem Statement

An entrance gas stream with high pressure is reduced using the expander to create an exit gas stream with low pressure and high velocity. An expander with a 40% efficiency is fed a combination of gas mixture (CH_4, C_2H_6, C_3H_8, CO_2, and N_2) with component mole percent of 40, 30, 20, 5, and 5, respectively, at 70 °F and 300 psia. The gas mixture is flowing at a rate of 200 lbmol/h, with an expander outlet pressure 60 psia. The outlet temperature of the natural gas and energy production should be determined. Use the Peng–Robinson as a fluid package.

3.2.2 Simulation

1) Open a new case and enter the following values:
   Components: CH_4, C_2H_6, C_3H_8, CO_2, and N_2
   Fluid Package: Peng–Robinson
2) Enter the **Simulation** environment.
3) Add a new Material stream with the following values.

Name	Gas_In
Temperature	70 °F
Pressure	300 psia
Molar Flow Rate	200 lbmol/h
Component Mole Fraction	$C_1 = 0.4$, $C_2 = 0.3$, $C_3 = 0.2$, $CO_2 = 0.05$, $N_2 = 0.05$

4) Add an Expander by clicking the **Expander** icon  from the Object Palette, and drag it to the simulation space.
5) Double-click the expander icon and from connections page, enter the following information:

Name	Expander
Inlet	Gas_In
Outlet	Gas_Out
Energy	Q

The window of the expander will appear as in Figure 3.7.
6) **Parameters page:** Enter the Adiabatic Efficiency equal to 40%.
7) Worksheet tab, Conditions page, Specify the pressure for Outlet stream = 60 psia, as shown in Figure 3.8.

What is the condition of the outlet stream? And what is the power produced of the expander?

Increase the efficiency of the expander to 60%, and see the outlet stream conditions and power of the expander. Reduce the pressure outlet, and see the conditions of the outlet stream and power produced.

Using case study tool, draw the effect of the expander efficiency on the power produced and temperature of the outlet stream.

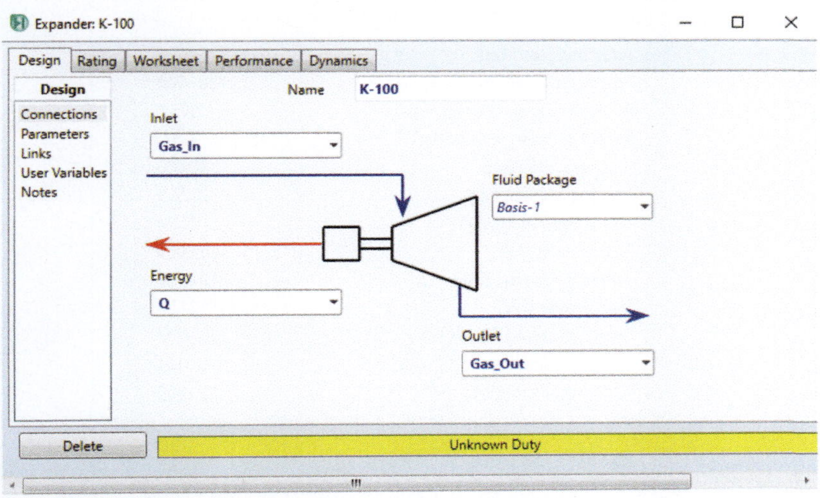

Figure 3.7 Window of the expander, Design tab, connection page.

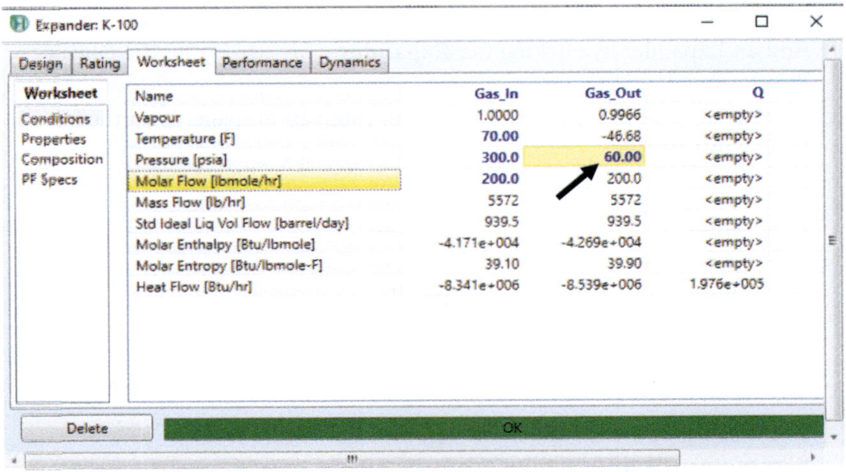

Figure 3.8 Entering the outlet pressure through worksheet tab.

3.2.3 Other Results

Similar to the pump, many results can be obtained from operation of expander, **Performance** tab**,** Results page as in Figure 3.9.

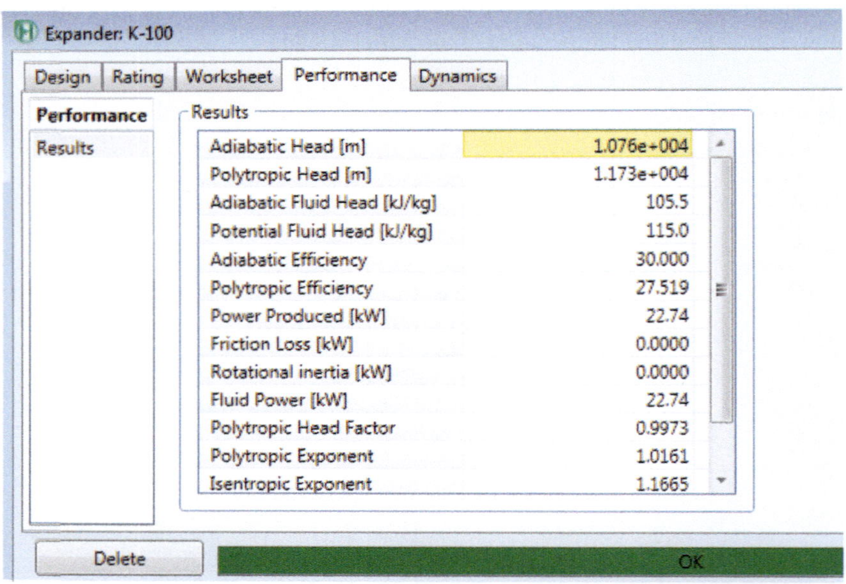

Figure 3.9 Results page - Performance tab of the expander.

3.2.4 Summary

This illustration demonstrates how the expander utilized the pressure drop of the gas to produce the work required. When the efficiency increased, the temperature of the outlet gas stream decreased. The work produced represents the actual work; according to Eq. (3.5), the ideal (isentropic) work is equal to actual work divided by the adiabatic efficiency of the expander.

3.3 Compressor

The Centrifugal Compressor operation is used to increase the pressure of an inlet gas stream with relatively high capacities and low compression ratios. Depending on the information specified, the Centrifugal Compressor calculates either a stream property (pressure or temperature) or a compression efficiency [1, 3]. For a Centrifugal Compressor, the isentropic efficiency is given as the ratio of the isentropic (ideal) power required for compression to the actual power required [2, 6]:

$$Efficiency\,(\%) = \frac{Fluid\ Power\ Produced_{isentropic}}{Fluid\ Power\ Produced_{actual}} \times 100\% \tag{3.8}$$

For an adiabatic Centrifugal Compressor, Aspen HYSYS calculates the centrifugal compression rigorously by following the isentropic line from the inlet to outlet pressure. Using the enthalpy at that point, as well as the specified efficiency, Aspen HYSYS then determines the actual outlet enthalpy. From this value and the outlet pressure, the outlet temperature is determined. For a polytropic Centrifugal Compressor, the path of the fluid is neither adiabatic nor isothermal. For a 100% efficient process, only the condition of mechanical reversibility is required. For an irreversible process, the polytropic efficiency is less than 100%. The work determined for the mechanically reversible process is divided by an efficiency to give the actual work [4].

The work for a mechanically reversible process can be determined from Eq. (3.6).

The actual power is equivalent to the heat flow (enthalpy) difference between the inlet and outlet streams [2].

$$Power\ Produced_{actual} = Heat\ Flow_{outlet} - Heat\ Flow_{inlet} \tag{3.9}$$

3.3.1 Problem Statement

A compressor with a 40% efficiency is fed a mixture of natural gas (N_2, CO_2, methane, ethane, propane, *i*-butane, *n*-butane, *i*-pentane, *n*-pentane, *n*-hexane, and C_7+, all with mole fractions of 0.02, 0.04, 0.31, 0.13, 0.10, 0.09, 0.08, 0.07, 0.06, 0.058, and 0.042, respectively) at 200 °F and 18 psia. Natural gas is flowing at a rate of 150 lbmol/h, with a compressor pressure drop of 60 psia. Find the natural gas outlet temperature and the energy required of the compressor. Use the Peng–Robinson as a fluid package [5].

3.3.2 Simulation

1) Open a new case and enter the following values:
   Components: N_2, CO_2, C_1, C_2, C_3, *i*-C_4, *n*-C_4, *i*-C_5, *n*-C_5, *n*-C_6, C_7+
   Component C_7+ is not available in the Component Library. Therefore, you need to create this component using ***Hypothetical***.
2) Define a New Component
   (2-1) From the **Select** Menu choose the **Hypothetical**, Figure 3.10.
   (2-2) From **Method** Menu choose **Create and Edit Hypos,** Figure 3.11.
   (2-3) Then click **New Hypo.**
   A hypothetical component can be used to model non-library components, defined mixtures, undefined mixtures, or solids. You will be using a hypothetical component to model the component in the gas mixture heavier than hexane.
3) In the view of the hypo component that you are creating, in the Component **Name** cell, type C_7+ and in the **Normal Boiling Point** cell, enter 110°C, Figure 3.12.
   Since you do not know the structure of the hypothetical component and you are modeling a mixture, the Structure Builder will not be used.
4) Click the **Estimate Unknown** button to estimate all the other properties and fully define the hypothetical component.

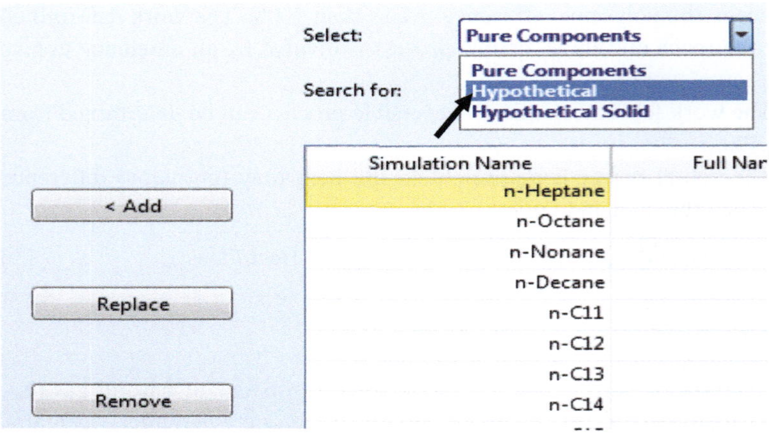

Figure 3.10 How you can add the Hypothetical component.

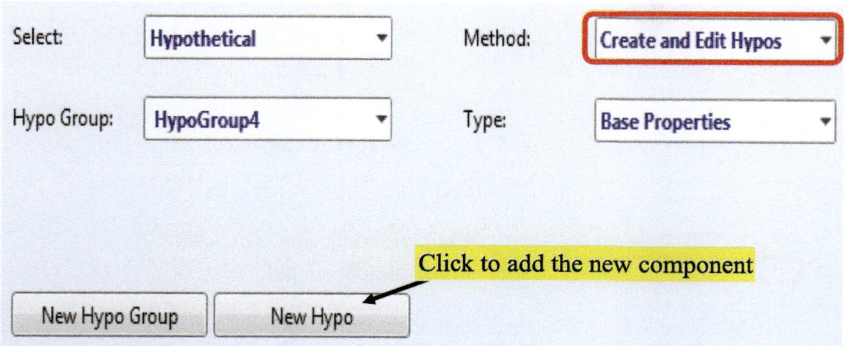

Figure 3.11 How you can add the Hypothetical component (continue).

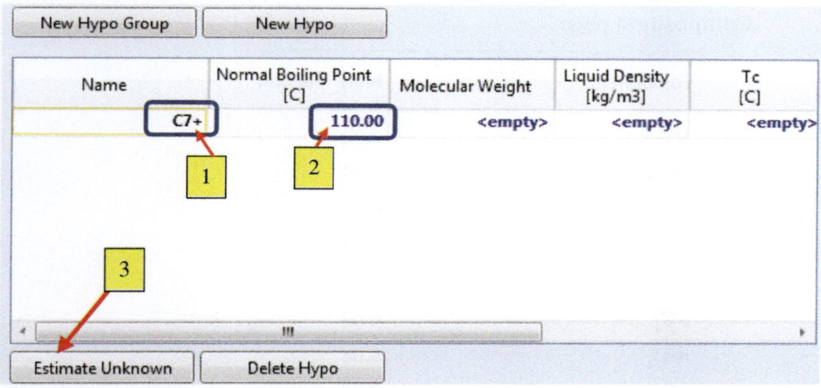

Figure 3.12 Entering the data required for hypothetical component.

5) After the properties of the new component are estimated, click the **Add** button to add the component to your list, Figure 3.13.
   Note: *The minimum information required for defining a hypo is the Normal Boiling or Molecular Weight and Ideal Liq Density.*
6) Property package: *Peng-Robinson*
7) Enter the **Simulation** environment.
8) Create a new Material stream with the following conditions:

Cell	**Enter**
Name	Gas_In
Temperature	200°F
Pressure	18 psia
Molar Flow	150 lbmol/h

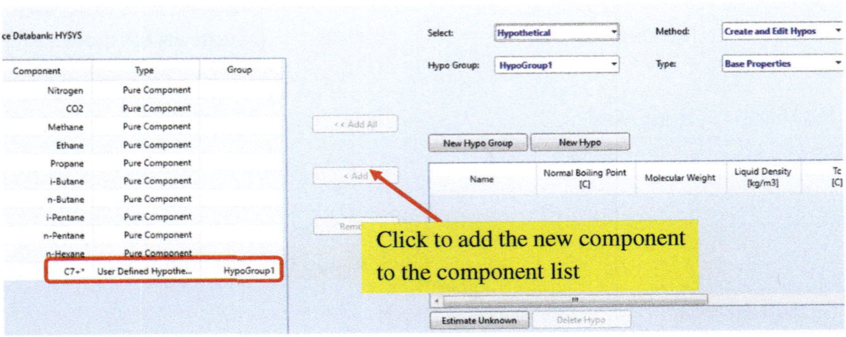

Figure 3.13 When the physical properties of C_7+ are estimated, it can be added to the component list.

Composition page:

Component	Mole Fraction
N_2	0.02
CO_2	0.04
C_1	0.31
C_2	0.13
C_3	0.10
i-C_4	0.09
n-C_4	0.08
i-C_5	0.07
n-C_5	0.06
n-C_6	0.058
C_{7+}	0.042

9) Add a Compressor from **Object Palette** by clicking the icon of compressor  and drag it to the simulation page.
10) **Connections tab:** Add the following information to the compressor and as shown in Figure 3.14:

Cell	Enter
Name	Compressor
Feed	Gas_In
Outlet	Product
Energy	W

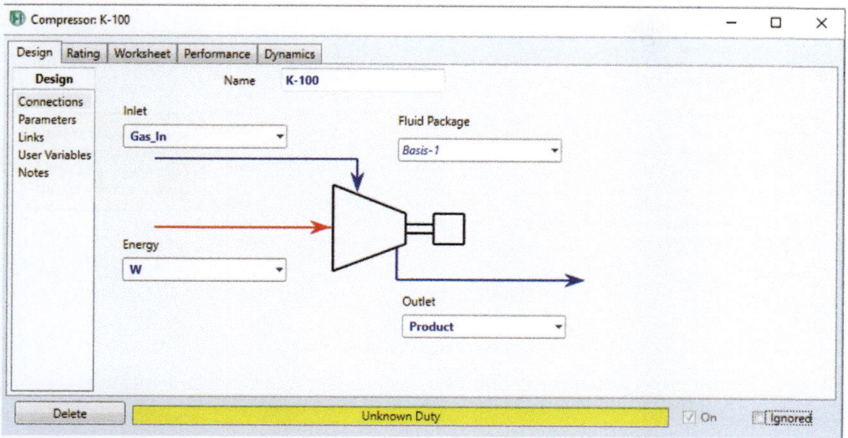

Figure 3.14 Connections page of the compressor.

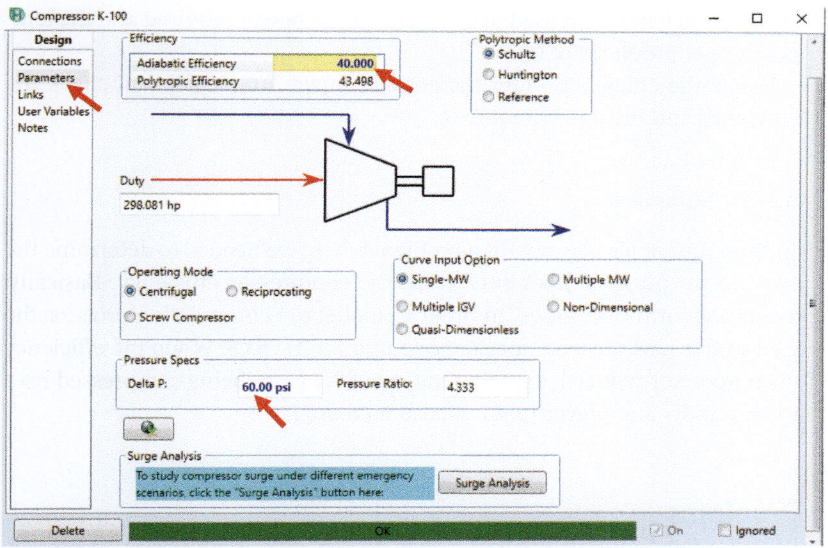

Figure 3.15 Parameters page of the compressor: Entering efficiency 40%, and pressure drop 60 psi.

11) **Parameters page:** Enter the Adiabatic Efficiency equal to 40%, and the pressure drop equal to 60 psi, Figure 3.15.
12) The condition of the outlet stream is illustrated in the **Worksheet** tab, Figure 3.16.

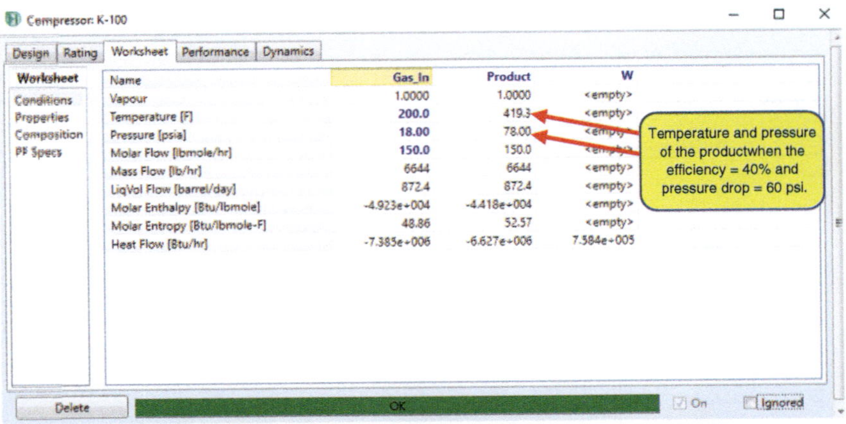

Figure 3.16 The results of the compressor after entering the efficiency and pressure drop.

If the efficiency increased to 60%, what is the power required and the temperature of product stream?

Delete the efficiency, enter the outlet temperature 400°F, and check the efficiency and the power required.

3.3.3 Summary

In this section, we began with a problem where we needed to determine the compressor outlet temperature given the compressor efficiency. Basically, a compressor moves gases. In order to model the compressing process, the user in this part ran a compressor operation in HYSYS. When the efficiency of compressor reduced, the temperature of the gases being compressed rises more rapidly and power required also increased.

Problems

P.3.1 What is the horse power required of the pump to increase the pressure of water from 1 to 7 atm, where the conditions of entrance to the pump are 20 °C and 1100 kg/h?

P.3.2 A steam turbine with rated capacity of 56,400 kw operates with steam at inlet conditions of 8600 kPa and 500°C and discharge into a condenser at a pressure of 10 kPa. Assuming a turbine efficiency of 75%, determine the state of the steam at discharge and the mass rate of the steam [3].

P.3.3 10,000 kg/h of saturated-vapor steam at 100 kPa is compressed adiabatically to 300 kPa. If the compressor efficiency is 75%, what is the work required and what are the properties of the discharge stream? [3]

P.3.4 A centrifugal compressor with an efficiency of 72% is used to compress 2000 kmol/h of natural gas with composition of C_1 0.92, C_2 0.03, C_3 0.02, N_2 0.025, and CO_2 0.005 between two compressor stations, and the pressure drop is compensated. The inlet temperature is 20°C, the inflow pressure is 55 bar, and the ratio of the compressor pressure is 1.6. What is the polytropic temperature of the gas at the compressor outlet, polytropic fluid head, and the power needed? [6]

References

1 Nash, D. and Banks, W.M. (2003). *Pressure Equipment Technology: Theory and Practice*. Wiley.
2 (2017). *Aspen HYSYS User Guide Version 10.0*. Aspen Technology.
3 Smith, J.M., Van Ness, H.C., and Abbott, M.M. (2001). *Introduction to Chemical Engineering, Thermodynamics*, 6e. McGraw-Hill Companies, Inc.
4 McCabe, W., Smith, J., and Harriott, P. (2004). *Unit Operations of Chemical Engineering*, 7e. McGraw Hill.
5 Mohd. Kamaruddin Abd Hamid, *HYSYS®: An Introduction to Chemical Engineering Simulation For UTM Degree++ Program*, 2007, University Teknology Malaysia.
6 Elliott, J.R. and Lira, C.T. (2012). *Introductory Chemical Engineering Thermodynamics*, 3e. Prentice Hall.

4

Heating and Cooling

In the majority of chemical processes, heat is either given out or absorbed, and fluids must often be either heated or cooled in a wide range of plants, such as furnaces, evaporators, distillation units, dryers, and reaction vessels, where one of the major problems is transferring heat at the desired rate. In addition, it may be necessary to prevent the loss of heat from a hot vessel or pipe system. The control of the flow of heat at the desired rate forms one of the most important areas of chemical engineering. Provided that a temperature difference exists between two parts of a system, heat transfer will take place in one or more of three different ways [1].

4.1 Heat Exchanger

This section starts with a problem to determine the temperature of the cold stream passing through the heat exchanger under the specified stream conditions. The process will be modeled in this part using the shell-and-tube heat exchanger from HYSYS. Calculations for the two-sided energy and material balance are done by the heat exchanger. Temperatures, pressures, heat flows (including heat loss and heat leak), material stream flows, or UA can all be solved using the heat exchanger's high degree of flexibility (overall heat transfer coefficient times area) [2].

In HYSYS, you can choose the Heat-Exchanger Model for your analysis. Your choices include a **Simple End-Point** analysis design model, an ideal ($F_t = 1$) counter-current **Weighted designed** model, a **Simple steady-state rating** model, a **Rigorous shell-&-tube** model, and a **dynamic rating** model for use in dynamic simulations. The dynamic rating method is available as either a Basic or Detailed model, and can also be used in Steady-State mode for Heat-Exchanger rating [3].

Chemical Process Simulations using Aspen Hysys, First Edition. Khalid W. Hameed.

In the following general relations, the hot fluid supplies the Heat-Exchanger duty to the cold fluid [1, 4]:

Balance Error $= (M_{cold}\,[H_{out} - H_{in}]_{cold} - Q_{leak}) - (M_{hot}[H_{in} - H_{out}]_{hot} - Q_{loss})$

Or the above equation can be re-written as follows [5]:

$$Q = (mC_P\Delta T)_{cold} = (mC_P\Delta T)_{hot} \tag{4.1}$$

where:

M is the fluid mass flow rate
H is the enthalpy
Q_{leak} is the heat leak
Q_{loss} is the heat loss
Balance Error is a Heat-Exchanger Specification that equals zero for most applications
hot and cold is the hot and cold fluids
in and out is the inlet and outlet stream
C_P is the specific heat
ΔT is the temperature difference.

The Heat-Exchanger operation allows the heat curve for either side of the exchanger to be broken into intervals. Rather than calculating the energy transfer based on the terminal conditions of the exchanger, it is calculated for each of the intervals and then summed to determine the overall transfer [1].

The total heat transferred between the tube-and-shell sides (Heat-Exchanger duty) can be defined in terms of the overall heat transfer coefficient, the area available for heat exchange, and the log mean temperature difference [5]:

$$Q = UA\Delta T_{lm}F_t \tag{4.2}$$

where:

U is the overall heat transfer coefficient
A is the surface area available for heat transfer
ΔT_{lm} is the log mean temperature difference (LMTD)
F_t is the LMTD correction factor.

The heat transfer coefficient and the surface area are often combined for convenience into a single variable referred to as UA. The LMTD and its correction factor are defined in the Performance section.

4.1.1 Forms of Tubes

1) **Fixed tubes:** This configuration has fixed the setting, straight pipes on both sides using welded stationary tube sheets. It is a low-cost

construction. Its major disadvantages are maintenance since being fixed to the shell cannot mechanically clean this part of the tube; so it is very difficult to change a damaged tube [5].

2) **U-tubes:** This configuration requires only a stationary tube sheet; however, the cost is equal because it is expensive to produce tubes that form. One advantage is that one can remove the tube bundle, thus facilitating cleaning of the shell side. However, cleaning the side of the tubes is very complicated because the U-shape is necessary to use cleaning equipment with moving head. Such tubes should not be used for services with very dirty fluids [5].
3) **Floating head:** It is the most versatile and also the most expensive configuration. In this design, one of the heads is relatively fixed to the shell, while the other end is left free to "float" inside the shell. Thus, thermal expansion joints do not allow, and likewise it is mechanically possible to clean both sides of the tubes. This design can be used for applications where both fluids (the tubes and the shell) are dirty [5].

4.1.2 Models of Heat Exchange

1) **Simple end-point model:** This model assumes that the overall heat transfer coefficient is always constant, as the specific heat of both end streams of the equipment. Here it is also assumed that both heat curves are linear. Similarly, assume that no phase changes and the specific heat is constant [1, 6].
2) **Simple weighted model:** The use of this model for systems where heat curves exhibit a nonlinear behavior, such as phase-change systems for pure components on either side of the equipment-recommended behavior. It is available only for counter-current exchangers [1].
3) **Simple steady-state rating model:** It is similar to end-point model but requires very detailed information on the equipment geometry; if you have this information, we recommend running the calculation with this model [6].
4) **Dynamic rating model**: Three models are available for Dynamic Rating using the Heat-Exchanger unit operation: Basic, Intermediate, and Detailed. If you specify three temperatures or two temperatures and a UA, you can rate the exchanger with the Basic model. If you want to perform zone analysis and model tube thermal inertia without providing geometric details such as tube ID (inside diameter), OD (outside diameter), length, shell diameter, and baffles, you can rate the exchanger using the Intermediate model. If you provide detailed geometry information, you can rate the exchanger using the Detailed model [6].

5) **Rigorous shell-and-tube model:** This mode accesses the Aspen Exchanger Design and Rating (EDR) program and lets you set up a rigorous shell-and-tube exchanger model within it. Using this model also allows advanced analysis and error checking through the EDR activated analysis environment. You must have Aspen Exchanger Design and Rating (EDR) installed to access this model [1].

4.1.3 Problem Statement

Cold water at 20°C and 150 kPa is used to cool a hot stream (equal amounts by mole of ethanol and 1-Pentanol) in a shell-and-tube heat exchanger. The inlet temperature and pressure of the hot stream are 65°C and 150 kPa, respectively. It is required to cool the hot stream to 35°C. If the flow rate of the hot and cold streams is 300 and 600 kg/h, respectively, determine the temperature of the outlet cold stream passing through the exchanger. Assume the pressure drop through the shell and tubes is 30 kPa each.

4.1.4 Simulation

1) Open the HYSYS software, open a new case and enter the following values:
   Components: Water, ethanol, 1-Pentanol
   Fluid package: NRTL, click **Binary Coeff** tab, then click **Unknowns only** button.
2) Enter the **Simulation** Environment.
3) Add two material streams with the following values:

Name	Water_In	Feed
Temperature, °C	20	65
Pressure, kPa	150	150
Mass flow, kg/h	600	300
Composition	100% Water	0.5 mol fraction for each of ethanol & 1-Pentanol

4) Adding a Heat Exchanger:
   The calculations of the two-sided energy and material balance are done by the heat exchanger. Temperatures, pressures, heat flows (including heat loss and heat leak), material stream flows, and UA can all be resolved by the heat exchanger.

   (4-1) From the **Object Palette**, click the **Heat Exchanger** icon 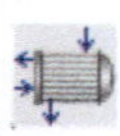, and drag it to the simulation area.

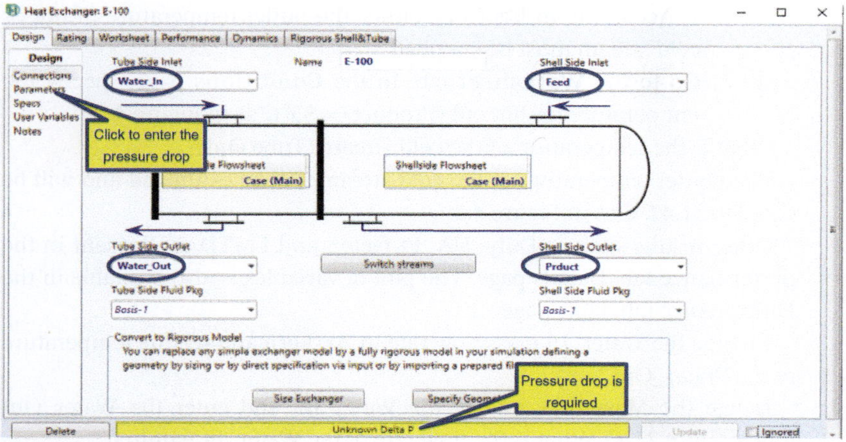

Figure 4.1 Heat- exchanger, connections page,- design tab.

(4-2) **Connections** page: connect the streams as in Figure 4.1.

(4-3) The yellow bottom bar of Figure 4.1 informs us about the pressure drop. Click the **Parameters** page: Enter the pressure drops for the shell-and-tube sides, each 30 kPa as shown in Figure 4.2.

The yellow bottom bar of Figure 4.2 informs us that some specific data is required.

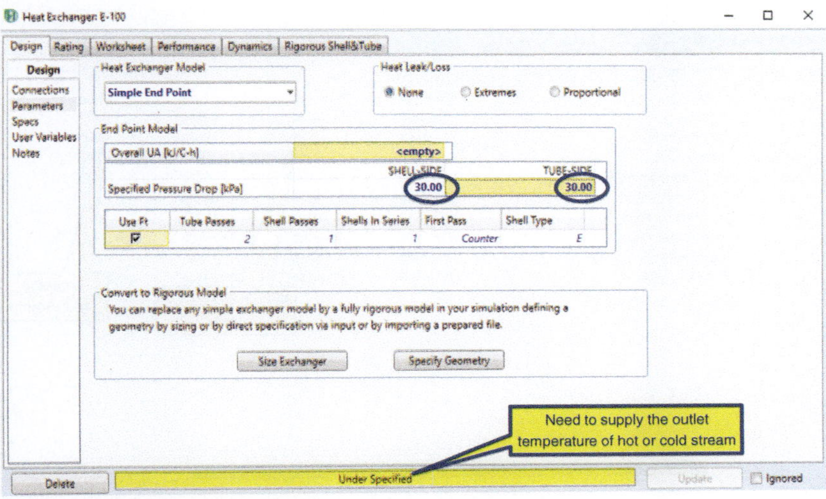

Figure 4.2 Heat exchanger, parameters page, design tab, entering the pressure drop.

According to Eq. (4.1), either the outlet temperature of hot or cold stream must be supplied.

(4-4) Go to the **Worksheet** tab. In the **Conditions** page, specify the temperature for hot out (Product) = 35°C as in Figure 4.3.

What is the temperature of the cold stream (Tube Out)?

The Outlet temperature of the cold stream will be estimated and will be equal to 31.47°C as in Figure 4.3.

Other results such as Duty, UA, Ft factor, and LMTD are present in the performance tab, Details page. The plot of variables is also available in the Performance tab, Plots page.

Change the Water_In mass flow rate to 700 kg/h and see the temperature of the Water_Out.

Delete the Mass flow rate of the Water_In, and enter the Water_Out temperature 27°C. What is the flow rate of the Water_In required?

The greater the temperature difference between the hot output and the cold output, the smaller the dimensions of the heat exchanger, but in return, the operating cost increases.

The default model of the heat exchanger that the software uses is "Simple End Point." This model is preferred for single phase, but when there is a phase change, it is better to use "Simple Weighted," because this model has the ability to calculate the C_P of gases with temperature [6, 7].

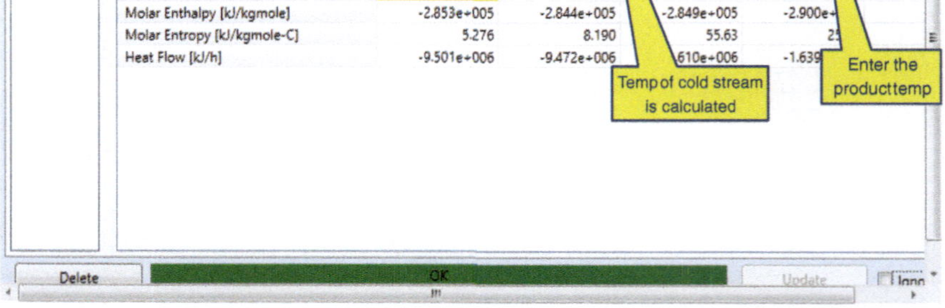

Figure 4.3 Heat exchanger, conditions page, worksheet tab, specify the shell Out Temp.

4.1.5 Design of Heat Exchanger

Note: *To achieve the design of heat exchanger successfully, the pressure drop across shell and tube* ***must be*** **> 0** [5, 7]. Initially, we can assume any expected value, and the simulator will compute the actual value. In this example, the pressure drops on entering the shell and tube were 30 kPa each.

Again, for the same condition of the problem and after executing the basic calculation, on the **Design** tab, **Parameters** page, click the **Size Exchanger** button. From the new message that appears, select Auto Size option, and click the **Convert** button. Wait few minutes for new message; click **Yes** then click **Rigorous Shell & Tube** tab, Figure 4.4, [8, 9].

To see the details of exchanger design, click the **View EDR Browser** button; the design of Heat Exchanger can be seen from Figure 4.5.

The mechanical design of heat exchanger is available from **Mechanical Summary → Setting Plan**, Figure 4.6. The Tube-sheet layout can be seen from **Tubesheet Layout** tab; more information about Cost/weight is also available in the page of **Cost/Weight.**

To view full details of the design, click **Model Details** button at the bottom of the form.

Notes

(1) *For more accurate results: In the* ***Process*** *tab* (Figure 4.5), *set the* ***Fouling resistance = 0.001 hF/Btu (0.00018 m²K/W)*** *for hot and cold side.* Where: 1 $m^2K/W = 5.67\ h.f^{2}{}^{\circ}F/Btu$ (hF/Btu).

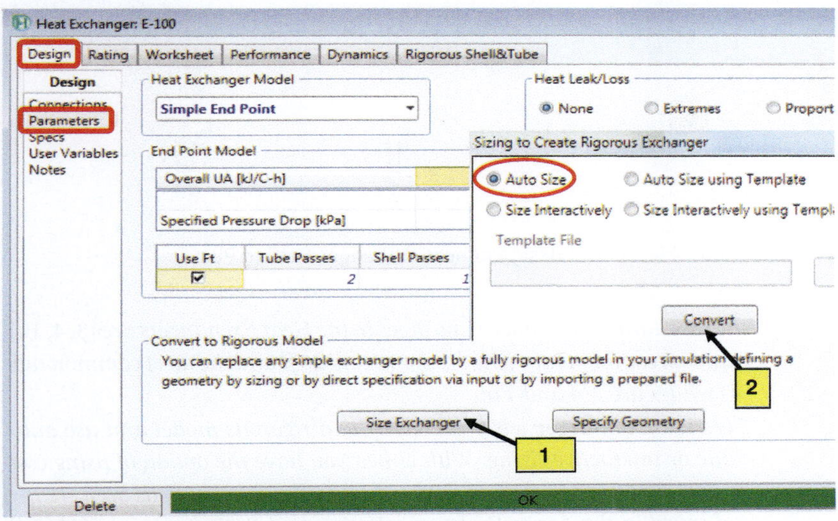

Figure 4.4 How you can arrive at the size exchanger.

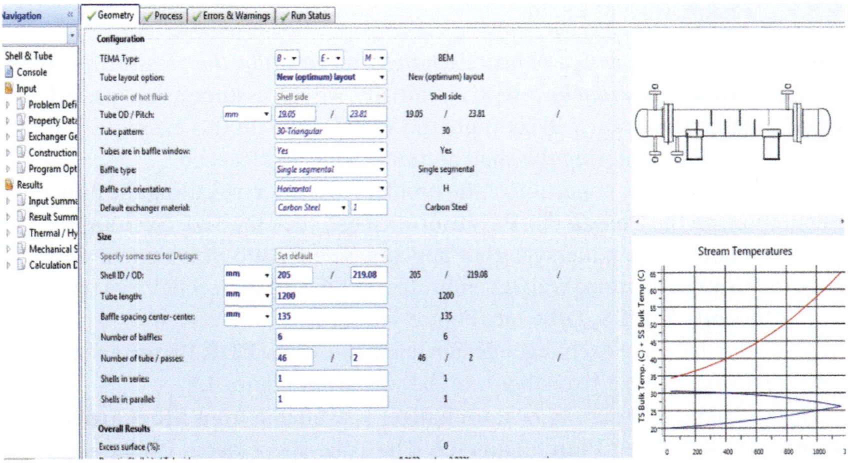

Figure 4.5 Design results of heat exchanger.

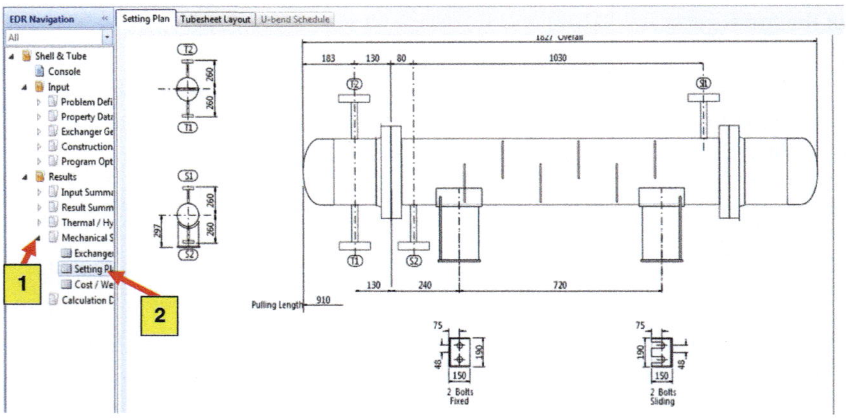

Figure 4.6 Mechanical design of heat exchanger/Setting Plan.

(2) *The standard dimensions of the pipes in the Heat Exchangers are* [3, 4, 10]*: The most common tube lengths are 16 and 20 ft and the most common tube OD values are 3/4 and 1 in.*

(3) *The* ***Size Exchanger*** *lets you convert to a rigorous model and use automatic or interactive sizing. With either, you have the option of using constraints contained within an EDR template file (.edt) that has been created and saved in the Aspen Exchanger Design and Rating program* [11–14]. *In reference to* Figure 4.4, Table 4.1 *shows the translation of each option.*

Table 4.1 Translation of each option (Selection) for exchanger design and rating.

Selection	Translation
Auto Size	Find the most economical design for the rigorous exchanger based on stream flows. When you click Convert, a progress report informs you of the number of designs that have been evaluated and the number that are "near" and "OK." When the process is completed, Activated EDR selects and uses the most economical design as the rigorous specification for the model.
Size interactively	When you click Convert, Activated EDR opens the EDR Sizing Console where you can specify the rigorous exchanger geometry and process properties relevant to sizing the exchanger. You can start the sizing routine repeatedly from this form and use Accept Design to pass the best EDR results back to the converted exchanger.
Auto Size Using Template	Find the most economical design for the rigorous exchanger based on stream flows in combination with constraints used in an EDR exchanger template (.edt) previously saved in the EDR program.
Size Interactively Using Template	Specify an .edt template file for the exchanger. The EDR Sizing Console opens with the template information loaded into it. You can edit the template specifications and use the Sizing button to run the auto sizing routine and examine the results. You can start the sizing routine repeatedly from this form, and use Accept Design to pass the best EDR results back to the converted exchanger.

4.2 Air Cooler

The Air Cooler unit operation uses an ideal air mixture as a heat transfer medium to cool (or heat) an inlet process stream to a required exit stream condition. One or more fans circulate the air through bundles of tubes to cool process fluids. The air flow can be specified or calculated from the fan rating information. The Air Cooler can solve for many different sets of specifications including [1]:

- Overall heat transfer coefficient, UA
- Total air flow
- Exit stream temperature.

An **Air Cooler** will be used to cool a liquid process stream by air as shown in Figure 4.7.

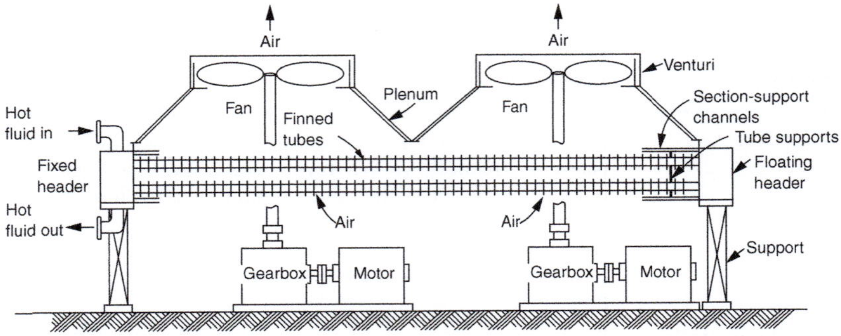

Figure 4.7 Side view of Air Cooler.

The Air Cooler uses the same basic equation as the Heat-Exchanger unit operation; however, the Air Cooler operation can calculate the flow of air based on the fan rating information. The Air Cooler calculations are based on an energy balance between the air and process streams. For a cross-current Air Cooler, the energy balance is calculated as follows [6]:

$$M_{air}(H_{out} - H_{in})_{air} = M_{process}(H_{in} - H_{out})_{process}$$
$$\text{OR } (mC_P\Delta T)_{Air} = (mC_P\Delta T)_{process} \tag{4.3}$$

where:

M_{air} is the air stream mass flow rate
$M_{process}$ is the process stream mass flow rate
H is the enthalpy.

The Air Cooler duty, Q, is defined in terms of the overall heat transfer coefficient (U), the area available for heat exchange (A), and the log mean temperature difference (ΔT_{lm}):

$$Q = -UA\Delta T_{lm}F_t \tag{4.4}$$

The parameters of Eq. (4.4) are the same as in Eq. (4.2).

The air flow through the fan is calculated using a linear relation [1]:

$$Fan\,Air\,Flow = \frac{Speed}{Design\quad Speed} \times Design\quad Flow \tag{4.5}$$

Each fan in the Air Cooler contributes to the air flow through the Cooler. The total air flow is calculated as follows:

$$Total\,Air\,Flow = \sum Fan\,Air\,Flow \tag{4.6}$$

4.2.1 Problem Statement

One tube row and one pass configuration of the Air Cooler is used to cool a hot stream of (0.5 mass fraction of ethanol and 1-Pentanol each) by cold air at 70°F and 101.3 kPa. The inlet temperature and pressure of the hot stream are 130°F and 16 psia, respectively. It is required to cool the hot stream to 85°F. If the flow rate of the hot stream is 12,000 lb/h, determine the duty of the Air Cooler and the temperature of the outlet cold air. The pressure drop of the process stream is 1.2 psi, while demand speed and design speed of the fan of Air Cooler are 42 and 60 rpm, respectively.

4.2.2 Simulation

1) Open a new case and create the following:
   Components: Ethanol, 1-Pentanol, Fluid Package: NRTL
   Don't forget to click the **Binary coefficient** button and click **Unknowns Only.**
2) **Simulation** Environment
   Material Stream: Feed,
   $T = 130°F$, $P = 16$ psia, Mass flow rate = 12,000 lb/h,
   Composition: 0.5 mass fraction for each of Ethanol and 1-Pentanol

   From the object palette add an **Air-cooler** 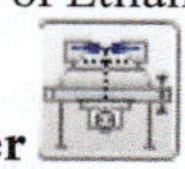 icon.
3) Specify the Air Cooler connections and parameters as shown below. The **Parameters** page contains the fan rating and speed specifications, Figure 4.8.

Tab (Page)	Input Area	Entry
Design (Connection)	Process Stream Inlet	Feed
	Process Stream Outlet	Product
Design (Parameters)	Process Stream Delta P	1.2 psi
	Configuration	1 tube row, 1 pass
	Air Intake Temperature	70°F

4) From **Worksheet** tab, enter the temperature of the product = 85°F.
5) Using the following specifications, HYSYS is able to calculate the actual volumetric flow of air.

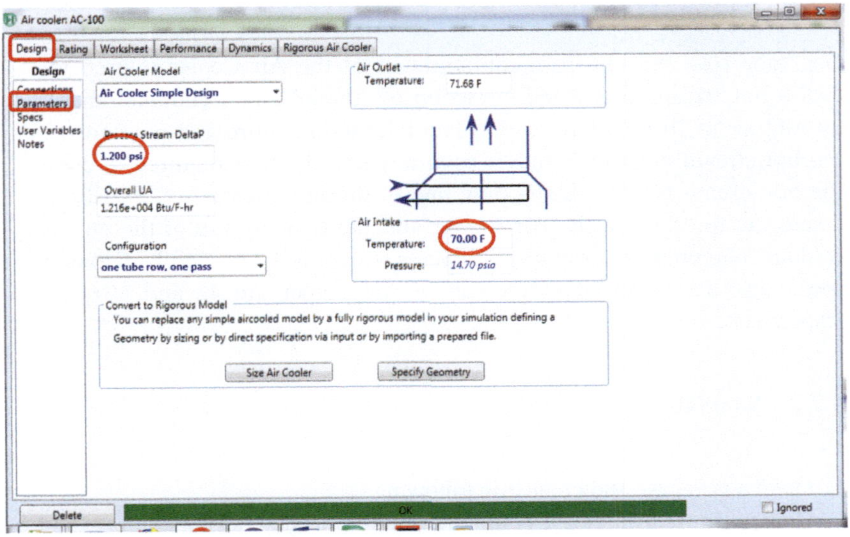

Figure 4.8 Parameters page of the Air Cooler.

Tab (Page)	Input Type	Entry
Rating (Sizing)	Number of Fans	1
	Demanded Speed	42 rpm
	Design Speed	60 rpm
	Design flow	2.12e5 [ACFM]

6) At this point, the **Air Cooler** unit operation is fully specified. The exit stream conditions are displayed in the **Conditions** page of the **Worksheet** table.
7) The **Air Cooler** unit operation parameters are shown in the **Results** page of the **Performance** tab, Figure 4.9.

4.2.3 Design of Air Cooler

To estimate the dimensions of the Air Cooler, return to the Design Tab > Parameters page: click the [Size Air Cooler] button, Figure 4.10. From the new window, click the **Convert** button to perform the design of the Air Cooler. Wait some minutes; from new window, select **Auto Size** and click "Yes," then Click **Rigorous Air Cooler** tab and click **View EDR Browser** button, Figure 4.11. The results can be seen in detail as in the Heat Exchanger.

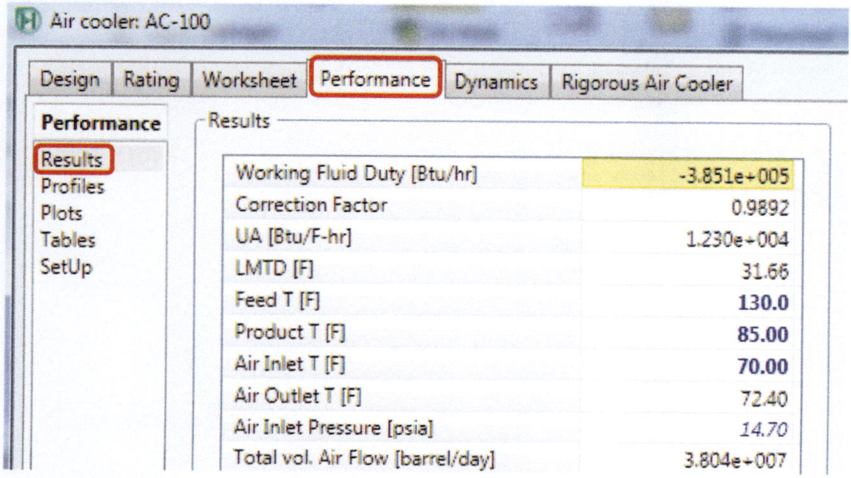

Figure 4.9 Performance results of Air Cooler.

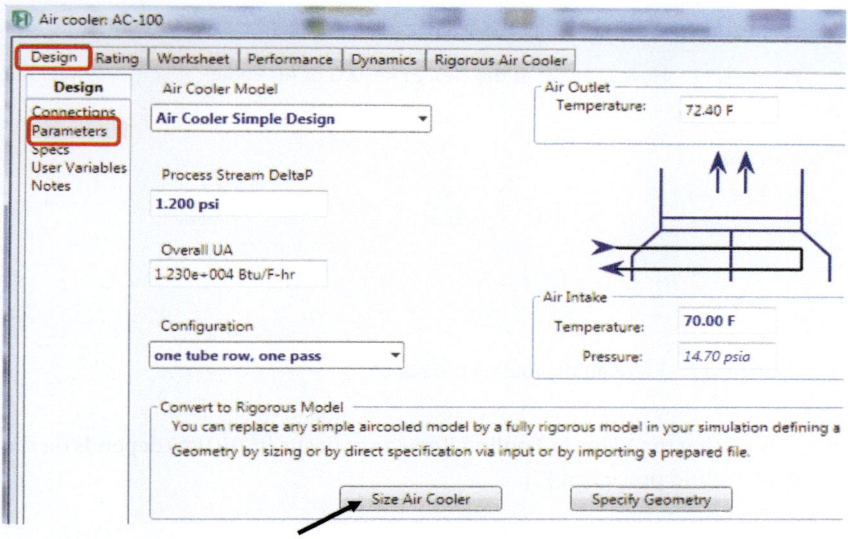

Figure 4.10 Estimating the size of Air Cooler, parameters page, design tab.

For more accurate results: In the **Process** tab, Figure 4.12, specify the fouling resistance of 0.00018 m^2K/W in the Hot Side, and 0.00035 m^2K/W in cold side, then click [Size Air Cooler] another time.

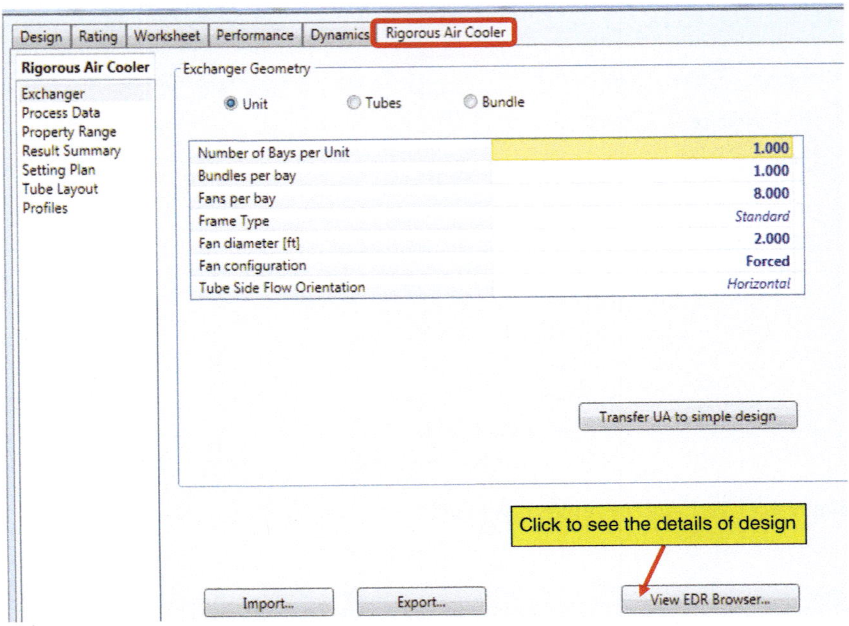

Figure 4.11 How to arrive at the Design Results of Air Cooler.

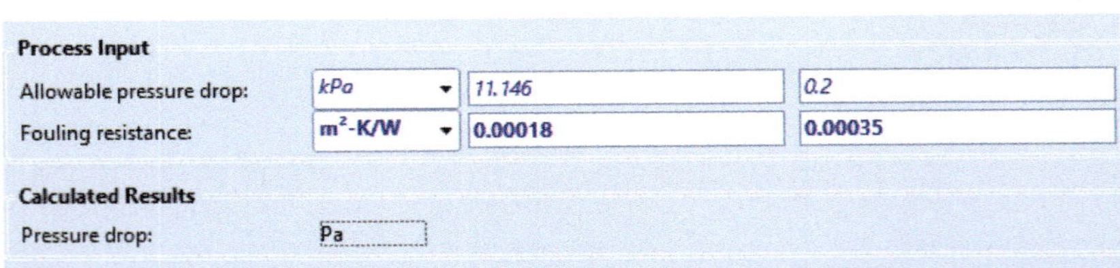

Figure 4.12 Entering the fouling resistance.

Selecting the Value of Fouling Resistance in (hr.ft^2°F/Btu) depends on the type of fluid processed [7].

4.3 Fired Heater

It performs energy and material balances in steady-state or dynamic modes to model a direct Fired Heater-type furnace. This type of equipment requires a large amount of heat input. Heat is generated by fuel combustion and

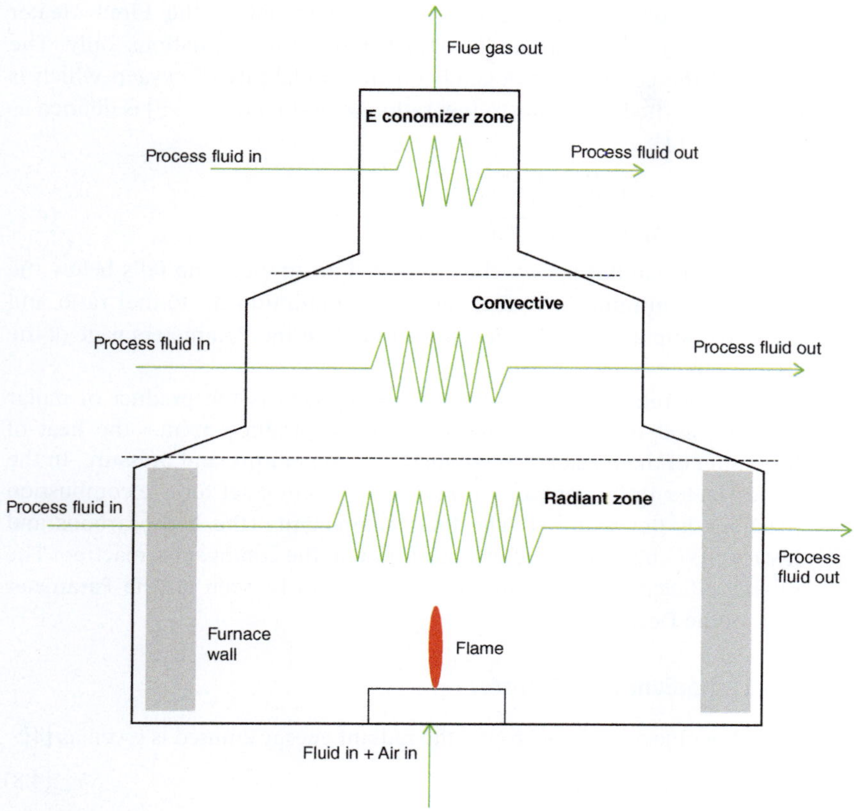

Figure 4.13 Schematic diagram of the Fired Heater.

transferred to process streams. A simplified schematic of a direct Fired Heater is illustrated in Figure 4.13 [6].

In general, a Fired Heater can be divided into three zones:

- Radiant zone
- Convective zone
- Economizer zone.

Note: *The dynamic mode does not support multiple fuel streams, but uses one combined fuel/air stream. If there are multiple fuel streams defined for steady state, when switching from steady state to dynamics a new stream is created from the sum of the fuel streams and the air stream is defined in steady state. When switching back to steady state, one fuel and one air stream is created from the dynamic settings* [1, 14].

The combustion reaction in the burner model of the Fired Heater performs pure hydrocarbon (CxHy) combustion calculations only. The extent of the combustion depends on the availability of oxygen which is usually governed by the air-to-fuel ratio. Air-to-fuel ratio (AF) is defined as follows [6, 14]:

$$AF = \frac{\left(\frac{Mass\ \ of\ \ flow\ \ O_2)}{\sum Mass\ \ flow\ \ of\ \ fuel}\right)}{Mass\ \ Ratio\ \ of\ \ O_2\ \ in\ \ Air} \tag{4.7}$$

The flame cannot light if the calculated air-to-fuel ratio falls below the specified minimum air-to-fuel ratio. The minimum air-to-fuel ratio and the maximum air-to-fuel ratio can be found on the Parameters page of the Design tab.

The heat released by the combustion process is the product of molar flowrate, and the heat of formation of the products minus the heat of formation of the reactants at combustion temperature and pressure. In the Fired Heater unit operation, a traditional reaction set for the combustion reactions is not required. The fuels components (the hydrocarbons and hydrogen) to be considered can be chosen in the combustion reaction. The mixing efficiency of each fuel component can be seen on the Parameter page of the Design tab.

4.3.1 Radiant Heat Transfer

For a hot object in a large room, the radiant energy emitted is given as [4]:

$$Q_{radiative} = \delta A \varepsilon \left(T_1^4 - T_2^4\right) \tag{4.8}$$

where:

δ = Stefan–Boltzmann constant, 5.669×10^{-8} W/m^{2}K^4
ε = emissivity, (0–1), dimensionless
A = area exposed to radiant heat transfer, m^2
T_1 = temperature of surface 1, K
T_2 = temperature of surface 2, K.

4.3.2 Convective Heat Transfer

The convective heat transfer, taking part between a fluid and a metal, is given in the following [4]:

$$Q_{convective} = UA(T_1 - T_2) \tag{4.9}$$

where:

U = overall heat transfer coefficient, W/m^{2}K
A = area exposed to convective heat transfer, m^2

T_1 = temperature of surface 1, K
T_2 = temperature of surface 2, K

The U actually varies with flow according to the following flow-U relationship if this Flow-Scaled method is used [3, 4]:

$$U_{used} = U_{specified} \frac{1}{2} \left(\frac{Mass\ flow\ at\ time\ \ t}{Reference\ \ Mass\ flow} \right)^{0.8} \tag{4.10}$$

where:

$U_{specified}$ = *U value at steady-state design conditions.*

The ratio of mass flow rate at time t to reference mass flow rate is also known as flow-scaled factor. The minimum flow-scaled factor is the lowest value, where the ratio is anticipated at low flow region. For the Fired Heater operation, the minimum flow-scaled factor can be expressed only as a positive value.

For example, if the minimum flow-scaled factor is +0.001 (0.1%), when this mass flow ratio is achieved, the U_{used} stays as a constant value. Therefore,

$$U_{used} = U_{specified}(0.001)^{0.8} \tag{4.11}$$

4.3.3 Conductive Heat Transfer

Conductive heat transfer in a solid surface is given as [2, 8]:

$$Q_{conductive} = -kA\frac{(T_1 - T_2)}{\Delta t} \tag{4.12}$$

where:

k = thermal conductivity of the solid material, W/mK
Δt = thickness of the solid material, m
A = area exposed to conductive heat transfer, m^2
T_1 = temperature of inner solid surface 1, K
T_2 = temperature of outer solid surface 2, K.

4.3.4 Problem Statement

The feed stream (molar flow rate in kmol/h: C_1 = 300, C_2 = 250, C_3 = 200, i-C_4 = 150, n-C_4 = 150, i-C_5 = 100, n-C_5 = 100) at a temperature of 50°C and a pressure of 5 atm is heated to 300°C by a Fired Heater using methane gas as a fuel (40°C, 200 kPa) with 100% excess air at 40°C and 200 kPa. What is the temperature of flue gas? What is the molar flow rate of the fuel? And what is the amount of air required if the efficiency of furnace is 60%?

4.3.5 Simulation

1) Open a new case and add the components as in problem above (C_1, C_2, C_3, i-C_4, n-C_4, i-C_5, n-C_5, O_2, and N_2).
2) **Fluid package:** Peng–Robinson.
3) Enter the Simulation environment.
4) Construct a **Feed stream** with conditions as mentioned in the problem above ($T = 50°C$, $P = 5$ atm, molar flow rate (kmol/h): $C_1 = 300$, $C_2 = 250$, $C_3 = 200$, i-$C_4 = 150$, n-$C_4 = 150$, i-$C_5 = 100$, n-$C_5 = 100$).
5) Construct the **Air stream** with molar composition (0.21 O_2 and 0.79 N_2) and with conditions as mentioned in the problem above ($T = 40°C$, $P = 200$ kPa).
6) Construct the **Fuel stream** with data as mentioned in the problem above ($T = 40°C$, $P = 200$ kPa, composition: 100% C_1).
7) From Object Palette, add Fired Heater  to the simulation area.
8) In the Connection page - Design tab, connect the previous streams above as shown in Figure 4.14.
9) Switch to the **parameters** page - **Design** tab, and specify the Excess Air = 100% and Efficiency = 60%, Figure 4.15.
10) From the **Worksheet,** specify the temperature of the product by 300°C.

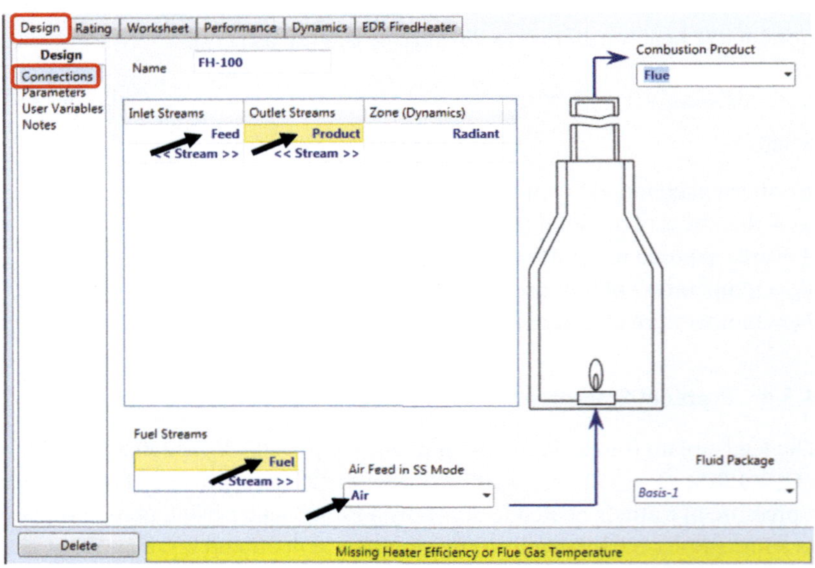

Figure 4.14 Fired Heater, connections page, design tab.

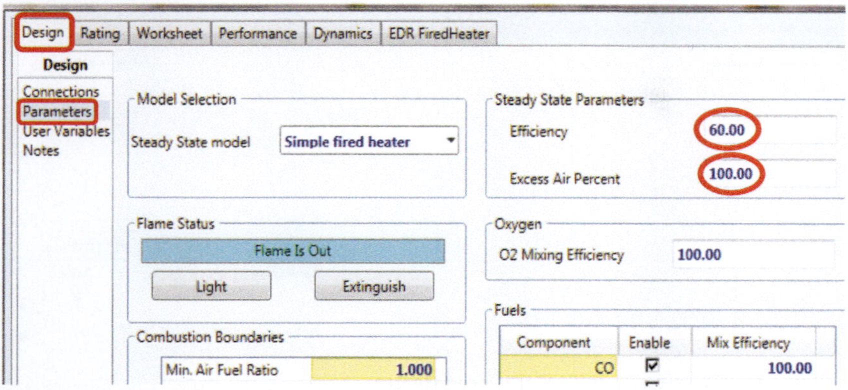

Figure 4.15 Fired Heater, parameters page, design tab, entering the efficiency and excess air percent.

Figure 4.16 Process flow diagram of Fired Heater after convergence.

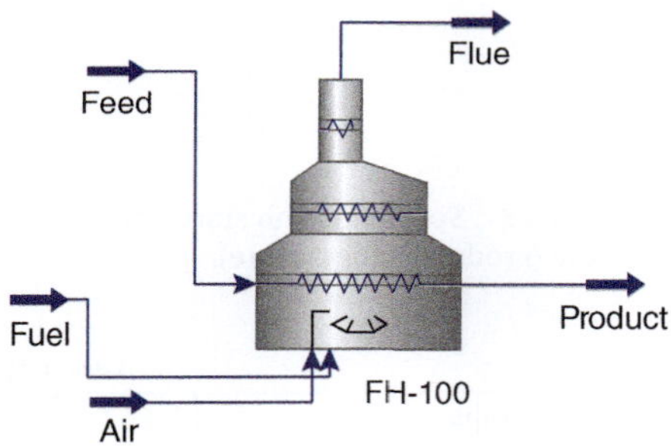

Figure 4.16 shows the final view of the Fired Heater as PFD. What is the molar flow rate of the fuel and air required?

4.3.6 Case Study

To study the effect of the temperature of the product on the molar flow of the fuel required to heat the feed to the 300°C, create a case study as follows:

1) Independent variable = Temperature of the product, Figure 4.17.
2) Dependent variable: Molar flow of the Fuel, Figure 4.17.
3) From **Case Study Setup,** Figure 4.17, Specify the range of the Fuel's Temperature: Start value = 250°C, End value = 400°C, and the step size = 10°C, Figure 4.18.
4) Click **Run** and then click plot to see the results, Figure 4.19.

It is seen from Figure 4.19 that the molar flow rate of the fuel is increased with increasing temperature of the product. When the amount of the fuel

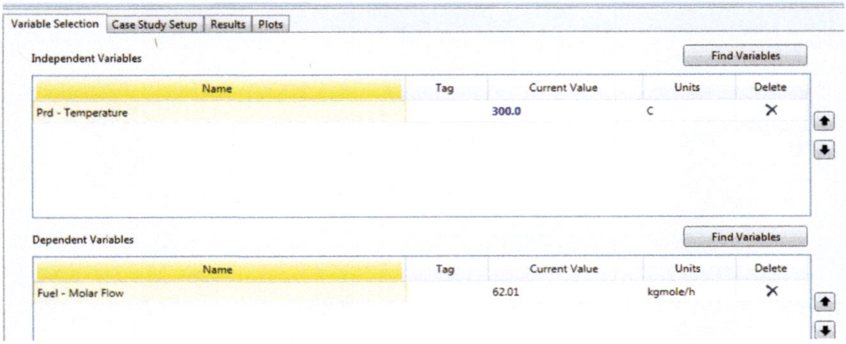

Figure 4.17 Case study of the Fired Heater.

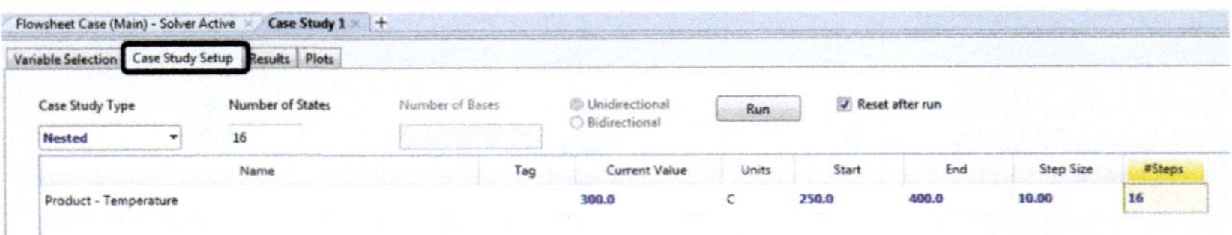

Figure 4.18 Specifying the start, end, and step size values of independent variable (Product Temperature).

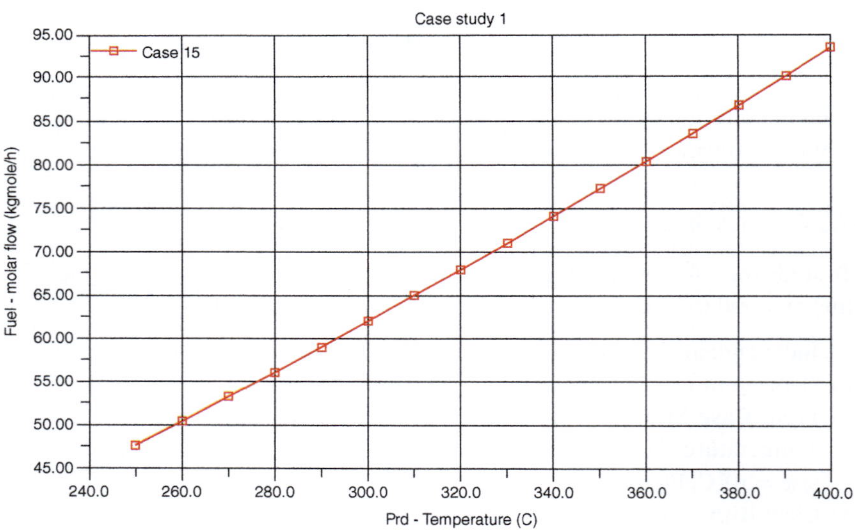

Figure 4.19 Results of case study of the Fired Heater, molar flow of fuel as a function of product temperature.

increased the amount of air required will also increase. The temperature of the product at 300°C requires molar flow of the fuel equal to 62.01 kmol/h.

So, it can be concluded from Figure 4.19, that the fuel required increases when the temperature of the product increased.

4.3.7 Design of Fired Heater

To obtain the sizing of the Fired Heater, follow the following procedure:

1) Delete the excess air and efficiency of the Fired Heater from **Parameters** page, because the simulator will calculate them by **EDR** (Exchanger Design and Rating) model.
2) In **Parameters** page, **Model Selection**, select **EDR FiredHeater** from the drop-down list, Figure 4.20.
3) Click the **EDR FiredHeater** tab; Figure 4.21 will appear.
4) There are some templates of a Fired Heater that can be imported. From Figure 4.21, **EDR FiredHeater** tab, Click **Import** button.

   The templates of Fired Heater are found in [C:\Program Files (x86)\AspenTech\Aspen Exchanger Design and Rating V9.0\Examples\FiredHeater]. Select **QFiredHeater1** and click Open. The **EDR Fired Heater** will be solved as shown in Figure 4.22, where the summary results are shown.

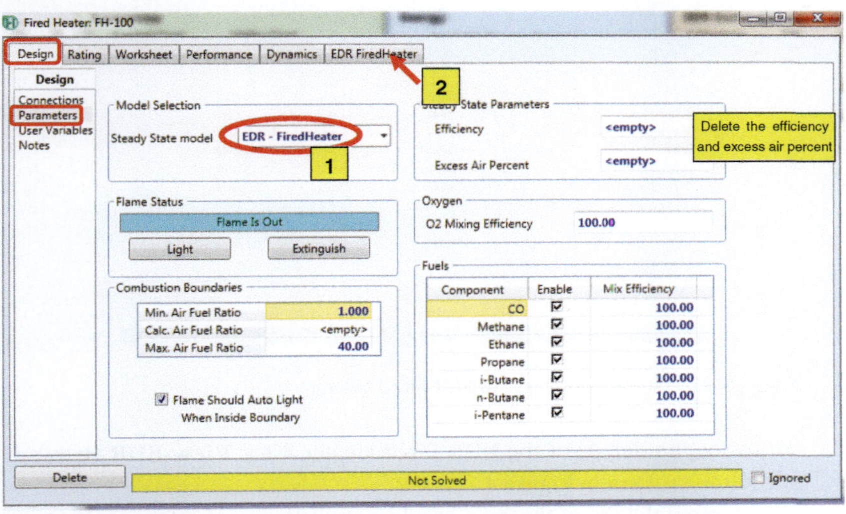

Figure 4.20 Convert the Fired Heater from Simple to EDR.

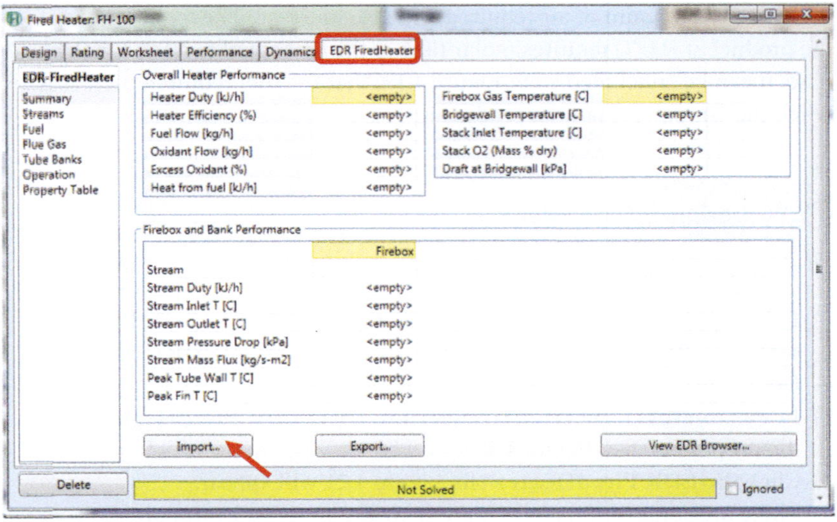

Figure 4.21 To import the templates of the Fired Heater, click Import button.

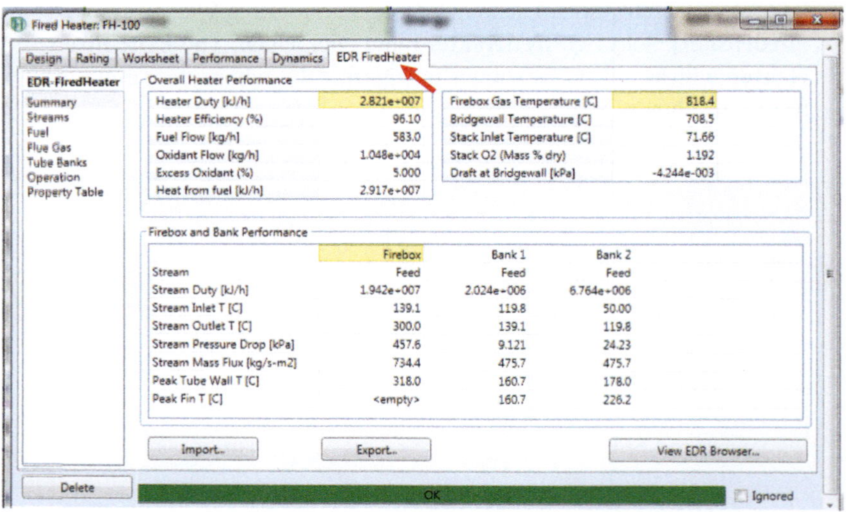

Figure 4.22 Summary results of EDR Fired Heater.

5) To see the details of the EDR Fired Heater, click **View EDR Browser** button, as seen in Figure 4.23.
6) For more details of the process and geometry, one can browse between the pages as in Figures 4.24–4.26.

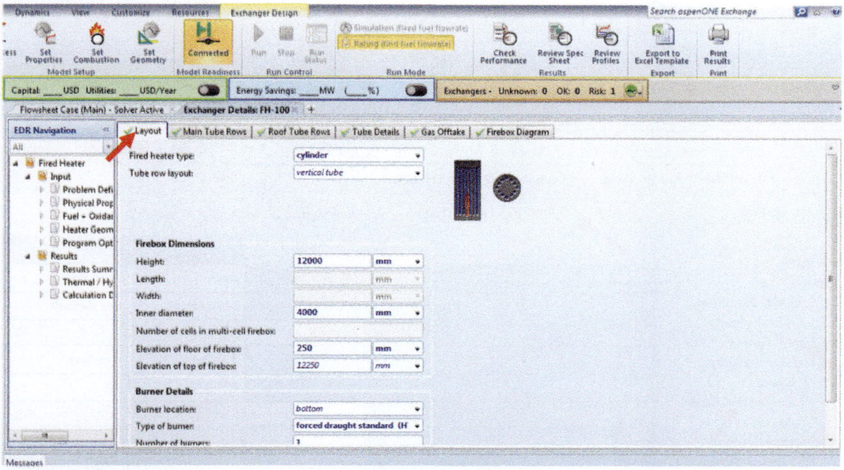

Figure 4.23 Window of details of EDR Fired Heater, Layout page.

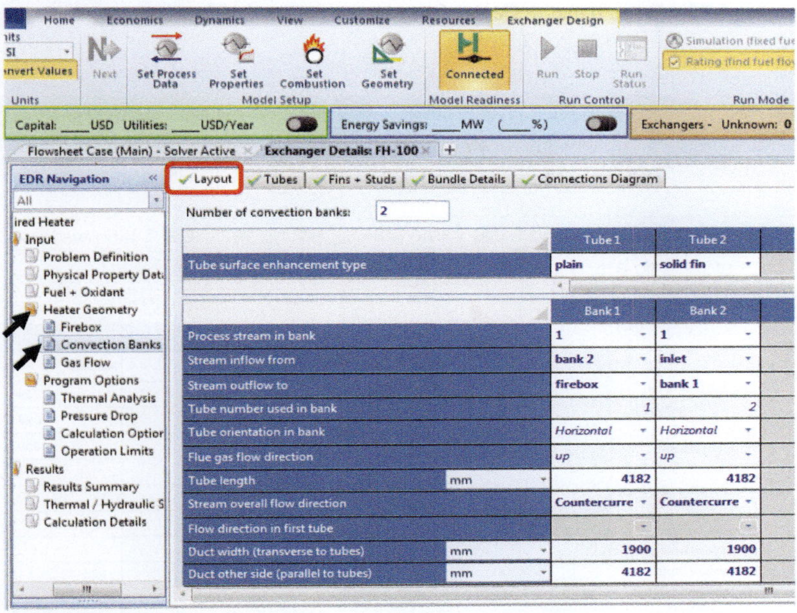

Figure 4.24 Layout page > Heater Geometry> Convective Banks.

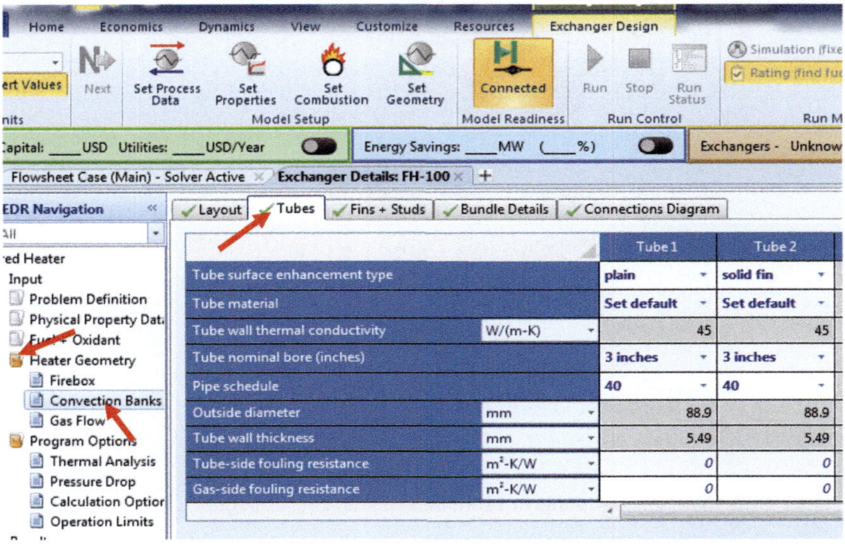

Figure 4.25 Tubes pages > Heater Geometry > Convective Banks.

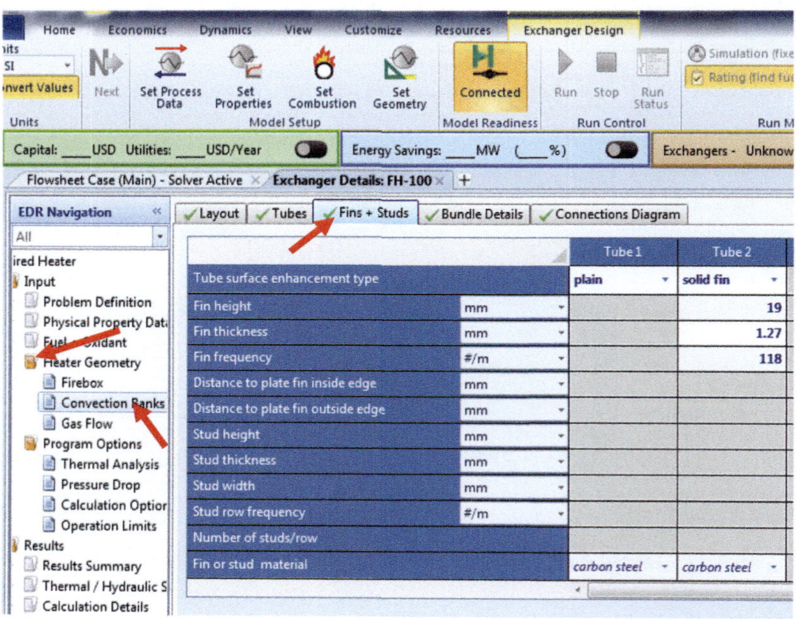

Figure 4.26 (Fins + Studs) pages > Heater Geometry > Convective Banks.

4.4 Summary

When different substances involve different sorts of processes, the heat-exchange modules are particularly versatile and allow for multiple calculation methods. Their computational models have been extensively researched and are being employed in the computation of various types, ranging from mechanically intricate designs to thermodynamic exchangers. When evaluating the performance of equipment in the early stages of the process, Aspen HYSYS is a great resource. This is especially true when using pre-existing equipment. On the other hand, the Aspen Exchanger Design and Rating® interface enables the performance of accurate and thorough estimations when a rigorous heat-exchanger design is necessary. In addition, it permits application-dependent settings where the equipment results acquired can be applied to building projects. A heat exchanger uses two streams to supply/absorb energy; it may either be that both of them are liquids or gases, or one of them liquid and the other is gas [15, 16].

An air cooler uses the air to cool or heat the process fluid, its efficiency is less than that of a heat exchanger, but it can increase its efficiency by increasing the flow rate of air or/and increasing the number of fans. Air coolers are mostly used as a pre-cooler in the chemical, petrochemical, and refining industries. The operation of air coolers depends on the convection and conduction of heat from a process fluid (liquid or gas) into the surrounding air. Heat transfer is facilitated by forcing air across the exterior surfaces of the tubes that the process fluid passes through in a loop or series of tubes.

In the Fired Heater, since the energy required is huge and high temperature of fluid output is needed, the source of energy supplying is the combustion of fuel in the present of air. It is widely used in the oil and gas, refining, chemical, and power generation industries. Fired Heaters are designed to maximize thermal efficiency, often achieving efficiencies of 80% or higher.

The design procedure for the above equipment is almost similar, and it is possible to control the dimensions and choose the appropriate design through the template saved in the simulator.

Problems

P.4.1 The mixture of toluene and benzene, which contains 0.4 and 0.6 by mass of benzene and toluene, respectively, is cooled using a heat exchanger. At 1.25 bar, 58°C, and a flow rate of 4400 kg/h, the feed enters the heat exchanger. It exits the heat exchanger at 32°C. Water serves as the coolant medium, and it enters the heat exchanger at a temperature of 23°C and 1.28 bar. The pressure drops across the tubes and shell are 12 and 7 kPa, respectively. Determine the mass flow rate of water required [14].

P.4.2 A heat exchanger is used to raise the temperature of a binary combination of acetic acid and ethanol from 20 to 50°C while using hot water at 75°C. Calculate the hot water needed if the feed's flow rate and pressure are 4800 kg/h and 125 kPa. Hot water enters the heat exchanger at a pressure of 125 kPa and exits at 58°C. Assume a 6 kPa pressure drop across the tubes and shell [14].

P.4.3 The feed stream (mole flows kmol/h: $C_1 = 300$, $C_2 = 250$, $C_3 = 200$, $i\text{-}C_4 = 150$, $n\text{-}C_4 = 150$, $i\text{-}C_5 = 100$, $n\text{-}C_5 = 100$) at a temperature of 50°C and pressure of 5 atm is heated to 300°C by a Fired Heater and using a methane gas as a fuel (40°C, 200 kPa) with 100% excess air at 40°C and 200 kPa. What is the temperature of the flue gas? What is the molar flow rate of the fuel? And what is the amount of air required if the efficiency of furnace is 60%? If the flue gases are used to generate steam in a heat exchanger, what is the mass flow rate of water when converted to steam? Assume the temperature of flue gases outlet from exchanger is 150°C and the conditions of cold water are ($P = 150$ kPa and $T = 20$°C). Assume the pressure drop in the exchanger is 20 kPa for shell and tube. The PFD of the process is shown in Figure 4.27.

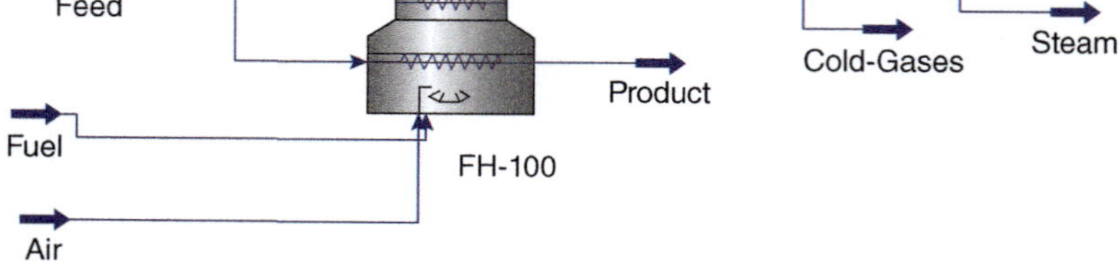

Figure 4.27 Process flow diagram (PFD) of the Problem (P.4.3).

References

1 (2017). *Aspen HYSYS User Guide Version 10.0*. Aspen Technology.
2 Backhurst, J.R. and Harker, J.H. (1973). *Process Plant Design*. Heinemann Educational Books Ltd.
3 AEA Technology, "*Process Modeling Using HYSYS with Chemical Industry Focus*", 2000.
4 Coulson, J.M. and Richardson, J.F. (2003). *Chemical Engineering Design*, Chemical Engineering, 3e, vol. 6 (ed. R.K. Sinnott).
5 Ludwig, E.E., Rouge, B., and Louisiana (1999). *Applied Process Design for Chemical and Petrochemical Plants*, Volumes 3, 3e. Butterworth-Heinemann.
6 (2005). *HYSYS® 2004.2, Tutorials and Applications*. Aspen Technology, Inc.
7 Michael E. Hanyak, Jr., "*Chemical Process Simulation and the Aspen HYSYS Software*", 2012.
8 Sittler, L. and Ralston, T. (2013). *Jump Start: Aspen Shell and Tube Exchanger V8.0*. Aspen Technology, Inc. AspenTech®, aspenONE®.
9 Sittler, L. (2013). *Jump Start: Activated Exchanger Design and Rating in Aspen HYSYS®*. Aspen Technology, Inc. AspenTech®, aspenONE®.
10 Rohsenow, W.M., Hartnett, J.P., and Cho, Y.I. (1998). *Handbook of Heat Transfer*, 3e. New York: McGraw-Hill.
11 Dimian, A.C., Bildea, C.S., and Kiss, A.A. (2014). *Integrated Design and Simulation of Chemical Processes*. Elsevier.
12 Foo, D.C.Y., Chemmangattuvalappil, N., Ng, D.K.S. et al. (2017). *Chemical Engineering Process Simulation*. Elsevier Inc.
13 Chaves, I.'n.D.'o.G. (2016). *Process Analysis and Simulation in Chemical Engineering*, ©. Springer.
14 Haydary, J. (2019). *Chemical Process Design and Simulation Aspen Plus and Aspen HYSYS Applications*. John Wiley and Sons, Inc.
15 McCabe, W., Smith, J., and Harriott, P. (2004). *Unit Operations of Chemical Engineering*, 7e. McGraw Hill.
16 Turton, R. and Shaeiwitz, J.A. (2017). *Chemical Process, Equipment Design*. Pearson Education, Inc.

5

Flash Separator

A flash separator, also known as a flash drum or flash vessel, is a type of process equipment used in the oil and gas, chemical, and petrochemical industries to separate a pressurized stream into its vapor and liquid components [1, 2]. The first issue in this chapter involves determining the composition and flowrate of the liquid and vapor outlet streams of the flash separator. The separator separates the vessel's contents into their individual liquid and vapor phases in steady-state operation. Before being separated, the vapor and the liquid in the vessel are given time to attain equilibrium. To ascertain the conditions and stages of the product, a flash separator is used. The lowest feed pressure minus the pressure drop throughout the vessel is the pressure at which the flash is executed. The aggregate feed enthalpy plus or minus the duty equals the enthalpy [3, 4].

The separator has the capacity to compute results in reverse. The separator can use a known product composition to determine the composition of the other product stream(s), as well as by balancing the feed composition, in addition to the usual application (fully defined feed stream(s) being separated at the vessel pressure and enthalpy) [4, 5].

There are two-phase separators (vapor and liquid) and three-phase separators (vapor, organic liquid, and aqueous liquid). Also the separator may be spherical cylinder, vertical cylinder, or horizontal cylinder [3].

Figure 5.1 shows the schematic diagram of a single-stage liquid–vapor continuous separation process of a flash separator. Assuming an equilibrium stage: $P_V = P_L$ and $T_L = T_V$.

The following equations represent the mathematical model of the process [4–6]:

$$n_F x_{Fi} = n_L x_i + n_V y_i \tag{5.1}$$

$$y_i = K_i x_i \tag{5.2}$$

Chemical Process Simulations using Aspen Hysys, First Edition. Khalid W. Hameed.

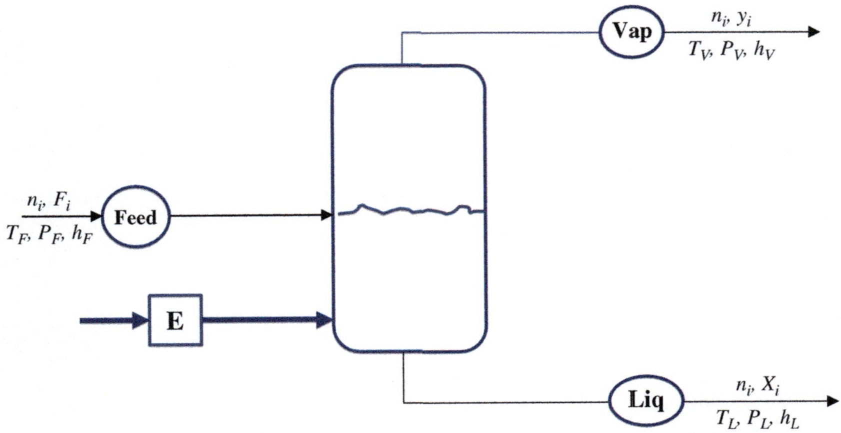

Figure 5.1 Scheme of a continuous single-stage liquid–vapor separation.

$$K_i = f\,(T, P, x_i, y_i) \tag{5.3}$$

$$n_F h_F + Q = n_L h_L + n_V h_V \tag{5.4}$$

$$\sum x_{Fi} = 1, \sum x_i = 1, \sum y_i = 1 \tag{5.5}$$

$$h_F = f\,(T_F, P_F, x_{Fi}), h_L = f\,(T_L, P_L, x_i), h_V = f\,(T_V, P_V, y_i) \tag{5.6}$$

where:

n_i is the moles of the stream i
y_i is the mole fraction in the vapor phase
x_i is the mole fraction in the liquid phase
K_i is the equilibrium constant of component i
h_i is the enthalpy of accompanying stream i
F, L, V are streams of feed, liquid, and vapor, respectively [7, 8].

5.1 Simulation of Flash Separator

5.1.1 Problem Statement

An equimolar mixture of a saturated liquid of benzene and toluene at 200 kPa and a flow rate of 100 kmol/h is subjected to flash distillation in the separator. The feed mixture is passed through a valve with a pressure drop of 50 kPa. Determine the composition of the liquid and vapor leaving the separator when the feed is 15% vaporized [9].

5.1.2 Simulation

1) Open a new case and enter the following components: Methane, ethane, and propane.
   Fluid package: Peng–Robinson
2) Enter the **Simulation** environment
3) Add a new **Material** stream with the following values.

Name	Feed
Vapor fraction	0.0
Pressure	200 kPa
Molar Flow	100 kmol/h
Molar Compositions	Benzene = 0.5, Toluene = 0.5

4) Add a Valve 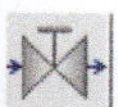, and a Separator to the simulation area.
5) Connect the process flow diagram as in Figure 5.2.
6) Double-click the Valve, Design tab, Parameters page, enter the pressure drop = 50 kPa.
7) Double-click the separator, Design tab, Parameters page, enter the pressure drop = 50 kPa, Figure 5.3.
8) Right click the Vap stream and select Show table, Figure 5.4.
9) As you see, the molar flow of Vap stream is 12.89 kmol/h (i.e. 12.89% of feed stream).

   To vaporize 15% of the feed, by trial and error, increase the pressure drop of the separator until the molar flow rate of Vap stream becomes 15 kmol/h.

   After the increase the pressure drops to 62.6 kPa, the flow of Vap stream reached 15 kmol/h as in Figure 5.5.

   Now, check the conditions and composition of the liquid and vapor streams.

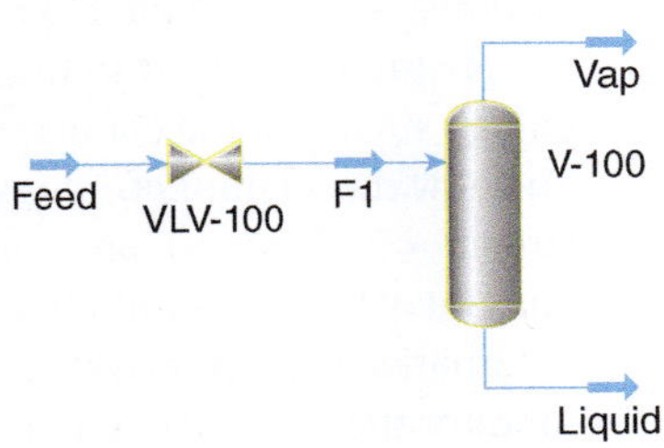

Figure 5.2 Connecting the material streams with the valve and separator.

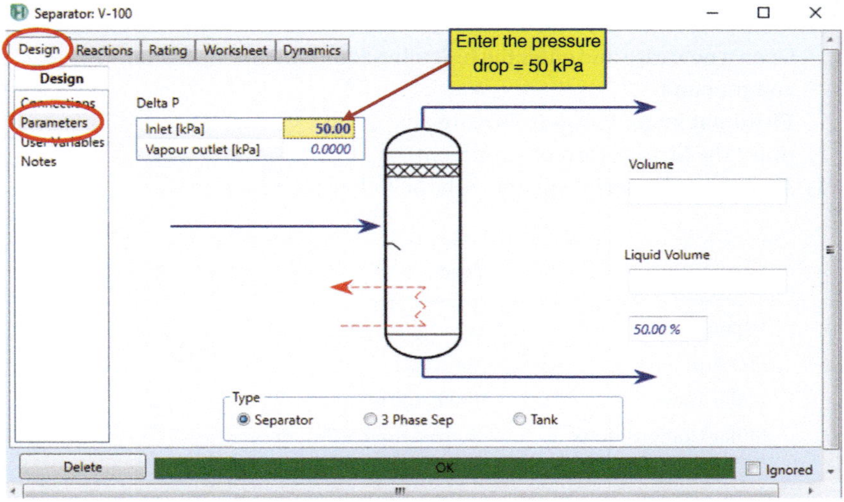

Figure 5.3 Window of separator, Design tab, Parameters page, entering the pressure drop = 50 kPa.

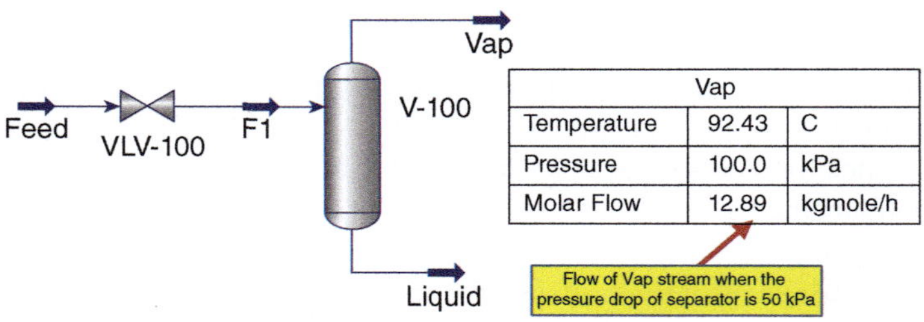

Figure 5.4 Process flow diagram of valve and separator with table of results of Vap stream.

Benzene in vapor phase = 67.8% and the rest is toluene.

If it is required to vaporize 20% of the feed, what is the pressure drop of the separator?

10) Another way to obtain 15 kmol/h of Vap stream. Instead of reducing the pressure, add heat to the separator. To execute this operation, add energy stream in the connection page of the separator, Figure 5.6.
11) To know the amount of energy required to vaporize 15% of the feed (Vap stream = 15 kmol/h), one can do that either by trial and error for the amount of Q-energy until the flow of the Vap stream reaches 15 kmol/h. Alternatively, enter the value of the flow of Vap stream by 15 kmol/h and monitor the value of Q-energy, Figure 5.7.

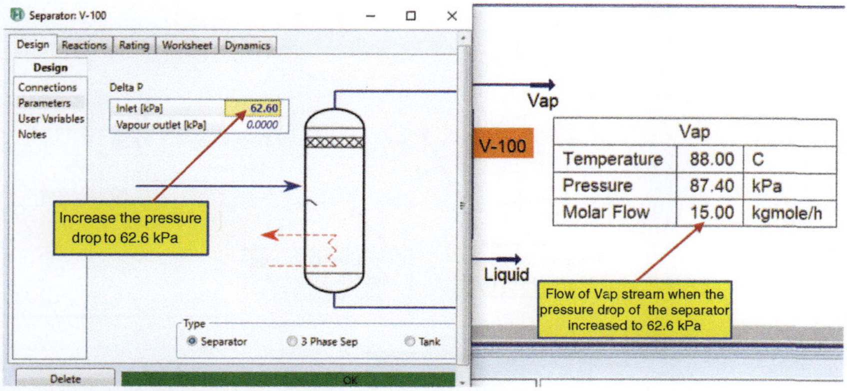

Figure 5.5 After increasing the pressure drop of separator to 62.6 kPa, the flow of Vap stream reached 15 kmol/h (15% of feed).

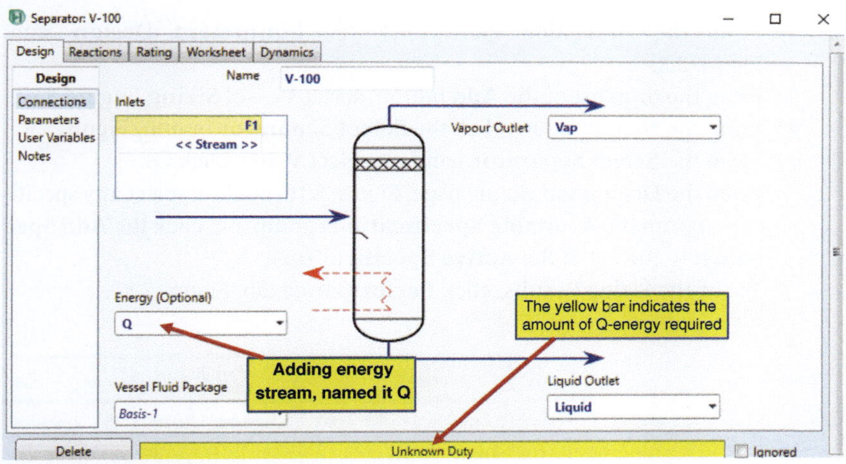

Figure 5.6 Adding energy stream to the separator.

Now, check the conditions and composition of liquid and vapor; are they different from the previous method?

The benzene in vapor phase = 67.6% and the rest is toluene.

Create a case study to see the effect of the amount of heat added on the vaporized feed.

Study the difference between the cost of a reduced pressure of separator and the heat added so that it vaporizes 15% of the feed.

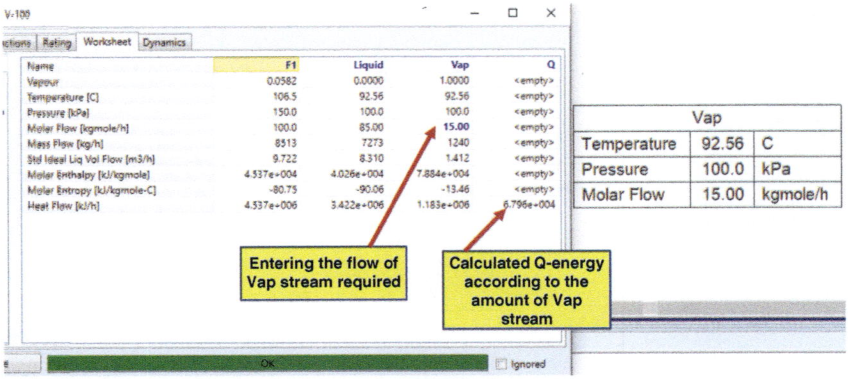

Figure 5.7 Entering the flow of Vap stream required from which the Q-energy will be calculated.

5.1.3 Vessel Design

1) From the Navigation Pane, click the **Equipment Design** icon, Figure 5.8.
2) From the drop list of the **Add** button, select **Vessel Sizing**, Figure 5.8.
3) From the New Window, click the **Select Separator** button, Figure 5.9.
4) From the **Select Separator** window, select V-100, click OK.
   From the **Design** tab, Sizing page, Figure 5.10, one can select any specification from the **Available Specifications** group and click the **Add Spec** button to the list of the **Active Specifications**.
5) To see the Sizing Results, click **Performance** tab, Figure 5.11.

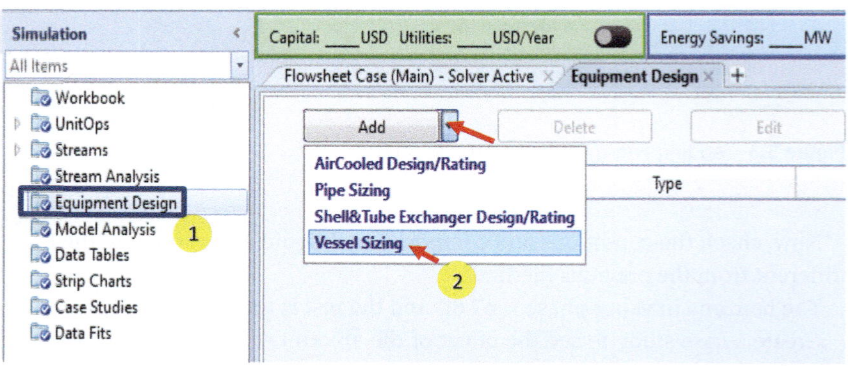

Figure 5.8 Selecting vessel/Separator for executing the design.

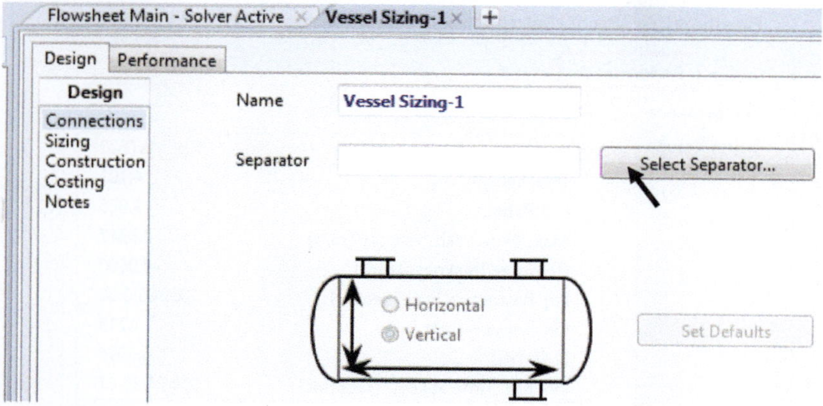

Figure 5.9 Window of Vessel Sizing, Design tab Connection page, Select Separator.

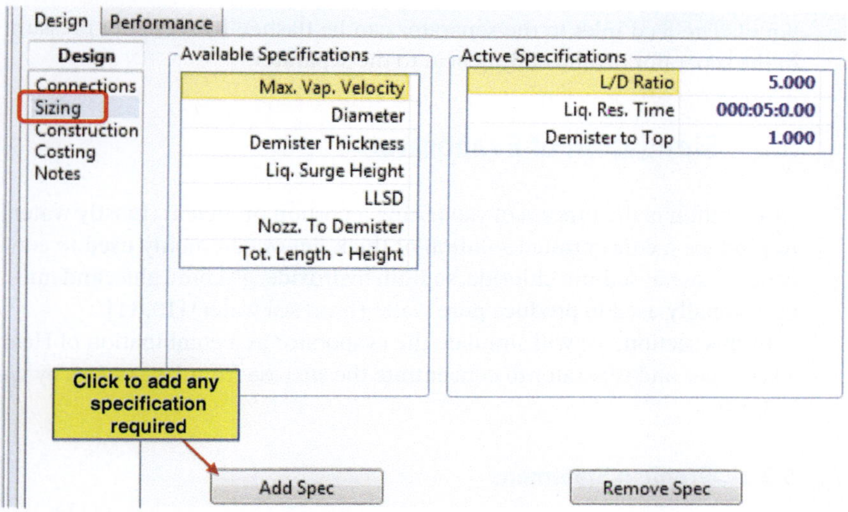

Figure 5.10 Specifications of vessel design.

5.1.4 Summary

Flash separator is an essential equipment in many industrial processes, contributing to the efficient separation and recovery of valuable components from complex mixtures. The pressure drop of the separator affects the vaporized fraction of the inlet feed and then the composition of vapor and

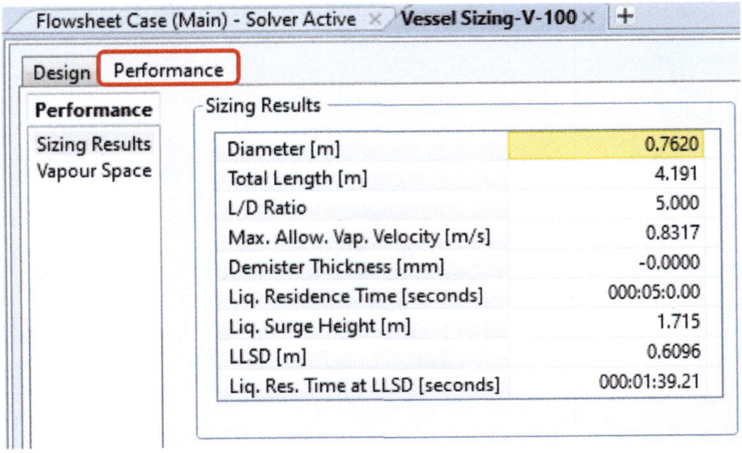

Figure 5.11 Sizing results of the separator in the performance tab.

liquid. The feed inlet to the separator can be flashed by reducing pressure in the separator or/and adding heat to the separator.

5.2 Simulation of Evaporator

Evaporation is the process of vaporizing a portion of solvent (mostly water) to produce a concentrated solution of thick liquor commonly used to concentrate sugar, sodium chloride, sodium hydroxide, glycerol, glue, and milk, occasionally used to produce pure water (from sea water) [10, 11].

In this section, we will simulate the evaporator as a combination of Heat Exchanger and separator to concentrate the sucrose from 10% to 50% by wt [12, 13].

5.2.1 Problem Statement

A 3000 kg/h sucrose solution of 10% by weight at 30°C and 110 kPa is to be concentrated to 50% using the latent heat of saturated steam at 152°C. The pressure of product in the shell is 70 kPa. Calculate

(1) The mass flow rate of steam required for this duty and the efficiency of the process.
(2) The effect of steam's temperature (120–170°C) on the mass flow of steam required.
(3) The effect of product's pressure (50–80 kPa) in the shell on the amount of steam required.

(4) The effect of product's concentration (40–60 wt%) on the steam (@152 °C) required.
(5) The effect of steam's temperature and pre-prod's pressure (shell out pressure) on the *UA* of the Heat Exchanger.

5.2.2 Simulation

Branch (1)

1) Open a new case and add the following components: Water and Sucrose. **Fluid package: *PRSV*.**
2) The evaporator will simulate as a combination of Heat Exchanger and separator. So, construct the process flow diagram as in Figure 5.12, where the feed stream enters the shell and the steam stream enters the tubes.
3) Specify the conditions of feed's stream: Temp = 30 °C, P = 110 kPa, mass flow rate = 3000 kg/h, and component fraction *by mass* = 0.1 for sucrose and 0.9 for water.
4) Specify the conditions of steam's stream: Vapor fraction = 1, Temp = 152 °C, mole fraction = 1 for water.
5) **Heat Exchanger:** Tube-side inlet = Steam, Tube-side outlet = Conds Shell-side inlet = Feed, Shell-side outlet = Pre-Prod.
6) Since the steam lost its latent heat only, set the pressure drop in the Tube-side = 0, and set the vapor fraction of Conds (Condensate) stream = 0.
   Set the pressure of Pre-prod stream = 70 kPa.
7) Now by trial and error enter the mass flow rate of the steam up to the sucrose-mass fraction in the product = 0.5 by mass.

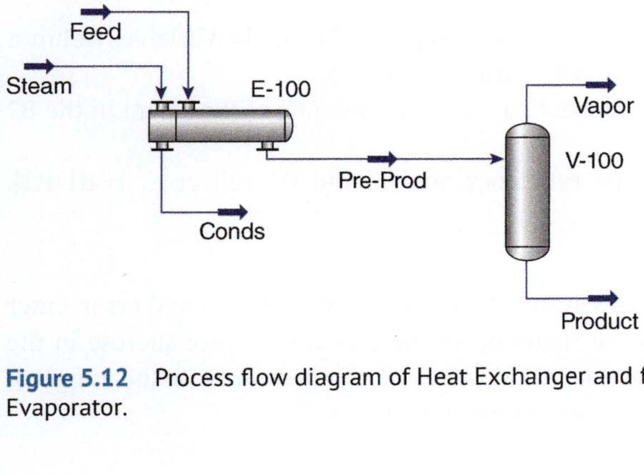

Figure 5.12 Process flow diagram of Heat Exchanger and flash separator as an Evaporator.

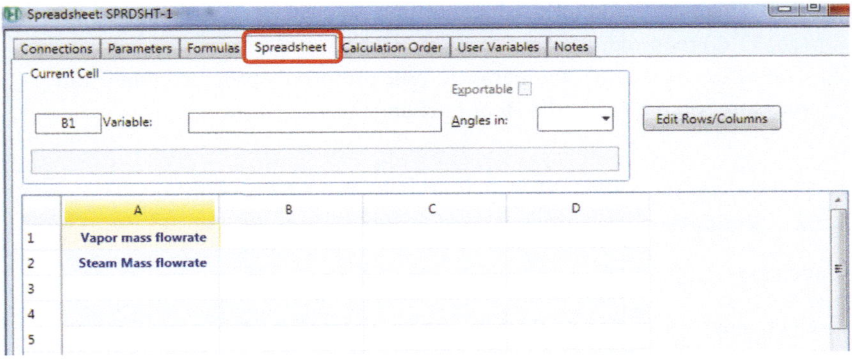

Figure 5.13 Typing the specified terms in the spreadsheet.

The required steam will be 2894 kg/h. The efficiency of the evaporator is estimated by the following equation [7, 8].

$$Efficiency = \frac{vapor \quad separated}{steam \quad required} \tag{5.7}$$

Vapor separated = Feed – Product = 3000–600 = 2400 kg/h [Estimated by material balance or taken from Vapor stream directly]

$$\rightarrow Efficiency = \frac{2400}{2894} = 0.83$$

Or it can be used in the **Spread-Sheet** as follows:

- Add the spread sheet  from Object palette and double-click it to open.
- Click the **Spread-sheet** tab, and in the A1 cell, type "Vapor mass flowrate" and in the A2 cell, type "Steam mass flowrate," Figure 5.13.
- Right click on the B1 cell and choose **Add Import Variable(s)**, Figure 5.14.
- From the Objects column select **Vapor** and from the Variables Column select **Mass Flow**, click **Done**, Figure 5.15.
- By the same way above add the mass flow rate of the steam in the B2 cell.
- In the C1 cell-type Efficiency, and in the D1 cell enter [=B1/B2], Figure 5.16.

Branch (2):

Enter the desired temperature of the steam, and by trial and error, enter the mass flow rate of the steam up to the concentration of sucrose in the product = 0.5. Repeat for other values of temperatures in the range of (120–170 °C). The results are shown in Figure 5.17.

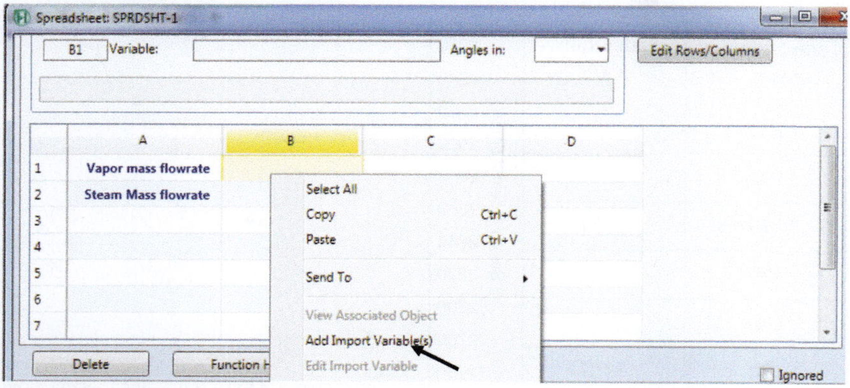

Figure 5.14 How to add import variables to the cells in the spreadsheet.

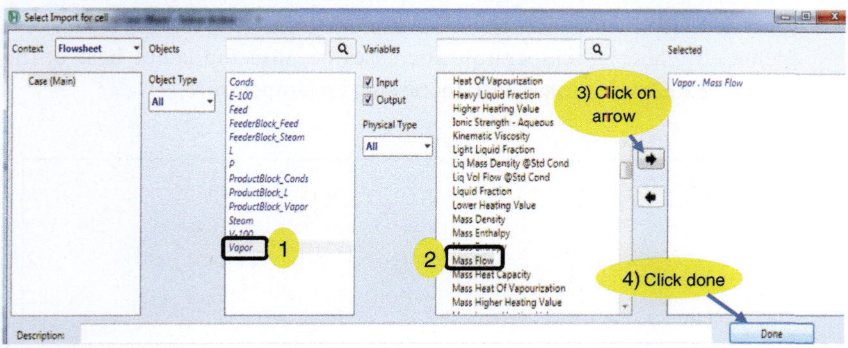

Figure 5.15 Selecting the required variables.

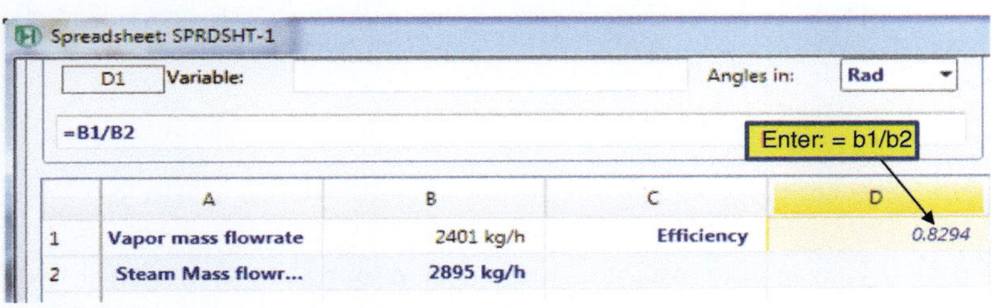

Figure 5.16 Estimating efficiency after entering the results in the cells of the spreadsheet.

Temperature of steam °C	Steam required Kg/h
120	2785
130	2817
140	2851
150	2887
160	2926
170	2967

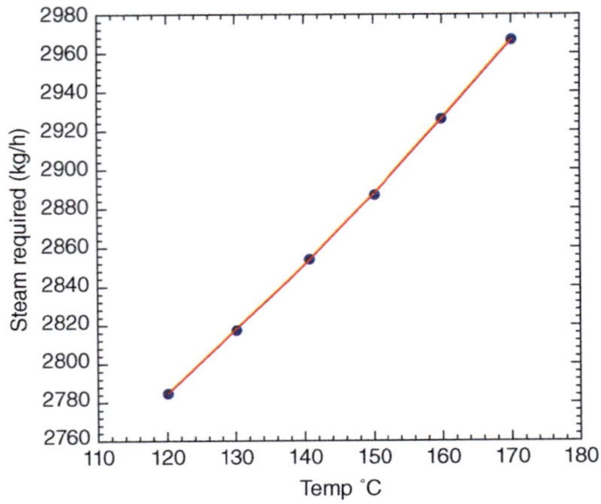

Figure 5.17 How the steam's temperature affects the steam required.

As can be seen from Figure 5.17, the required mass flow rate of the saturated steam to concentrate the sucrose from 10% to 50% by mass is increased when its temperature increases because the latent heat of the saturated steam decreases with increasing its temperature.

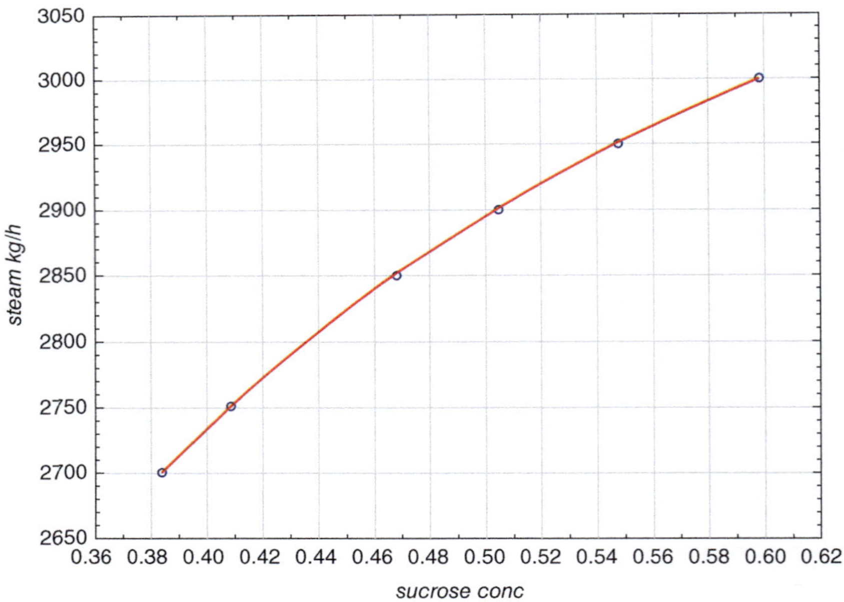

Figure 5.18 Effect of sucrose concentration in the product on the amount of steam required, branch 4.

Branch (3):

Similar to branch 2, enter the pressure of Pre-prod (50–80 kPa, step 10 kPa), and by trial and error, enter the mass flow rate of the steam at each pressure up to the mass fraction of sucrose in the product = 0.5.

Branch (4):

Here, a case study can be used in the software by changing the amount of steam (2700–3000 kg/h), and the results will be as shown in Figure 5.18.

From Figure 5.18, the mass flow rate of the steam increased when the concentration of the sucrose required increased because the amount of water separated from the solution increased.

Branch (5):

Change the temperature of the steam from 120 to 200°C, and change the pressure of the pre-prod from 50 to 80 kPa. For each point, calculate the steam required and see the value of the UA of the Heat Exchanger:

i.e. $UA = f$(steam's temp, pre-prod's pressure), we achieve the concentration of sucrose 50% [11, 14].

Problems

P.5.1 In Figure 5.19, steam is generated (basis = 100 kg/h) in power plants at a pressure of 8600 kPa and a temperature of 500°C and is fed to a turbine. Exhaust from the turbine enters a condenser at 10 kPa where it is condensed into saturated liquid which is then pumped to the

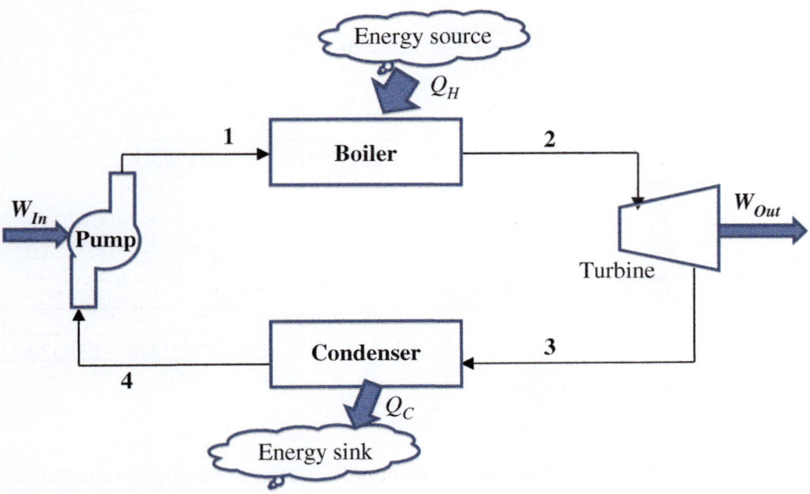

Figure 5.19 Steam power plant system.

boiler. (a) Determine the thermal efficiency of a practical cycle operating under these conditions if both the turbine efficiency and pump efficiency are 75%. (b) If the rating of the power cycle of part (a) is 80,000 kW, what is the steam rate and what are the heat-transfer rates in the boiler and condenser? [15] (Fluid Package = ASME steam or NBS steam). *Use Spreadsheet in the calculation of efficiency*

$$\textit{Where the efficiency of the cycle} = \frac{W_{out} - W_{in}}{Q_H} \tag{5.8}$$

Repeat the problem above using a Fired Heater instead of the heater (Boiler), and Shell-and-Tube Heat Exchanger instead of the cooler (Condenser). Use methane as fuel in the fired heater and water as coolant in the Heat Exchanger, both components are at atmospheric conditions.

It will simulate this problem to explain how the steam power plant is simulated:

1) Open the software, open a new case, and add the water from the components library and NBS steam as a fluid package.
2) Enter the simulation environment, and construct the PFD as in Figure 5.20.
3) Enter the given conditions ($T = 500\,°C$, $P = 8600\,kPa$, mass flowrate = 100 kg/h, and mole fraction of water = 1) in stream 2.
4) Enter the pressure drop for heater and cooler = 0 in the parameters page of each unit.
5) Enter the vapor/phase fraction = 0 in the stream 4.
6) *The efficiency of the cycle* = $\frac{W_{out}-W_{in}}{Q_H}$

$$= \frac{9.56^{*}10^{4} - 1157}{3.188^{*}10^{5}} = 0.296$$

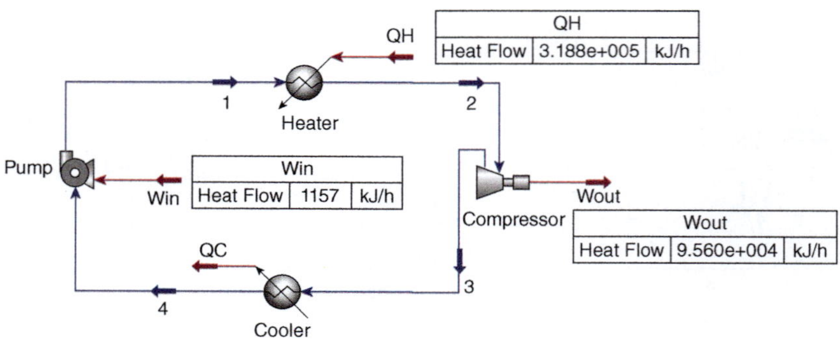

Figure 5.20 PFD of the steam power plant of P.5.1 after executing the simulation.

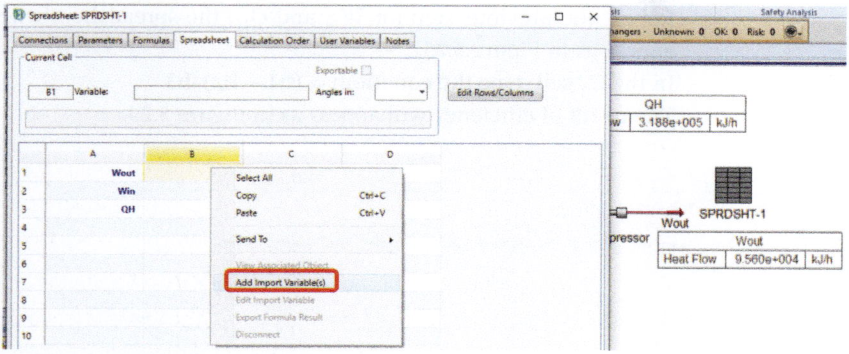

Figure 5.21 Spreadsheet; typing the energy's streams and how to import the variables in the B's cells by right clicking the required cell and select "Add Import Variable(s)."

Since W_{in} *is* $\ll$ W_{out}, *one can neglect the* W_{in}.

One can use a spreadsheet to execute these calculations as follows:

- Open the Spreadsheet by double-clicking on it from the object palette, 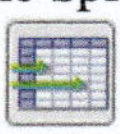
- Click the **Spreadsheet** tab, in the A's cells, type the energy's streams as in Figure 5.21.
- Right click in the B1 cell, and select "Add Import Variable(s)," select "W_{Out}" as objects, and select "*Heat Flow*" as variables, click the select arrow, then click **Done**, Figure 5.22.

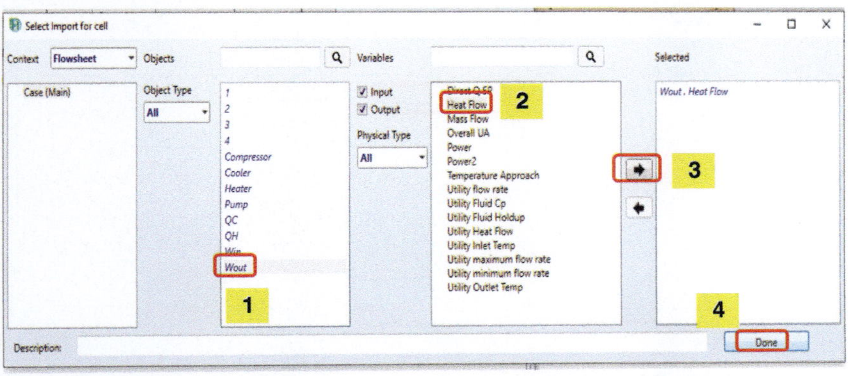

Figure 5.22 Steps to import the required variables to the spreadsheet.

- Repeat the previous step for W_{in} and Q_H, the spreadsheet will appear as in Figure 5.23.
- In the C_4 cell, type the equation: = **(b1 − b2)/b3**
- The result of efficiency will appear as in Figure 5.24.

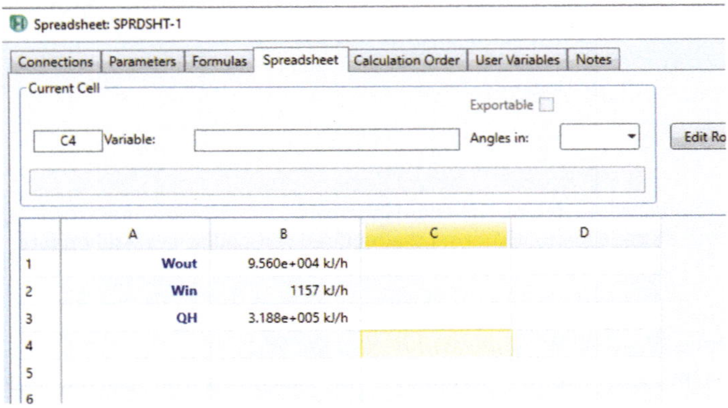

Figure 5.23 Spreadsheet after importing the energy's streams.

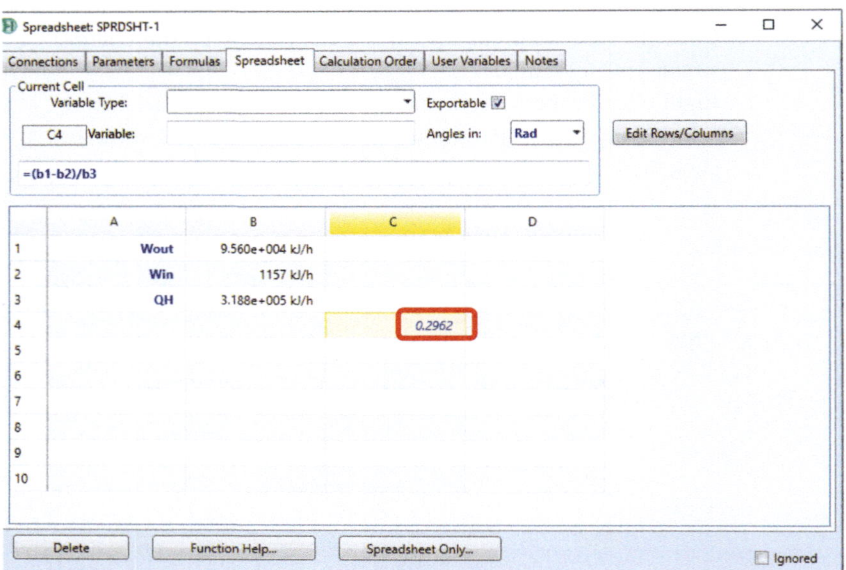

Figure 5.24 Result of efficiency of the steam power plant after calculation using a spreadsheet.

The solution of branch (2) of this problem is to delete the mass flow rate of steam from point 2 and enter the power of the expander equal to 80,000 kW. The amount of steam required will be estimated.

P.5.2 A gas stream with a flow rate of 80 kmol/h, a composition of 20% C_2, 25% C_3, 50% i-C_4, and 5% n-C_4, is available at 10°C and 1 atm. This stream needs to be cooled to 0°C after being compressed to 4 atm, Figure 5.25. It is necessary to split the resultant liquid and vapor into two product streams. What are the two streams' compositions and flow rates? [16]

P.5.3 A vapor-compression refrigeration system operates on the cycle of Figure 5.26. The refrigerant is water. Given that the evaporation temp = 277.15 K, the condensation temp = 307.15 K, efficiency (compressor) = 76%, and the refrigeration rate = 1200 kW, determine the circulation rate of the refrigerant, the heat-transfer rate

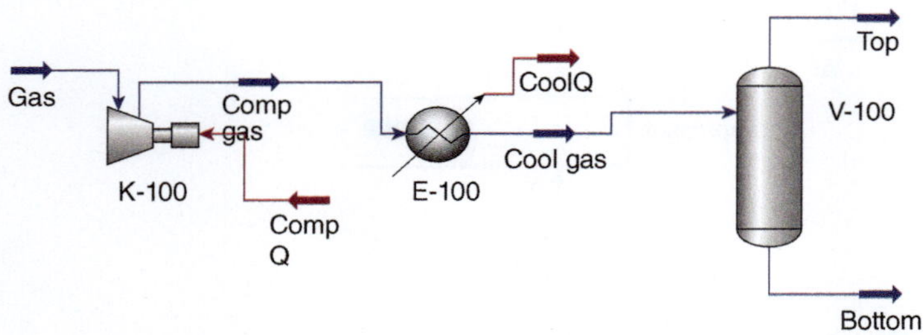

Figure 5.25 Process flow diagram of the compressor, cooler, and flash separator.

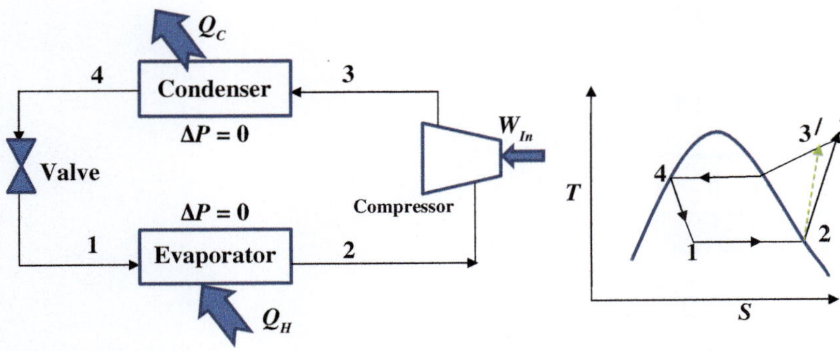

Figure 5.26 Refrigeration cycle.

in the condenser, the power requirement, and the coefficient of performance of the cycle [15].

P.5.4 Dry methane is supplied by a compressor and precooling system to the Heat Exchanger of a Linde–Hampson liquid-methane system (Figure 5.27) at 180 bar and 300 K. The high-pressure methane leaves the Heat Exchanger at a temperature of 230 K. The separator operates at 1 bar, and the product is saturated liquid at this pressure. What is the maximum fraction of the methane entering the Heat Exchanger that can be liquefied? What is the enthalpy of the high-pressure stream that enters the valve? The makeup gas enters at 5°C and 150 kPa [15]. (**Note:** Take *Basis of the gas input to the compressor* = 100 kg/h).

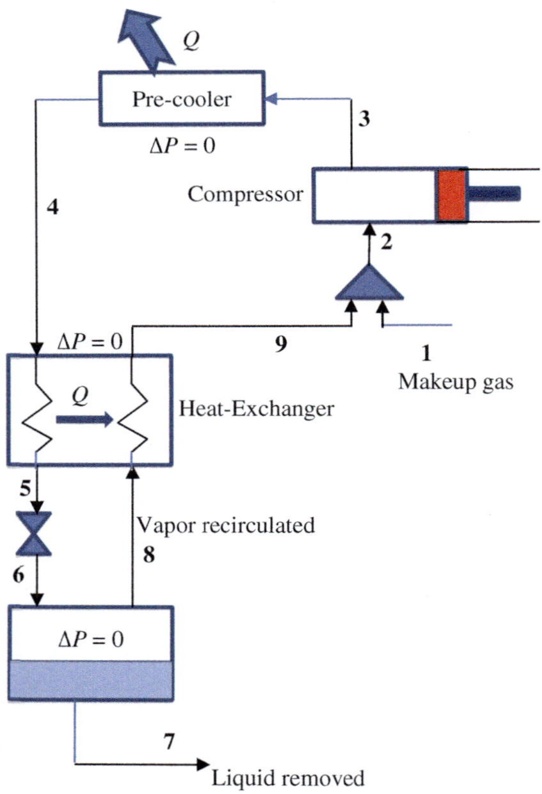

Figure 5.27 Linde–Hampson liquid-methane system.

P.5.5 Natural gas, assumed here to be pure methane, is liquefied in a Claude Process, Figure 5.28, and compressed to 60 bar. The expander and throttling exhaust to a pressure of 1 bar. The recycled methane at this pressure leaves the exchanger system (point 15) at 295 K. Assume no heat leaks into the system from the surrounding, an expander efficiency of 75%, and an expander exhaust of saturated vapor. For a draw-off to the expander of 25% of the methane entering the exchanger system ($x = 0.25$), what fraction of the methane is liquefied (z), and what is the temperature of the high-pressure stream leaving the cooler? The temperature of stream 7 is −75°C, the pressure of stream 1 is 2 bar, and the temperature and flow rate of stream 2 are 20°C and 100 kg/h, respectively [17].

P.5.6 A dilute solution of dextrose in water, where the mass percent of dextrose is 10%. It is required to concentrate the dextrose to 55% using an evaporator. The liquid product would be at 76°C by reducing the pressure in the separator (or the pressure of the outlet). Simulate this case study and assume the separator is the evaporator. The solution enters the separator at 1000 kg/h, 25°C, and 1 atm. At what pressure does the liquid product have to achieve this target?
Fluid package: PRSV

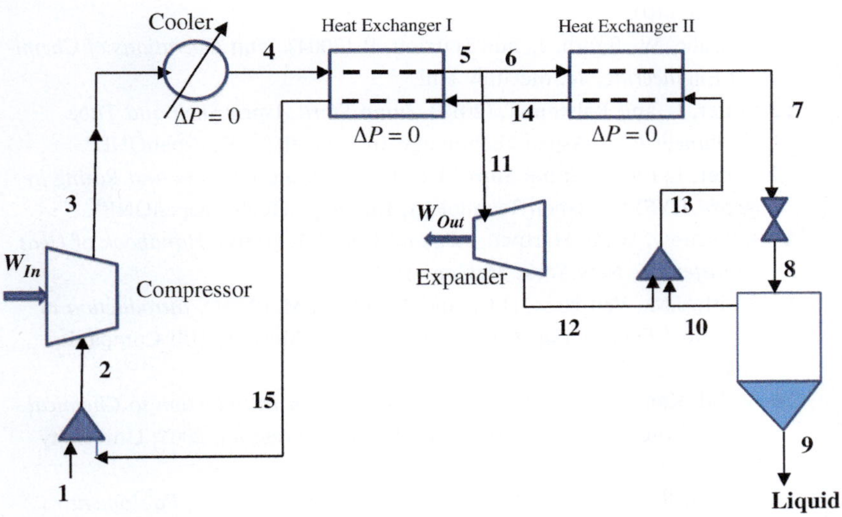

Figure 5.28 Liquefaction Process (Claude system).

References

1 Backhurst, J.R. and Harker, J.H. (1973). *Process Plant Design*. Heinemann Educational Books Ltd.
2 Coulson, J.M. and Richardson, J.F. (2003). *Chemical Engineering Design*, Chemical Engineering, 3e, vol. 6 (ed. R.K. Sinnott).
3 Dimian, A.C., Bildea, C.S., and Kiss, A.A. (2014). *Integrated Design and Simulation of Chemical Processes*. Elsevier.
4 (2017). *Aspen HYSYS User Guide version 10.0*. Aspen Technology.
5 AEA Technology, *"Process Modeling Using HYSYS with Chemical Industry Focus"*, 2000.
6 Foo, D.C.Y., Chemmangattuvalappil, N., Ng, D.K.S. et al. (2017). *Chemical Engineering Process Simulation*. Elsevier Inc.
7 Ludwig, E.E., Rouge, B., and Louisiana (1999). *Applied Process Design for Chemical and Petrochemical Plants*, Volumes 3, 3e. Butterworth-Heinemann.
8 Chaves, I.'n.D.'o.G. (2016). *Process Analysis and Simulation in Chemical Engineering*. Springer.
9 Haydary, J. (2019). *Chemical Process Design and Simulation Aspen Plus and Aspen HYSYS Applications*. John Wiley and Sons, Inc.
10 Hameed, K.W., Khadom, A.A., and Mahood, H.B. (2024). Simulation of single-effect and triple-effect evaporator for fruit juice concentration using Aspen HYSYS. *Chemical Product and Process Modeling*, vol. 19, issue 3, (10).
11 McCabe, W., Smith, J., and Harriott, P. (2004). *Unit Operations of Chemical Engineering*, 7e. McGraw Hill.
12 Sittler, L. and Ralston, T. (2013). *Jump Start: Aspen Shell and Tube Exchanger V8.0*. Aspen Technology, Inc. AspenTech®, aspenONE®.
13 Sittler, L. (2013). *Jump Start: Activated Exchanger Design and Rating in Aspen HYSYS®*. Aspen Technology, Inc. AspenTech®, aspenONE®.
14 Rohsenow, W.M., Hartnett, J.P., and Cho, Y.I. (1998). *Handbook of Heat Transfer*, 3e. New York: McGraw-Hill.
15 Smith, J.M., Van Ness, H.C., and Abbott, M.M. (2001). *Introduction to Chemical Engineering, Thermodynamics*, 6e. McGraw-Hill Companies, Inc.
16 Mohd. Kamaruddin Abd Hamid, *HYSYS®: An Introduction to Chemical Engineering Simulation For UTM Degree++ Program*, 2007, University Teknology Malaysia.
17 Turton, R. and Shaeiwitz, J.A. (2017). *Chemical Process, Equipment Design*. Pearson Education, Inc.

6

Logical Operations

A logical operation is a special symbol that connects two or more phrases of objectives. It is most often used to test whether a certain relationship between the phrases is true or false. These types of operations are called Boolean operations [1, 2].

6.1 Set Operation 

The Set is an operation used to set the value of a specific Process Variable (PV) in relation to another PV. The relationship is between the ***same PV in two like objects***; for instance, the temperature of two streams, or the UA of two exchangers [1].

The Set unit operation can be used in both Dynamic and Steady-State modes, see Figure 6.1.

Figure 6.2 shows the Connection tab, from which the Target Variable and Source can be selected.

The dependent or target variable is defined in terms of the independent, or source variable according to the following linear relation and Figure 6.3:

$$Y = MX + B \tag{6.1}$$

where:

Y is the dependent (target) variable
X is the independent (source) variable
M is the multiplier
B is the offset (intercept).

- **Target object:** The stream or operation to which the dependent variable belongs. This is chosen by clicking the **Select Var** button. This brings up the Variable Navigator property view.

Chemical Process Simulations using Aspen Hysys, First Edition. Khalid W. Hameed.

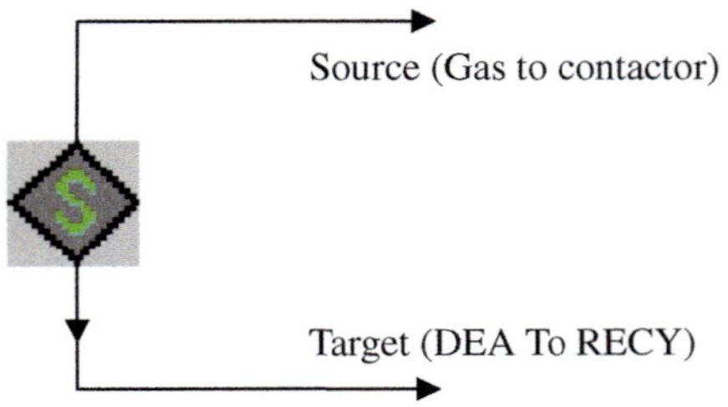

Figure 6.1 Connection of set operation.

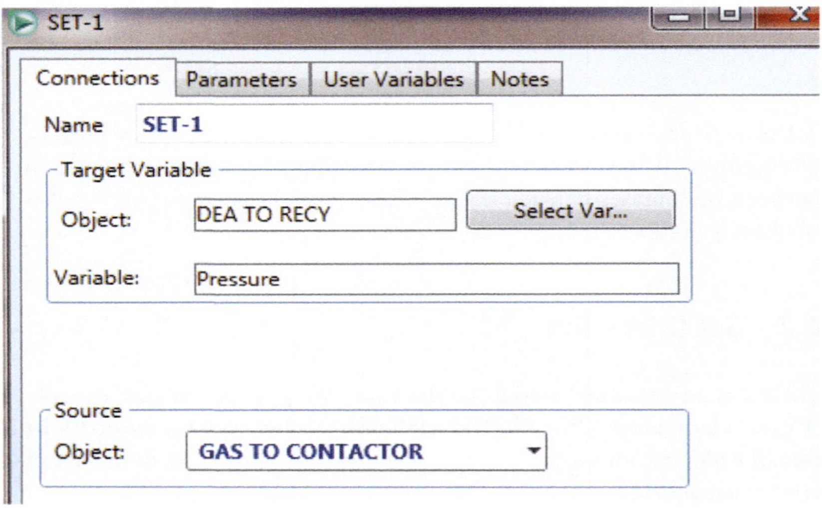

Figure 6.2 Connection tab of set operation.

- **Target variable:** The type of variable you want to set, for example, temperature, pressure, and flow. The available choices for Variable are dependent on the Object type (stream, heat exchanger, and so forth). Your choice of Variable is automatically assigned to both the Target and Source object.
- **Source object:** The stream or operation to which the independent variable belongs.

6.2 Adjust Operation 

The Adjust operation varies the value of one stream variable (the independent variable) to meet a required value or specification (the dependent variable) in another stream or operation, Figure 6.4.

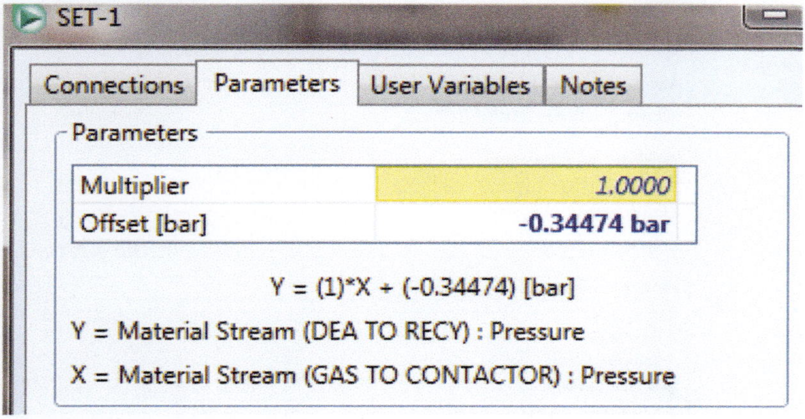

Figure 6.3 Parameters of set operation.

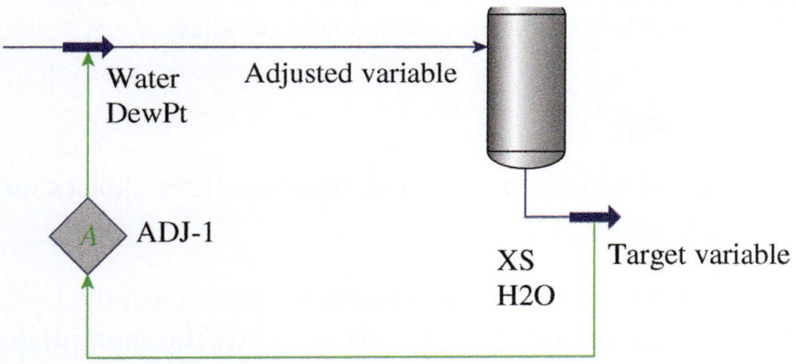

Figure 6.4 Connection of Adjust-operation.

The Adjust is a steady-state operation; HYSYS ignores it in dynamic mode.

It may be necessary to combine specifications in a flowsheet that cannot be resolved immediately. Trial-and-error methods must be used to solve these kinds of issues. This type of flowsheet problem can be solved rapidly by using the Adjust operation, which will carry out the trial-and-error iterations automatically [1, 2].

- **Adjusted variable:** It is the variable that is used to adjust the Target.
- **Target variable:** It is the variable that is required to change it.

Figure 6.5 shows the Connection tab, from which the Adjusted Variable, Target Variable, and Target Value can be selected.

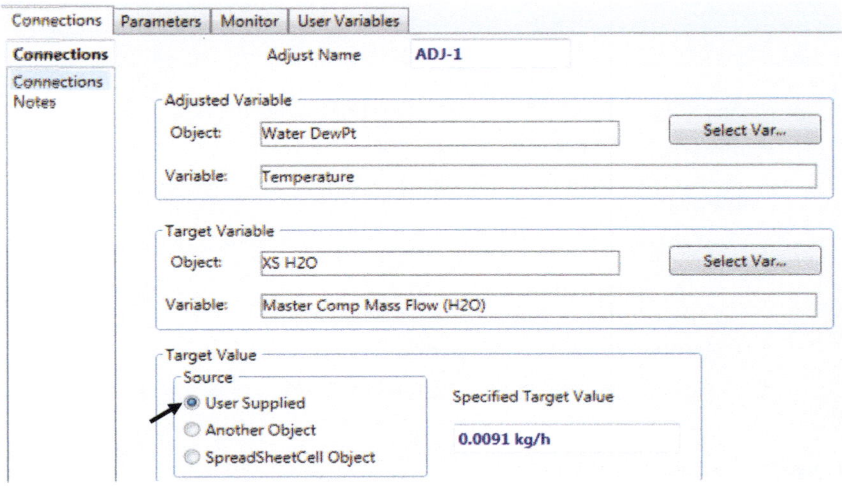

Figure 6.5 Connection page of adjust-operation, select the User Supplied radio button when the target variable is to meet a certain numerical value.

6.2.1 Target Value Group

Once the target object and variable are defined, there are three choices for how the target is to be satisfied:

- If the target variable is to meet a certain numerical value, select the User **Supplied** radio button (as shown in Figure 6.5), and enter the appropriate value in the Specified Target Value field.
- If the target variable is to meet the value (or the value plus an offset) of the same variable in another stream or operation, select the **Another Object** radio button (as shown in the Figure 6.6), and select the stream or operation of interest from the Matching Value Object drop-down list. If applicable, enter an offset in the available field, Figure 6.6.
- If the target variable is to meet the value (or the value plus an offset) of the same variable specified in the spreadsheet, select the **SpreadSheetCell Object** radio (as shown in Figure 6.7), and select the cell that you want from the Matching Value Object drop-down list. This allows the SpreadSheetCell to be attached as an adjusted variable and source to the target variable. You can also specify the offset in the available field, Figure 6.7.

Once you have chosen the dependent and independent variables, the convergence criteria must be defined. Table 6.1 shows the description of the solving parameter of the Adjust operation.

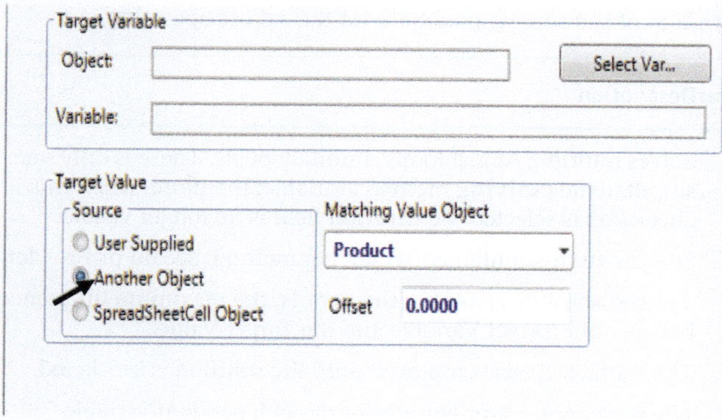

Figure 6.6 Connection page of adjust-operation, select the Another Object radio button when the target variable is to meet the value (or the value plus an offset) of the same variable in another stream or operation.

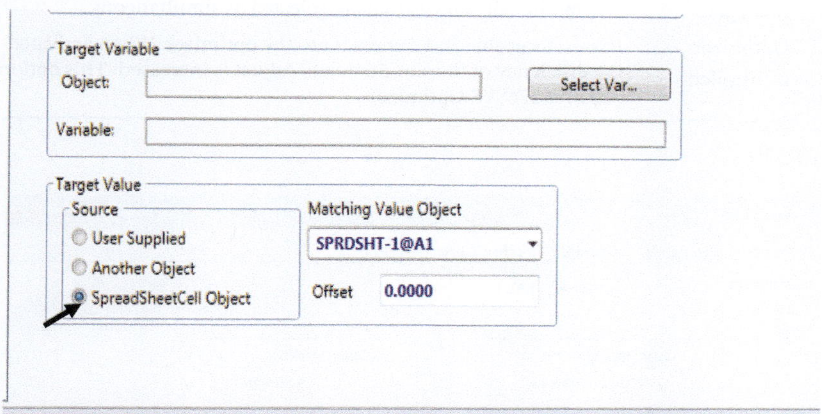

Figure 6.7 Connection page of adjust-operation, select the SpreadSheetCell Object radio button when the target variable is to meet the value (or the value plus an offset) of the same variable in spreadsheet.

6.3 Balance 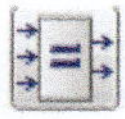

The Balance operation provides a general-purpose heat and material balance facility. The names of the streams entering and exiting the operation are the only data needed by the Balance. The window of connection page, connection tab, for the balance action are displayed in Figure 6.8. Component ratios for the General Balance can also be specified. The Balance can

Table 6.1 Description of the solving parameter of the Adjust operation.

Solving Parameter	Description
Simultaneous Solution	Solves multiple Adjust loops simultaneously. There is only one simultaneous **solving** method available; therefore, when this check box is selected the **Method** field is no longer visible.
Method	Sets the (non-simultaneous) solving method: Secant or Broyden.
Tolerance	Sets the absolute error. In other words, the maximum difference between the Target Variable and the Target Value.
Step Size	The initial step size employed until the solution is bracketed.
Maximum/Minimum	The upper and lower bounds for the independent variable (optional) are set in this field.
Maximum Iterations	The number of iterations before HYSYS quits calculations, assuming a solution has not been obtained.
Sim Adj Manager	Opens the Simultaneous Adjust Manager allowing you to monitor and modify all Adjusts that are selected as simultaneous.
Optimizer Controlled	Passes a variable and a constant to the optimizer. When activated the efficiency of the simultaneous Adjust is increased. This option requires HYSYS Optimizer.

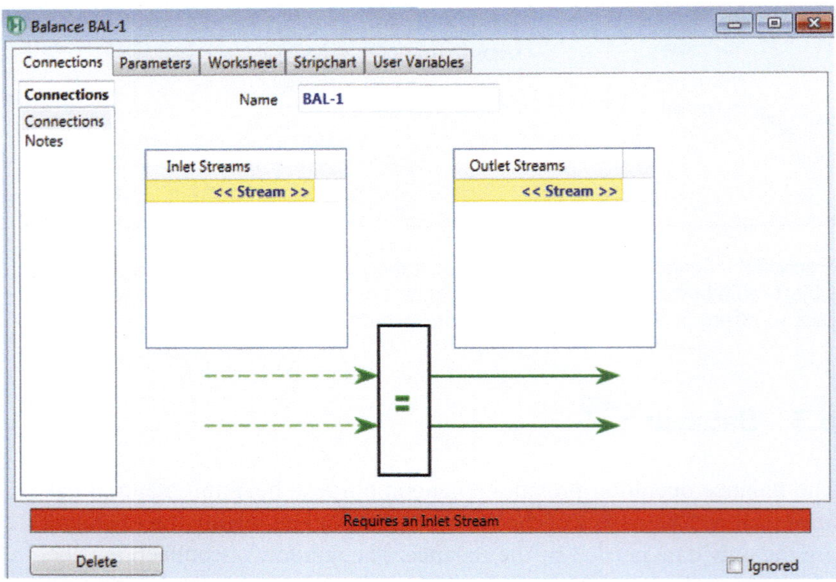

Figure 6.8 Connection page of Balance-operation.

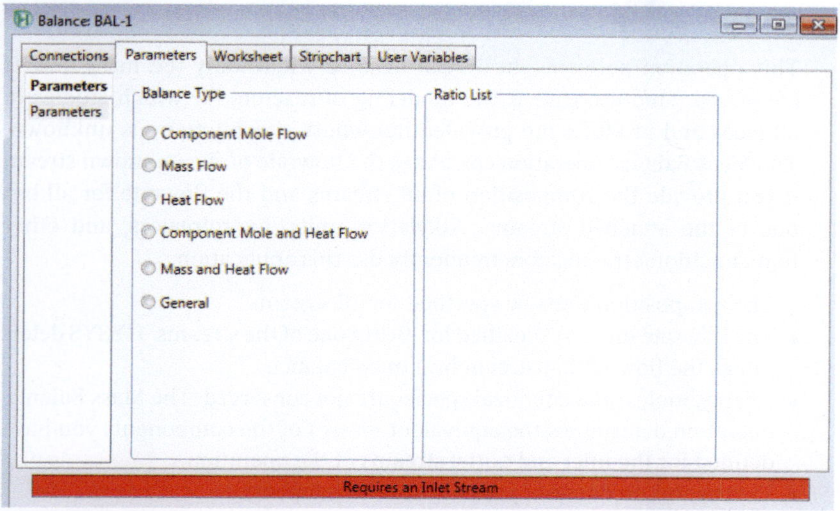

Figure 6.9 Parameters that can be selected in balance-operation.

be used in parallel with other units for overall material and energy balances because HYSYS allows streams to enter or exit multiple operations [2].

The Balance overrides the filtering of streams that HYSYS typically performs.

The Parameters tab contains two groups as shown in Figure 6.9:

- Balance Type
- Ratio List

The Balance Type group contains a series of radio buttons, which allow you to choose the type of Balance you want to use. The radio buttons are shown in Figure 6.9, Parameters tab.

These buttons are explained as follows.

6.3.1 Component Mole Flow

This operation performs an overall mole balance on selected streams; no energy balance is made. It can be used to provide material balance envelopes in the flowsheet or to transfer the flow and composition of a process stream into a second stream.

- The composition does not need to be specified for all streams.
- The direction of flow of the unknown is of no consequence. HYSYS calculates the molar flow of a feed to the operation based on the known products, or vice versa.
- This operation does not pass pressure or temperature.

6.3.2 Mass Flow

This operation performs an overall balance where only the mass flow is conserved. One use case is the modeling of reactors for which studies of all feeds and products are provided but whose stoichiometry is unknown. The Mass Balance operation calculates the flowrate of the unknown stream if you provide the composition of all streams and the flowrate for all but one of the attached streams. Alkylation units, hydrotreaters, and other non-stoichiometric reactors frequently use this application.

- The composition must be specified for all streams.
- The flowrate must be specified for all but one of the streams. HYSYS determines the flow of that stream by a mass balance.
- Energy, moles, and chemical species are not conserved. The Mass Balance operation determines the equivalent masses of the components you have defined for the inlet and outlet streams of the operation.
- This operation does not pass pressure or temperature.

6.3.3 Heat Flow

This operation performs an overall heat balance on selected streams. It can be used to provide heat balance envelopes in the flowsheet or to transfer the enthalpy of a process stream into a second energy stream.

- The composition and material flowrate must be specified for all material streams. The heat flow is not passed to streams which do not have the composition and material flowrate specified, even if there is only one unknown heat flow.
- The direction of flow for the unknown stream is of no consequence. HYSYS calculates the heat flow of a feed to the operation based on the known products, or vice versa.
- This operation does not pass the pressure or temperature.
- You cannot balance the heat into a Material Stream.

6.3.4 Component Mole and Heat Flow

The most common application for this balance is to perform overall material (molar basis), and energy balance calculations of selected process streams to either check for balances or force HYSYS to calculate an unknown variable, such as a flowrate.

- The Mole and Heat Balance independently balance energy and material.
- The Mole and Heat Balance calculate ONE unknown based on a total energy balance, and ONE unknown based on a total material balance.

- The operation is not directionally dependent for its calculations. Information can be determined about either a feed or product stream.
- The balance remains a part of your flowsheet and as such defines a constraint; whenever any change is made, the streams attached to the balance always balance with regard to material and energy. As such, this constraint reduces by one the number of variables available for specification.
- Since the Mole and Heat Balance work on a molar basis, it should not be used in conjunction with a reactor where chemical species are changing.

6.3.5 Mass and Heat Flow

Similar to the Mass balance mode, this balance mode performs a balance on the overall mass flow. In addition, however, energy is also conserved.

- The composition must be specified for all streams.
- Flow rate must be specified for all but one of the streams. HYSYS determines the flow of that stream by a mass balance.
- Enthalpy must be specified for all but one of the streams. HYSYS determines the enthalpy of that stream by a heat balance.
- Moles and chemical species are not conserved.

6.4 Recycle

The capability of any flowsheet simulator to solve recycles reliably and efficiently is critical. HYSYS has inherent advantages over other simulators in this respect. It has the unique ability to back-calculate through many operations in a non-sequential manner, allowing many problems with recycle loops to be solved explicitly. For example, most heat recycles can be solved explicitly (without a Recycle operation). Material recycles, where downstream material mixes with upstream material, require a Recycle operation. The Recycle installs a theoretical block in the process stream. The stream conditions can be transferred either in a forward or backward direction between the inlet and outlet streams of this block. In terms of the solution, there are assumed values and calculated values for each of the variables in the inlet and outlet streams. Depending on the direction of transfer, the assumed value can exist in either the inlet or outlet stream. For example, if the user selects Backward for the transfer direction of the Temperature variable, the assumed value is the Inlet stream temperature and the calculated value is the Outlet stream temperature. Figure 6.10 shows a picture of the Recycle operation [1, 2].

Figure 6.10 Picture of the Recycle operation.

The following steps take place during the convergence process:

1) HYSYS uses the assumed values and solves the flowsheet around the recycle.
2) HYSYS then compares the assumed values in the attached streams to the calculated values in the opposite stream.
3) Based on the difference between the assumed and calculated values, HYSYS generates new values to overwrite the previous assumed values.
4) The calculation process repeats until the calculated values match the assumed values within specified tolerances.

6.5 Spreadsheet

The Spreadsheet applies the functionality of Spreadsheet programs to flowsheet modeling. With essentially complete access to all process variables, the Spreadsheet is extremely powerful and has many applications in HYSYS.

Variables in the flowsheet can be customized or altered using the Spreadsheet. Since it's an operation, computations are done automatically, and when variables in the flowsheet change, spreadsheet cells are updated as well. The Spreadsheet cells are updated while the integrator is operating in Dynamics mode. Calculating the pressure drop during a heat exchanger's dynamic operation is one use for the spreadsheet. Regardless of flow, the pressure drop in the HYSYS Heat Exchanger is constant on both sides. Nonetheless, the real pressure drop on one or both sides of the exchanger could be computed as a function of flow utilizing the Spreadsheet [2].

The HYSYS Spreadsheet has standard row/column functionality. You can import a variable, or enter a number or formula anywhere in the spreadsheet.

Sophisticated mathematical expressions can be produced with a syntax akin to that of traditional spreadsheets. The Spreadsheet has mathematical functions such as arithmetic, logarithmic, and trigonometric functions. The Spreadsheet's extensive mathematical functions are complemented with logical programming. Boolean logic is available, enabling you to use

Table 6.2 Description of the two methods of importing and exporting variables to and from the Spreadsheet.

Method	Description
Using the Variable Navigator	Do one of the following: • On the Connections tab, click the Add Import or AddExport button. • On the Spreadsheet tab, right-click the cell you want and select export/import command from the object inspection menu. Then using the Variable Navigator, select the variable you want to import or export.
Dragging Variables	Simply right-click the variable value you want to import, and drag it to the desired location in the Spreadsheet. If you are exporting the variable, drag it from the Spreadsheet to an appropriate location. When using the Dragging Variables method, the property views have to be non-modal.

logical operators to compare the values of two or more variables and take the necessary action based on the outcome.

You can import virtually any variable in the simulation into the Spreadsheet, and you can export a cell's value to any specifiable field in your simulation. There are two methods of importing and exporting variables to and from the Spreadsheet, see Table 6.2.

It is important to take the entire computation sequence into account when applying the Spreadsheet to deliver a result back to the Flowsheet, especially when recycling is involved. There is a chance of inconsistency if the Spreadsheet calculates something and sends the results back upstream since the previous Recycle loop's full impact may not have spread across the flowsheet. You can reduce the likelihood of issues of this kind by utilizing the Calculation Sequencing feature.

Figure 6.11 shows the property view of the spreadsheet.

6.5.1 Spreadsheet Functions

The HYSYS Spreadsheet has extensive mathematical and logical function capability. To view the available Spreadsheet Functions and Expressions, click the Function Help button to open the Available Expressions and Functions property view.

All functions must be preceded by "+" (straight math) or "@" (special functions like logarithmic, trigonometric, logical, and so forth).

Examples are "+A4/B5" and "@ABS(A4-B5)."

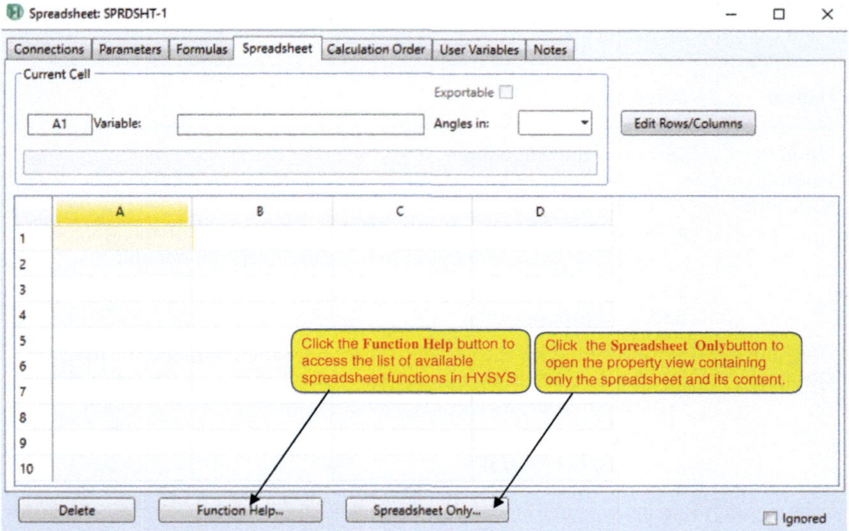

Figure 6.11 The property view of Spreadsheet.

The application of logical operations will be presented in the next chapters.

To explain more details, see "Aspen HYSYS Unit operation V10 (or other version) Reference guide."

References

1 (2002). *HYSYS 3.2 Operation Guide*, © Hyprotech, a subsidiary of Aspen Technology, Inc.

2 (2017). *Aspen HYSYS Unit Operations V10, Reference Guide*, Aspen Technology, Inc.

7

Piping Equipment

7.1 Pipe Segment

The *Pipe Segment* is used to simulate a wide variety of piping situations ranging from single or multiphase plant piping with rigorous heat transfer estimation to a large-capacity looped pipeline problem [1].

The *Pipe Segment* operation contains four calculation modes:

Pressure Drop, Length, Flow, and Diameter.

The mode is automatically assigned, depending on what information is specified.

Regardless of which mode you use, you must specify the number of increments in the pipe. Calculations are performed in each increment; for example, to determine the pressure drop, the energy and mass balances calculations are performed in each increment, and the outlet pressure in that increment is used as the inlet pressure to the next increment. The calculation continues down the length of the pipe until the pipe outlet pressure is determined.

The *Pipe Segment* can solve in either direction. The solution procedure generally starts at the end where the temperature is known (temperature is typically unknown on both ends). HYSYS then begins stepping through the pipe from that point, using either the specified pressure, or estimating a starting value. If the starting point is the pipe outlet, HYSYS steps backward through the pipe. At the other end of the pipe, HYSYS compares the calculated solution to other known information and specifications, and, if necessary, restarts the procedure with a new set of starting estimates.

Aspen HYSYS calculates the total pressure loss across the pipe length as a sum of the pressure gradient caused by the gravitational force, pressure

Chemical Process Simulations using Aspen Hysys, First Edition. Khalid W. Hameed.

losses caused by fluid friction, and a kinetic or acceleration component of the pressure drop [2, 3]:

$$\frac{dp}{dL} = \rho g \sin\theta + \left(\frac{dP}{dL}\right)_{fr} + \rho v \frac{dv}{dL} \tag{7.1}$$

where:

P is the pressure,
ρ is the fluid density,
g is the gravity acceleration,
L is the pipe length,
v is the liquid velocity,
fr is the friction factor, and
θ is the inclined pipe.

The power of the fluid, P_w, is estimated by

$$P_w = \rho g h Q \tag{7.2}$$

where Q is the volumetric flow rate of the fluid and h is the head.

7.1.1 Problem Statement

In Figure 7.1, 700 kg/h of Water at 60 °F and one atmosphere is being transferred with a 2 hp pump that is 50% efficient. All the piping is 1 in. diameter, schedule 40, and cast-iron pipe except for the last section, which is 0.5 in.

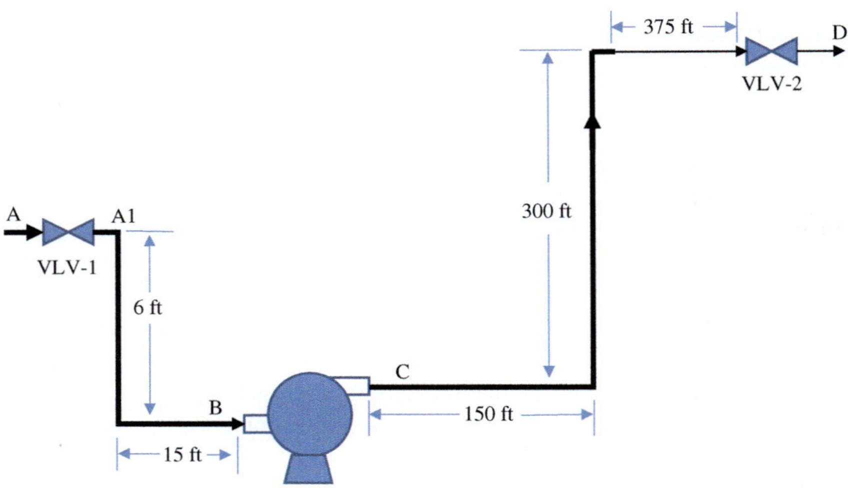

Figure 7.1 Schematic diagram of a fluid flow system.

schedule 40 cast-iron pipe. The elbows are standard 90 ° and one reducer to connect the 0.5 in. pipe. Add VLV-1 separately, but VLV-2 is including the Pipe Segment, type of VLV-2: Gate valve, Half.

(a) What is the pressure at point D? Plot the pressure of point D versus the inlet flow rate.
(b) What is the *hp* of the pump if the pressure at Point D is 3.2 atm and its temperature is 60 °F?

7.1.2 Simulation: (a)

1) Open a new case, Add the Component: Water. Fluid package: *ASME Steam.*
   Enter the **Simulation** Environment.
2) Construct the material stream (A) of feed water with the following conditions:
   Flow rate = 700 kg/h, $T = 60\,°\text{F}$, $P = 1$ atm, composition = 100% water
3) Add a valve  from the object palette.
   Inlet = A, Outlet = A1
   From **Parameters** page, activate "Use sizing methods to calculate delta P" as in Figure 7.2.
   To enable the valve to estimate the pressure drop, the size of the valve must be first given.
4) Click the **Rating** tab, the size of valve is determined by entering the specific value of C_v.

$$C_V \text{ [USGPM (60°F, 1 psi)]} = \text{flow (m}^3/\text{h)} \times 246.2/60$$

   So for current case, the volumetric flow rate = 0.7014 m^3/h (*taken from worksheet*)

$$\rightarrow C_V = 0.7014 \times (246.2/60) = 2.878 \text{ USGPM (60°F, 1 psi).}$$

   Enter this value in the C_V [USGPM (60 °F, 1 psi)] cell, Figure 7.3.
   The pressure drop is calculated to be 31.7 kPa. This is for 50% opening valve.
   Note: *If the size of valve is required, the pressure drop instead of activate (use sizing methods to calculate delta p) radio must be entered in the Parameters Page. The C_v will be estimated, then click **Size Valve** button.*
   The pipe section from (A1 – B) is created as follows:
5) Add **Pipe Segment** 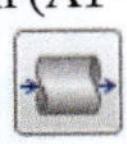 from the object palette. Connect the Pipe Segment with required streams as in Figure 7.4, where inlet = *A*1, Outlet = *B*, and Energy = *Q*-1.

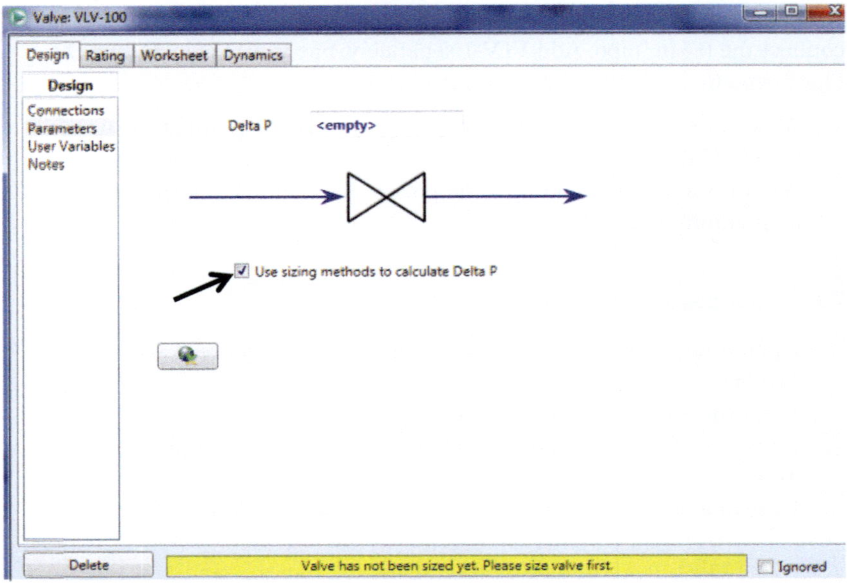

Figure 7.2 Activating the "Use sizing methods to calculate delta P" radio in the valve.

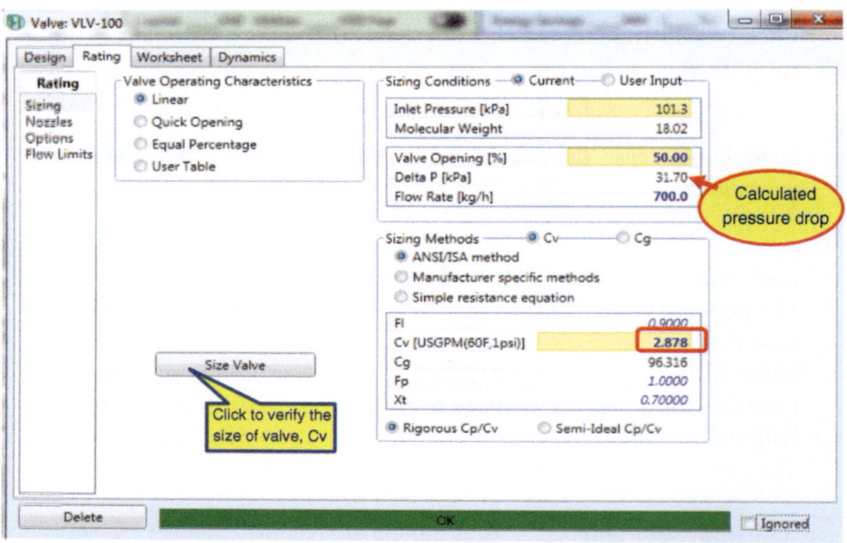

Figure 7.3 On entering the C_v value, the delta P will be estimated.

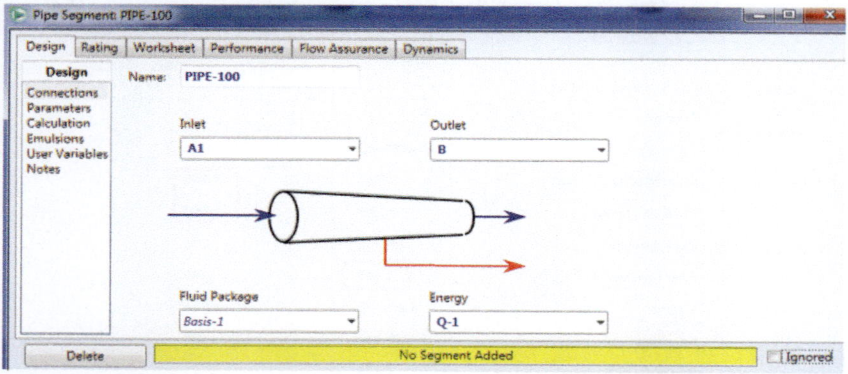

Figure 7.4 Window of pipe segment when opening first time.

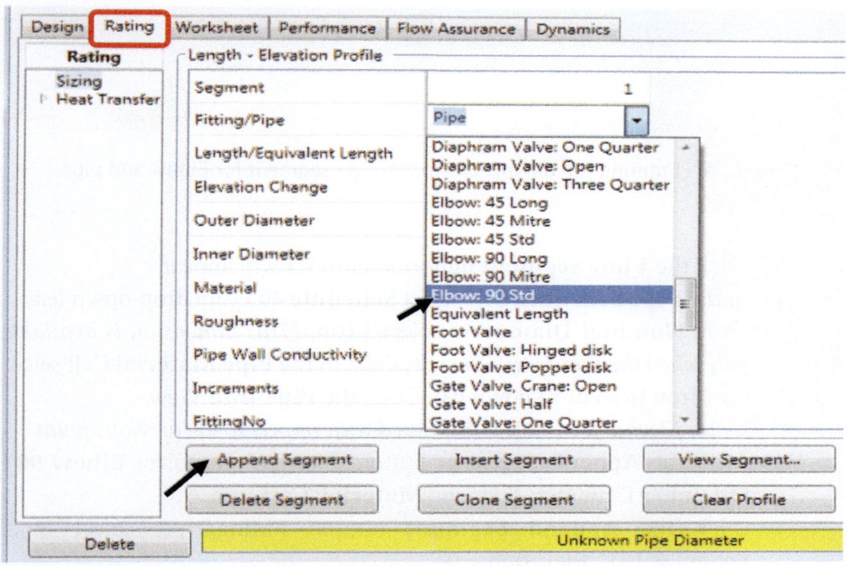

Figure 7.5 How you can append segment in the piping system.

6) From the **Rating** tab, **Sizing** page, Click the **Append Segment** button, Figure 7.5.
   For Fitting/Pipe, Select **Elbow: 90 Std** from drop list. Enter the inner diameter = 1.315 in. (= outer diameter of the next pipe), Material = Cast Iron.
7) Click **Append Segment**, Fitting/Pipe: Select Pipe, Enter the Length = 6 ft, Elevation change = −6 ft, Figure 7.6.

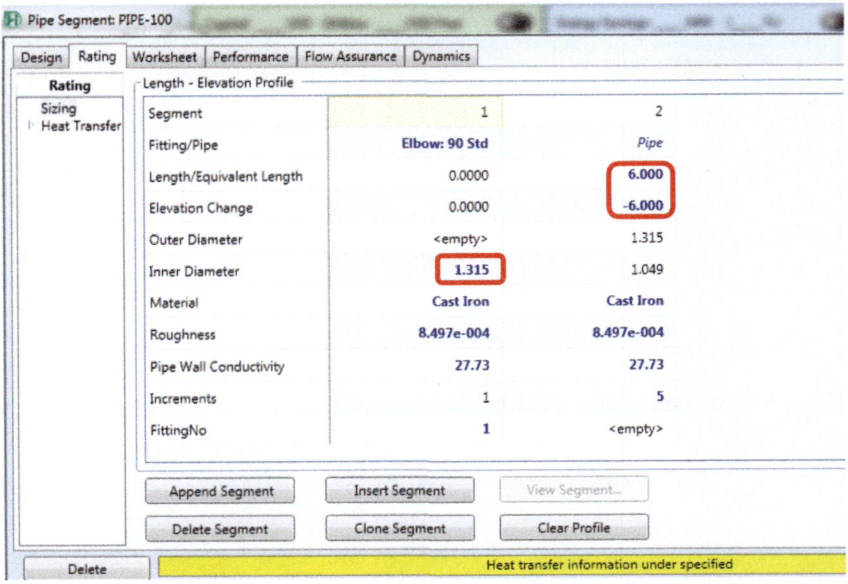

Figure 7.6 Entering the dimensions in the pipe segment for Elbow and pipe.

8) Click the **View Segment** button, Figure 7.7 will appear.
9) In the **Pipe Schedule** Cell select **Schedule 40** From drop-down list.
10) From **Nominal Diameters** select 1.0 in. (*This dimension is available only when the **Field** units are selected.*). In the **Pipe Material** Cell select Cast Iron from drop-down list. Close the **Pipe Info** View.
    Note: *Elevation change is estimated with respect to the previous point.*
11) Click the **Append Segment** button: Fitting/Pipe: select **Elbow 90° Std**, Inner Diameter: 1.315 in., **Material**: Cast Iron.
12) Click the **Append Segment** button: Fitting/Pipe: select Pipe, Length = 15 ft, Elevation = 0.
    Click the **View Segment** button: In the **Pipe Schedule** Cell, select **Schedule 40** From the drop-down list.
    From the **Available Nominal Diameters** Group select 1.0 in. In the **Pipe Material** Cell select Cast Iron from drop-down list. The segments added for Pipe-100 are shown in Figure 7.8.
13) Select the **Heat Transfer** Page: In the **Heat Loss** Cell: Enter 0.0 Btu/h, Figure 7.9.
    Close the **Pipe segment** view.
    Note: *To estimate the **heat transfer coefficient**; Instead of entering 0.0 in the Heat loss, select **Estimate HTC** button radio and enter the*

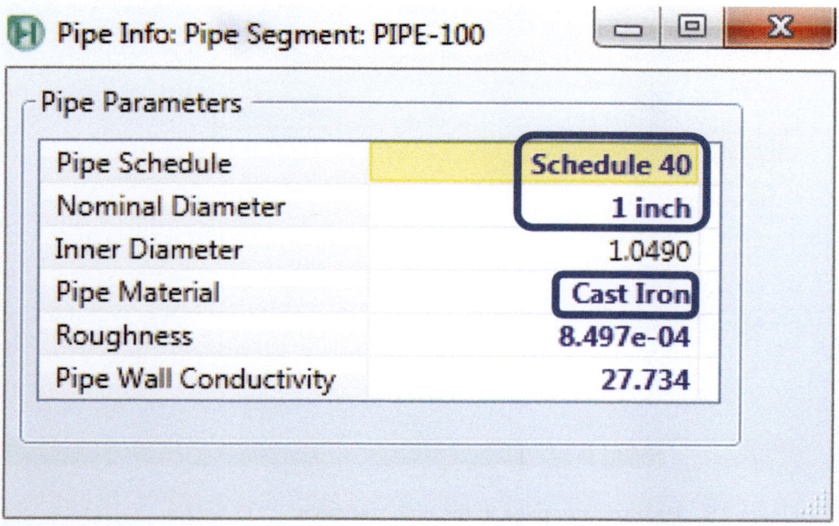

Figure 7.7 Selecting the diameter of pipe segment and its material from Pipe standard schedules.

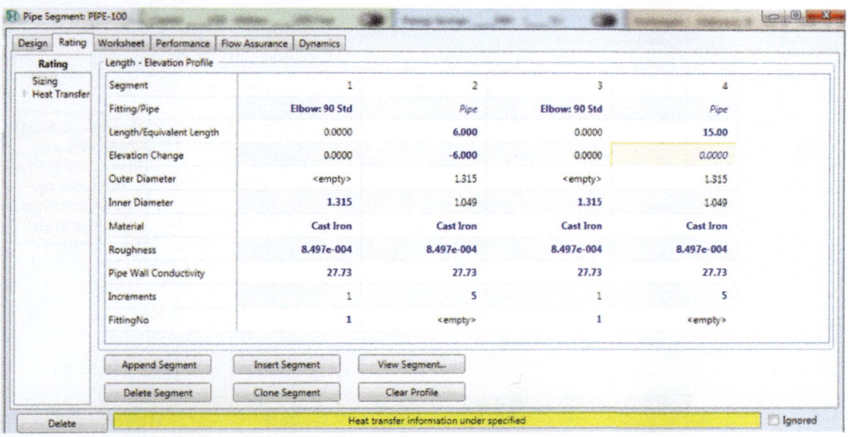

Figure 7.8 Segments of (Pipe-100).

Ambient temperature, then determine the parameters that are ***required*** *to be estimated, Figure 7.10*
Then click ***Overall HTC*** *button, all the types of heat transfer coefficient will be estimated.*

14) Add a **Pump**
Design tab, **Connection** page: Inlet: *B*, Outlet: *C*, Energy: *Q*-2
Parameters tab, enter the efficiency 50%

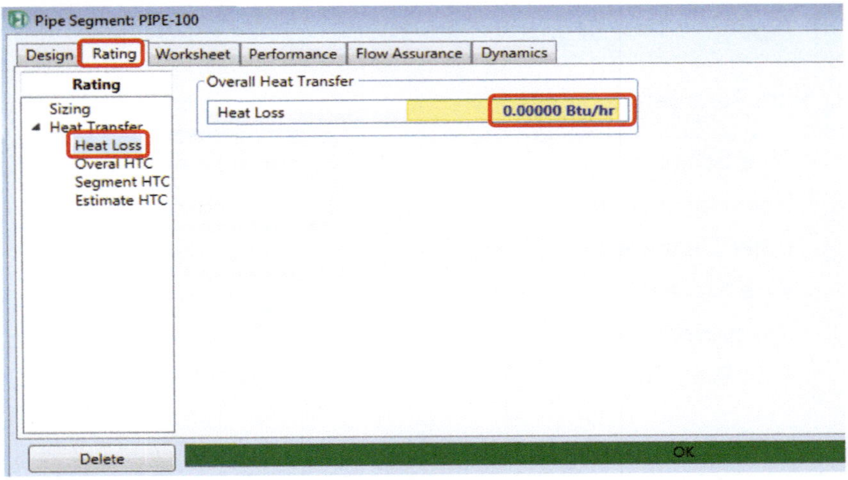

Figure 7.9 Heat transfer page in the pipe- segment.

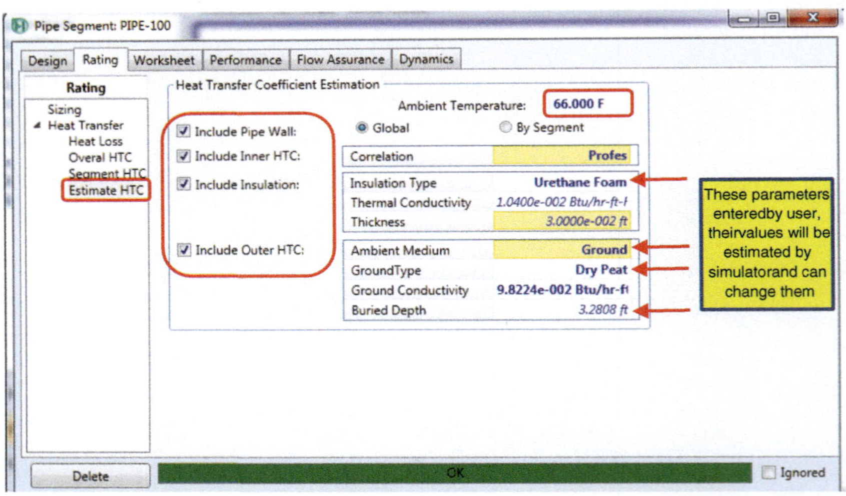

Figure 7.10 Estimating the overall heat transfer coefficient in the pipe segment in case of Heat loss ≠ 0.

Enter the **Duty** = 2 hp. Close the view of the pump.

15) Add **Pipe Segment; Design** tab, **connection** page; Inlet: *C*, Outlet: *D*, Energy: *Q-3*.
16) Add the segments required (section PIPE-101) as shown in schematic diagram of Figure 7.1 and as the previous procedure,

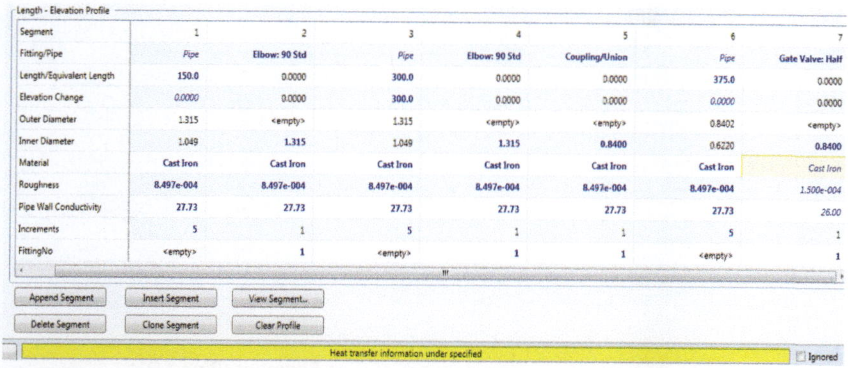

Length - Elevation Profile

Segment	1	2	3	4	5	6	7
Fitting/Pipe	Pipe	Elbow: 90 Std	Pipe	Elbow: 90 Std	Coupling/Union	Pipe	Gate Valve: Half
Length/Equivalent Length	150.0	0.0000	300.0	0.0000	0.0000	375.0	0.0000
Elevation Change	0.0000	0.0000	300.0	0.0000	0.0000	0.0000	0.0000
Outer Diameter	1.315	<empty>	1.315	<empty>	<empty>	0.8402	<empty>
Inner Diameter	1.049	1.315	1.049	1.315	0.8400	0.6220	0.8400
Material	Cast Iron	Cast Iron	Cast Iron	Cast Iron	Cast Iron	Cast Iron	Cast Iron
Roughness	8.497e-004	8.497e-004	8.497e-004	8.497e-004	8.497e-004	8.497e-004	1.500e-004
Pipe Wall Conductivity	27.73	27.73	27.73	27.73	27.73	27.73	26.00
Increments	5	1	5	1	1	5	1
FittingNo	<empty>	1	<empty>	1	1	<empty>	1

Figure 7.11 completing all parts of the pipe segment (Pipe-101).

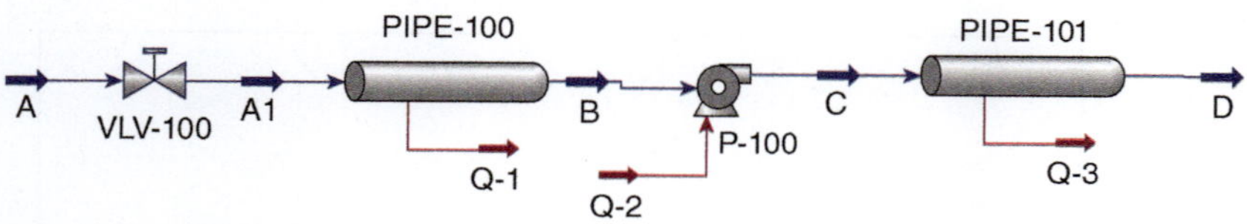

Figure 7.12 Process flow diagram of the piping system in HYSYS.

after completing all the parts, the window of the pipe will appear as in Figure 7.11.

Note: *The valve can be added to the piping system separately from the object palette as in the step 3 or from the pipe segment as in Figure 7.11, branch segment 7.*

17) **Heat Transfer** page: enter the Heat Loss = 0

Figure 7.12 shows the PFD of the piping system of schematic diagram of Figure 7.1.

From **Workbook** check the pressure at point D.

From the **Performance** tab, PIPE-101, click the **View Profile** button to see the results, Figure 7.13.

The results can be seen by table or graph. Figure 7.13 shows the most results that may be required such as pressure, temperature, heat transfer, and Reynolds number with length of pipe.

18) Now, to plot the pressure of point *D* as a function of mass flow rate of water inlet, use the case steady, where the independent variable is (mass flow rate in point *A*), and the dependent variable is (pressure at point *D*). Take the range of mass flow rate (600–1000 kg/h) with step size equal to 20 kg/h. The results of pressure of point *D* as a function of mass flow rate of point *A* are shown in Figure 7.14.

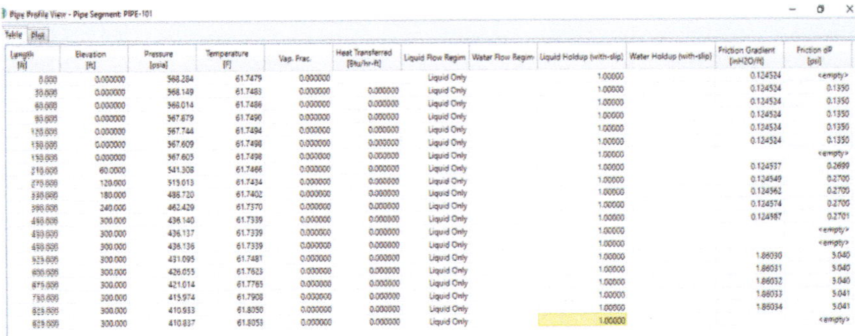

Pipe Profile View - Pipe Segment: PIPE-101

Table | Plot

Length [ft]	Elevation [ft]	Pressure [psia]	Temperature [F]	Vap. Frac.	Heat Transferred [Btu/hr-ft]	Liquid Flow Regim	Water Flow Regim	Liquid Holdup (with-slip)	Water Holdup (with-slip)	Friction Gradient [inH2O/ft]	Friction dP [psi]
0.000	0.000000	568.284	61.7479	0.000000		Liquid Only		1.00000		0.124524	<empty>
30.000	0.000000	568.149	61.7483	0.000000	0.000000	Liquid Only		1.00000		0.124524	0.1350
60.000	0.000000	568.014	61.7486	0.000000	0.000000	Liquid Only		1.00000		0.124524	0.1350
90.000	0.000000	567.879	61.7490	0.000000	0.000000	Liquid Only		1.00000		0.124524	0.1350
120.000	0.000000	567.744	61.7494	0.000000	0.000000	Liquid Only		1.00000		0.124534	0.1350
150.000	0.000000	567.609	61.7498	0.000000	0.000000	Liquid Only		1.00000		0.124524	0.1350
150.000	0.000000	567.605	61.7498	0.000000	0.000000	Liquid Only		1.00000			<empty>
210.000	60.0000	541.308	61.7466	0.000000	0.000000	Liquid Only		1.00000		0.124537	0.2699
270.000	120.000	513.013	61.7434	0.000000	0.000000	Liquid Only		1.00000		0.124549	0.2700
330.000	180.000	486.720	61.7402	0.000000	0.000000	Liquid Only		1.00000		0.124562	0.2700
390.000	240.000	462.429	61.7370	0.000000	0.000000	Liquid Only		1.00000		0.124574	0.2700
450.000	300.000	436.140	61.7339	0.000000	0.000000	Liquid Only		1.00000		0.124587	0.2701
450.000	300.000	436.137	61.7339	0.000000	0.000000	Liquid Only		1.00000			<empty>
450.000	300.000	436.136	61.7339	0.000000	0.000000	Liquid Only		1.00000			<empty>
525.000	300.000	431.095	61.7481	0.000000	0.000000	Liquid Only		1.00000		1.86030	5.040
600.000	300.000	426.055	61.7623	0.000000	0.000000	Liquid Only		1.00000		1.86031	5.040
675.000	300.000	421.014	61.7765	0.000000	0.000000	Liquid Only		1.00000		1.86032	5.040
750.000	300.000	415.974	61.7908	0.000000	0.000000	Liquid Only		1.00000		1.86033	5.041
825.000	300.000	410.933	61.8050	0.000000	0.000000	Liquid Only		1.00000		1.86034	5.041
825.000	300.000	410.837	61.8053	0.000000	0.000000	Liquid Only		1.00000			<empty>

Figure 7.13 The results of Pipe Network of Pipe Segment (Pipe-101).

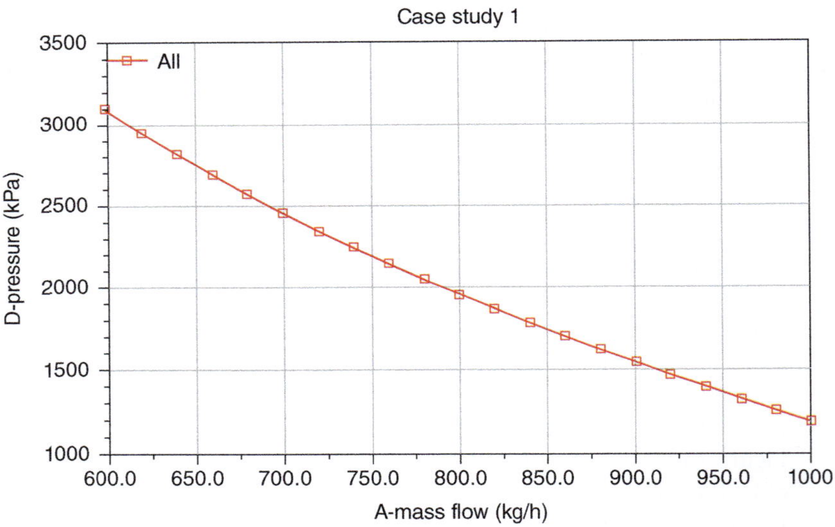

Figure 7.14 The results of pressure of point *D* as a function of mass flow rate of water.

It can be seen from Figure 7.14, that the pressure of point *D* decreases with increasing mass flow rate of water, since when increasing the amount of water, it will require more energy to deliver it and a portion of the energy of the pump will be lost on the weight of the water.

(b) To find the *hp* of the pump:

Delete the value of energy of the pump (*Q*-2) that is given in step (14). In point *D* enter pressure = 3.2 atm and temperature = 60 °F. See the result of the energy of the pump.

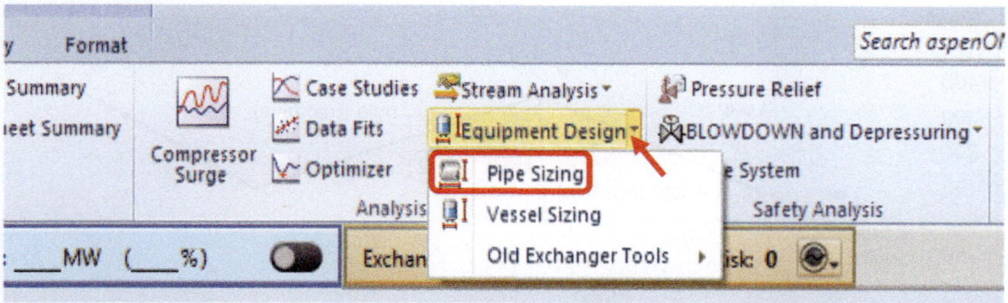

Figure 7.15 Another method to estimate the pressure drop or pipe diameter.

Note: *The sizing of the pipe can be estimated from the* **Home** *tab of the Ribbon;* **Equipment Design** *→ Pipe Sizing, Figure 7.15, then click the* **Select Stream** *button to choose the required stream, and determine the pressure drop to estimate the diameter or determine diameter required to estimate the pressure drop.*

7.2 PIPESYS

A pipeline must transport fluids over diverse topography and under varied conditions. Ideally this would be done efficiently with a correctly sized pipeline that adequately accounts for pressure drop, heat losses and includes the properly specified and sized inline facilities, such as compressors, heaters, or fittings [4, 5].

Pipesys has far-reaching capability to accurately and powerfully model pipeline hydraulics. It uses the most reliable single- and multi-phase flow technology available to simulate pipeline flow. Functioning a seamless extension to HYSYS, **Pipesys** has access to HYSYS features such as component database and fluid properties. Pipesys includes many inline equipment and facility options relevant to pipeline construction and testing. The extension models pipeline that stretch over varied elevation and environments. Pipesys enables you to:

- Rigorously model single-phase and multi-phase flows.
- Compute detailed pressure and temperature profiles for pipelines that traverse irregular terrain, both onshore and off.
- Perform forward and reverse pressure calculations.
- Model the effects of inline equipment such as compressors, pump, heaters, coolers, regulators, and fittings including valves and elbows.
- Perform special analyses including pigging slug prediction, erosion velocity prediction, and severe slugging checks.

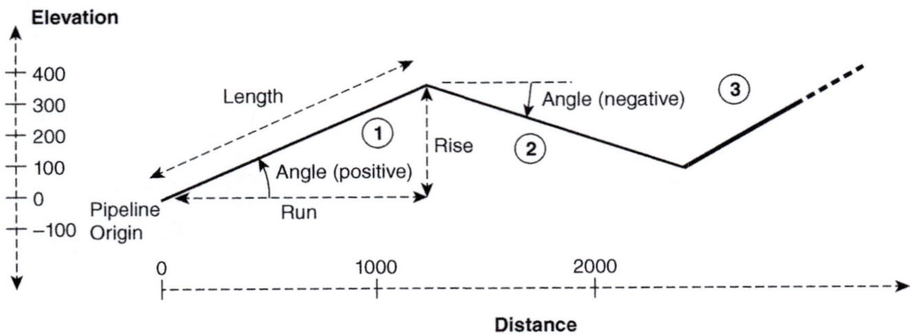

Figure 7.16 Simplified diagram about estimations of Pipesys.

- Model single pipelines or networks of pipelines in isolation or as part of a Hysys process simulation.
- Perform sensitivity calculations to determine the dependency of system behavior on any parameter.
- Quickly and efficiently perform calculations with internal calculation optimizer, which significantly increases calculations speed without loss of accuracy.
- Determine the possibility of increasing capacity in existing pipelines based on compositional effects, pipeline effects, and environmental effects.

This system is similar to the Pipe Segment in Section 7.1, but this system is often used for oil and gas pipelines.

Figure 7.16 shows the important terms used in the PIPESYS.

where:

$$\text{Distance} = \sum \text{Runs} \tag{7.3}$$

$$\text{Rise} = \text{Elevation of each } \textit{pipe segment} \text{ from the } \textit{previous point.} \tag{7.4}$$

$$\text{Elevation} = \text{Elevation of overall } \textit{Pipe system} \text{ from } \textit{origin point.} \tag{7.5}$$

7.2.1 Problem Statement

Figure 7.17 shows the physical configuration of the performance of a small gas-condensate-gathering system. The system consists of three wells distributed over an area of approximately 1.0 mile^2 connected to a gas plant via a network of pipelines [4].

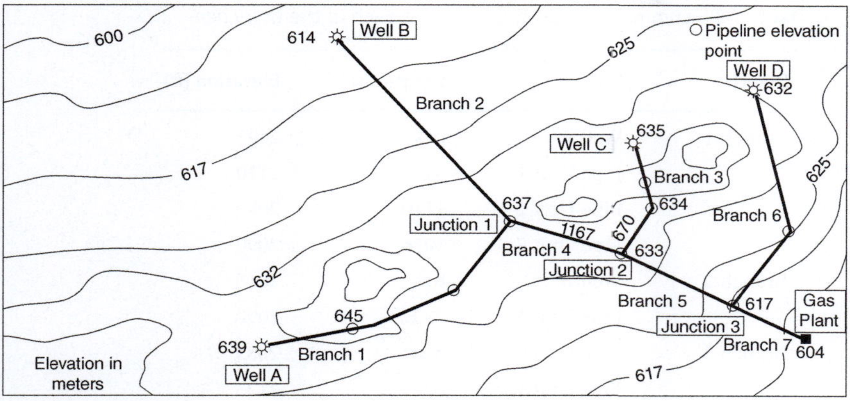

Figure 7.17 Small gas-condensate-gathering system.

Field data shows that the wells are delivering the following rates:

Well	MMSCFD
A	8.6
B	7.4
C	10.1

All three wells have the same composition. The residue of all the heavier components in the condensate has a molecular weight of 122 and a density of 760 kg/m^3. The characteristics of this component will be accounted for by using the hypothetical component facility in HYSYS. The compositional analysis of the gas condensate resulted as in Table 7.1.

Table 7.1 compositional analysis of the gas condensate.

Component	Composition	Component	Composition
Methane	0.623	*n*-Pentane	0.00405
Ethane	0.280	*n*-Hexane	0.00659
Propane	0.0163	C_7+	0.00992
i-Butane	0.00433	Nitrogen	0.00554
n-Butane	0.00821	Carbon Dioxide	0.0225
i-Pentane	0.00416	Hydrogen Sulfide	0.0154

Table 7.2 summary of elevation data for each of the branches.

Branch	Unit	Length (ft)	Elevation (ft)
Branch 1	Well A	n/a	2095
	Pipe Unit 1	945	2110
	Pipe Unit 2	1110	2089
	Pipe Unit 3	1056	2090
Branch 2	Well B	n/a	2015
	Pipe Unit 1	2822	2090
Branch 3	Well C	n/a	2085
	Pipe Unit 1	528	2125
	Pipe Unit 2	334	2080
	Pipe Unit 3	670	2077
Branch 4	Pipe Unit 1	1167	2077
Branch 5	Pipe Unit 1	2110	1980

Pipe diameters for each of the branches are:

Branch Diameter	
Branch 1	3″
Branch 2	3″
Branch 3	3″
Branch 4	4″
Branch 5	6″

A schedule 40 steel pipe is used throughout, and all branches are buried at a depth of three feet. All pipes are uninsulated. Table 7.2 summarizes the elevation data for each of the branches. The elevation given for the pipe units is for the endpoint of the pipe (i.e. the downstream end).

The flow rate at each well is specified and is independent of the flow rate at each of the other wells. In cases such as this, the system can be modeled with only one pressure drop determination per branch. Simultaneous pressure and temperature calculations can be performed if the temperature at each wellhead is also known.

The PFD generated by HYSYS for the completed case will be shown as in Figure 7.18.

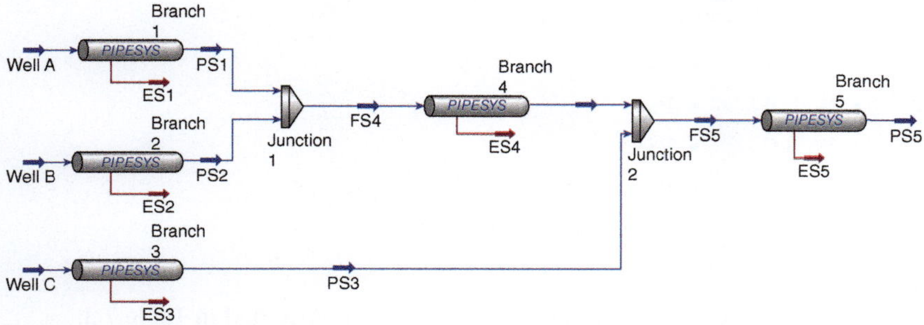

Figure 7.18 PFD generated by HYSYS for the completed small gas-condensate-gathering system.

A pressure of 1060 psia will be specified for Well *A*. PIPESYS will then determine the pressures elsewhere in the network that are consistent with this specification.

The fluid temperatures at wells *A*, *B*, and *C* are known and must be entered as fixed conditions. PIPESYS will then perform an iterative pipeline calculation in the branches, where the upstream temperature and downstream pressure are known. Temperatures of the blended fluids will be computed on a mass basis downstream of the junctions of two or more streams.

7.2.2 Simulation

1) Start the Aspen HYSYS and open a new case. Create the following data: **Pure Components:** C_1, C_2, C_3, *i*-C_4, *n*-C_4, *i*-C_5, *n*-C_5, C_6, N_2, CO_2, H_2S. Create a hypothetical component, C_7+, with the following user-defined properties.

Name	C_7+
Molecular Weight	122
Ideal Liquid Density (lb/ft^3)	47.45

Click the **Estimate Unknown** button, then click the **Add** button to add the component C_7+ to the component's list.

Fluid Package: Peng–Robinson.

2) Enter the **Simulation** Environment, open the Workbook, and add the **10 Material Streams** listed below:

- **Well A**
- **Well B**
- **Well C**
- **PS1**
- **PS2**
- **PS3**
- **PS4**
- **PS5**
- **FS4**
- **FS5**

3) Enter the compositional data for Well *A* as specified in Table 7.3:
   The composition of the streams Wells *A*, *B*, and *C* is identical
4) Use the **Define from Other Stream**... button on Well *B*'s stream view to define its composition as shown in Figure 7.19.
   Repeat the same procedure for well C.
5) Open the Energy Streams tab on the Workbook view. Enter the names of five energy streams:
   ES1, ES2, ES3, ES4, ES5
6) Adding First PIPESYS Extension
   (6-1) Add a PIPESYS Extension to the flowsheet (by clicking F12, select **Extension** and double-click the PIPESYS), Or from Object Palette, **Custom** group, PIPESYS Extension icon .

Table 7.3 Composition of Well A.

Component	(mole frac.)
Methane	0.623
Ethane	0.280
Propane	0.0163
i-Butane	0.00433
n-Butane	0.00821
i-Pentane	0.00416
n-Pentane	0.00405
n-Hexane	0.00659
C_7+	0.00992
Nitrogen	0.00554
Carbon Dioxide	0.0225
Hydrogen Sulfide	0.0154

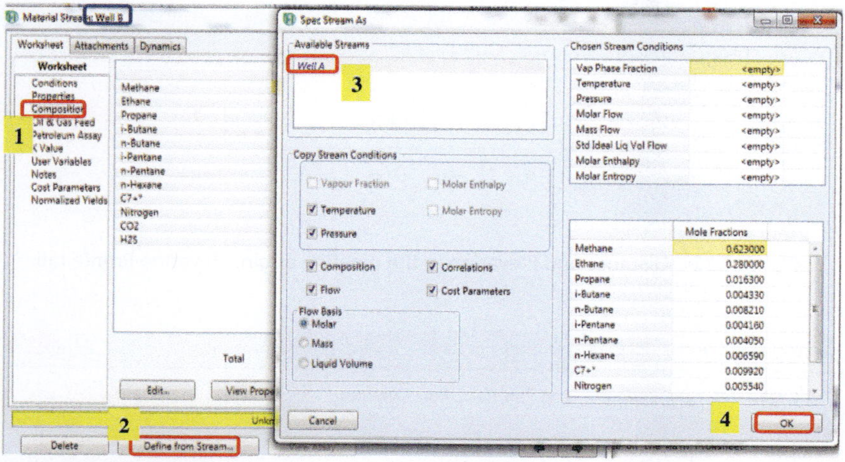

Figure 7.19 How to copy the composition of Well *B* from Well *A*.

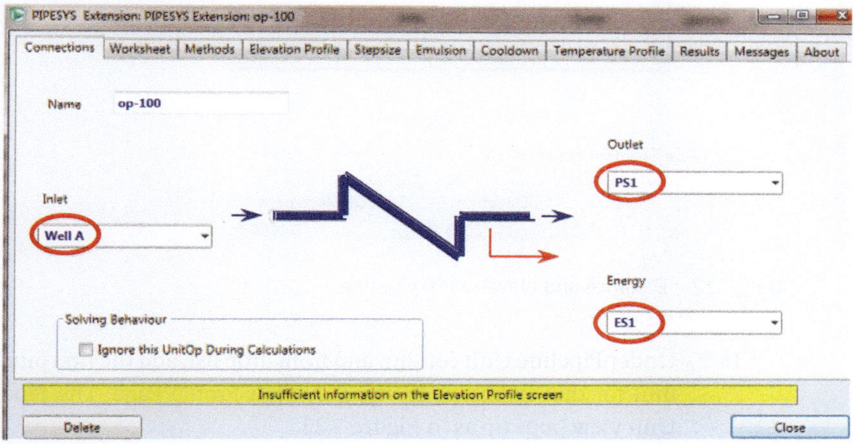

Figure 7.20 Connections tab of the Pipesys.

(6-2) Complete the Connections tab as shown in Figure 7.20.

(6-3) Go to **the Elevation Profile** table

(6-4) In the **Pipeline Origin** Enter 0.0 ft into the **Distance** field, and 2095 ft into the **Elevation** field as shown in Figure 7.21.

Note: *A starting point for the profile must be specified at the* ***top*** *of the tab in the Pipeline Origin group, using the Distance and Elevation input cells.*

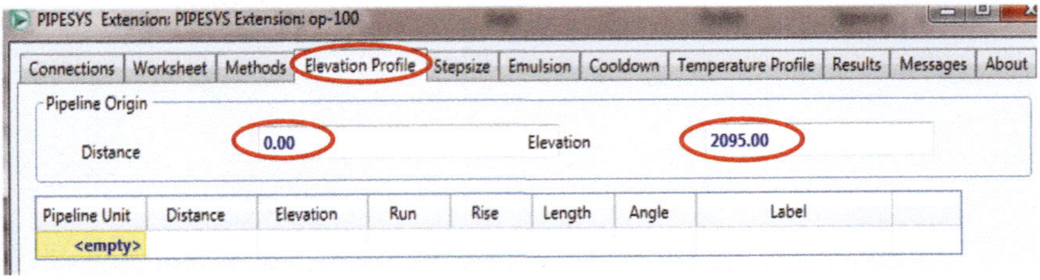

Figure 7.21 Distance and Elevation in the pipeline origin, Elevation Profile tab.

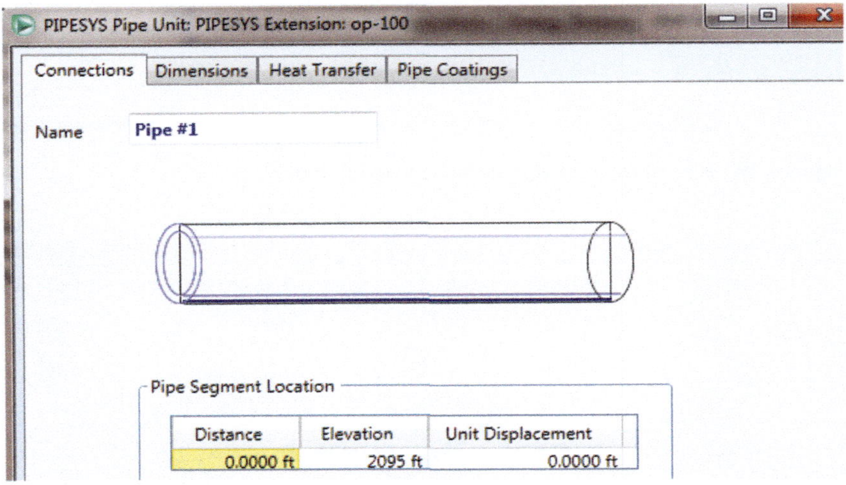

Figure 7.22 Distance and elevation in pipeline.

(6-5) Under Pipeline Unit column and from drop list, add the first pipe unit for this extension on the Elevation Profile Page. The Pipe Unit view pops up as in Figure 7.22.

(6-6) Go to the Dimensions tab, Figure 7.23 will appear.

(6-7) Specify the Nominal diameter = 3 in., and select Pipe Schedule = 40 as shown in Figure 7.23.

(6-8) Go to the Heat Transfer tab:

Click **Centre Line Depth** cell and enter 3.0 ft.

Leave all other parameters at their default values as shown in Figure 7.24.

Close the Pipe Unit view.

Figure 7.24 shows the details of circumstance of the pipe, such as whether the pipe is underground (Buried), Submerged, Above

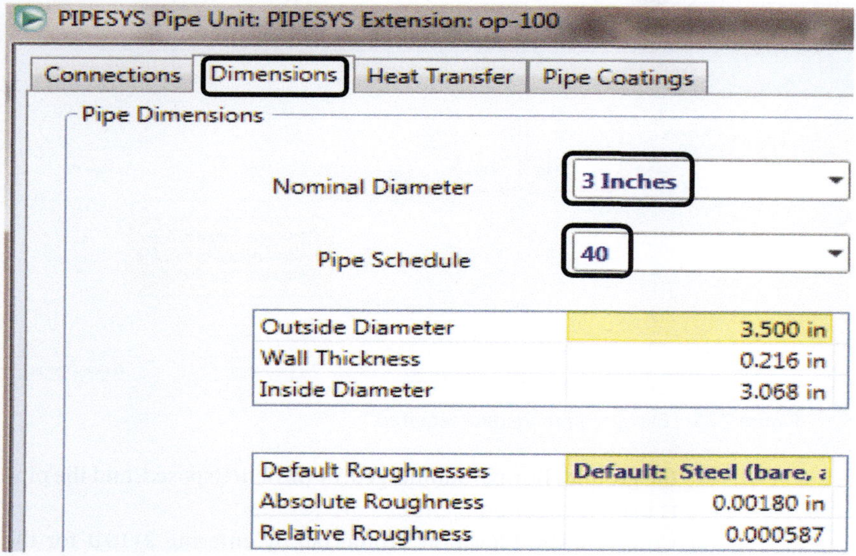

Figure 7.23 Selection of nominal diameter and pipe schedule.

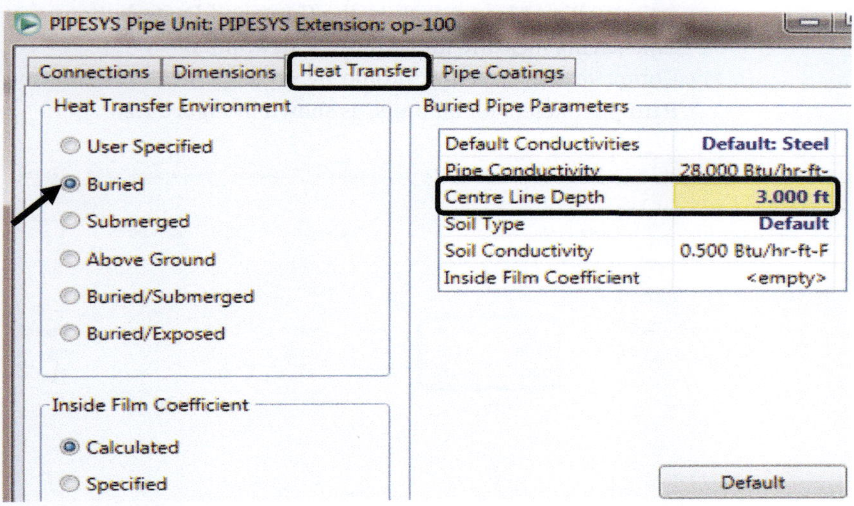

Figure 7.24 Selection of heat transfer environment and entering the value of centerline depth.

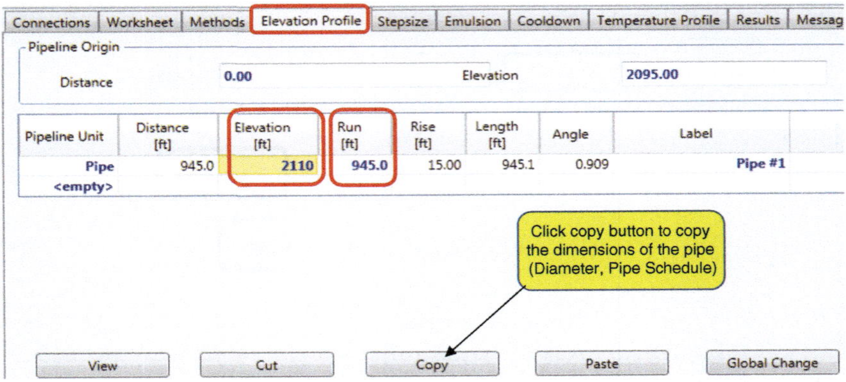

Figure 7.25 Elevation profile data required.

the ground, Buried/Submerged, or Buried/Exposed, and the pipe is insulated or not.

(6-9) Complete the Elevation Profile tab by entering 2110 ft for the **Elevation** parameter and 945 ft for the **Run** parameter as in Figure 7.25.

(6-10) Add the remaining two pipe units.
Because all the pipe units for the extension have identical properties to Pipe #1, you can use the **Copy** and **Paste** buttons as a time-saving measure for adding the new units.

(6-11) Complete the elevation profile by adding the **Elevation** and **Run** parameters for all units, as shown in Figure 7.26.

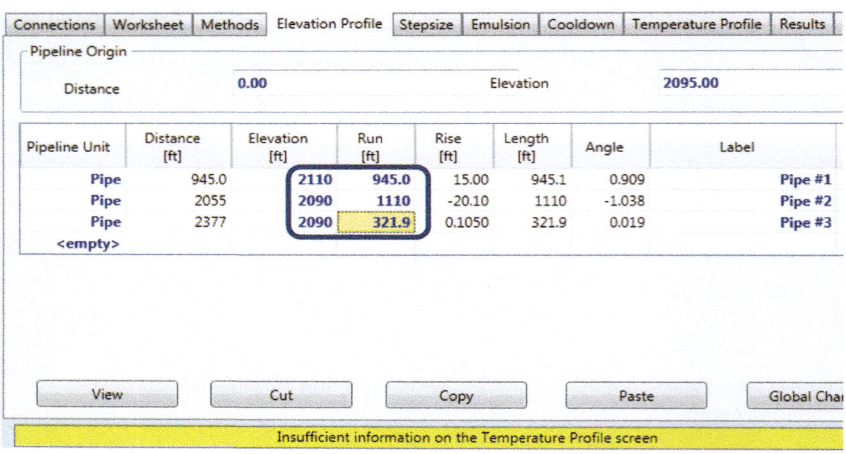

Figure 7.26 Results of the elevation data.

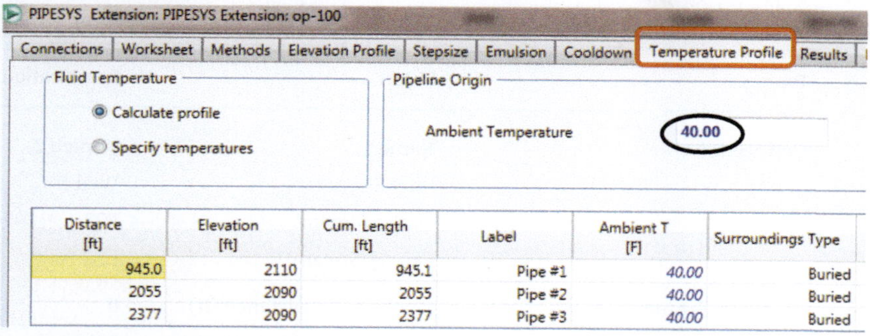

Figure 7.27 Temperature profile page in the Pipesys.

(6-12) Go to the **Temperature Profile** tab and enter 40 °F in the Ambient Temperature field (in the **Pipeline Origin** group), Figure 7.27.
This completes the first PIPESYS extension for your case.

7) Add the Second PIPESYS Extension
The second PIPESYS extension consists of a single Pipe Unit.
Enter the required information for the second extension as defined in Table 7.4.

8) Add a **Mixer** to your simulation and named Junction 1.
Input: PS1 and PS2.
Output = FS4
On the **Parameters** page of the Mixer, select the **Equalize All** radio button in the Automatic Pressure Assignment group.
Notes
- *In the* ***parameters*** *page of the* ***Mixer*** *there are two options in the* ***Automatic Pressure assignment****:*
- ***Equalize All****: Selected in the case of one stream has known pressure and the other is unknown.*
- ***Set Outlet to Lower Inlet****: Selected in the case of two streams are known pressures.*

9) Add the Third PIPESYS Extension
The third PIPESYS extension is used to represent Branch 4. It consists of a single pipe unit.
Enter the data for the third PIPESYS extension as defined in Table 7.5.

10) Add the Fourth PIPESYS Extension

Branch 3 of this pipeline system is represented by the fourth PIPESYS extension.

Table 7.6 contains the required information to complete the fourth PIPESYS extension.

Table 7.4 Information of the second extension.

Tab		Object		Specification
Connections		Name		Branch 2
		Inlet Stream		Well B
		Outlet Stream		PS2
		Energy Stream		ES2
Elevation Profile		Pipeline Origin	Distance (ft)	0
			Elevation (ft)	2015
		Pipeline Unit		Pipe # 1
		Elevation (ft)		2090
		Run (ft)		2822
click **View** button	Dimension	Nominal Diameter (Inch)		3
		Pipe Schedule		40
	Heat Transfer	Centerline Depth		Default
Temperature Profile		Ambient Temperature (°F)		40

Table 7.5 Information of the third extension.

Tab		Object	Specification	
Connections		Name	Branch 4	
		Inlet Stream	FS4	
		Outlet Stream	PS4	
		Energy Stream	ES4	
Elevation Profile		Pipeline Origin	Distance (ft)	0
			Elevation (ft)	2090
		Pipeline Unit	Pipe # 1	
		Elevation (ft)	2077	
		Run (ft)	1167	
click **View** button	Dimension	Nominal Diameter (Inch)	4	
		Pipe Schedule	40	
	Heat Transfer	Centerline Depth	Default	
Temperature Profile		Ambient Temperature (°F)	40	

Table 7.6 Information of the Fourth extension.

Tab		Object		Specification
Connections		Name		Branch 3
		Inlet Stream		Well C
		Outlet Stream		PS3
		Energy Stream		ES3
Elevation Profile		Pipeline Origin	Distance (ft)	0
			Elevation (ft)	2125
		Pipeline Unit		Pipe # 1
		Elevation (ft)		2077
		Run (ft)		528
		Pipeline Unit		Pipe # 2
		Elevation (ft)		2080
		Run (ft)		334
		Pipeline Unit		Pipe # 3
		Elevation (ft)		2077
		Run (ft)		670
click **View** button	Dimension	Nominal Diameter *for all 3 pipes above* (Inch)		3
		Pipe Schedule		40
	Heat Transfer	Centerline Depth		Default
Temperature Profile		Ambient Temperature (°F)		40

11) Add the Fifth PIPESYS Extension
    In this segment, the total gas flows from Wells *A*, *B*, and *C* are merged, and the endpoint of the extension is the gas plant.
    Enter the information for the final extension as defined in Table 7.7.
12) Add a **Mixer** named Junction 2 to your simulation.
    Inlet = PS3 and PS4, and the Outlet = FS5.
    On the **Parameters** page of the Junction 2, select the **Equalize All** radio button.
13) Define the Well Streams
    Finish specifying streams Wells *A*, *B*, and *C* with the data in Table 7.8.
    Save your case, and view the results from **Material streams** tab in the workbook.

Table 7.7 information of the fifth extension.

Tab		Object		Specification
Connections		Name		Branch 5
		Inlet Stream		FS5
		Outlet Stream		PS5
		Energy Stream		ES5
Elevation Profile		Pipeline Origin	Distance (ft)	0
			Elevation (ft)	2090
		Pipeline Unit		Pipe # 1
		Elevation (ft)		1980
		Run (ft)		2110
click **View** button	Dimension	Nominal Diameter (Inch)		6
		Pipe Schedule		40
	Heat Transfer	Centerline Depth		Default
Temperature Profile		Ambient Temperature (°F)		40

Table 7.8 Conditions of the well's stream.

Stream	Object	Specification
Well A	Temperature (°F)	105
	Pressure (psia)	1060
	Molar Flow (MMSCFD)	8.6
Well B	Temperature (°F)	115
	Molar Flow (MMSCFD)	7.4
Well C	Temperature (°F)	110
	Molar Flow (MMSCFD)	10.1

14) Optimize Gas-Condensate-Gathering System
    The PFD of the Pipesys in HYSYS is shown in Figure 7.28. It will attempt to increase production from the wells by adding a compressor to the fifth PIPESYS extension.
    The adjust will be used to maintain a constant pressure of 1000 psia at the gas plant with the pressure at **Well *A*** being the adjusted variable.

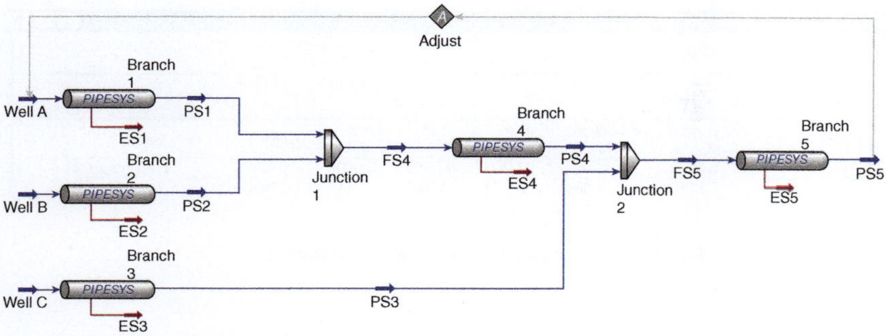

Figure 7.28 Process flow diagram of the small gas condensate in Hysys after adding adjust.

Table 7.9 specifications of Adjust operation.

Object		Specification
Connection	Adjusted Variable: Object	Well A
	Variable	Pressure
	Target Variable: Object	PS5
	Variable	Pressure
	Specified Target Variable	1000 psia
	Method	Secant
Parameters	Tolerance	0.1 psi
	Step Size	1 psi
	Max. Iteration	1000

(14-1) Add an ADJUST operation with the specifications of Table 7.9.
(14-2) Click the **Start** button at the bottom of the Adjust property view to begin the adjust calculates. *(HYSYS may need several minutes to reach a solution.)*
(14-3) Upon convergence, the **well** pressures should be:

Well	Pressure (psia)
A	1093
B	1077
C	1052

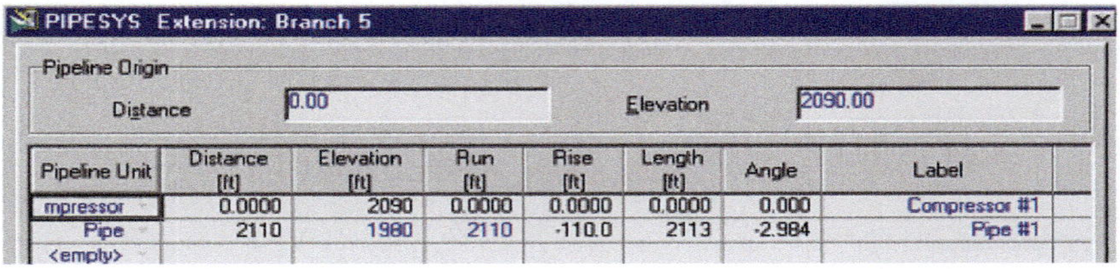

Figure 7.29 Adding compressor to the Pipesys, branch 5.

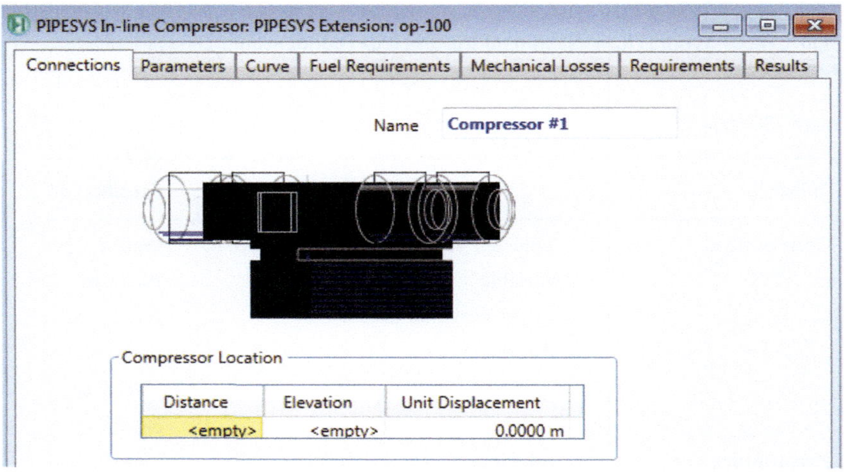

Figure 7.30 Window of the compressor in the pipesys.

15) Add an Inline Compressor: The next step will be adding an inline compressor at the upstream end of PIPESYS Branch 5.
   (15-1) Go to **the Elevation Profile** tab of Branch 5.
   (15-2) Modify the pipe extension, so that the **Compressor** is placed before the Pipe. See Figure 7.29 for placement of the **pipe** units.
   (15-3) Open the Inline Compressor view, Figure 7.30.
   (15-4) Use the information in Table 7.10 to complete the tabs as specified on the Inline Compressor view, see Figure 7.31.
   The solution process may take several minutes to perform the iterative calculation for the PIPESYS network and converge.
   When the process is complete, the well pressures should be as follows:

Table 7.10 Specification of Inline compressor.

Object	Specification
On the Parameters tab	
Brake Power-Specified	1000 hp
Max. Discharge Temp	100 °F
Max. Interstage Temp	100 °F
Number of Stages	2
Adiabatic Efficiency	0.73
Interstage delta P	10 psi
On the Mechanical Losses tab	
Overall Efficiency	0.95

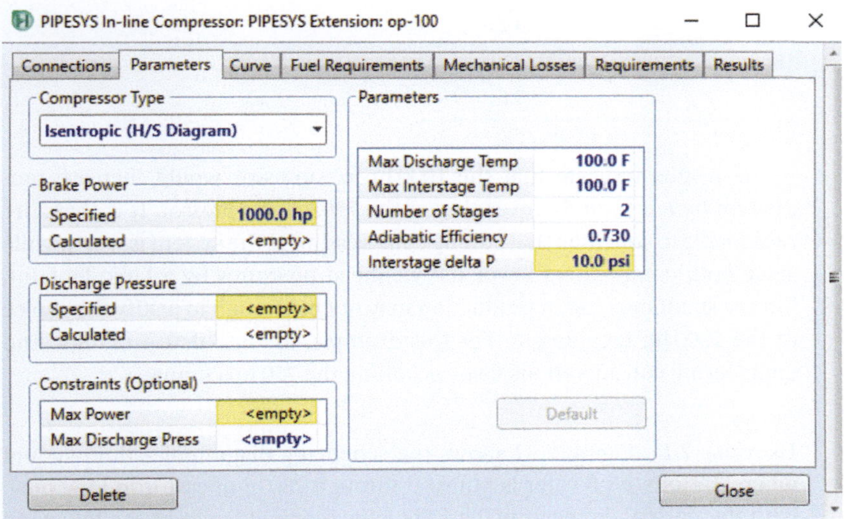

Figure 7.31 Inline Compressor view, parameters tab.

Well	Pressure (psia)
A	686.7
B	655.9
C	619.5

To compare the performance between two compressors, the same calculations are repeated using a **750 hp** compressor.

(15-5) On the Inline Compressor view, go to the Parameters table
(15-6) In the Brake Power group, change the 1000 hp in the Specified cell to 750 hp.

When HYSYS completes the iteration, the new pressures are:

Well	Pressure (psia)
A	753.2
B	726.3
C	693.5

The results indicate that the 1000 hp compressor would increase production by *less than 5%,* over that of the 750 hp compressor. It is therefore reasonable to conclude that adding compression to the system is worthwhile since both compressors lower the wellhead pressures by a large amount, but the small increase in production may not be enough to justify the choice of the 1000 hp compressor. For this example, assume that economic and engineering considerations favor installing the 750 hp compressor.

Exercise 7.1 Figure 7.32 shows the schematic diagramof a fluid system where 3.8 tons/h of water is pumped through 280 m of cast-iron long pipe, with the normal diameter of 2 in. The pipe segment contains the following fittings: an open gate valve, a half-open gate valve, a 2 standard 90 ° elbows, a disk water meter, and an open globe valve. The inlet pressure and temperature of water are 6.5 bar, 26 °C respectively. Ambient temperature is 21 °C, and the average value of the heat transfer coefficient is 20 Wm^{-2}/K. Using Pipe Segment and PIPESYS, estimate the pressure drop, and pressure and temperature profile across the pipe [6].

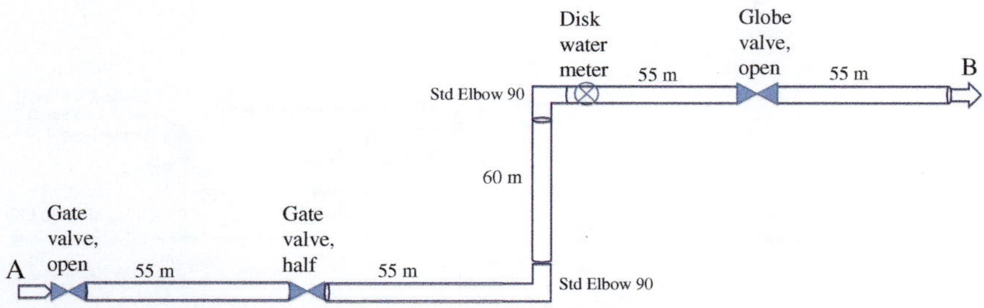

Figure 7.32 Schematic diagram of Exercise 7.1.

7.3 Sizing of Pressure Relieve Valve

When designing a chemical process and its equipment, safety is of substantial concern. All possible strategies should be considered to eliminate the risk of accident and injury when implementing a process design.

One safety concern that occurs frequently in a plant is the over-pressurization of equipment or piping. In order to prevent dangerous bursts, explosions, and fires, pressure relief valves are designed and installed to bleed out excess liquid or vapor causing pressure build-up.

Aspen HYSYS has a pressure safety valve (PSV)-sizing feature, which uses data from a process simulation to help automate PSV calculations. It will be taking an example from which the PSV has been designed [7, 8].

7.3.1 Problem Statement

In this example, the methane is oxidized in the Fired Heater using excess air to ensure complete oxidation. The gases from the fired heater are used to generate the steam in the Heat Exchanger. To maintain the pressure of steam under control, safety valve must be used to avoid sudden excess pressure and then ensure safety operation.

The process flow diagram and its conditions are shown in Figure 7.33. Design a pressure safety valve for this purpose. Assume the efficiency of the Fired Heater = 60%, and the excess Air = 100%.

7.3.2 Simulation

1) Open a new case in the HYSYS and add the following components: C_1, C_2, C_3, i-C_4, n-C_4, i-C_5, n-C_5, O_2, N_2, and H_2O.
   Fluid package: Peng–Robison.

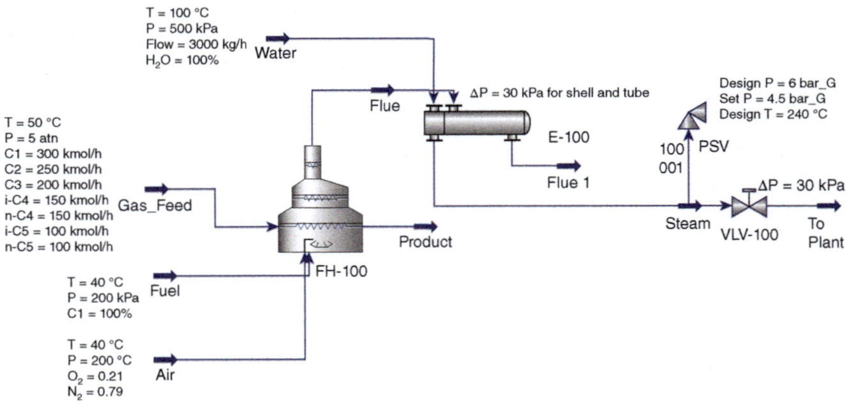

Figure 7.33 PFD for Fired Heater, Heat Exchanger, and Pressure Safety Valve with their conditions.

2) Enter the Simulation environment and add streams with the conditions as in the process flow diagram, PFD, Figure 7.33.
3) Add a Fired Heater and connect the streams as in PFD, Figure 7.33. Enter the Efficiency of Fired Heater = 60%, and Excess Air = 100%.
4) Add a Heat Exchanger, Tube in = Water, Tube out = Steam, Shell in = Flue, Shell Out = Flue1,
ΔP for Shell and tube sides = 20 kPa, and the temperature of Product and Flue1 = 300 °C.
5) Add a valve: Input = Steam, Output = to Plant, ΔP of valve = 30 kPa.
6) From the Navigation Pane, click the **Safety Analysis** button, Figure 7.34.
This environment provides a setting for relief scenario setup and PSV sizing and documentation.
7) From the new window, click the **Add PSV** button, Figure 7.35.
This switches the cursor into attach mode. Hover over the stream or unit operation on which a PSV is necessary. When the cursor arrow changes and displays a white box, click to attach a relief valve over **Steam** stream.
8) Double-click a pressure safety valve in the flowsheet opens the "Valve Design" tab, as shown in Figure 7.36.
Enter the Design Temperature = 240 °C and Design Pressure = 6 barG, Figure 7.36.
9) Click the **PRD Data** tab in Figure 7.36, and change the **Discharge to** Atmosphere (*because it is steam*), and **Set Pressure** (4.5 barG) after deactivating the radio of **Design P** Figure 7.37.

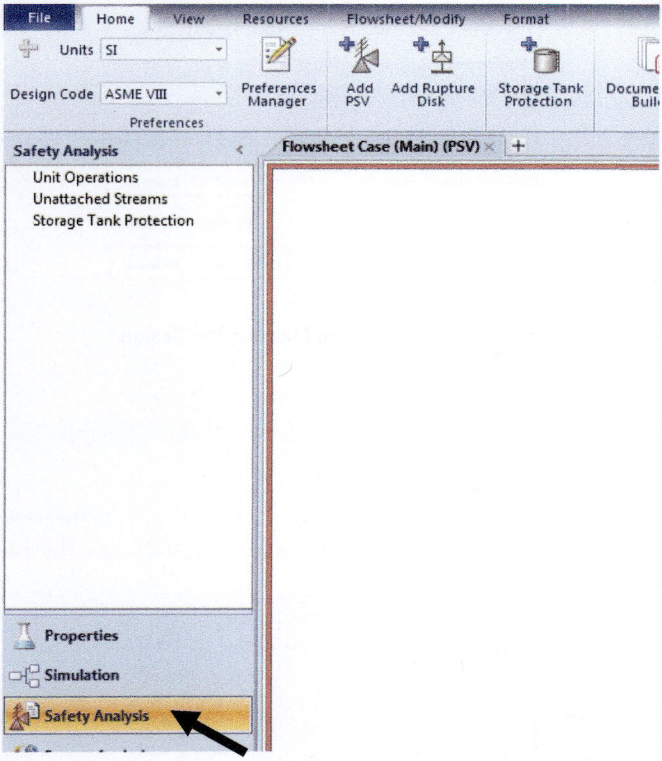

Figure 7.34 Entering the Safety Analysis by clicking the Safety Analysis button.

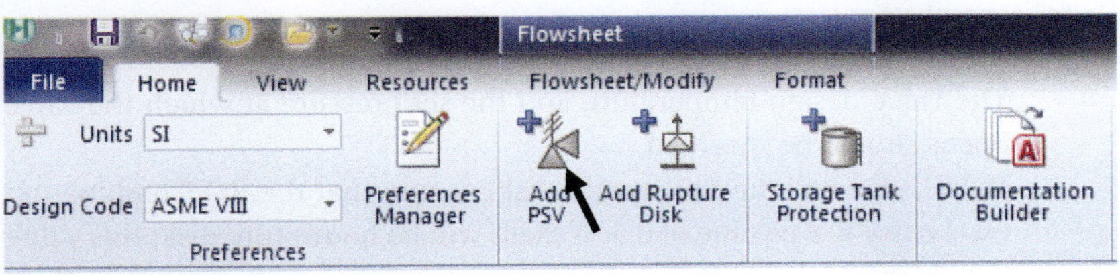

Figure 7.35 Adding PSV Option in Safety Analysis Environment.

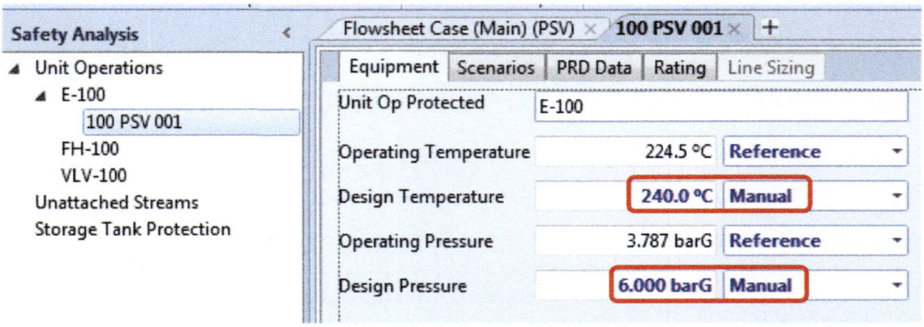

Figure 7.36 Window of pressure safety valve and Entering the Design Temperature and Design Pressure.

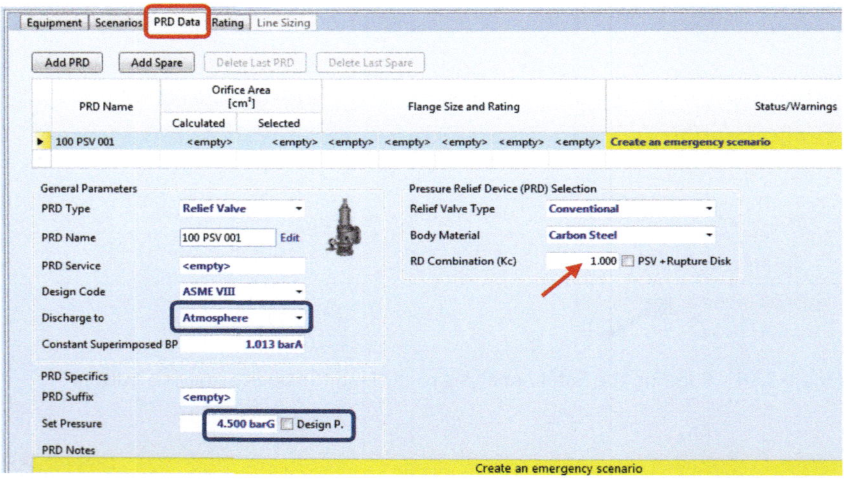

Figure 7.37 Window of PRD Data and Entering Set Pressure and Set "Discharge to" *Atmosphere.*

The valve design temperature and the set pressure at which the valve opens should be specified.

If the valve will have a rupture disk, ensure that the RD Combination (K_C) entry has a value of 0.9. If there will be no rupture disk, this value should be 1.

10) Click the **Scenario** tab, then click **"Create Scenario,"** Figure 7.38 will show.

    Check the **Sizing Case** radio.

11) From Figure 7.38, Under **Scenario** Group, **Type** column, click the drop list of **User Defined,** Figure 7.39 will show. Select the

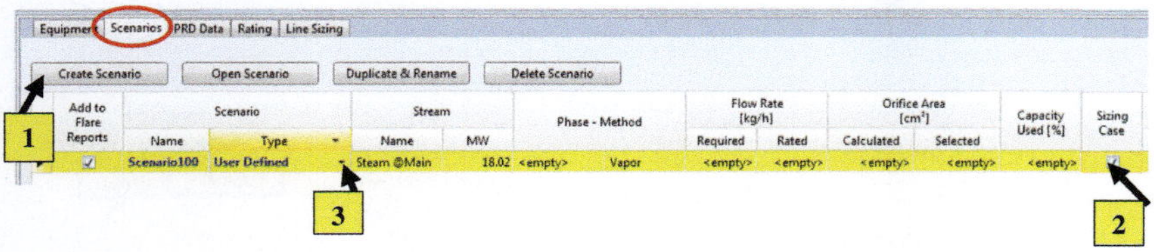

Figure 7.38 Scenarios tab of PSV.

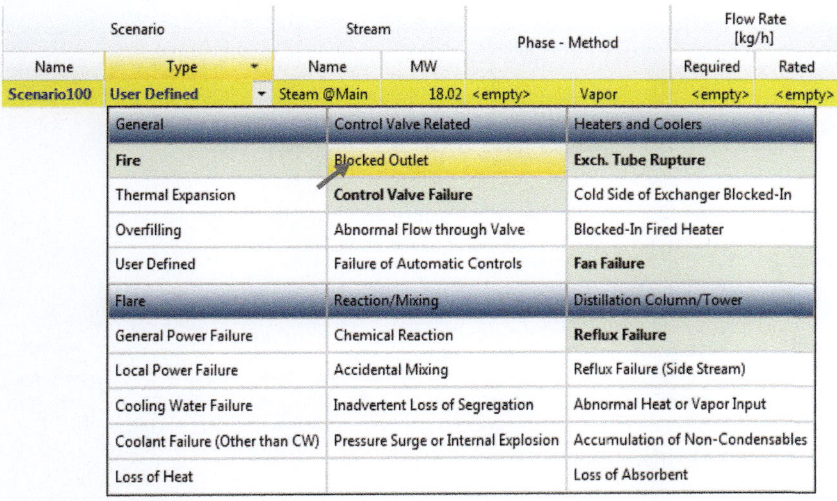

Figure 7.39 Relief Load Scenarios Available in the scenario Type list.

contingency/emergency scenario that causes overpressure in the system. There are number of different of scenarios that could have upon at different places of plant. For this case, select **"Blocked Outlet."**

12) Under **Stream** group, Figure 7.39, double-click the **Steam @Main**, Figure 7.40 will show.
    For Relieving Temperature, from drop list, select **Reference.**
    For Relieving Flow Method, check the **Reference** radio.
13) Now, under **Sizing Case,** it can be seen the desired value (Calculated) of orifice area is (10.17 cm^2); click the **"Select"** to enter the standard value of Orifice area.
14) Return to the (**100 PSV 001**) tab, **Scenario** page to see other results as shown in Figure 7.41.

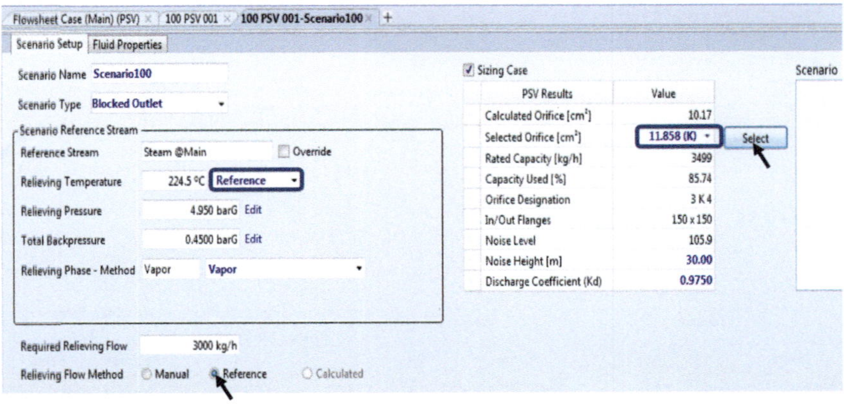

Figure 7.40 Valve Results Table inside "Scenario Setup" tab.

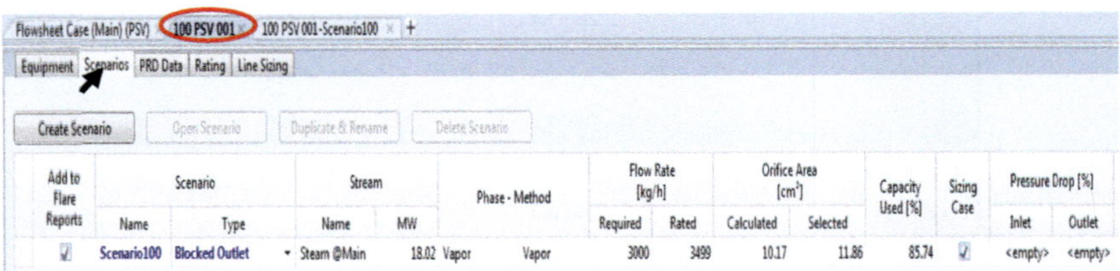

Figure 7.41 Results of PSV.

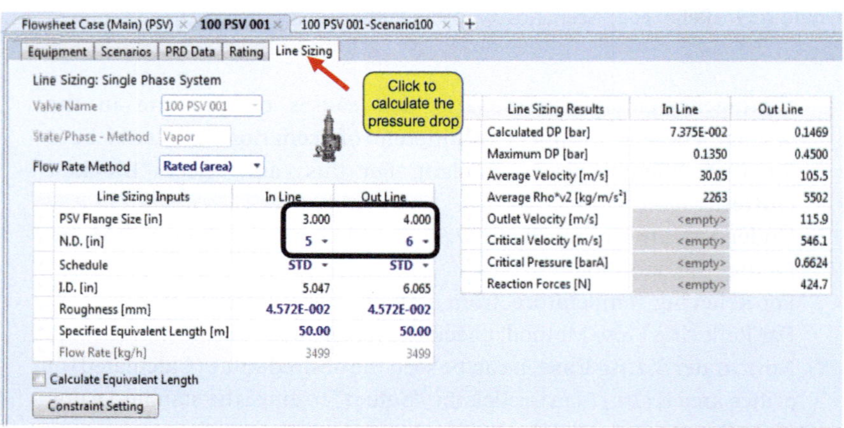

Figure 7.42 Entering the nominal diameter for In Line and Out Line and Results of Line Sizing.

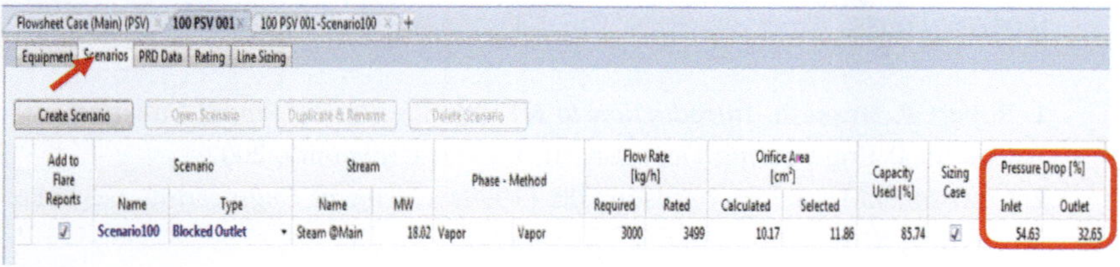

Figure 7.43 Calculating pressure drop in the PSV.

15) To calculate the **Pressure Drop (%),** click **Line Sizing** tab, and enter the nominal diameter (N.M. [in]) for Inlet Line = 5 in, and Outlet Line = 6 in, and select **STD** for Schedule, Figure 7.42.
    - The Line Sizing will be calculated.
16) Return to the **Scenario** tab to see the Pressure Drop (%), Figure 7.43.

7.4 Summary

There are two types of piping systems, Pipe Segment and Pipesys, as mentioned in Sections 7.1 and 7.2, respectively. Both of these can be used in all areas of industries, and the calculation methods for both systems are very similar, but the first is preferred to be used for chemical industries, and the second for oil and gas industries, especially for large distances. In the first system, at least three variables must be specified: the conditions of the inlet stream, the length, and the diameter of the pipe. The simulator will calculate the pressure difference, the outlet pressure, and the outlet temperature. In the second system, some variables must be specified: inlet stream conditions, pipe diameter, pipe length and displacement, and the simulator will calculate the angle of deflection, pressure difference, and pipe height. The (Adjust) operation has been used in this system to maintain the plant pressure (PS5) by changing the pressure of Well *A*.

In Section 7.3, the sizing of the pressure relief valve is evaluated on the output stream of the generated steam from the Heat Exchanger where the energy supplied to the Heat Exchanger is from flue gases out of the fired heater. Pressure relief valve design and sizing can be a tedious and lengthy process. Relief valve sizing in Aspen HYSYS allows the user to design pressure relief valves quickly using a small amount of process information at the location of the valve. Aspen HYSYS now encompasses all facets of the valve design process in one convenient location. In addition, various scenarios involving multiple valves can be explored to find the most optimal relief schedule for a unit.

References

1 Robert P. Hesketh, *Introduction to HYSYS Simulation Piping Calculations*, Chemical Engineering Department, Rowan University, 2002.

2 Coulson, J.M. and Richardson, J.F. (1999). *Chemical Engineering*, 6e, vol. 1. (ed. J. R. Backhurst and J. H. Harker).

3 Holland, F. and Bragg, R. (1995). *Fluid Flow for Chemical Engineers*, 2e. Butterworth-Heinemann.

4 (2010). *Aspen HYSYS Pipeline Hydraulics – PIPESYS*. Aspen Technology, Inc. AspenTech®, aspenONE®.

5 (2003). *PIPESYS, User Guide*. Aspen Technology, Inc. AspenTech®, aspenONE®.

6 Haydary, J. (2019). *Chemical Process Design and Simulation Aspen Plus and Aspen HYSYS Applications*. Wiley.

7 Qassam, A., Dyment, J., and Mofor, W. (2014). *Jump Start: Relief Sizing in Aspen HYSYS® and Aspen Plus® V8.6*. Aspen Technology, Inc. AspenTech®, aspenONE®.

8 Brownrigg, N. and Mofor, W. (2013). *Jump Start: Pressure Safety Valve Sizing in Aspen HYSYS®*. Aspen Technology, Inc. AspenTech®, aspenONE®.

8

Reactors

In Aspen HYSYS, there are two basic reactor styles and several subgroups. These reactors are the two ideal reactor models of a CSTR (continuous stirred tank reactor) and a PFR (plug flow reactor) [1]. The CSTR model is a standard algebraic model that has been in simulation packages for a number of years. The order differential equations (ODE's) of the PFR are a recent addition to simulation packages and are solved by dividing the volume into small segments and finding a sequential solution for each volume [2]. In these more recent models, these reactors not only include energy balances, but also pressure drop calculations are a standard feature for packed bed reactors.

With the exception of the PFR, all of the reactor operations share the same basic property view. The primary differences are the functions of the reaction type (conversion, kinetic, equilibrium, heterogeneous catalytic, or simple rate) associated with each reactor [3]. As opposed to a separator or general reactor with an attached reaction set, specific reactor operations can only support one particular reaction type. For instance, a conversion reactor only functions properly with conversion reactions attached. If you try to attach an equilibrium or a kinetic reaction to a conversion reactor, an error message appears. The Gibbs reactor is unique in that it can function with or without a reaction set.

Table 8.1 shows the summary of the reactions in the HYSYS [1, 2].

8.1 Reactor Types

Table 8.2 shows the type of reactors and their suitable reactions.

Chemical Process Simulations using Aspen Hysys, First Edition. Khalid W. Hameed.

Table 8.1 Summary of Reactions in HYSYS.

Reaction Type	Description
Conversion	Conversion% ($X\% = C_o + C_1T + C_2T^2$)
Equilibrium	$K_{eq} = f\ (T)$; equilibrium based on reaction stoichiometry, K_{eq} predicted or specified
Gibbs	minimization of Gibbs free energy of all components
Kinetic	$r_A = k_f C_A^{\alpha} C_B^{\beta} - k_{rev} C_R^{\varphi} C_S^{\gamma}$ where the reverse rate parameters must be thermodynamically consistent and rate constants are given by $k = AT^n \exp(-E/RT)$
Heterogeneous Catalytic	Yang and Hougen form: $-r_A = \frac{k\left(C_A^a C_B^b - \frac{C_R^r C_S^s}{K}\right)}{1+\sum K_i C_i^{\gamma_i}}$ This form includes Langmuir–Hinshelwood, Eley–Rideal, and Mars–van Krevelen.
Simple Rate	$r_A = -k_f\left(C_A^{\alpha} C_B^{\beta} - \frac{C_R^{\varphi} C_S^{\gamma}}{K_{eq}}\right)$ in which K_{eq} is predicted from equilibrium data.

Table 8.2 Types of reactors and their suitable reactions.

HYSYS Reactor Name	Reaction Types (See above in Table 8.1)
Conversion Reactor	Conversion% ($X\% = C_o + C_1T + C_2T^2$)
CSTR	Simple Rate, Heterogeneous Catalytic, Kinetic
Equilibrium Reactor	$K_{eq} = f(T)$; equilibrium based on reaction stoichiometry, K_{eq} predicted from Gibbs Free Energy, K_{eq} specified as a constant or from a table of values.
Gibbs Reactor	minimization of Gibbs free energy of all specified components, (Option 1) no the reaction stoichiometry is required (Option 2) reaction stoichiometry is given
Yield Shift Reactor	supports efficient modeling of reactors by using data tables to perform shift calculations. The operation can be used for complex reactors where no model is available, or where models are too computationally expensive.
Plug Flow Reactor (PFR)	The suitable reactions for this reactor type are: Simple Rate, Heterogeneous Catalytic, Kinetic.

8.2 Parameters in the Reactors

8.2.1 Pressure Drop

The pressure drop across the vessel is defined as [2]:

$$\Delta P = P_{feed} - P_{vapor} = P_{feed} - P_{liquid}$$
$$P = P_{vapor} = P_{liquid} \tag{8.1}$$

where:

P is the vessel pressure,
P_{vapor} is the pressure of vapor product stream,
P_{liquid} is the pressure of liquid product stream,
P_{feed} is the pressure of feed stream (assumed to be the lowest pressure of all the feed streams),
ΔP is the pressure drop in vessel (Delta P).

The default pressure drop across the vessel is zero. The vessel pressure is used in the reaction calculations.

8.2.2 Duty

The steady-state reactor energy balance is as [1, 2]:

$$Duty = H_{vapor} + H_{liquid} - H_{feed} \tag{8.2}$$

where:

$Duty$ is the heating (+ve) or cooling (−ve) by the optional energy stream,
H_{vapor} is the heat flow of the vapor product stream,
H_{liquid} is the heat flow of the liquid product stream,
H_{feed} is the heat flow of the feed stream(s).

The enthalpy basis used by HYSYS is equal to the ideal gas enthalpy of formation at 25°C and 1 atm. As a result, the heat of reaction calculation is amalgamated into any product/reactant enthalpy difference.

8.2.3 Volume

The total volume of the vessel is user specified. While not necessarily required for solving Conversion, Gibbs or Equilibrium reactors in Steady-State mode, this value must be entered for CSTR and PFR. The vessel volume, together with the liquid level set point, defines the amount of holdup in the vessel. The amount of liquid volume or holdup in the vessel at any time is given by Eq. (8.3):

$$Holdup = Vessel\ volume \times \frac{PV(\%Full)}{100} \tag{8.3}$$

where:

$PV(\%Full)$ = liquid level in the vessel

The vessel volume is necessary when modeling reactors in steady state, as it determines the residence time.

8.3 Conversion Reactor

Conversion Reactor is used when reaction stoichiometry and conversion or molar extent for each reaction is known, and reaction kinetics are unknown or unimportant. ***Conversion Reactor*** can model reactions occurring simultaneously or sequentially. In addition, these models can calculate the heat of reaction.

There is no need for thermodynamic expertise for this kind of reaction. The base reactant's stoichiometry and conversion must be entered. The conversion cannot go above 100%. The reaction will continue until either the desired conversion is attained or the supply of a limiting reagent is depleted [2, 3].

In a reaction set, conversion reactions cannot be grouped with any other type of reaction. They can be prioritized to work sequentially or simultaneously and grouped with other conversion reactions. First place finishes last (may start with either 0 or 1). Simultaneous reactions cannot total more than 100% conversion of the same basis, just like single reactions. PFRs and CSTRs cannot be used with conversion reactions. They ought to be utilized exclusively in conversion reactors, in general.

8.3.1 Problem Statement

CO at 10°C is completely burnt at 1 atm pressure with 50% excess air that is fed to a burner at a temperature of 540°C. The combustion products leave the burner chamber at a temperature of 425°C. Calculate the heat evolved, Q, from the burner. The reaction that represents the burning of CO is [4]:

$$CO + 1/2O_2 \rightarrow CO_2 \tag{8.4}$$

Conversion 100% with respect to CO.

The feed consisting of 100 kmol/h of CO reacts with air (21% O_2, 79% N_2). According to Eq. (8.4), each mole of CO needs 0.5 mol of O_2 or $0.5/0.21 = 2.381$ mol of air plus 50% excess, i.e. each 100 mol of CO will require $1.5 \times 238.1 = 357.15$ mol of air.

Create a model that illustrates the conversion of carbon oxide to carbon dioxide.

8.3.2 Simulation

1) The choice of a suitable fluid package is the initial stage in the simulation of CO oxidation. Open a new case and enter the following information into the specified fluid package:
   Components: O_2, N_2, CO, and CO_2.
   Fluid package: Peng–Robinson
2) **Specifying the reactions:** Similar to how components are introduced to a simulation, HYSYS adds reactions in a similar way:
   (2-1) In the Navigation Pane, Click **Reactions** folder, then add **Reaction Set** by clicking the **Add** button, Figure 8.1.
   (2-2) To specify the type of reaction into reaction set, select the **Add Reaction** button, Figure 8.2.
3) Choose **Conversion** as the type from the displayed list and click **Add Reaction**, Figure 8.2.
4) Under the **Active Reaction**: Double-click the **Rxn-1**; Figure 8.3 will show the window of the Conversion Reaction.

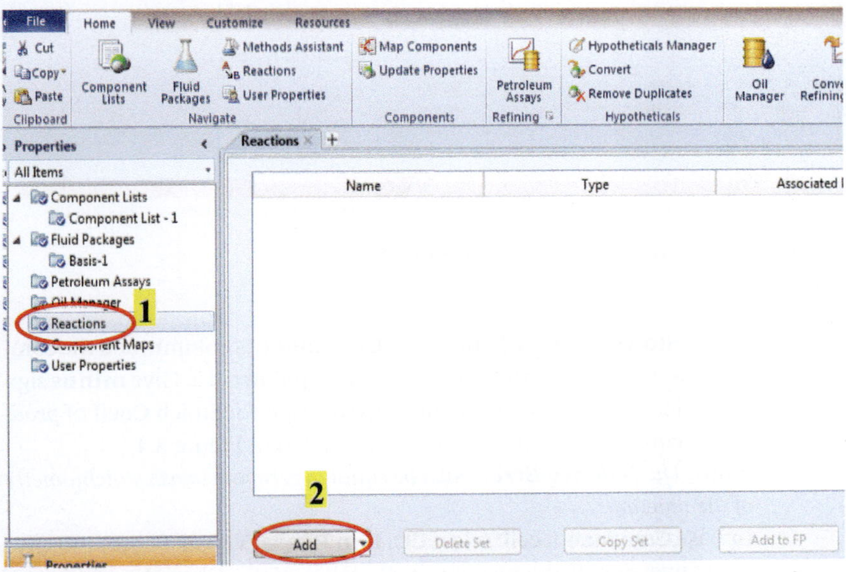

Figure 8.1 How to add a reaction set.

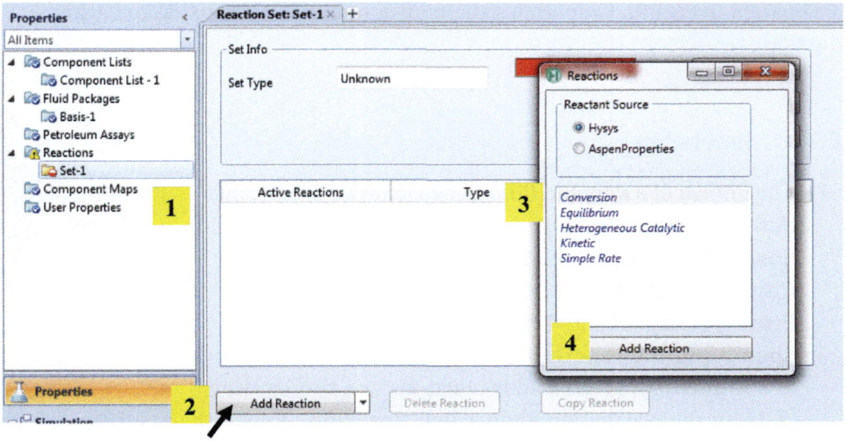

Figure 8.2 Adding a reaction to the Reaction Set-1.

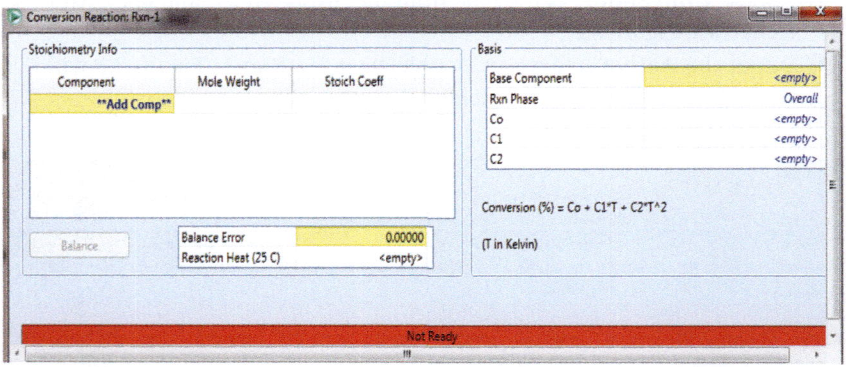

Figure 8.3 Window of conversion reaction type.

5) Under **Stoichiometry Info** group, **Components** column, add the components according to the reaction from drop-down list. Give **minus** sign for Stoich Coeff of *reactants* and **positive** sign for Stoich Coeff of *products* according to reaction (8.4) and as shown in Figure 8.4.
Note: *The* ***Balance Error*** *must be equal to zero for correct stoichiometry of the reaction.*
6) In Base Component cell: select CO, **Rxn Phase**: VapourPhase, and Give the conversion of the reaction in $\mathbf{C_0}$ cell = 100%. Close the window of the Conversion reaction: Rxn-1, Figure 8.4.

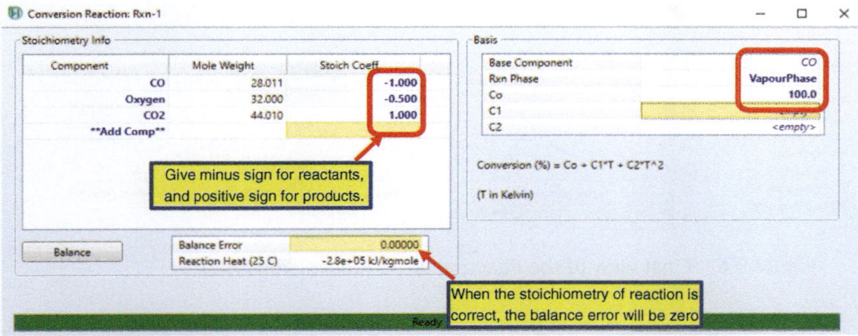

Figure 8.4 Stoichiometric and conversion information of reaction (8.4).

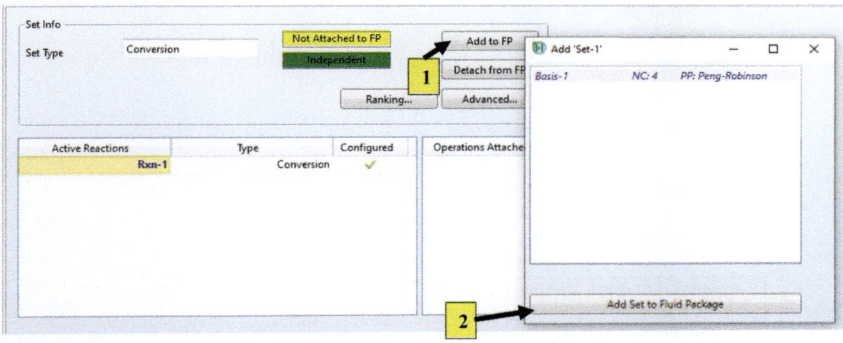

Figure 8.5 How to add the Reaction Set to the fluid package.

7) Attach the **Reaction Set** to the Fluid Package by clicking the **Add to FP** button. From the new window, click **Add Set to Fluid Package**, Figure 8.5.
   The final view of the **Reaction Set-1** must be as in Figure 8.6.
8) When the reaction set is added to the **Fluid Package**, enter the **Simulation** environment and start creation the simulation.
9) Add two Material streams with the following conditions:

Stream Name	**CO**	**Air**
Temp	10°C	540°C
Pressure	1 atm	1 atm
Molar flow	100 kmol/h	357.15 kmol/h
Composition by mole	100% for CO	21% O_2, 79% N_2

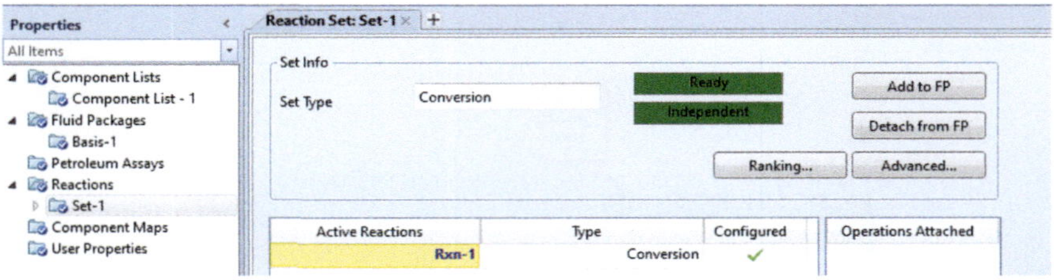

Figure 8.6 Final view of the Reaction Set-1 (Conversion type).

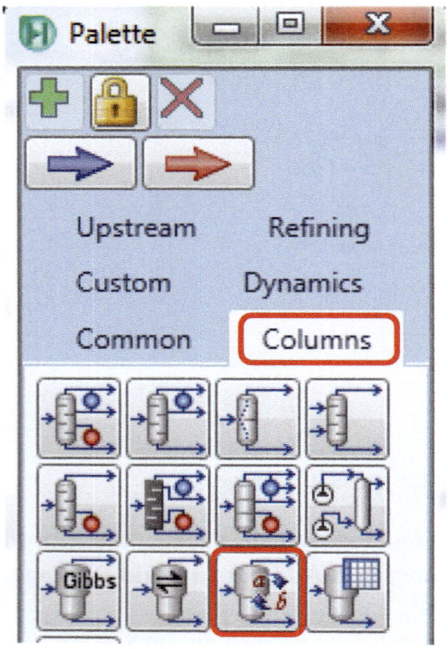

Figure 8.7 Conversion reactor in the object palette, Columns group.

10) Add the Conversion Reactor from the Object Palette by clicking the **Columns** group, Figure 8.7. Select the **Conversion Reactor**, and drag it into the simulation area.
11) Connect the **CO** and **Air** streams as feeds. Name the vapor outlet **Vap** and even though the liquid product from this reactor will be zero, it must attach the liquid stream and name it **Liq**, Figure 8.8.
12) Add stream energy (*Q*) in the Energy cell.
    Note: *If the Reaction is adiabatic, the* **Energy** *cell must be* **empty,** *and if the Reaction is at a specified temperature, it must* **add** *Energy stream. In this example, the Reaction is specified at 425°C.*

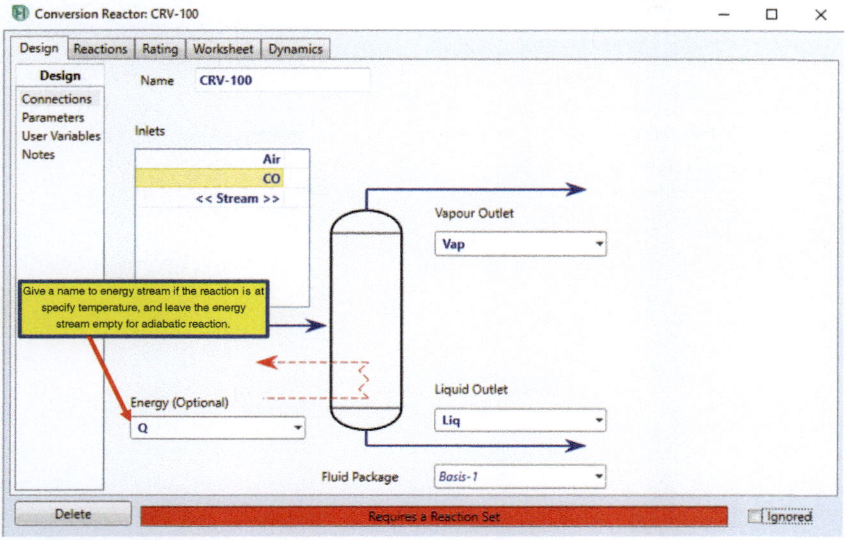

Figure 8.8 Connection page of the conversion reactor.

13) Switch to the **Parameters** Page of the reactor; if there is a pressure drop, its value must be entered; in this example the $\Delta P = 0$.
14) **Reactions** tab: **Details** page, from the **Reaction Set**, select **Set-1**. This will automatically connect the proper reactions to this reactor, Figure 8.9.
    From Figure 8.9, the bottom yellow bar indicates either specifying the outlet temperature, or specifying the amount of heat added to the reactor. From the Worksheet tab, enter the Vap stream temperature equal to 425°C. The data required now is complete, and the reactor is converged, see Figure 8.10.
    The value of Q is (-2.845×10^7 kJ/h), and since its sign is negative, that means the reaction is exothermic, and this amount of heat must be removed during the reaction.
    One can see the conversion of the reaction from the **Reactions** tab, **Results** page. Select Reaction Balance radio to see the amount of each component input and output from the reactor, Figure 8.11.
15) Switch to the **Worksheet** tab, **Composition** page, in the **Vap** stream, analyze the composition.
    Specify the molar flow rate of the following components: N_2, O_2, CO, and CO_2.

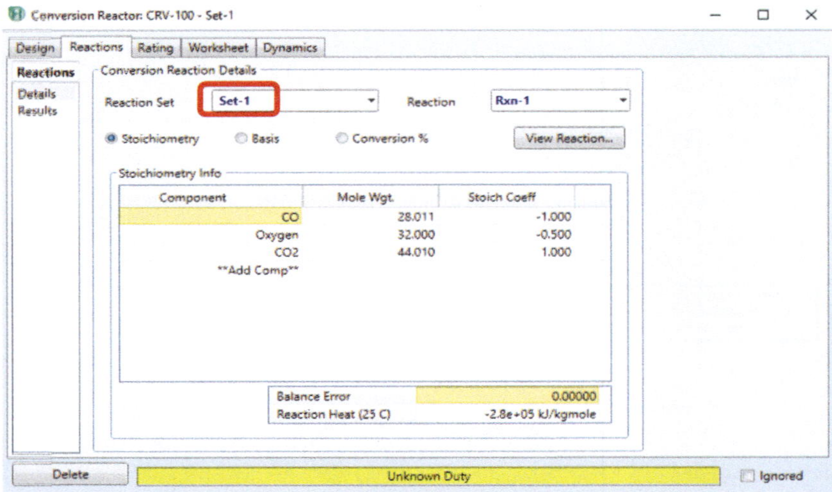

Figure 8.9 Adding the reaction set to the conversion reactor.

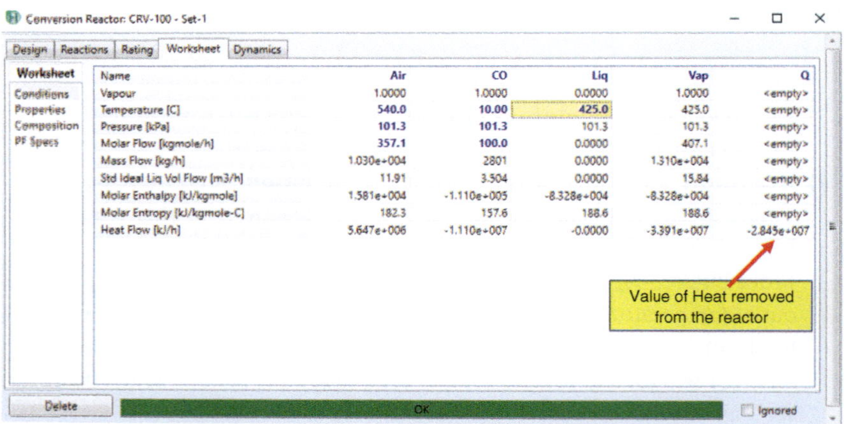

Figure 8.10 Worksheet tab of the conversion reactor, through which the heat removed is presented.

Save this case by **Conversion.** The process flow diagram (PFD) of the conversion reactor is presented in Figure 8.12.

8.3.3 Summary

In this section, the goal is to create a model that illustrates how CO undergoes a total oxidation reaction. In the total oxidation process of CO to CO_2,

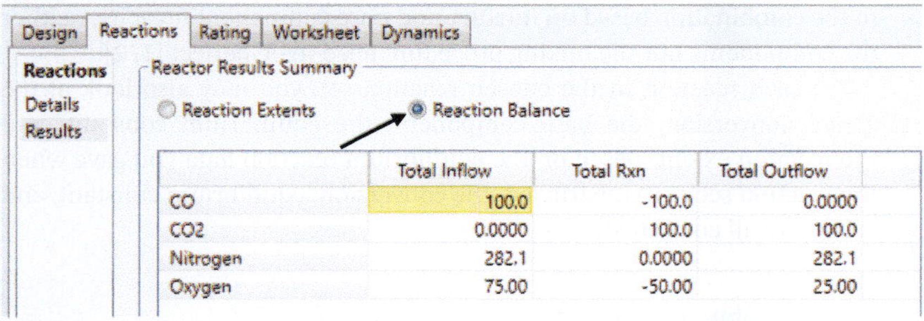

	Total Inflow	Total Rxn	Total Outflow
CO	100.0	-100.0	0.0000
CO2	0.0000	100.0	100.0
Nitrogen	282.1	0.0000	282.1
Oxygen	75.00	-50.00	25.00

Figure 8.11 Reaction tab, Results page, Reaction Balance radio, explaining the flow rate of each component that inputs/outputs to the reactor.

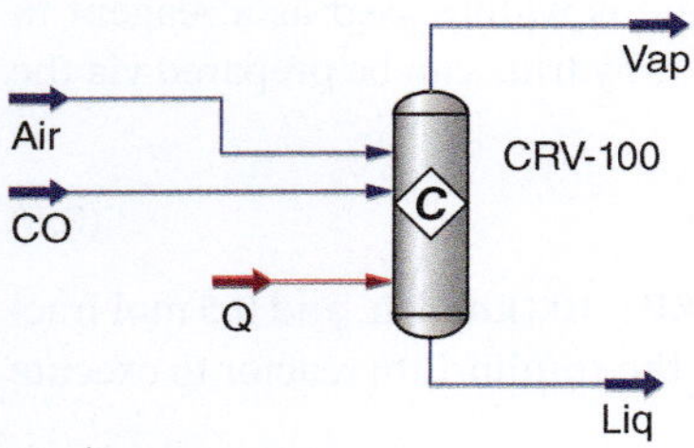

Vap		
Temperature	425.0	C
Pressure	101.3	kPa
Molar Flow	407.1	kgmole/h
Master Comp Molar Flow (CO)	0.0000	kgmole/h
Master Comp Molar Flow (CO_2)	100.0000	kgmole/h
Master Comp Molar Flow (Oxygen)	25.0015	kgmole/h

Figure 8.12 PFD of the conversion reactor.

air is used as a source of oxygen. The main data required in the conversion reactor is the stoichiometry of the reaction and the conversion of the reactant. Additionally, the user gains knowledge of how to add conversion reactions and reaction sets in HYSYS. In the case of specifying the temperature of the reactor or outlet product, it must add a stream of energy to calculate the heat removed (for an exothermic reaction) or the heat added (for an endothermic reaction), while for an adiabatic reaction, the energy cell must be empty. These connections of energy streams are applied for all types of reactors.

8.4 Equilibrium Reactor

The equilibrium reactor is a device that simulates equilibrium reactions. The exit streams from the reactor are in an equilibrium state with respect to both chemistry and physics. There is no limit to the number of equilibrium reactions that can be included in the reaction set that you attach to the equilibrium reactor; these reactions can be addressed simultaneously or sequentially. Since HYSYS can calculate the chemical activity of each component in the combination based on mixture and pure component fugacity, neither the components nor the mixing procedure need to be perfect [1, 2].

For each reaction in the chosen reaction set, you may also look at the actual conversion, the base component, the equilibrium constant, and the reaction extent. Based on the equilibrium reaction data you gave when the reaction set was constructed, the conversion, equilibrium constant, and extent are all computed.

8.4.1 Problem Statement

Acetic anhydride, or ethonic anhydride, is a chemical compound with the formula $(CH_3CO)_2O$. Commonly abbreviated as Ac_2O, it is the simplest isolable anhydride of a carboxylic acid and is widely used as a reagent in organic synthesis. It is known that acetic anhydride can be prepared via the reaction of acetic acid and ketene:

$$C_2H_2O + CH_3COOH \rightarrow C_4H_6O_3 \tag{8.5}$$

The feed enters the reactor at 50°C, 20 kPa, 100 kmol/h, and 0.5 mol fraction of each of ketene and acetic acid. Use the equilibrium reactor to execute the reaction with adiabatic condition.

8.4.2 Simulation

1) Open a new case of HYSYS, add the components: Ketene, Aceticacid, and aceticanhydride.
   Fluid package: PRSV.
2) From the Navigation Pane, Click the **Reactions** Folder view.
3) In order to add a new Reaction Set, Click the **Add** button, then click **Add Reaction**, Figure 8.13. Select **Equilibrium** as the type from the displayed list, then click **Add Reaction**, Figure 8.14.
4) Under the **Active Reactions**, double-click the **Rxn-1.**
5) From the Components column, add the components of the reaction (8.5) with its stoichiometry coefficients as shown in Figure 8.15.
6) Attach the Reaction Set to the Fluid Package by clicking the **Add to FP** button, then click the **Add set to the fluid package** button.
   When the reaction set is added to the Fluid Package, Enter the **Simulation** environment.
   Note: *The conversion of the equilibrium reaction depends on the Keq Source.*
   Click ***Keq*** *tab and see the source of the Keq,* Figure 8.16.
   Keq Radio Button

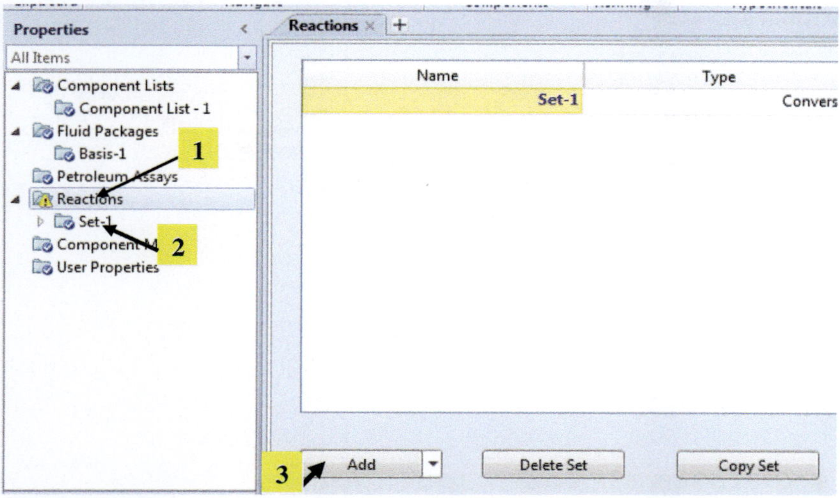

Figure 8.13 How to add the new Reaction Set.

Figure 8.14 Choosing equilibrium reaction type.

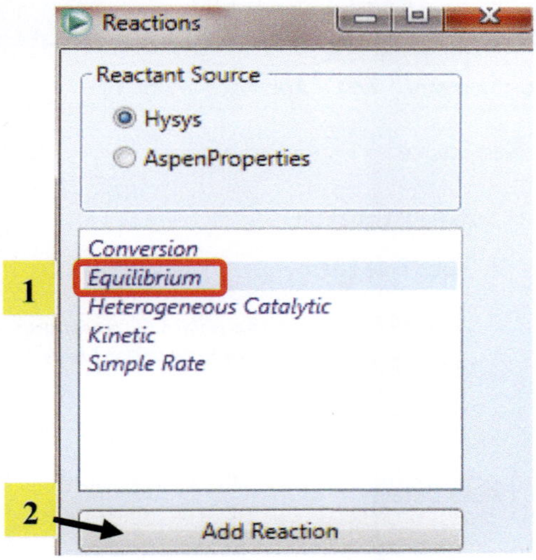

When you select the Keq radio button, the Ln(keq) group and K Table appear. The Ln(keq) group displays the Ln(Keq) relationship which may vary depending upon the Ln(K) Source value selected for the reaction. When you select the Ln(Keq) Equation radio button in the Ln(K) Source group, the parameters of the equilibrium constant equation appear. These values are either specified when the reaction was created or calculated by

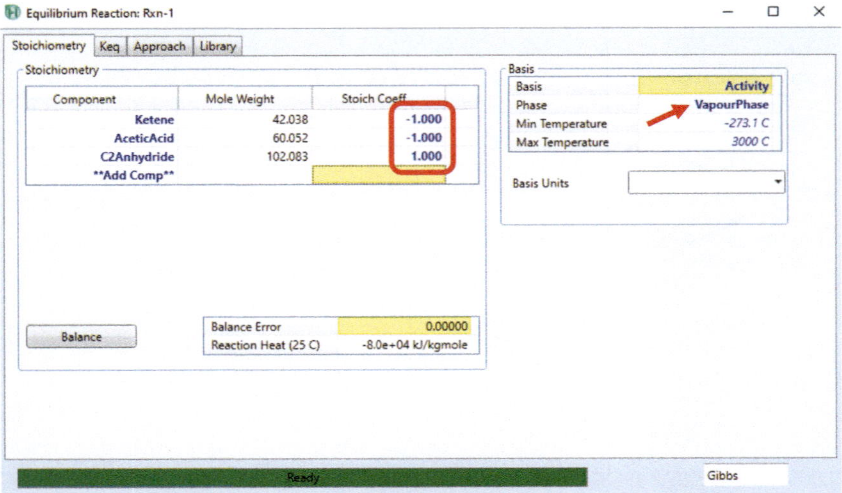

Figure 8.15 Stoichiometry of reaction (8.5) in the equilibrium reaction.

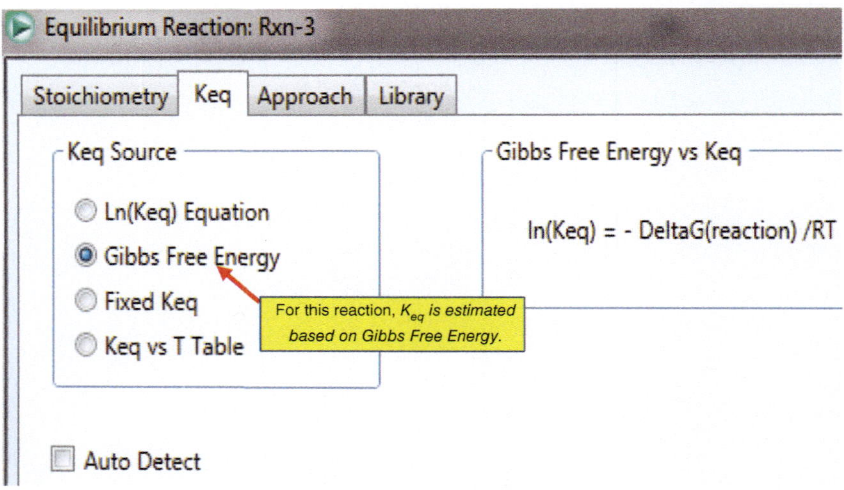

Figure 8.16 Specifying the type of equilibrium constant of the reaction.

HYSYS. If a fixed equilibrium constant was provided, it is shown here. Any of the parameters in the Ln(K) Equation group can be modified on this page. Changes made to the parameters only affect the selected reaction in the current reactor. After a change has been made, you can have HYSYS return the original calculated value by selecting the appropriate ***Use Default*** *check box.*

In this example, the Keq depends on Gibbs Free Energy which is taken from the database of components that are saved in software.

7) In the Simulation environment, add a new **Material** stream with the following conditions.

Name	Feed
Temp	50°C
Press	20 kPa
Molar flow	100 kmol/h
Composition by mole	0.5 for ketene and aceticacid

8) From the Object Palette, **Column** Group, Add an **Equilibrium** Reactor  to the simulation area.
9) Attach the feed, Vap, and Liq as in Figure 8.17. Even though the liquid product from this reactor will be zero, it must be attached to the liquid stream, which the simulator will estimate to be zero.
10) On Parameters page, since there is no data about pressure, and the simulator assumed zero, let it be the default value.

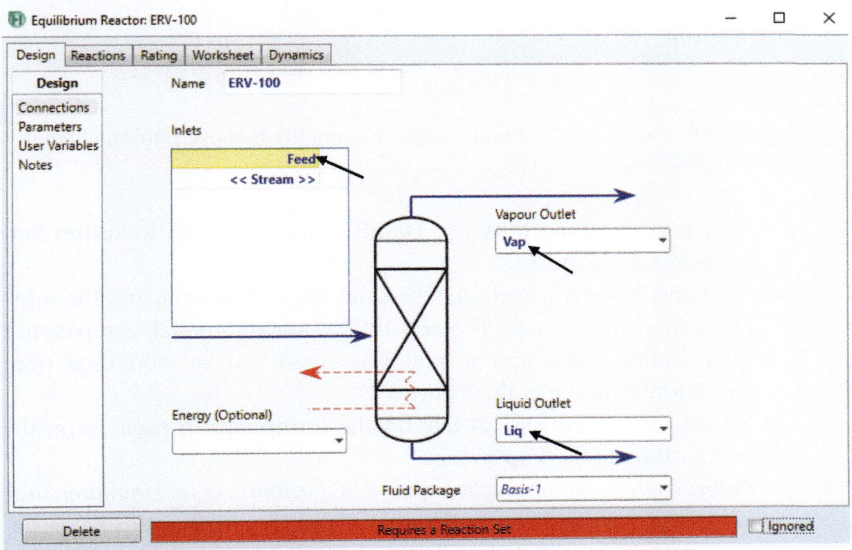

Figure 8.17 Window of equilibrium reactor, Design tab, Connection page.

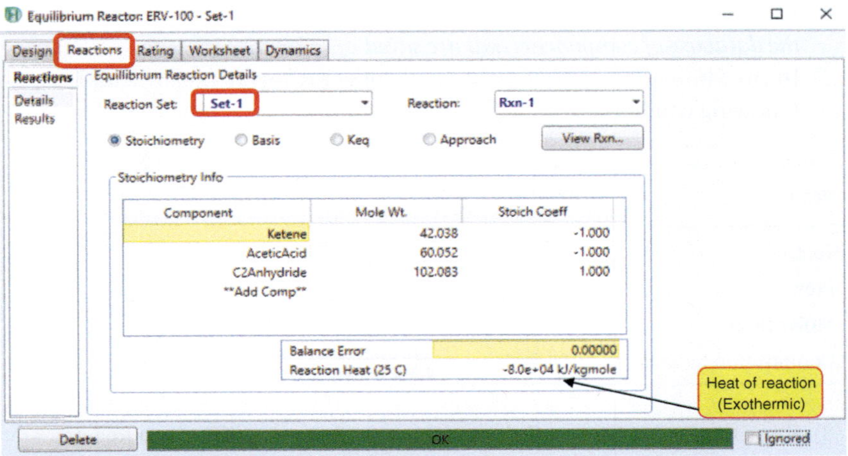

Figure 8.18 Adding the reaction set to the equilibrium reactor.

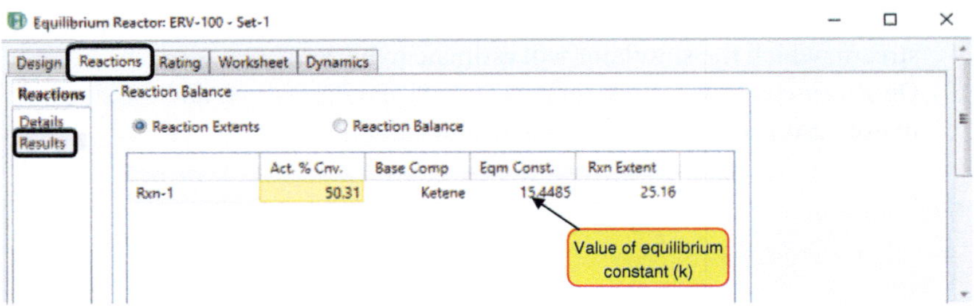

Figure 8.19 Reactions tab, Results page: showing the % conversion, equilibrium constant, and reaction extent.

11) Switch to the **Reactions** tab, **Details** page, and from **Reaction Set**, select **Set-1**, Figure 8.18.
12) Click the **Results** page to see the conversion of reaction and the value of equilibrium constant, Figure 8.19. The amount of each component's inlet, outlet, consumption, and generation can be monitored from **Reaction Balance** radio, Figure 8.20.
13) Switch to the **Worksheet** tab; on the **Composition** page, exam the composition of the **Vap** stream.

Note: *There are some reactions found in the library of the simulator, such as water gas shift reaction and ammonia production. Before using the stoichiometry, we can click the* ***Library*** *button, select the reaction, then click* ***Add Library Rxn*** *as shown in* Figure 8.20, *and the* K_{eq} *will be estimated*

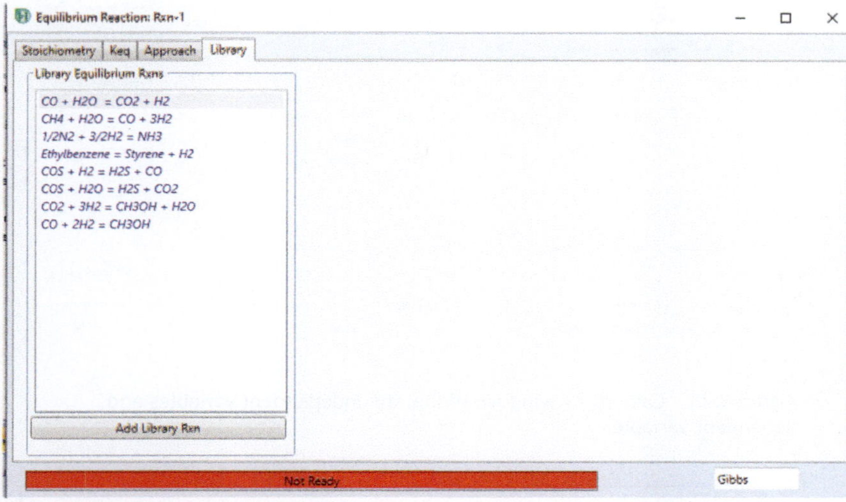

Figure 8.20 List of available reactions in the library, e.g. we can add the shift gas reaction from library of the simulator.

using (K_{eq} versus T Table). But for the current example, the reaction is not found; it must give the data for the K_{eq} or use the stoichiometry for adding the information of the reaction and using the Gibbs free energy to estimate the K_{eq}.

8.4.3 Effect of Feed's Pressure and Product's Temperature on the Reaction Conversion

1) Return to Figure 8.17, add the energy stream, *Q*, and enter the temperature of the product at 225°C. (This is the same temperature of the product when the reaction is adiabatic.)
2) Open a case study and add the feed's pressure, and product's temperature as independent variables, and add the ERV-100-% conversion (Equilibrium Reaction-1) as dependent variable, Figure 8.21.
3) Switch to Case study setup, and give the range of pressure (20–40 kPa) step 5 kPa, and range of temperature (100–250°C) step 25 °C.
4) Click **Run** button and draw the results using Excel; the graph will appear as in Figure 8.22.

It can be seen from Figure 8.22, the conversion of acetic acid to acetic anhydride increases with increasing product's pressure at high temperature of product, but almost the feed's pressure has no effect on the conversion at lower product's temperature. On the other hand, the reaction conversion

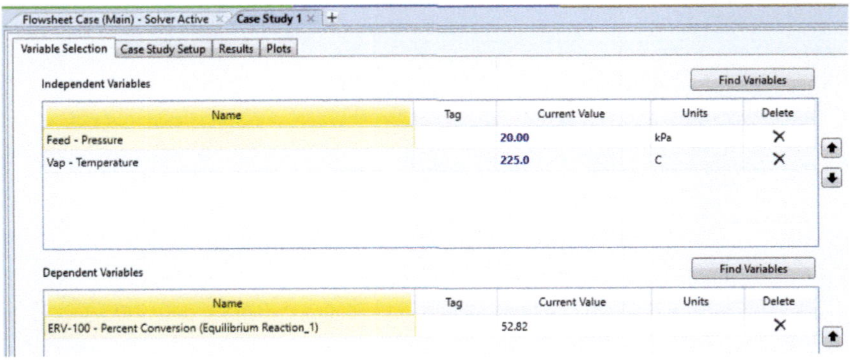

Figure 8.21 Case study window, adding the independent variables and dependent variable.

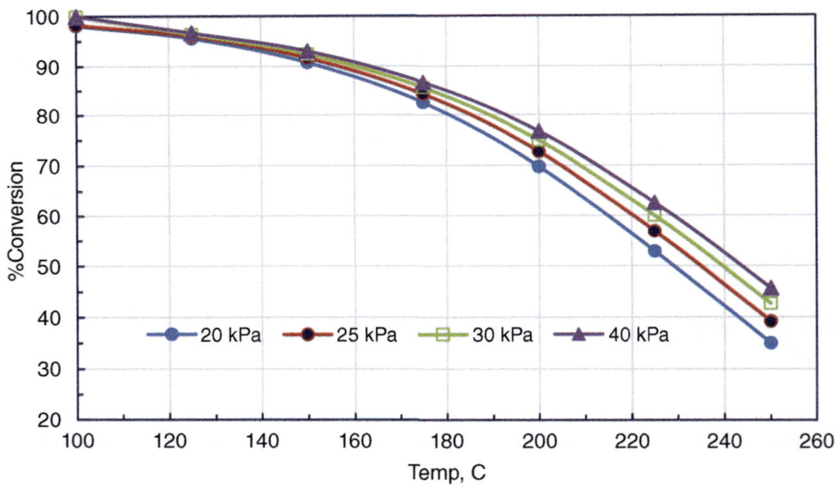

Figure 8.22 Effect of feed's pressure and product's temperature on the reaction conversion.

decreases with increasing product's temperature, since the reaction is exothermic (see Figure 8.18).

8.4.4 Summary

In this section, the goal is to create a model that would illustrate the production of acetic anhydride from the reaction of acetic acid and ketene using Gibbs free energy as a source of equilibrium constant (K_{eq}) estimation. There are many ways to estimate K_{eq} depending on the data available. In the equilibrium reactor, the data required is the stoichiometry of the reaction, and if there is data available about K_{eq}, it can be dependent, if not, it is possible to suffice for the stoichiometry of the reaction. Additionally, the user gains knowledge on how to add equilibrium reactions and sets of reactions in HYSYS.

8.5 Miscellaneous Examples

You can use the HYSYS as a robust tool to address any issues relating to processes. However, you cannot use it effectively unless you have a solid understanding of the process and solution methods because you have to present a variety of circumstances and options in order to solve an issue. To improve their capacity to utilize the software, the user must use several examples.

8.5.1 Reaction and Separation-Based Process

Toluene is produced by dehydrogenation of *n*-heptane over a Cr_2O_3 catalyst as follows:

$$CH_3CH_2CH_2CH_2CH_2CH_2CH_3 \rightarrow C_6H_5CH_3 + 4H_2 \tag{8.6}$$

The first step in the manufacturing of toluene is to heat n-heptane in a heater from 20 to 427°C. It is fed to a catalytic reactor, which converts 17 mol% of the n-heptane to toluene isothermally. Its effluent is fed to a separator after being cooled to 20°C. Find the species flow rates in each stream, assuming that all of the equipment worked at 14.7 psia pressure (take the basis flow rate of feed = 100 kmol/h and the composition = 100% *n*-heptane), [5].

8.5.1.1 Simulation

1) Open a new case; Components: toluene, *n*-heptane, and hydrogen.
2) Property Package: **Peng–Robinson**.
3) Add a Reaction of Conversion type with stoichiometry as in Reaction (8.6):
   Rxn Phase: Vapourphase, Conversion = 17%, add the Reaction Set to the fluid package.
4) Enter the simulation environment and Construct the flow sheet as in Figure 8.23.

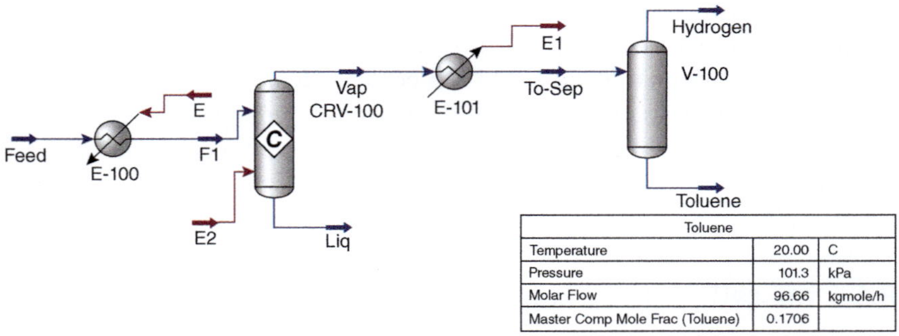

Figure 8.23 PFD of the toluene production.

5) Double-click the reactor, Choose Set-1 as a Reaction set, and close the windows.
6) Open the worksheet, and enter the known conditions for the streams. Make sure that the pressure drop across all units is zero. Enter the temperatures of the Vap stream and To-Vap stream, equal to 427, and 20°C, respectively.
   Record the values of energy for *E* and *E1*.
7) Improvement through Process Modification
   The results of the calculation reveal that the cooling and heating duties are comparable, indicating that the utility demand may be decreased by preheating the feed stream with hot reactor product.
   Add a **Heat Exchanger** before the heater to vary the procedure, as seen in Figure 8.24. This can be done by cutting the stream Feed from heater, and cutting the stream Vap from cooler, and re-arranging the PFD as in Figure 8.24.
8) From the Parameters page, Design tab of the Heat Exchanger, set $\Delta p = 0$ for both tube side and shell side.

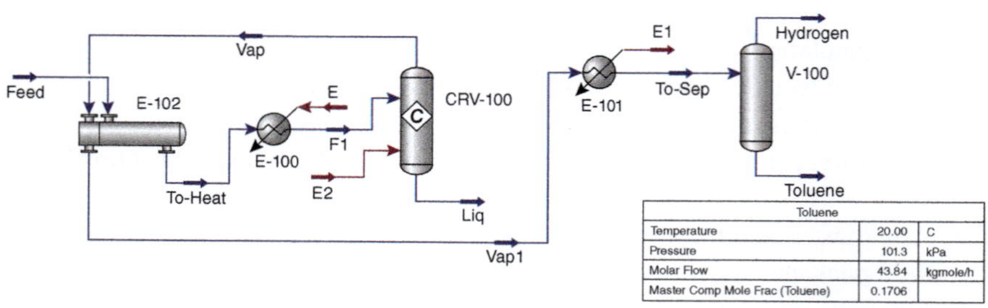

Figure 8.24 Modified PFD of the toluene production.

9) One additional condition needs to be specified. Open the Worksheet and set the Pre-Feed stream's temperature to 180°C. You can alter this temperature to observe how the Heat-duty and C-Duty are affected.
10) You can alter the temperature of the Pre-Feed stream to observe how it impacts the H-Duty and UA (heat transfer coefficient times interfacial area) of the Heat Exchanger. It can lower the H-Duty of the heater (E-100) by raising the To-Heat stream's temperature, resulting in a higher interfacial area of the Heat Exchanger and vice versa. By adjusting the temperature and observing how other values change, you can observe this impact. The Case study function can be used for this. The method is best explained as follows:
    (10-1) In the **Home** tab of the Ribbon, open the **Case Study** function, Click the **Add** button.
    (10-2) **Independent variables:** click the **Find Variables** button, select **To-Heat** as object, **Temperature** as Variable.
    (10-3) **Dependent variables: E-102** as object, **UA** as Variable. Close the window.
    (10-4) Do the same for ***E*** as object, **Heat Flow** as Variable.
    (10-5) Click the **Case Study Setup** button: **Start** value = **150 °C**, **End** value = **200 °C**, and **5 °C** for the Step Size.
    (10-6) Click **Run**. After a few seconds, plot the results by Excel; the graph is shown in Figure 8.25.

From Figure 8.25, the heat duty of the heater (E-100), required to raise the temperature of the feed, decreases by increasing the To-Heat stream's temperature, which leads to a reduction in operating cost, but the combined overall heat transfer coefficient (*UA*) of the Heat Exchanger (E-102) increases by increasing the To-Heat stream's temperature, and then the capital cost, so one must choose the optimum value of the To-Heat stream's temperature. From this analysis, the desired temperature of To-Heat stream can be evaluated from Figure 8.25 which is equal to 172°C.

8.5.2 Steam–Methane Reformer (Hydrogen Production)

Methane reformation using steam is frequently done in conjunction with other processes that use a lot of hydrogen. Thermodynamics and catalysts are used in subsequent reaction stages to increase the production of hydrogen at the expense of the byproduct gases carbon monoxide and dioxide. To reduce CO and CO_2 carryover into the downstream process, the next step is to entirely convert any leftover carbon oxides back into methane. Figure 8.26 shows the PFD of Steam–Methane Reformer [6].

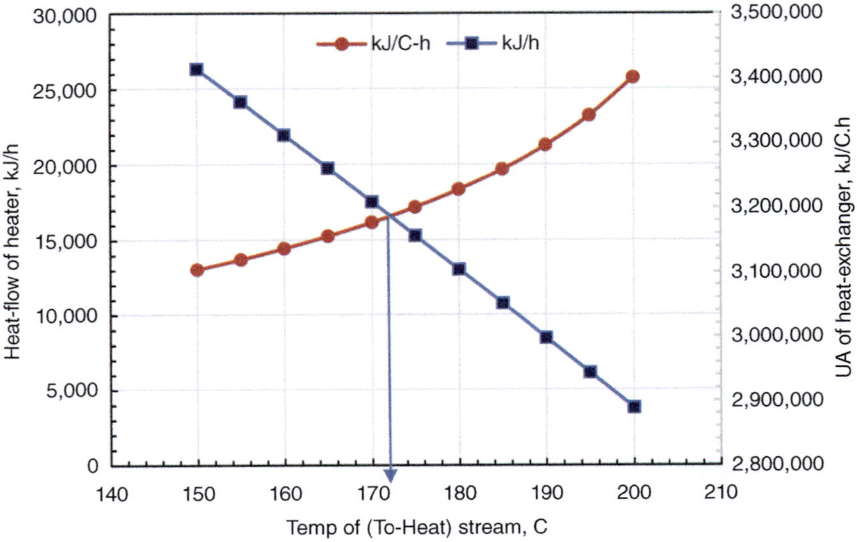

Figure 8.25 Obtaining the optimum value of the To-Heat temperature using a case study.

8.5.2.1 Simulation

1) Open a new case of HYSYS, and add the components: methane, carbon monoxide, carbon dioxide, hydrogen, and water. Fluid Package: Peng–Robinson
2) **Adding the Reactions:** The reactions which take place in this simulation are:

Reaction Name	Reaction	Reaction Type
Reform1	$CH_4 + H_2O \rightarrow CO + 3H_2$	Equilibrium
Reform2	$CO + H_2O \rightarrow CO_2 + H_2$	Equilibrium
Shift1	$CO + H_2O \leftrightarrow CO_2 + H_2$	Equilibrium
Meth1	$CO + 3H_2 \rightarrow H_2O + CH_4$	Conversion

Click the **Reaction** folder; click the **Add** button to add a Set of reactions, click the **Add Reaction** button, select **Equilibrium;** click the **Add Reaction** button, and enter the necessary information that is equivalent to the following reaction:

$$CH_4 + H_2O \rightarrow CO + 3H_2 \tag{8.7}$$

Rename this reaction to: **Reform1**

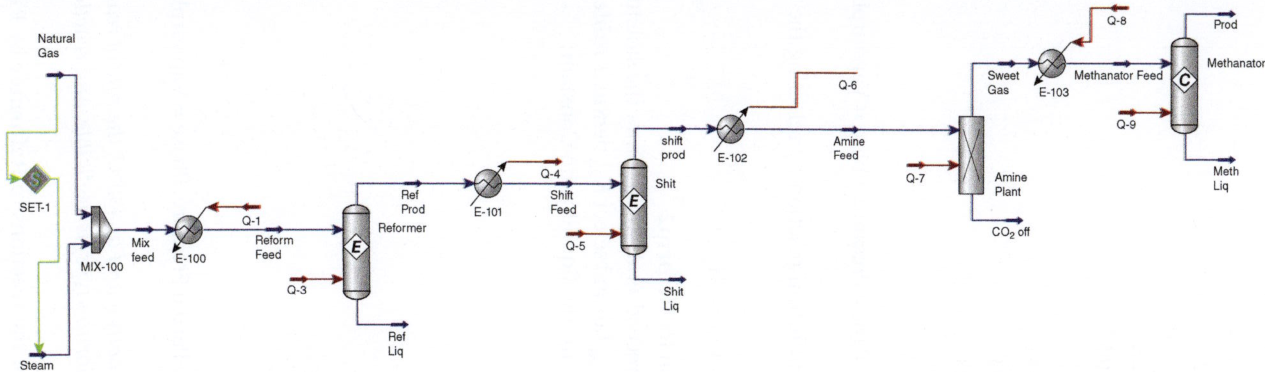

Figure 8.26 PFD of the Steam–Methane Reformer (Hydrogen Production).

3) Click K_{eq} tab, select the K_{eq} **versus** T **table** radio, and enter the following values:

Temperature °C	K_{eq}
595	0.5
650	3
705	14
760	63
815	243
870	817

Basis group: Basis = Partial Pressure, Phase = Vapourphase, Basis Units = atm.

4) Add the second Equilibrium reaction by selecting the reaction type as Equilibrium:

$$CO + H_2O \rightarrow CO_2 + H_2 \tag{8.8}$$

Rename this reaction to: **Reform2**

5) For reaction (8.8), proceed as above (by enter the stoichiometry for each component), click K_{eq} tab and select K_{eq} ***versus*** T ***table*** radio, and enter the following values for the Equilibrium Constant:

Temperature °C	K_{eq}
675	1.7
705	1.5
730	1.3
760	1.2
790	1.1
815	1.0

Basis group: Basis = Partial Pressure, Phase = Vapourphase, Basis Units = atm.

In the absence of a catalyst and at 430°C, the rate of reaction (8.7) in the Shift Reactor (Equilibrium type) is negligible, and reaction (8.8) becomes the only reaction.

Rename the **Set-1** of the reactions to **Reformer** by Right clicking the Set-1 in the Navigation pane.

6) Add **Reformer** Set of reactions to **FP** by clicking **Add to FP**, then click **Add Set to Fluid Package.**
HYSYS contains a library of some of the most commonly encountered chemical reactions with their Equilibrium Constants. For the Shift Reactor, the library values for the Equilibrium Constant will be used.
7) Add a second Set of reactions by clicking the **Reaction** Folder in the Navigation Pane, then click the **Add** button and rename this set to **Shift**.
8) Add the Equilibrium reaction to the **shift**-set by selecting the reaction type as **Equilibrium**. On the Library tab, highlight the reaction with the form:

$$CO + H_2O = CO_2 + H_2 \tag{8.9}$$

9) Press the **Add Library Rxn** button. This adds the reaction and all of the reaction's data to the simulation. Rename the reaction to **Shift1**
10) Add the **Shift** Set of reaction to the fluid package (**FP**).
11) Add a third Set of reactions and rename it to **Methanation** (this set will contain the conversion reactions).
12) Add a **Conversion** reaction for the reverse of reaction (8.7). The reaction is:

$$CO + 3H_2 \rightarrow H_2O + CH_4 \tag{8.10}$$

Enter the stoichiometry of the components as in the reaction.
Rename the reaction to **Meth1**
13) Move to the Basis tab: **Base Component** = CO, *Co%* = 100%, Phase = Vapourphase.
Add the **Methanation** Set of reactions to the fluid package (**FP**).
14) Enter the **Simulation** Environment.
15) Create **two** new material streams with the following information:

Name	Temp °C	Press kPa	Mass Flow kg/h	Composition
Natural Gas	20	520	800	$CH_4 = 1$
Steam	180	965		$H_2O = 1$

16) Add a Mixer with its streams as follows

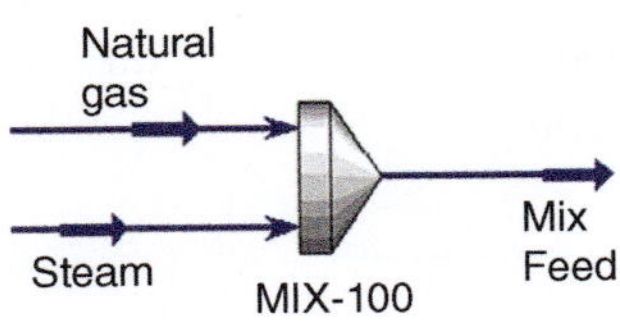

17) Add a **Set Operation**: A **Set** operation is needed to fix the steam rate relative to the methane feed rate.
The Set operation has the following information:

Connections Page/Design Tab	
Name	**SET-1**
Target Object	Steam, Molar Flow
Source	Natural Gas
Parameters	
Multiplier	2.5
Offset	0.0 kgmol/h

18) Add a Heater with the following information:

Pressure Drop	**10 kPa**
Reform Feed Temp	760°C

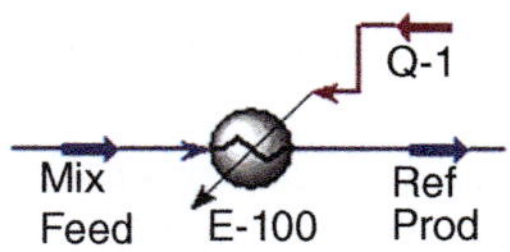

19) Add the steam **Reformer**: An Equilibrium Reactor will be used to simulate the steam Reformer.
From the Object Palette, Column Operations, Add the **Equilibrium** Reactor to the PFD. Make the following connections:

Connections Page/Design Tab	
Name	**Reformer**
Inlet	Reform Feed
Vapor Outlet	Reform Prod
Liquid Outlet	Reform Liq
Energy	Reform Q
Parameters	
Delta P	70 kPa

20) On the **Worksheet** tab, set the temperature of **Reform Prod** to 760°C
21) On the **Reactions** tab, select the **Reformer** as the Reaction Set.
    This will automatically connect the proper reactions to this Reactor and the Reactor will solve.
    What is the % conversion of Methane?
    How much CO and H_2 were produced in the reaction, i.e. what is the molar flowrate of these two compounds in the reactor's product stream?
22) Add a Cooler: Enter the streams with the following information:

ΔP	24 kPa
Shift Feed Temp	427°C

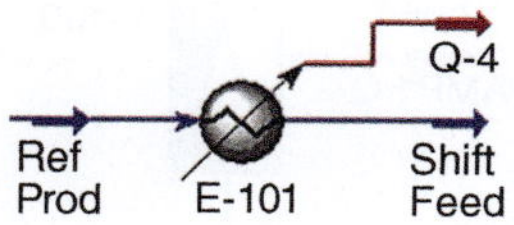

23) Add an **Equilibrium** Reactor (Name it: **Shift**) with the following data:

ΔP	**70 kPa**
Shift Prod Temp	427°C

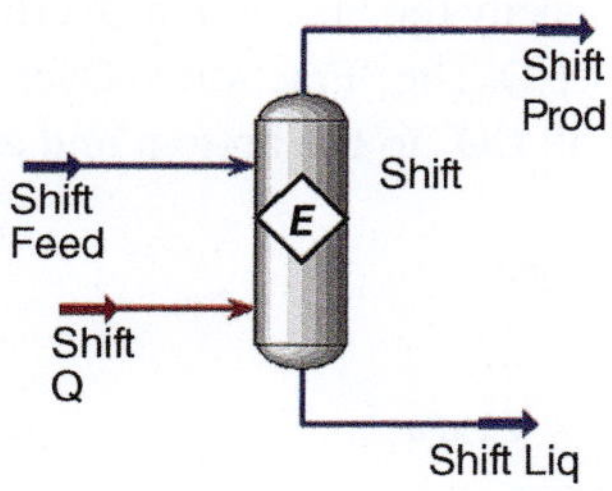

24) On the **Reactions** tab, select **Shift** as the Reaction Set.
    This will automatically connect the proper reactions to this reactor.
    What is the % conversion of CO in this reactor?
25) Add a Cooler with its streams as follows:

ΔP	35 kPa
Amine Feed Temp	38 °C

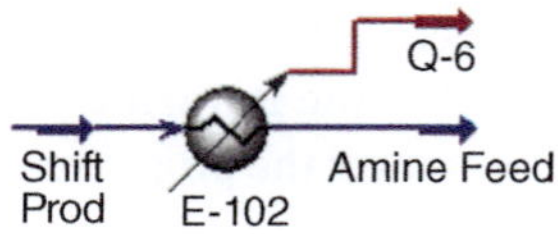

26) From Columns group, add Component Splitter 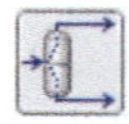 to model the Amine Plant. The purpose of this Splitter is only to remove the CO_2 present in the flow. The connections are shown below:

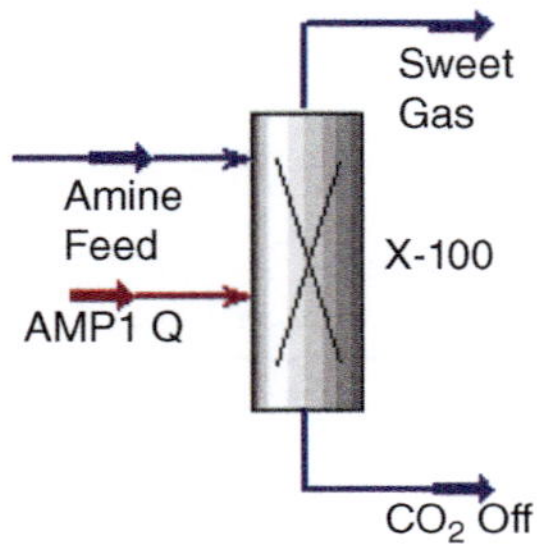

Sweet Gas Pressure	297 kPa
CO_2-Off Pressure	297 kPa
Sweet Gas Temp	138°C
CO_2-Off Temp	138°C

27) On the Splits page, specify the Fraction to Overhead as 1.0 for Methane, CO, Hydrogen, and H_2O. The "Fraction to Overhead" for the CO_2 must be 0; this will force all CO_2 to the bottom and all other components to the top.

28) Add a Heater as follows:

ΔP	**21 kPa**
Methanator Feed Temp	260°C

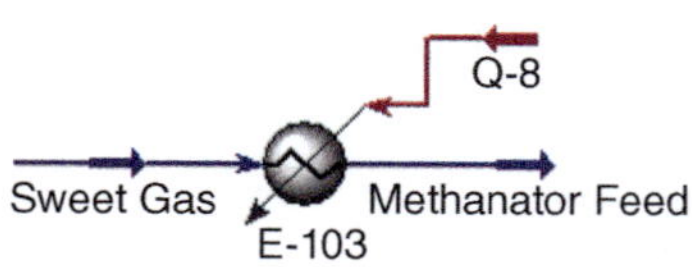

29) Add a **Conversion** Reactor (Name it: **Methanator**) with the following connections:

Pressure Drop	21 kPa
Temp of Meth Liq	260°C

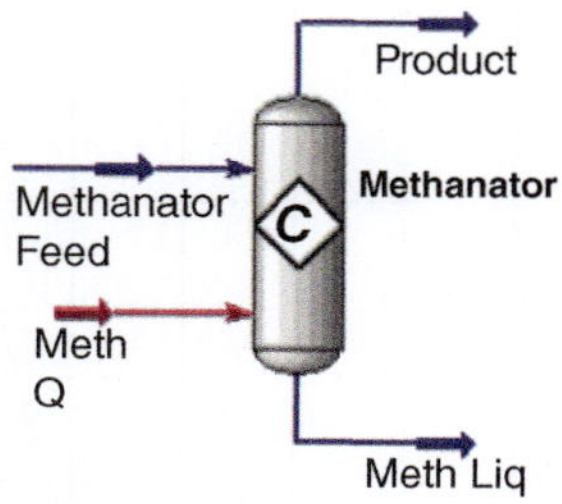

30) On the Reactions tab, choose the **Methanation** from the Reaction Set drop-down menu.
What is the % conversion of CO?
How much Methane is produced in this reactor?
What is the effect of varying the amount of Steam to the process on the flow of CO_2 entering the Amine Plant and the flow of methane leaving the Methanator Reactor?
   - Use case study to compare the steam flow rates of 1.5, 2.0, 2.5, 3.0, 3.5, and 4.0 times those of the natural gas:
     1) From the **Home** tab of the Ribbon, click the Case Study Case Studies and click the **Add** button.
     2) For the **Independent variables**, click **Find Variables**, Select SET-1 as Object, and Multiplier as Variable, click the Add  button.
     3) **Dependent Variables**: Select **Shift Prod** as an object, **Master Comp Molar Flow** as a variable, and $\mathbf{CO_2}$ as the variable's specific, then click **Add.**
     4) Add another dependent variable: **Product**; Master Comp Molar Flow (Methane).
     5) In the **Case Study Setup** button, Enter values for **Start** = 1.5, **End** = 4, and **Step Size** = 0.25.
     6) Click the **Run** button to begin the case study analysis. Press the **Plots** button to view the calculations as in Figure 8.27.

From Figure 8.27, the optimum value of the Set-1, Multiplier = 2.22.

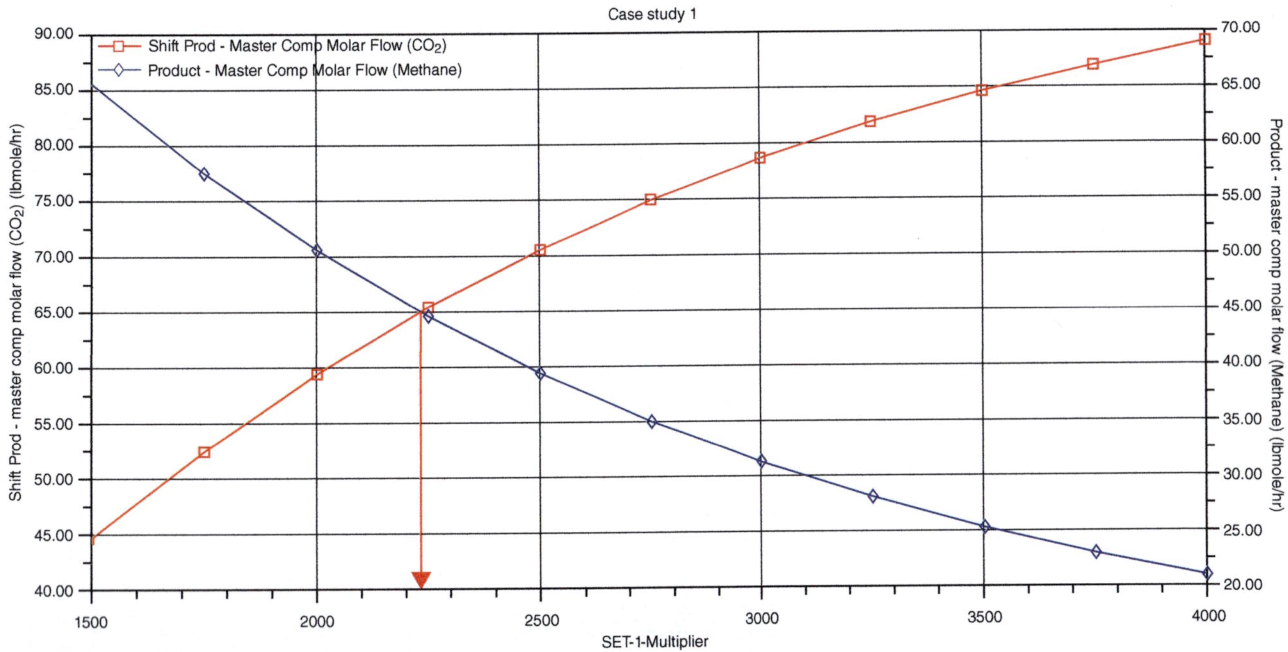

Figure 8.27 Optimum value of multiple steam flow rate times natural gas flow for maximum CO_2 and CH_4 moles production.

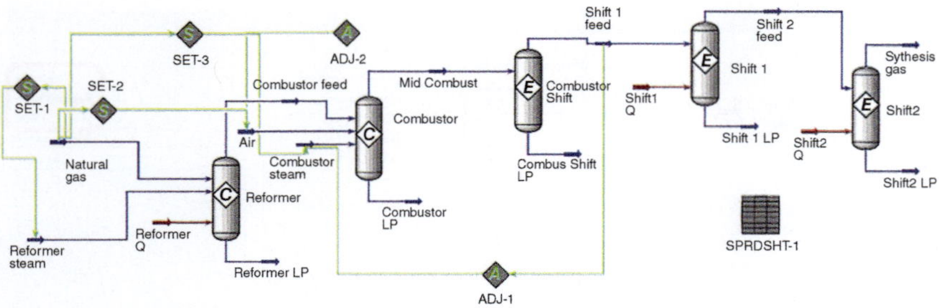

Figure 8.28 PFD of Synthesis Gas production.

8.5.3 Synthesis Gas production

This example will replicate the generation of Synthesis Gas (synthesis gas is important in Ammonia synthesis where the molar ratio of H_2 to N_2 is 3:1). The primary function of the synthesis gas plant is to produce hydrogen from natural gas, typically methane. The majority of synthesis gas facilities employ four reactors. However, five reactors will be employed in our simulation to represent this process. This is so that the combustor, which is a single vessel, can be treated as two reactors connected in series, each with a different sort of reaction. Conversion reactors make up the first reactor, and equilibrium reactors make up the second. Figure 8.28 displays the PFD for the creation of Synthesis Gas [6].

8.5.3.1 Simulation

1) Opening a new case and enter the components: CH_4, H_2O, CO, CO_2, H_2, N_2, and O_2.
   Fluid package: Peng–Robinson
2) Add a Reaction Set and rename it to **Reformer**
3) Add a Conversion Reaction as follows:

$$CH_4 + H_2O \rightarrow CO + 3H_2 \tag{8.11}$$

   vapor phase, conversion = 35%.
   Move to the **Basis** group and enter the information as shown in Figure 8.29.
4) Repeat the steps above for another Conversion Reaction. Use the following data:

Name	Reaction	Base Component	Phase Rxn	Co%
Rxn-2	$CH_4 + 2H_2O \rightarrow CO_2 + 4H_2$ ………. (8.12)	Methane	Vapor phase	65%

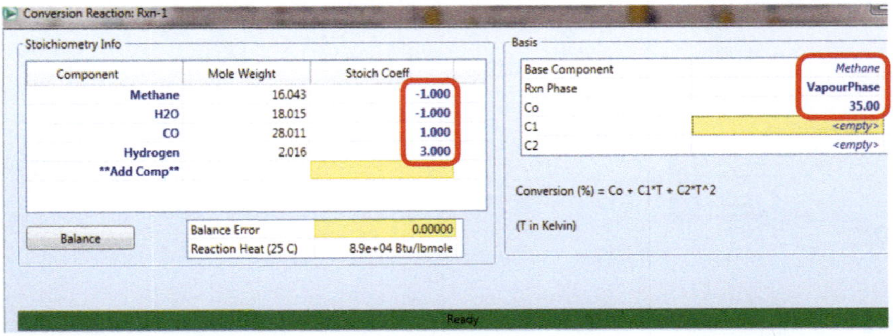

Figure 8.29 Stoichiometry and conversion percent for the reaction (8.11).

5) Add the Reaction Set to the Fluid Package by clicking the Add to FP button, then click Add Set to Fluid Package.
6) Add another Reaction Set and rename it to Combustor that included the following reactions:
   (6-1) Add two reactions that are constructed in steps 3 and 4 to the Combustor Reaction Set by clicking the arrow beside the **Add Reaction** button and selecting Existing Reaction, Figure 8.30.
   Then highlight the Rxn-1 and click **Add Reaction** button.
   By the same way above, add Rxn-2.
   Close the Existing Reaction view.
   (6-2) Add a 3rd reaction to the Combustor Set by clicking the **Add Reaction** button with the following information:

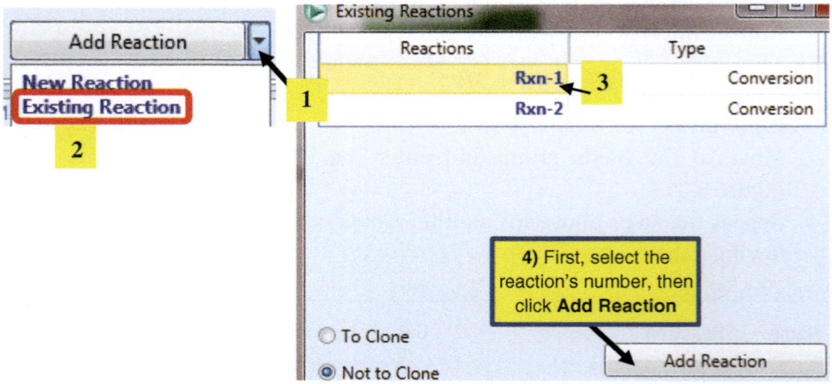

Figure 8.30 Choosing an existing reaction from the previous steps.

Name	Reaction	Base Component	Phase Rxn	Co%
Rxn-3	$CH_4 + 2O_2 \rightarrow CO_2 + 2H_2O$ .... (8.13)	Methane	Vapor phase	100%

Add the Combustor Set to the fluid package.

7) Add a new Reaction Set and rename it to **Shift** which contains the following reaction:
Add an **Equilibrium** reaction:

$$CO + H_2O = CO_2 + H_2 \tag{8.14}$$

Where this reaction exists in the library of simulator, click the **Library** button, select the reaction (8.14), and click the **Add Library Rxn** button. This adds the reaction and all of the reaction's data to the simulation, Figure 8.31.
Add Shift-Set to the fluid package.

8) Enter the **Simulation** Environment and create 4 material streams with the following information:

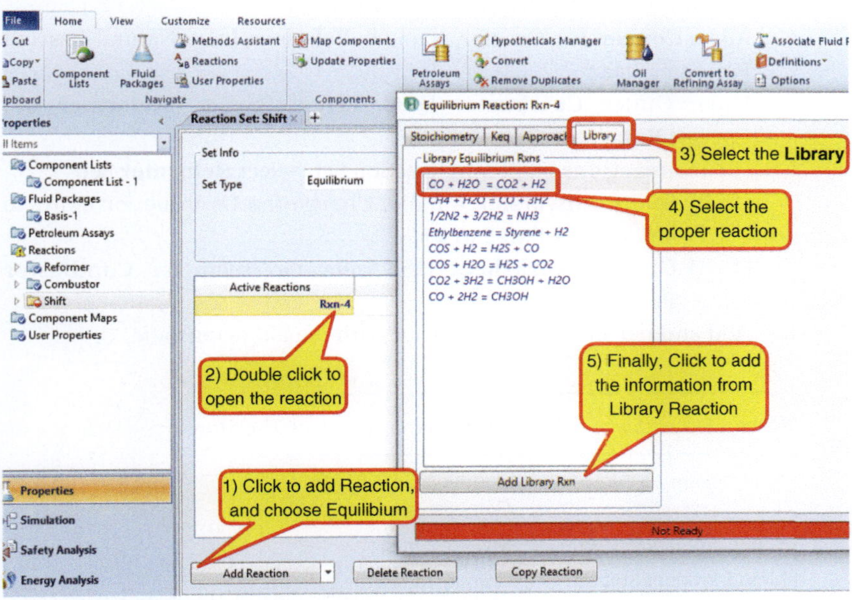

Figure 8.31 Adding reaction (8.14) from the list of library.

Name	Natural Gas	Reformer Steam	Air	Combustor Steam
Temp, °C	370	250	16	250
Pressure, kPa	3500	Empty	Empty	Empty
Molar Flow, kmol/h	90	240	90	140
Molar composition	100% CH_4	100% H_2O	79% N_2, 21% O_2	100% H_2O

9) Add a **Set Operation** from object palette; Set-1:
   Target Variable: **Object** = Reformer Steam, **Variable**: Pressure.
   Source: Object = Natural Gas,
   Parameter tab: **Multiplier** = 1, **Offset** = 0.
10) Add another Set Operation; Set-2:
   Target Variable: **Object** = Combustor Steam, **Variable**: Pressure.
   Source: Object = Natural Gas,
   Parameter tab: **Multiplier** = 1, **Offset** = 0.
11) Add another Set operation: Set-3:
   Target Variable: **Object** = Air**, Variable**: Pressure.
   Source: Object = Natural Gas,
   Parameter tab: **Multiplier** = 1, **Offset** = 0.
12) Add a **Conversion** Reactor, Name it: Reformer, Inlet: Natural Gas and Reformer Steam,
   Vapor Outlet: **Combustor Feed**, Liquid Product: **Reformer LP**, Energy: **Reformer Q**
   (12-1) **Reactions** tab, From Reaction Set: Select **Reformer** Set.
   (12-2) Select **Conversion%** radio, Change the ***Co*** value for Rxn-1 to 40% and for Rxn-2 to 30%
   (12-3) On the **Worksheet** tab, Enter the Temp for **Combustor Feed** = 930°C.
13) Add another **Conversion** Reactor with the following data:

Name	Combustor
Feed Stream	Combustor Feed, Air, Combustor Steam
Vapor Product Stream	Mid-Combust
Liquid Product Stream	Combustor LP
Reaction Set	Combustor Set

14) Add a Shift Reactor: The shift Reactor would be modeled as **Equilibrium** Reactor

Name	Combustor Shift
Feed Stream	Mid-Combust
Vapor Product Stream	Shift1 Feed
Liquid Product Stream	Combustor Shift LP
Reaction Set	Shift Set

15) Add another **Equilibrium Reactor** with the following data:

Name	Shifter 1
Feed Stream	Shift1 Feed
Vapor Product	Shift2 Feed
Liquid Product	Shift1 LP
Energy Stream	Shift1 Q
Shift2 Feed Temp	450°C
Reaction Set	Shift Set

16) Add the third **Equilibrium Reactor** with the following data:

Name	Shifter 2
Feed Stream	Shift2 Feed
Vapor Product	Synthesis Gas
Liquid Product	Shift2 LP
Energy Stream	Shift2 Q
Synthesis Gas Temp	400°C
Reaction Set	Shift Set

- What is the mole fraction of Hydrogen in the Synthesis Gas stream?
- What is the ratio of H_2/N_2 in the Synthesis Gas Stream?

17) Adding an Adjust Operations: In order to control the temperature of the product leaving the Combustor (2nd Conversion reactor), the flow rate of the steam to this reactor is controlled. It is desired to have an outlet temperature from the first shift reactor of 930°C.

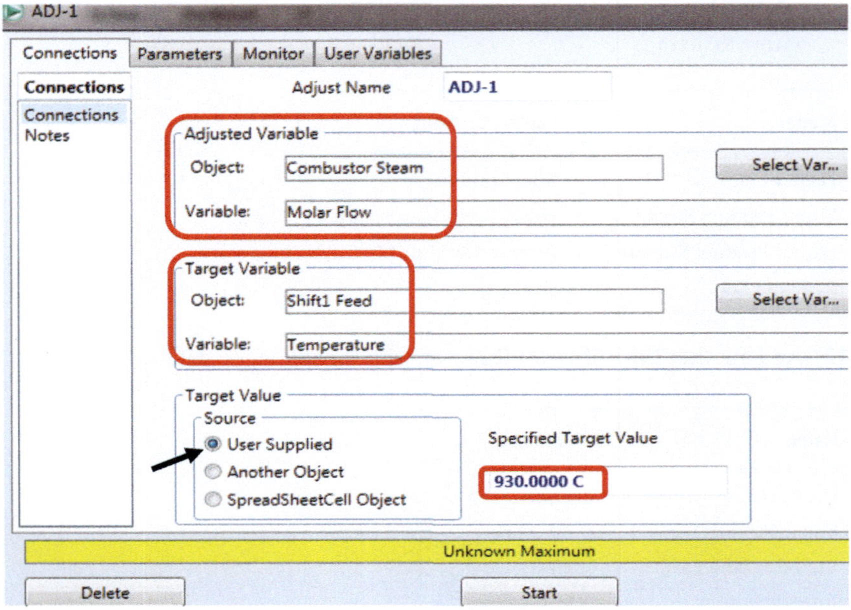

Figure 8.32 Adjust operation for adjusting the temperature of shift1 feed.

(17-1) Add an **Adjust Operation** from the object palette, and enter the data as in Figure 8.32:
Adjusted Variable: Click **Select variable;** Object = Combustor Steam, Variable = Molar flow.
Target Variable: Click **Select variable;** Object = Shift1 Feed, Variable = Temperature.

(17-2) Target Value; Source: User Supplied, Specified target value = 930 °C, Figure 8.32.

(17-3) **Parameters** tab, Tolerance = 0.1°C, Step Size = 1.0 Kmol/h, Maximum Iteration = 1000, Figure 8.33.

(17-4) Move to the **Monitor** tab and click Start, Figure 8.34. The molar flow rate of the Combustor steam that required to enable the temperature of shift1 feed (930°C) is 25.44 kmol/h.

18) Add a **Spreadsheet** operation 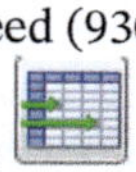 to the PFD and double-click it to open.

(18-1) Click the **spreadsheet** tab, in the cell A1, right click by mouse and select **Add Import Variable:**
Object = Synthesis Gas, **Variable** = Master Comp Molar Flow, **Variable Specify** = Hydrogen.

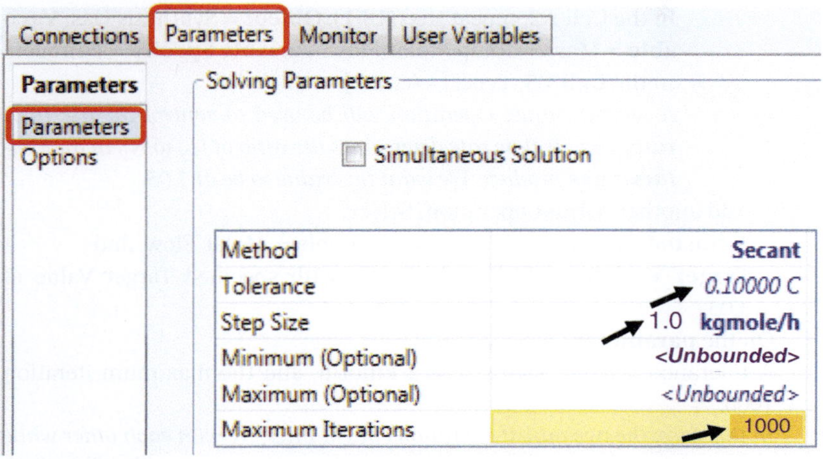

Figure 8.33 Parameters page of the adjust operation (Adj-1).

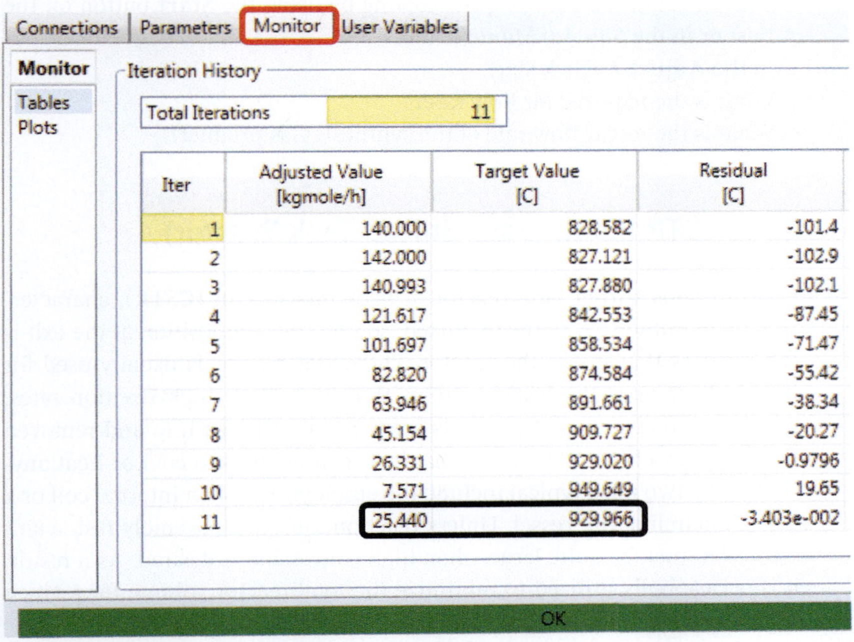

Iter	Adjusted Value [kgmole/h]	Target Value [C]	Residual [C]
1	140.000	828.582	-101.4
2	142.000	827.121	-102.9
3	140.993	827.880	-102.1
4	121.617	842.553	-87.45
5	101.697	858.534	-71.47
6	82.820	874.584	-55.42
7	63.946	891.661	-38.34
8	45.154	909.727	-20.27
9	26.331	929.020	-0.9796
10	7.571	949.649	19.65
11	25.440	929.966	-3.403e-002

Figure 8.34 Monitor page of the adjust operation (Adj-1).

(18-2) In the Cell A2, repeat step (18-1), **Object** = Synthesis Gas, **Variable** = Master Comp Molar Flow, **Variable Specify** = Nitrogen.

(18-3) In the **Cell B3**, Type: [=A1/A2]

A second Adjust Operation will be used to control the ***air flow rate****. The air flow rate determines the ratio of H_2 to N_2 in the* ***synthesis gas*** *product. We want this value to be at 3.05.*

19) Add another **Adjust** operation, Select:

Adjusted Variable: Object = Air, Variable = Molar Flow, and

Target Variable: SPREADSHT-1-B3 with specified Target Value of 3.05.

On the **parameters** page, choose:

A tolerance = 0.001, step size = 1 kmol/h, and the maximum iteration = 1000.

In this case, the two adjust operations might interfere with each other while they are solving. This is because changing either adjusted variable affects both target variables. To prevent this interference, the Adjusts can be set to solve simultaneously.

On the **Parameters** tab of the **ADJ-1 and ADJ-2** operations, select the **Simultaneous Solution** checkbox.

Start the simultaneous Adjusts solving by using the **Start** button on the Adjust or in the *Sim Adj Manager* tab.

In the **Adjust 2**, Click start.

What is the required air flow rate?

What is the molar flow rate of the synthesis Gas product?

8.6 CSTR (Continuous Stirred Tank Reactor)

A continuous stirred tank reactor or back-mix reactor (CSTR), characterized by its content, is perfectly mixed, so that the composition at the exit is the same as that inside the reactor. The CSTR reactor is usually used for liquid-phase or multiphase reactions that have fairly high reaction rates. Streams of reactants and products are continuously fed into and removed from the vessel [7]. Various mechanisms can be used to cool or heat anything. The two most typical include the employment of an internal coil or a jacket encircling the vessel. Unless reaction rates are extremely fast, a single CSTR must be quite large when high conversion is desired. As a result, it is occasionally utilized to minimize the total reactor volume for a given conversion by connecting numerous CSTRs in series [8].

In this section, a triethylamine synthesis flowsheet is provided. In a constantly stirred tank reactor (CSTR), ethanol and diethylamine are mixed to create triethylamine. The feed stream of ethanol and diethylamine is fed into a reactor that produces triethylamine under atmospheric pressure.

8.6.1 Problem Statement

Find the reactor volume that causes a 94% conversion of ethanol in the following liquid-phase reaction in an isothermal CSTR. 50 kgmol/h of ethanol, 50 kmol/h of diethylamine, and 100 kgmol/h of water make up the feed. The reaction is second order with respect to ethanol.

$$\text{Ethanol} + \text{Diethylamine} \rightarrow \text{Water} + \text{Triethylamine} \tag{8.15}$$

$$r_A = -kC_{Etoh}2, k = A\,e^{E/RT}$$

E = activation energy = 1×10^4 J/mol, and A = Pre-exponential factor = 4775 L/mol h.

The conditions of feed are: $P = 3.5$ atm, and $T = 50°\text{C}$.

8.6.2 Simulation

1) Open a new case of HYSYS and enter the following values:
   Components: Ethanol, Diethylamine, Triethylamine, and H_2O,
   Fluid package: NRTL
   In the Activity Model Interaction Parameters group, the Aij interaction table appears by default. HYSYS automatically inserts the coefficients for any component pairs for which library data is available. You can change any of the values provided by HYSYS if you have data of you own.
2) Next, you will use the **UNIFAC VLE** estimation method to estimate the unknown pair.
   - Click **Unknowns Only** button. HYSYS provides values for the unknown pair.
     The final Activity Model Interaction Parameters table for the Aij coefficients appears in Figure 8.35.
   - To view the Bij coefficient table, select the Bij radio button. For this case, all the **Bij** coefficients will be left at the default value of zero.
3) Click the **Reactions** folder in the Navigation Pane.
   The reaction between Ethanol and Diethylamine to produce Triethylamine and Water is shown in Eqs. (8.15) and (8.16):

$$C_2H_5OH + C_4H_{11}N \rightarrow H_2O + C_6H_{15}N \tag{8.16}$$

   These steps will be followed in defining our reaction:

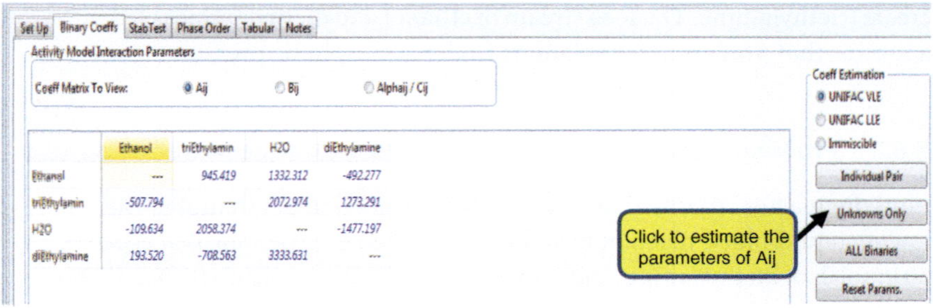

Figure 8.35 Final Activity Model Interaction Parameters table for the Aij coefficients.

4) In the Navigation pane, click the Reactions group, then click the **Add** button to Add Reaction Set. The Reaction Set property view appears.
5) Click the **Add Reaction** button. From the mini-window, Select **Kinetic** type and click the **Add Reaction** button, close the **Reaction** property view.
6) Under Active Reactions, Double-click the **Rxn-1**, add the components with its Stoichiometry according to the reaction (8.16), Figure 8.36.
7) Change the reaction order to 2 with respect to ethanol, and 0.0 with respect to diethylamine for the forward reaction order. Since the reaction is irreversible, under (Rev Order), type zero for all components.

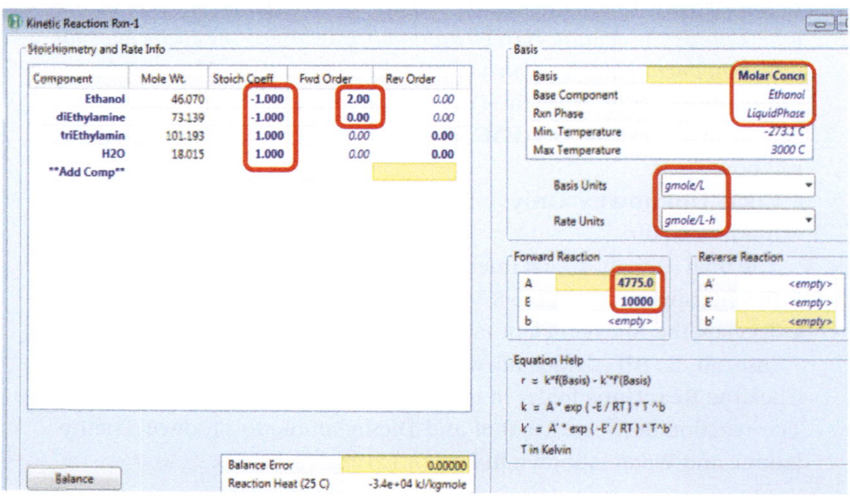

Figure 8.36 Entering the Stoichiometry, order of the reaction, and parameters of the Arrhenius equation of reaction (8.16).

8) On the Basis page, ethanol is the basis component and the basis unit is gmol/L. The units of the reaction rate are gmol/L.h. The reaction takes place in the liquid phase, Figure 8.36.
9) The **Arrhenius** parameters for the kinetic reaction must be given. In this case, there is no **Reverse Reaction** occurring, so you only need to supply the **Forward Reaction** parameters. Enter the values of $A = 4775$ L/mol h and $E = 10\,000$ J/mol, Figure 8.36.
10) The final task is to add the reaction set available to the Fluid Package, which also makes it available in the flowsheet. Add the Reaction Set to the Fluid Package by clicking the **Add to FP** button, then click the **Add set to Fluid Package** button.
11) Enter the **Simulation** Environment.
12) Add a new Material stream with the following values.

Name	Feed
Temp	50°C
Pressure	3.5 atm
Molar flow rate	200 kmolh
Molar composition	Ethanol = 0.25, Diethylamine = 0.25, Water = 0.5

13) The feed stream now is known; your next task is to install the necessary unit operations for producing the Triethylamine.
14) Installing the Reactor, where from the object palette (Figure 8.37), add the **CSTR** to the PFD.
15) Attach the **Feed** to the inlet of the reactor.
16) Name the vapor outlet as **CSTR Vent** and the liquid product stream as **CSTR Product** as in Figure 8.38.
17) On the **Reactions** tab, **Details** page, from the **Reaction set**, select **Set-1**. This will automatically connect the proper reaction to this reactor.
18) The next task is to specify the **Vessel Parameters**. From **Dynamics** tab, give the reactor volume = 4.433 m^3 and Liq Volume Percent = 85%, Figure 8.39.

    Reactor volume is obtained by trial and error until the conversion of reaction to 94% is attained.
19) The pressure drop is assumed to be zero. Since it is assumed that the reactor is functioning under isothermal circumstances, enter 50°C at the temperature cell in the CSTR Product column, Figure 8.40.

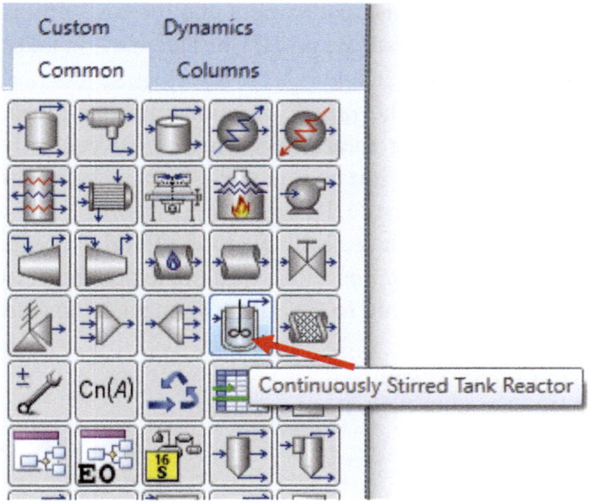

Figure 8.37 CSTR in the object palette.

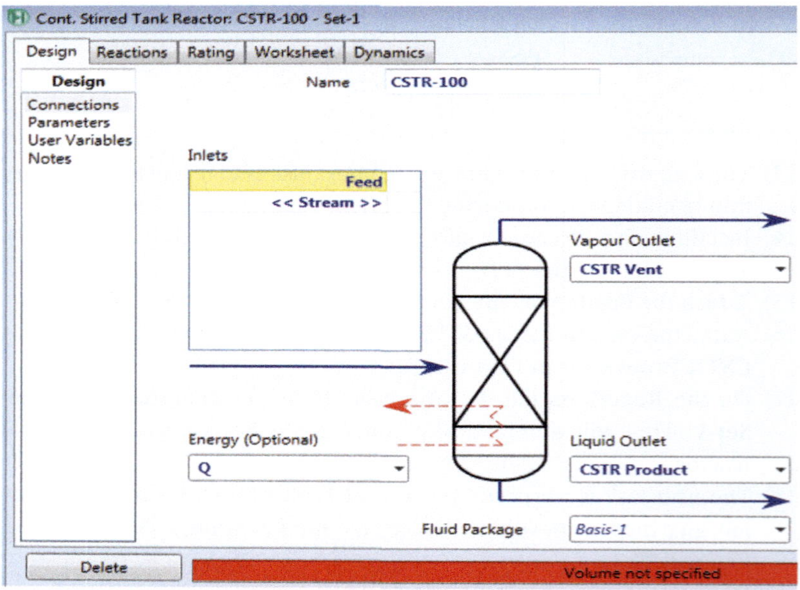

Figure 8.38 Design tab, Connections page in the CSTR and attached streams.

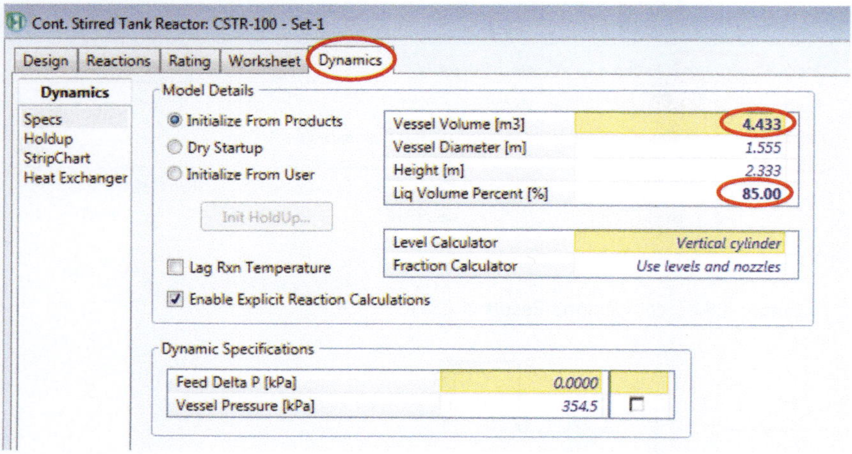

Figure 8.39 Entering the dimensions of CSTR in the Dynamics tab of the CSTR.

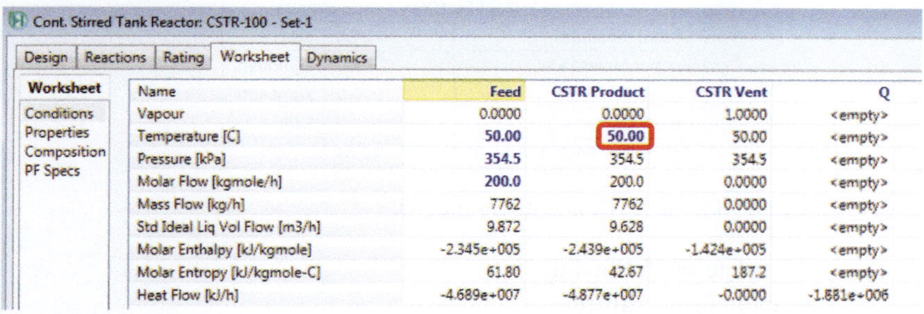

Figure 8.40 Worksheet tab, Conditions page, Entering the temperature in the CSTR Product.

20) There is no phase change in the reactor under isothermal circumstances since all the components (reactants and products) are liquid at 50°C and 3.5 atm, so the flow of the vapor product stream CSTR Vent will be zero. The calculated necessary cooling duty is also represented by the coolant stream's heat flow. Next, we'll look at reactor conversion as a function of temperature.
21) Select the Results page/Reactions tab. The conversion can be seen in Figure 8.41, the Reactor Results Summary table. The actual percent conversion (Act.% Cnv.) in the reactor at the moment is 93.34%.
Figure 8.42 shows the process flow diagram (PFD) of the CSTR reactor and its condition.

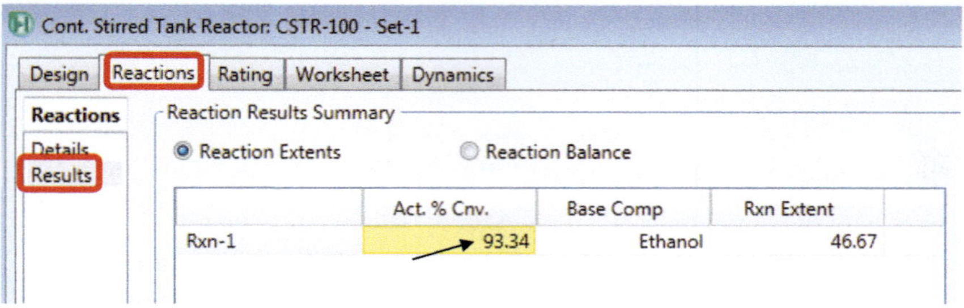

Figure 8.41 conversion's Result of CSTR.

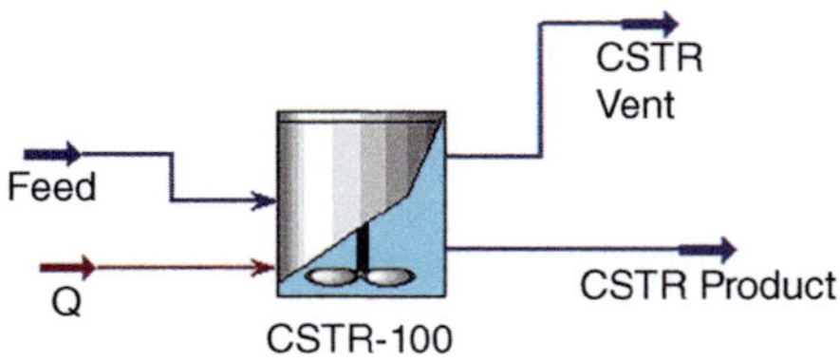

CSTR-100		
Vessel Pressure	354.5	kPa
Vapour Molar Flow	0.0000	kgmole/h
Liquid Molar Flow	200.0	kgmole/h
Heat Flow	−1.881e+006	kJ/h
Act. % Cnv. (Act. % Cnv._1)	93.34	

Figure 8.42 PFD of CSTR and its conditions.

22) Complete the following:
    Reactor Temperature ......
    Molar flow rate of components in the outlet flow (CSTR Product) ....

8.6.3 Summary

Continuous Stirred Tank Reactor (CSTR) is the main one of the two reactors in the simulator; it behaves as a well-mixed reactor; the reacting system is perfectly mixed, so that the temperature, pressure, and concentration are the same in all the points of the reactor. The suitable reaction types with CSTR are Simple Rate, Heterogeneous Catalytic, Kinetic; other types of reaction are not suitable with CSTR. As we see from Section 8.6.1, one of the parameters that affect the conversion is the volume of reactor as well as the operating conditions of reactor such as temperature, concentration of reactants, and pressure.

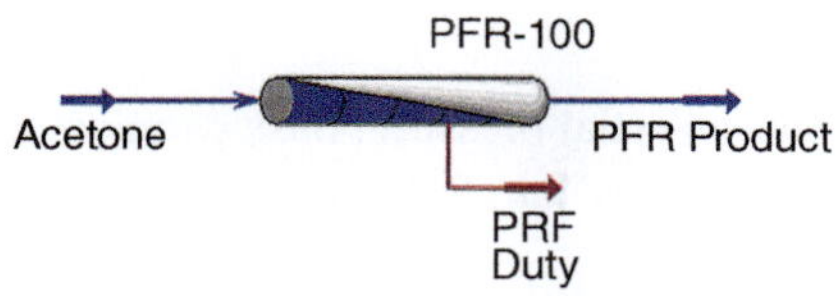

Figure 8.43 PFD of the plug flow reactor (PFR).

8.7 Plug Flow Reactor

The plug flow reactor (PFR) behaves as an ideal reactor when the reactants move in the axial direction without mixing with any element ahead or behind, and the residence time in the reactor is the same for all the elements of the fluid. It is assumed that no mixing occurs between adjacent fluid volume elements either radially or axially [7, 9]. The PFR as shown in Figure 8.43, also known as the Tubular Reactor, generally consists of a bank of cylindrical pipes or tubes. The flow field is modeled as plug flow, implying that the stream is radially isotropic (without mass or energy gradients). This also implies that axial mixing is negligible [8, 9].

As the reactants flow along the length of the reactor, they are continually consumed; hence, there is an axial variation in concentration. Since the reaction rate is a function of concentration, it also varies axially (except for zero-order reactions). To obtain the solution for the PFR (axial profiles of compositions, temperature, and so forth), the reactor is divided into several sub-volumes. Within each sub-volume, the reaction rate is considered to be spatially uniform.

8.7.1 Problem Statement-1: Simple Rate and Kinetics Reaction in One Set

In this process, styrene is manufactured by the dehydrogenation of ethylbenzene. Reaction occurs at high temperature (800–950 K) and low pressure (0.4–1.4 bar) in order to shift the equilibrium to the right to favor styrene production. The raw material is ethylbenzene, which is preheated to a saturated vapor and then mixed with steam to provide the heat of reaction and to serve as an inert diluent to help shift the reaction to the right. Steam also tends to limit side reactions and helps to extend catalyst life by reducing coke formation on the catalyst. In the reactor, the process uses a proprietary iron catalyst that minimizes (but does not eliminate) side reactions at higher temperatures.

Styrene is produced by the **simple rate** reaction [3, 10]:

$$\text{E} - \text{Benzene} \rightarrow \text{Styrene} + \text{Hydrogen} \tag{8.17}$$

where:

Basis: Partial pressure, Base component: E-Bz, Rxn phase: VapourPhase, Basis Units: bar, Rate units: $kmol/m^3s$.

The forward reaction equation is:

$$-r = k \times [\mathrm{f(Basis)} - \mathrm{f}^{/}(\mathrm{Basis})/K^{/}] \tag{8.18}$$

where: $k = Ae^{[-E/RT]*T\wedge B}$, $A = 7.3e7$, $E = 9.132e4\ kJ/kmol$, $B = 0$,

$$\mathrm{Ln}(K^{/}) = -5.5 - 1.3e4/T + 3.2\ln(T) - 2.4e - 3T \tag{8.19}$$

In addition to the above reaction, the following undesired **kinetic** reaction also occurred:

$$\mathrm{E - Benzene} \rightarrow \mathrm{Bz + Ethylene} \tag{8.20}$$

where:

Basis = Partial pressure, Base component: E-Bz, Rxn Phase: VapourPhase, Basis Units: bar, Rate units: $kmol/m^3s$.

The forward reaction equation is:

$$-r = kC_{EB} \tag{8.21}$$

where $k = Ae^{-E/RT}$, $EB = ethyl\text{-}benzene$

Data of kinetic reaction: $A = 1.01e13$, $E = 2.08e5\ kJ/kmol$.

The feed enters the adiabatic PFR at conditions of (607°C, 138 kPa, 10,180 kmol/h and mass fractions: E-Bz = 0.33, H_2O = 0.67), where the reactor has a volume of 5 m^3, a length = 2, a void fraction = 0.445, a single pipe, and a pressure drop across the reactor of 5 kPa. Find the mass flow rate of the styrene in the product.

8.7.1.1 Simulation

1) Open a new case and enter the components: E-Benzene, Styrene, Hydrogen, Benzene, ethylene, and water.
2) Fluid package: Peng–Robinson.
3) Add a Reaction Set, and Add a reaction of **Simple rate** type, reaction (8.17).
4) Enter the data required as given in the problem and as shown in Figure 8.44.
5) For the reaction (8.20), add the reaction of **kinetics** type in the *same* Reaction Set, and enter the data as in Figure 8.45.
6) Add the reaction Set to the **FP**.

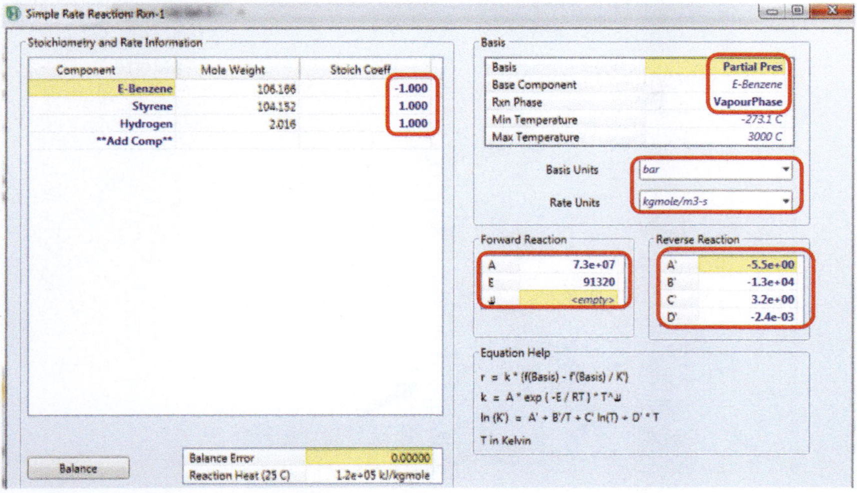

Figure 8.44 Window of the simple rate reaction of (8.17).

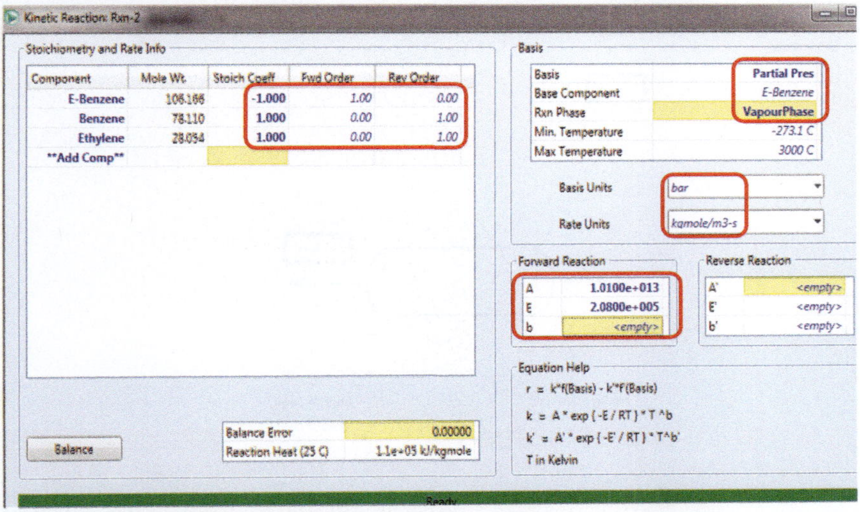

Figure 8.45 Stoichiometry and rate information of the kinetics reaction of (8.20).

7) Enter the **Simulation** Environment. Construct the Feed stream, Enter the data of the feed as mentioned in the problem.
8) Add the PFR reactor from objects palette, Figure 8.46.
9) Connect the Feed and Product streams with the reactor as shown in Figure 8.47.

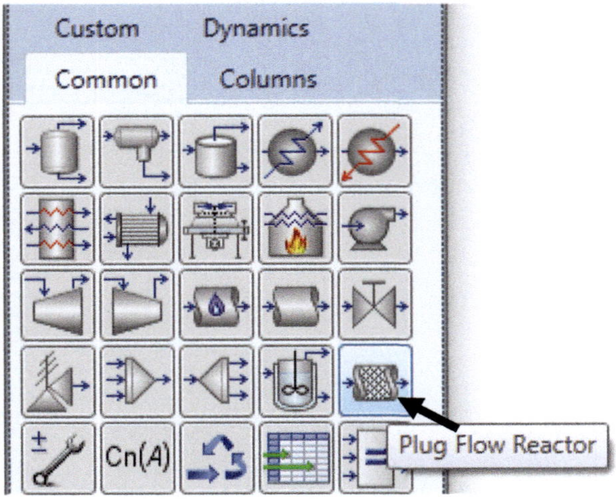

Figure 8.46 PFR in the object palette.

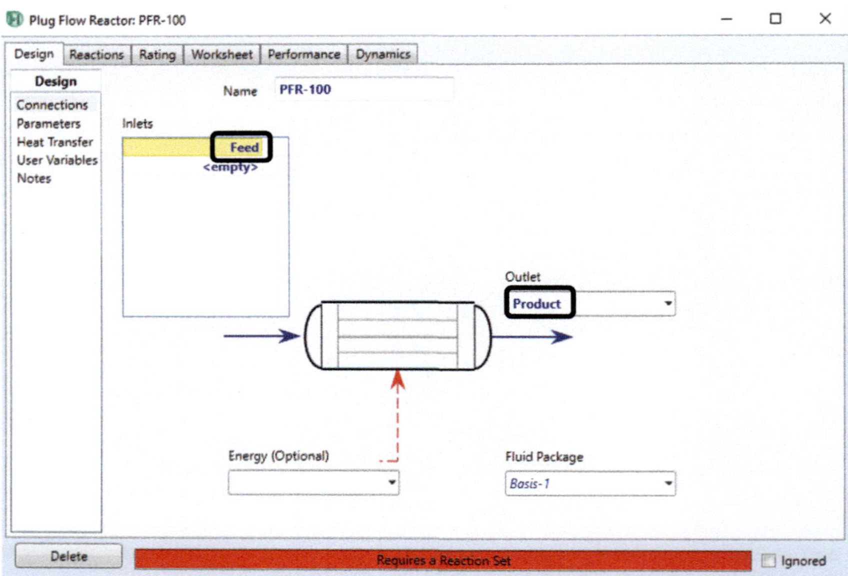

Figure 8.47 Attaching the streams with PFR in the Connection page of the PFR.

10) Parameters page: $\Delta P = 5\,\text{kPa}$.
11) Reaction tab: Insert **Set-1.**
12) Rating tab: add the data required as follows: Volume = $5\,\text{m}^3$, length = 2, Void Fraction = 0.445.

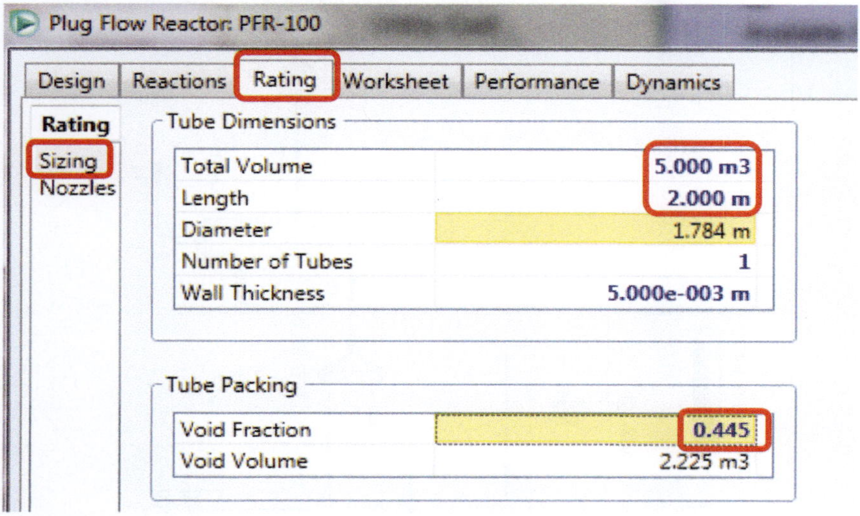

Figure 8.48 Rating tab, Sizing page in the PFR.

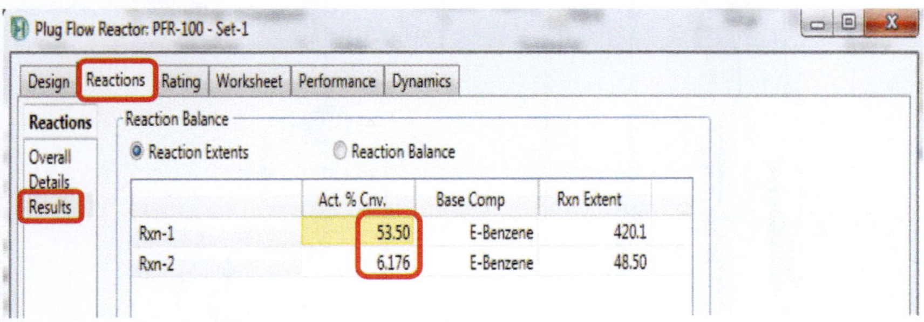

Figure 8.49 Conversion's result of the reactions in the PFR.

The window will look like that in Figure 8.48.

13) Go to the **Reaction** tab, **Results** page, and see the conversion of each reaction as shown in Figure 8.49.

8.7.2 Problem Statement-2: Ammonia Production Plant

Ammonia is produced in a process known as the Haber process, in which nitrogen and hydrogen react in the presence of an iron catalyst to form ammonia. The hydrogen is formed by reacting natural gas and steam at high temperatures, and the nitrogen is supplied from the air. The PFD of the ammonia production plant is shown in Figure 8.50.

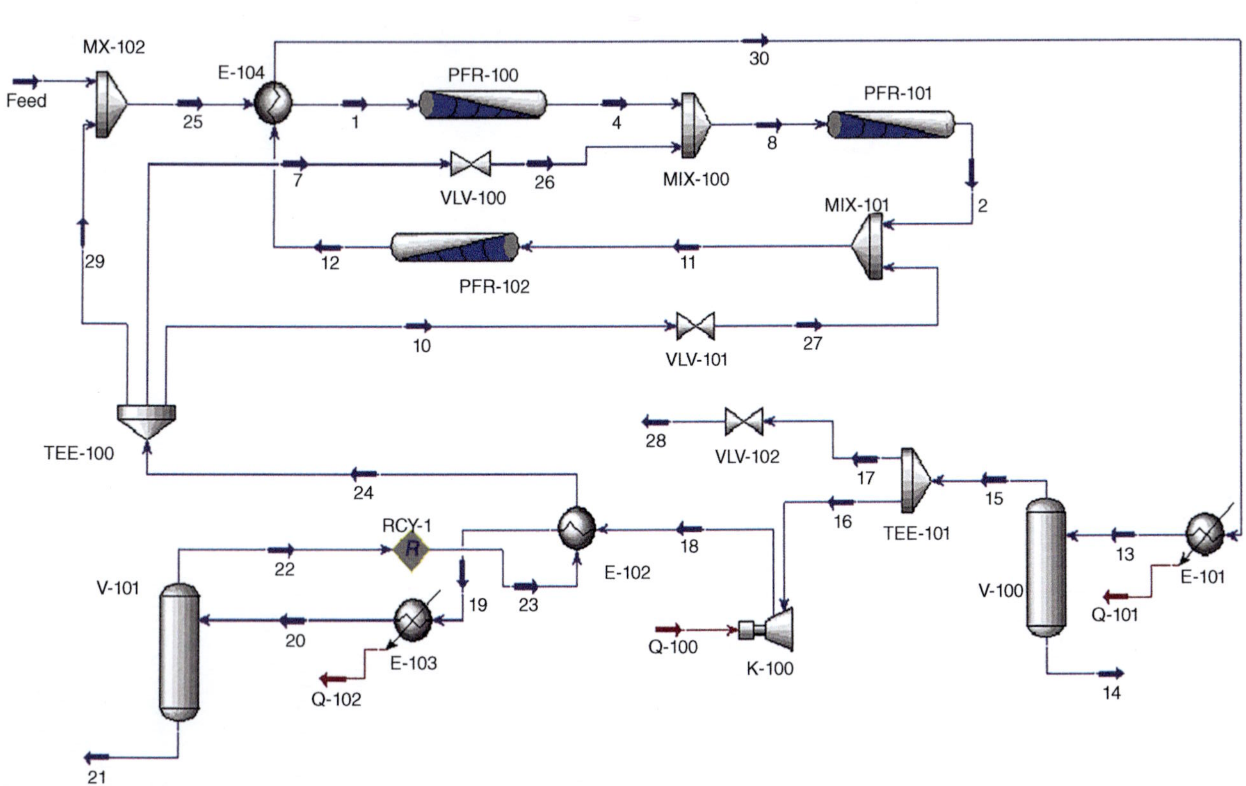

Figure 8.50 PFD of the Ammonia Production Plant.

Ammonia is produced from the reaction of nitrogen and hydrogen, as follows [4, 10, 11]:

$$0.5N_2 + 1.5H_2 = NH_3 \tag{8.22}$$

8.7.2.1 Simulation

1) Open a new case and add the components: Methane, H_2, N_2, NH_3, and Ar.
2) **Fluid package:** Peng–Robinson.
3) Add a Reaction Set: **Set-1, Reaction type:** Kinetics, Figure 8.51.
   Rxn-1: $0.5N_2 + 1.5H_2 = NH_3$
   Order of the Forward Reaction: $N_2 = 0.5$, $H_2 = 1.5$, $NH_3 = 0$
   Order of the Reverse reaction: $N_2 = 0$, $H_2 = 0$, $NH_3 = 1$
   Basis: Partial Pressure, **Base Component:** N_2, **Rxn Phase**: Vapor phase,
   Basis Unit: atm, **Rate Units**: kmol/m^{3}s
   Forward Reaction: $A = 10{,}000$, $E = 91{,}000$ kJ/kgmol,
   Reverse Reaction: $A' = 1.3e10$, $E' = 1.41e5$ kJ/kgmol, see Figure 8.51.
4) Add the Reaction Set to the **FP**.
5) Enter the Simulation environment.
   Create a material stream, Feed: $T = 200°C$, $P = 15\,000$ kPa.
   The composition in Molar Flow is shown below:

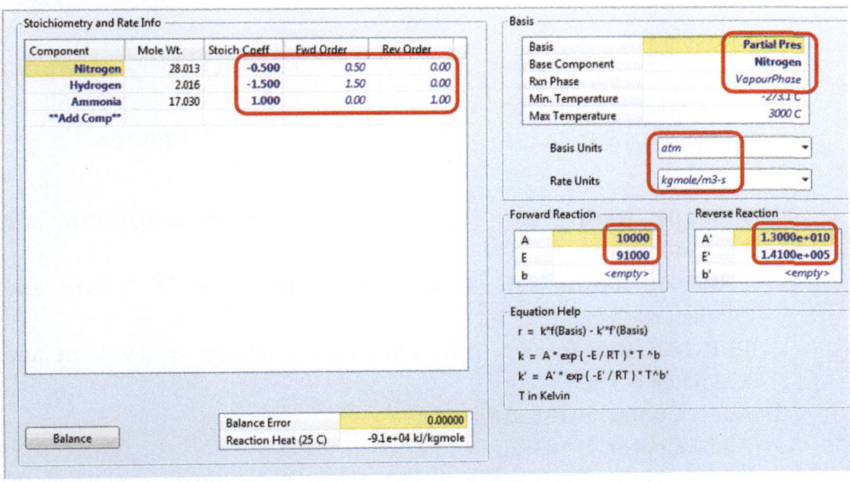

Figure 8.51 Window of kinetics reaction for ammonia kinetics.

Component Name	Comp Mole flow, kmol/h
Methane	294.44
Nitrogen	3417.5
Hydrogen	10,253
Ammonia	0.0
Argon	134

6) Add a **Heat Exchanger, E-104**:
   Tube side; inlet-stream: 12, outlet-stream: 30
   Shell side; inlet-stream: 25, outlet-stream: 1,
   Enter the Temp of stream-1 = 270°C,
   Parameter: $\Delta P = 0$ for **Tube** and **Shell**,
   Design tab, Parameters page, Heat-Exchanger model: **Simple Weighted**.
7) Add the **PFR-100, PFR-101, and PFR-102**:
   All reactors above have the following specifications:
   Parameters: $\Delta P = 15$ kPa, Adiabatic,
   Number of tubes = 1, Wall thickness = 0.005 m,
   Tube Packing: Void fraction = 0.5,
   Reaction tab: Overall Page: Reaction Set: Set-1.
   Specify the Integration Information as follows:

   Integration Information

Number of Segments	5
Minimum Step Fraction	1.0e-06

   PFR-100: Rating tab; Tube Dimensions: length = 0.9692 m and Diameter = 3 m.
   PFR-101: Rating tab; Tube Dimensions: length = 1.08 m and Diameter = 3 m.
   PFR-102: Rating tab; Tube Dimensions: length = 0.9512 m and Diameter = 3 m.
8) Add a **Cooler**, **E-101:** $\Delta P = 15$ kPa
9) Add a **Cooler**, **E-103:** $\Delta P = 15$ kPa
10) Add a **Heat Exchanger, E-102:** Tube Side: Inlet = Stream 18, Outlet = Stream 19, Shell Side: Inlet = Stream 23, Outlet = Stream 24.

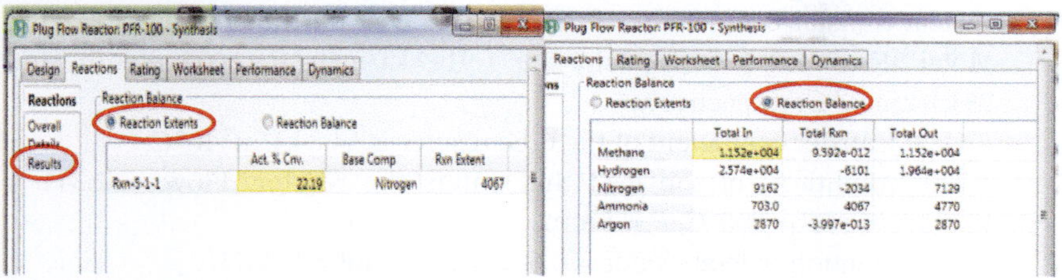

Figure 8.52 Reaction Results of plug flow reactor, PFR-100.

Parameters Page: $\Delta P = 15$ kPa for tube side,
Temp of Stream 20 = −40°C,
Temp of Stream 24 = 25°C,
Temp of Stream 13 = 35°C,
Pressure of stream 18 = 15 050 kPa.

11) TEE-101: Parameters Page: Stream 16 = 0.992.
12) TEE-100: Parameters Page: Stream 10 = 0.0005, and Stream 7 = 0.0005.
13) Stream 23: molar flow rate = 35,940 kmol/h.
    Composition (mole fraction): $C_1 = 0.3128$, $H_2 = 0.4314$, $N_2 = 0.16$, $NH_3 = 0.0196$, and Ar = 0.0762.
14) Add the mixers: **Mix-100, Mix-101, and Mix-102:** Parameters Page: Activate the radio **Equalize all** for each mixer.
15) **Recycle:** Parameters tab; Variables page: click [Use Component Sensitivities]
16) Add a valve, VLV-102: $\Delta P = 5$ kPa.
    The results of the reactions in the plug flow reactor PFR-100 are shown in Figure 8.52.

8.7.3 Problem Statement-3: Heterogeneous Catalytic Reaction

One method to produce the hydrogen is the reaction of methanol and water at 200 °C and 5 atm in the PFR in the vapor phase, where [3, 9]:

$$CH_3OH + H_2O \rightarrow CO_2 + 3H_2 \tag{8.23}$$

The dimension of reactor is: $L = 0.4$ m, $D = 0.04$ m.
The catalyst is cylindrical pellets: $H = D = 0.5$ cm.
Solid density of catalyst is 1312 kg/m^3, sphericity is 1.
The reaction is of second order (1st for methanol and 1st for water),

$$r = kC_{Etoh}C_{H_2O} \qquad k = Ae^{-E/RT}$$

$$A = 9.4868e9, \qquad E = 1.0794e5\ \text{kJ/kmol}$$

Bed void = 0.15.

Heat transfer coefficients of the medium used in the heating of reactor, Molar flow, and heat capacity are 1.91e7 kJ/m^{2}h·°C, 100 kmole/h, and 75 kJ/kmol·°C, respectively.

Heat transfer coefficient of the tube reactor is 64.8 kJ/m^2·h·°C.

The conditions of the feed that enters the reactor: Flow = 2.341 × 10^{-6} kmol/s, T = 200°C, P = 5 atm,

Mole fraction of feed: water = 0.6429, methanol = 0.3571.

Study the effect of reactor length on the hydrogen production.

8.7.3.1 Simulation

1) Open a new case and enter four components: CH_3OH, H_2O, CO_2, and H_2.
2) Fluid package: UNQAUC, click **Binary Coeff** tab, and click the **Unknowns only** button.
3) Add a Reaction Set, and add the Reaction type of Heterogeneous Catalytic.
4) Add reactants with the stoichiometry mentioned in Eq. (8.23) and other data as in Figure 8.53.

   In the right bottom of the window of the Heterogeneous Catalytic Reaction, click **Kinetics** Help. The formula for the Heterogeneous Catalytic reaction will be shown in Figure 8.54.

   Click the **Reaction Rate** tab, and enter the data as in Figure 8.55, where units of E are in (kJ/kgmole)

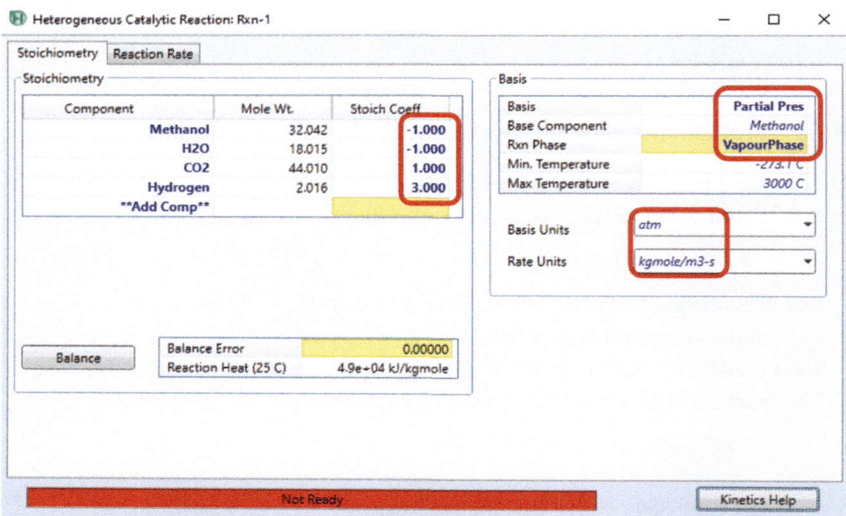

Figure 8.53 Window of Heterogeneous Catalytic reaction.

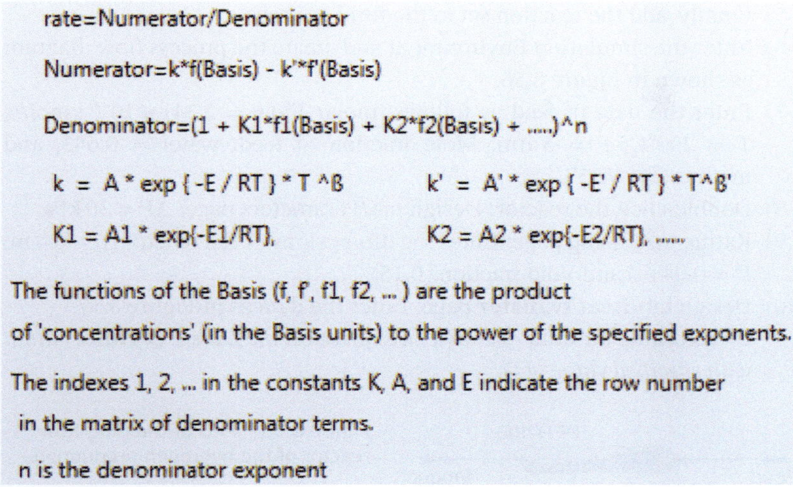

rate=Numerator/Denominator

Numerator=k*f(Basis) - k'*f'(Basis)

Denominator=(1 + K1*f1(Basis) + K2*f2(Basis) + ...)^n

k = A * exp { -E / RT } * T ^ß   k' = A' * exp { -E' / RT } * T^ß'

K1 = A1 * exp{-E1/RT},   K2 = A2 * exp{-E2/RT}, ...

The functions of the Basis (f, f', f1, f2, ...) are the product of 'concentrations' (in the Basis units) to the power of the specified exponents.

The indexes 1, 2, ... in the constants K, A, and E indicate the row number in the matrix of denominator terms.

n is the denominator exponent

Figure 8.54 A picture of the formula of Heterogeneous Catalytic reaction.

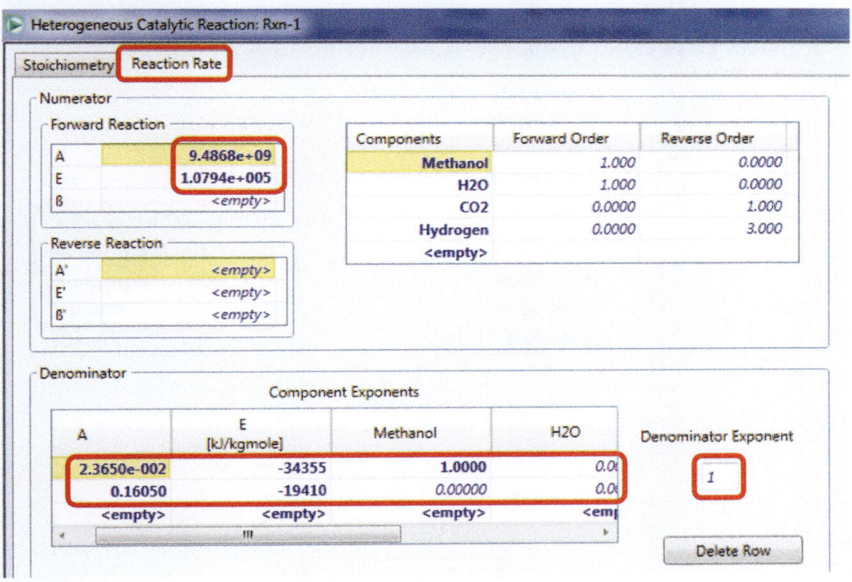

Figure 8.55 Reaction Rate of Heterogeneous Catalytic reaction.

5) Finally, add the reaction set to the fluid package.
6) Enter the simulation Environment and create the process flow diagram as shown in Figure 8.56.
7) Enter the data of feed as follows: molar Flow = 2.341×10^{-6} kmol/s, T = 200°C, P = 5 atm, Mole fraction of feed: water = 0.643, and methanol = 0.357.
8) Double-click the reactor: Design tab/Parameters page: ΔP = 30 kPa.
9) Rating tab/Sizing page: Enter the dimensions of the reactor (L = 0.4 m, D = 0.04 m) and Void fraction (0.15).
10) Design tab/**heat transfer** page: Enter the data as in Figure 8.57. *Or according the data available, we can select the* ***Direct Q Value*** *button with specified value of Q.*

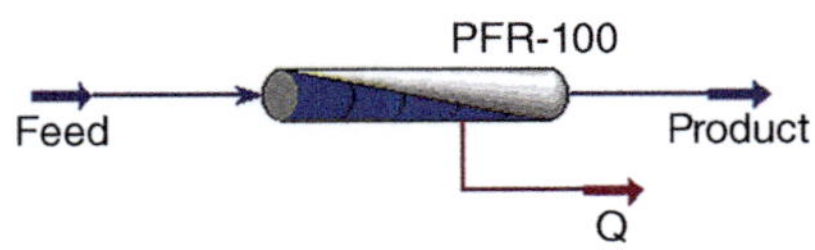

Figure 8.56 PFD of the plug flow reactor of the hydrogen production.

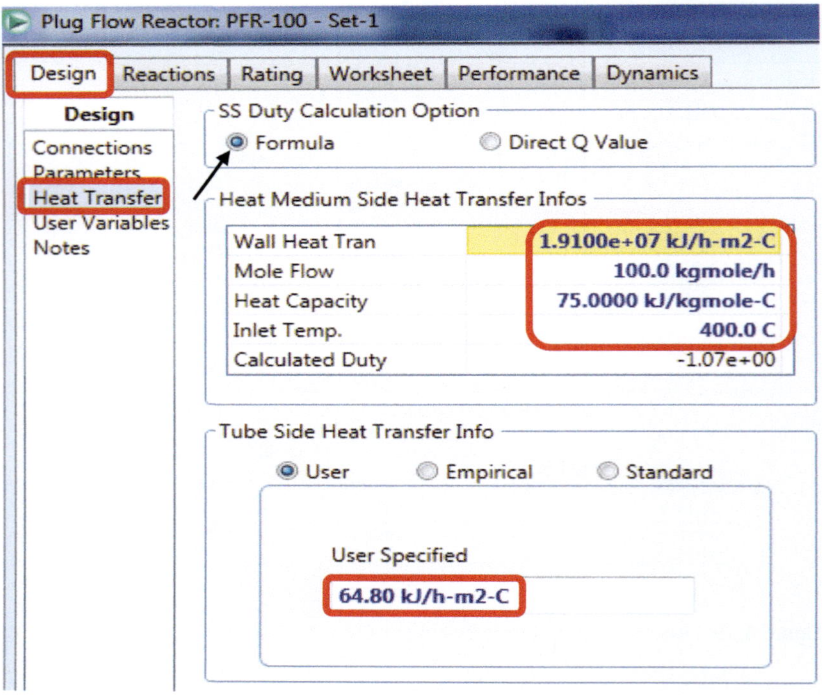

Figure 8.57 Heat transfer page in PFR.

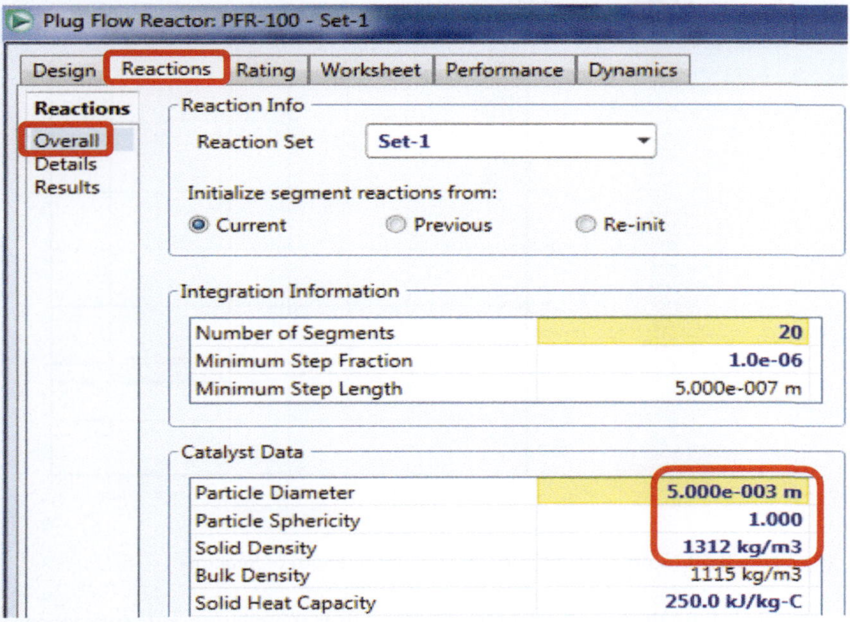

Figure 8.58 Data required of the Heterogeneous Catalytic reaction.

11) **Reaction** tab, **Overall** page: Reaction Set: Select Set-1,
Enter the remaining data as in Figure 8.58.
The results of reaction can be seen from **Performance** tab, **Compositions** page; select plot to show the composition with length as in Figure 8.59.
12) To study the effect of length of reactor on the reaction conversion and hydrogen product, we will use the **Case Study:** The independent variable is the length of reactor: 0.05–0.5 m, and the dependent variable is the mole fraction of hydrogen in the product stream.
The results will appear as in Figure 8.60.
It is clear that from Figure 8.60, the mole fraction of hydrogen in the product increases with increasing reactor length since the reactants will take more time for reaction, and that leads to decreasing methanol fraction in the product.

8.7.4 Summary

Plug flow reactor (PFR) is the main one of the two reactors in the simulator. The reaction takes place through the reactor, so that the concentration

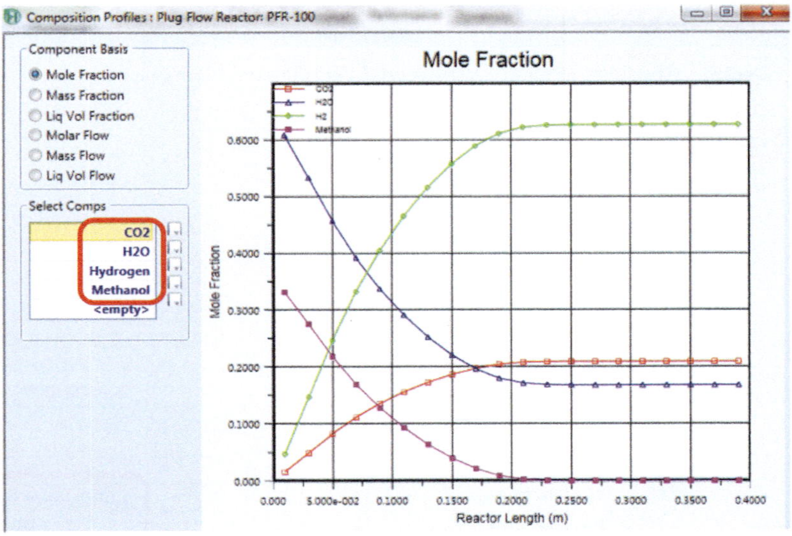

Figure 8.59 Component mole fraction along length of Reactor.

varies along the length of the reactor and thus the reaction rate (unless zero order). It has been applied to three types of reactions during the examples: Simple Rate, Heterogeneous Catalytic, and Kinetic. Other types of reactions are not suitable for PFR. One of the parameters that affect the conversion is the length of the reactor as well as the operating conditions of the reactor such as temperature, concentration of reactants, and pressure. PFR is suitable for gas, liquid, and two-phase instantly.

8.8 Gibbs Reactor

The Gibbs Reactor calculates the exiting compositions such that the phase and chemical equilibria of the outlet streams are attained. However, the Gibbs Reactor does not need to make use of a specified reaction stoichiometry to compute the outlet stream composition. The condition that the Gibbs free energy of the reacting system is at a minimum at equilibrium is used to calculate the product mixture composition. As with the Equilibrium Reactor, neither the pure components nor the reaction mixture are assumed to behave ideally [3, 10, 12].

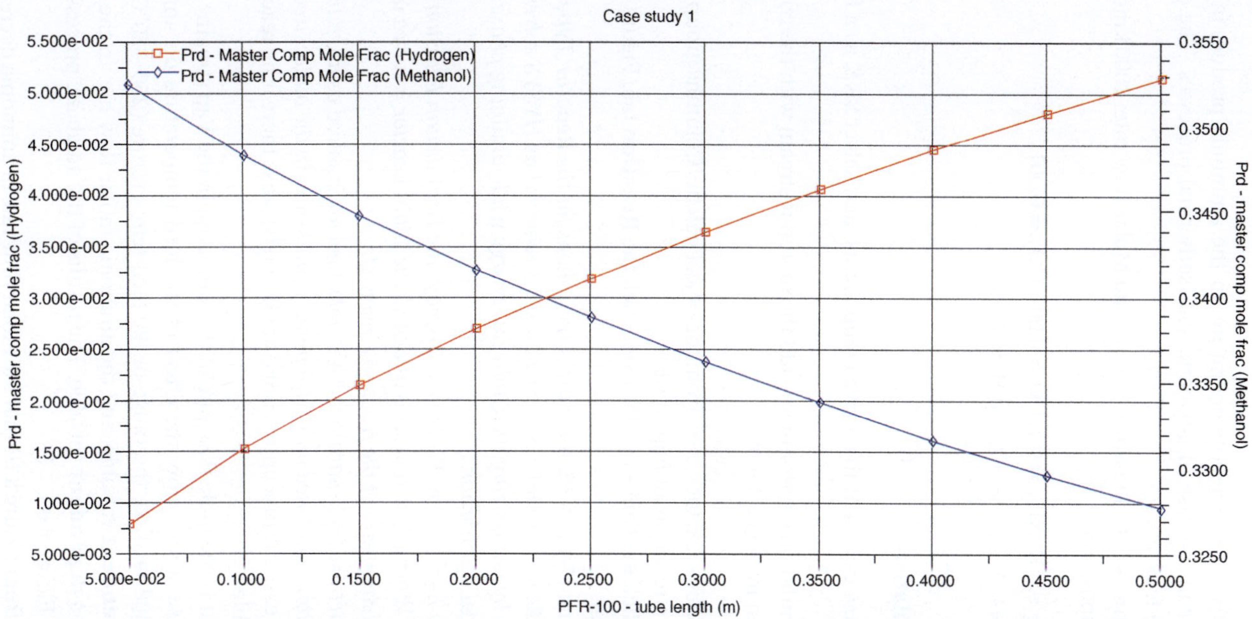

Figure 8.60 Effect of tube length on the compositions of the components.

8.8.1 Problem Statement

Study the effect of the feed temperature on the ammonia product in the Gibbs reactor in the case of adiabatic and isothermal processes. The conditions of the feed are:

Temperature: 300°C, Pressure: 80 atm, and Molar flow rate: 1000 kmol/h
Molar composition:

$$CH_4 = 0.001, NH_3 = 0.1, CO = 0.001, CO_2 = 0.001,$$
$$H_2O = 0.001, N_2 = 0.2, \text{and } H_2 = 0.696.$$

8.8.2 Simulation

1) Open a new case, add the above components, and select **SRK** as a fluid package.
2) In the simulation environment, add the material stream with the conditions given in the problem.
3) Add the Gibbs reactor 

from the object palette/Columns group and attach its streams as in Figure 8.61.
4) Double-click the Gibbs reactor icon, click the **Reaction** tab, Figure 8.62 will appear.
Note: *From Figure* 8.62, *there are three options for the* ***Reactor Type****:*
   * *Gibbs Reactions Only: No reaction set is required as HYSYS solves the system by minimizing the Gibbs free energy while attaining phase and chemical equilibrium.*
   * *Specify Equilibrium Reactions: Displays the Equilibrium Reaction Sets group. When a reaction set is attached, the Gibbs Reactor is solved using the stoichiometry of the reactions involved.*
   * *No Reactions (= Separator): The Gibbs Reactor is solved as a separator operation, concerned only with phase equilibrium in the outlet streams.*
5) So, in our case the first option must be chosen because there is no reaction set available.
Now, what is the mole flow rate of the ammonia in the Vap stream?
6) Using Case study, show the effect of the feed temperature on ammonia production. Take the range of the feed temperature (250–350°C) as an **independent** variable, and the ammonia molar flow rate in the Vap stream as a **dependent** variable. The results of the adiabatic process are shown in Figure 8.63.
It is clear from Figure 8.63 that the molar flow rate of ammonia decreases with increasing feed temperature because the reaction is exothermic.

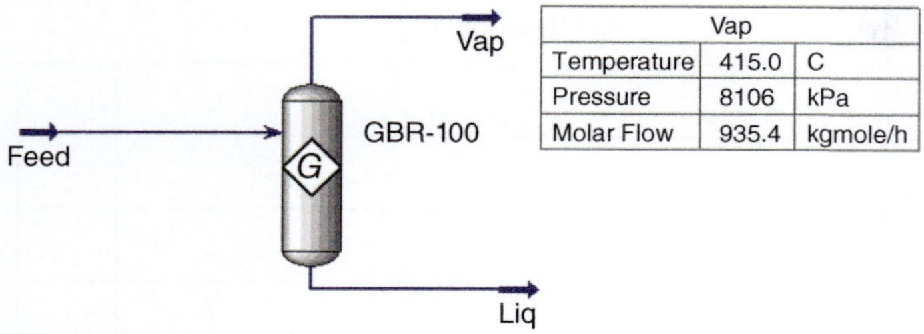

Figure 8.61 Gibbs reactor and its connected streams.

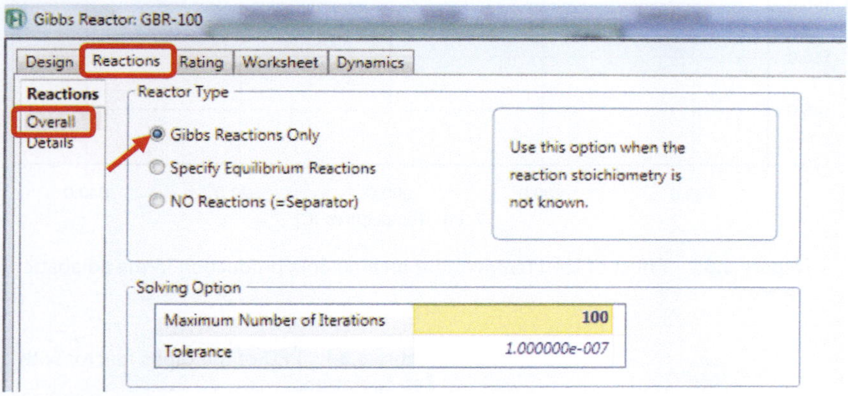

Figure 8.62 Selecting a suitable option of the Gibbs Reactor.

7) In the case of an isothermal process, adding the energy stream to the reactor and entering the temperature of the Vap stream are equal to 300°C. What is the mole flow rate of the ammonia in the Vap stream?
8) Now delete the temperature of the Vap stream and add a **Set**-operation to the reactor so that the final PFD will look like Figure 8.64. Target variable: Vap, Object = Temperature.
Source: Feed. Parameters: Multiplier = 1, Offset = 0.
9) Using the case study, the effect of the temperature (250–350°C) on the molar flow rate of ammonia in the product is shown in Figure 8.65.
Also from Figure 8.65, it can be seen that the molar flow rate of ammonia decreases with increasing feed's temperature because the reaction is exothermic, but the production rate of ammonia in the isothermal process is greater than that of the adiabatic process.

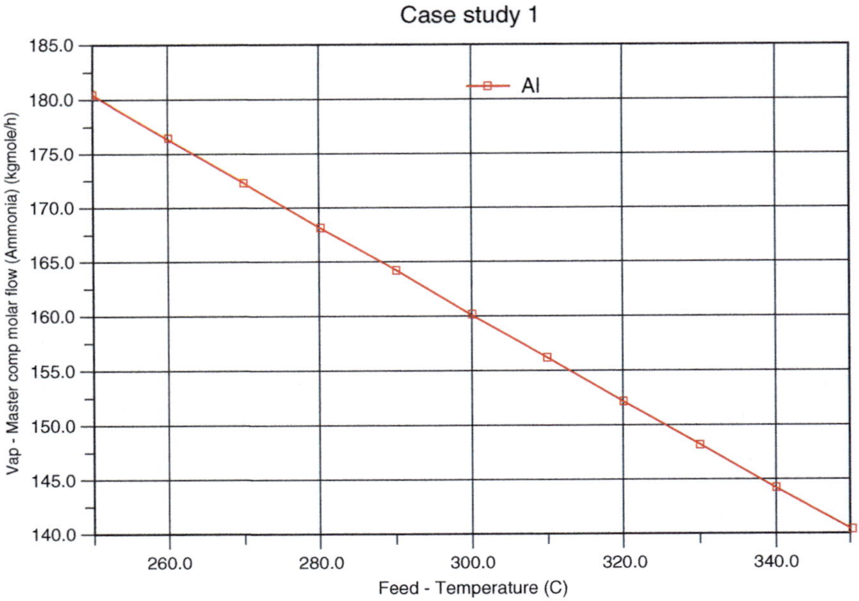

Figure 8.63 Effect of feed temperature on ammonia production in the adiabatic process.

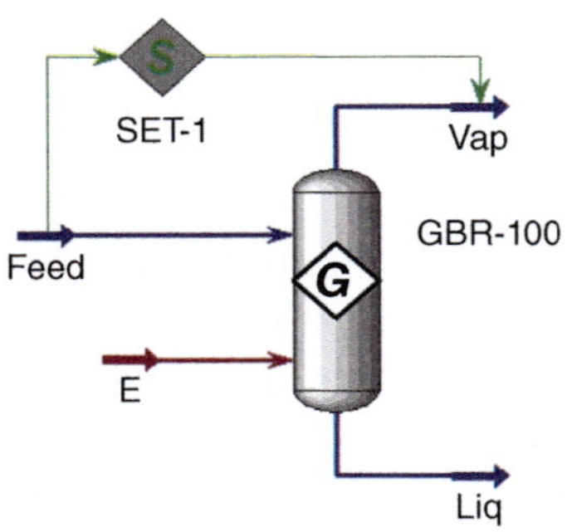

Figure 8.64 PFD of the Gibbs reactor with Set-operation.

8.8.2.1 Summary

The Gibbs Reactor evaluates the equilibrium composition of the outlet stream by minimizing the total Gibbs free energy of the effluent mixture. The Gibbs Reactor can be used to represent both equilibrium and constrained equilibrium systems. This module does not require the provision of reactions and therefore does not calculate the chemical equilibrium constants. Ammonia production from nitrogen and hydrogen is affected by the temperature of the feed and product. Since the reaction is exothermic,

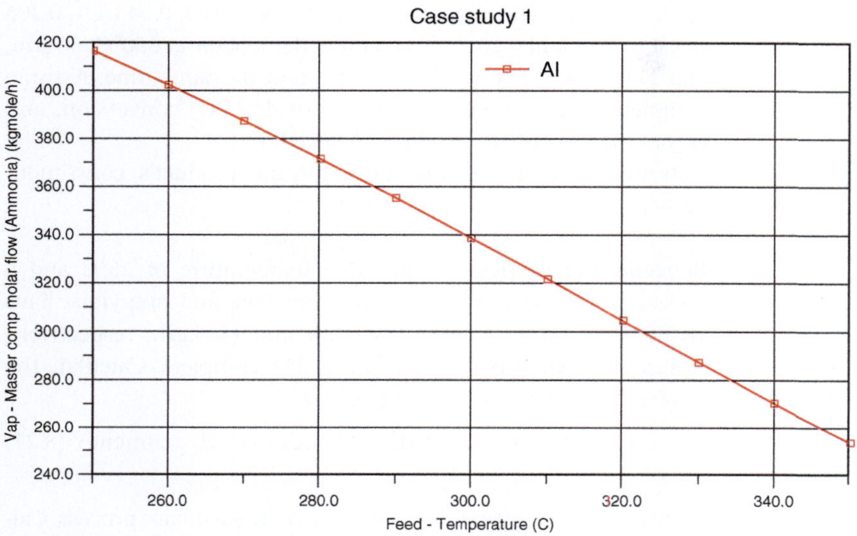

Figure 8.65 Effect of the feed temperature on the ammonia production in the isothermal process.

the lower the reactor temperature, the greater the ammonia production in two cases of adiabatic isothermal processes. The isothermal process gave more production rate of ammonia than the adiabatic because in the first one, the heat that evolved from the reaction is removed.

Problems

P.8.1 Ammonia is burned to make nitric oxide. Fifty percent of the limiting reactant converts. Inlet molar flow rates for both NH_3 and O_2 are 5 kmol/h. 25°C and 1 atm are the operating temperature and pressure, respectively. Calculate the exit component molar flow rates. The reactor is meant to operate adiabatically [13].

$$4NH_3 + 5O_2 \rightarrow 4NO + 6H_2O \quad (8.24)$$

P.8.2 Consider the following two reactions taking place in an isothermal reactor:

$$CH_4 + H_2O \rightarrow 3H_2 + CO \quad (8.25)$$

$$CO + H_2O \rightarrow CO_2 + H_2 \quad (8.26)$$

With feed mole fractions of 0.098 CO, 0.307 H_2O, 0.04 CO_2, 0.305 H_2, 0.1 CH_4, and 0.15 N_2, feed enters the reactor at 350°C, 30 atm, and 2110 mol/s. The responses happen at the same time. Assume complete methane and carbon monoxide (CO) conversion, and ignore the pressure drop across the reactor.
Determine the molar flow rates for the product's constituent parts [5].

P.8.3 Benzene and propylene react at a temperature of 25°C and a pressure of 1 atm to form cumene. Benzene and propylene have an inlet mass flow rate of 1000 kg/h and 180 kg/h, respectively. Assume that the limiting reactant is 45% complete. Calculate the product stream's molar flow rates [14].

$$C_6H_6(\text{benzene}) + C_3H_6(\text{propene}) \rightarrow C_9H_{12}(\text{cumene}) \quad (8.27)$$

P.8.4 Eighty percent conversion is achieved in the gas-phase process. Calculate the required amount of heat if the gases enter at 400 °C and 3 atm and exit at 500 °C. The following event transpires [3]:

$$CO_2 + 4H_2 \rightarrow 2H_2O + CH_4 \quad (8.28)$$

Take basis flow rate of gases = 100 kmol/h with molar compositions: 20% CO_2 and 80% H_2.

P.8.5 Partial oxidation can be used to convert the fuel to hydrogen. The partial oxidation technique relies on fuels, like methane, reacting with air to create carbon and hydrogen oxides.

$$CH_4 + 1/2O_2 \rightarrow CO + 2H_2 \quad (8.29)$$

Conversion 40% with respect to CH_4

$$CH_4 + O_2 \rightarrow CO_2 + 2H_2 \quad (8.30)$$

Conversion 60% with respect to CH_4
The feed consisting of pure methane (100% CH_4) at a temperature of 25°C, a pressure of 2 atm, and a molar flow rate equal to 100 kmol/h reacts with air (21% O_2, 79% N_2) at $T = 25°C$, $P = 2$ atm, and a Flow rate = 260 kmol/h according to the above two reactions. Create a model that illustrates the conversion of methane to hydrogen through partial oxidation.
The gases from the conversion reactor contain a high amount of CO, H_2O vapor used to convert the CO to CO_2 and H_2 in the equilibrium reactor, as shown in Figure 8.66. This reaction is called the

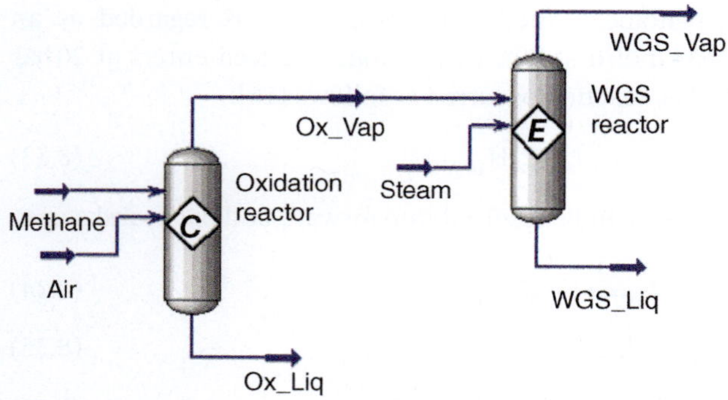

Figure 8.66 PFD of conversion and equilibrium reactor of problem P.8.5.

water gas shift reaction. A water gas shift reaction (WGS) makes up the purifying system that is most technologically possible. The reaction is

$$CO + H_2O \rightarrow CO_2 + H_2 \tag{8.31}$$

has been employed for H_2 production from liquid and gaseous hydrocarbons for 40 years in the industrial process. The purpose of the WGS reaction is to increase H_2 production while lowering CO levels, which can be poisonous to some catalysts. The steam conditions are: 100°C, 2 atm, 100 kmol/h, and 100% water [5, 10].

P.8.6 At 2 atm and 650°C, pure acetone reacts isothermally to produce ketene and methane. Acetone converts at a rate of 70%. Determine the acetone, ketene, and methane molar flow rates at the reactor's exit. Determine the amount of heat the reactor receives or loses. Take 100 kmol/h of pure acetone as a starting point. The reaction between 1 mol of acetone and 1 mol each of ketene and methane is noted. Package of properties: PRSV [15].

$$CH_3COCH_3 \rightarrow CH_2CO + CH_4 \tag{8.32}$$

P.8.7 In a plug-flow reactor (PFR), normal butane (n-C_4H_{10}) will be reversibly isomerized to isobutane (i-C_4H_{10}). The production of gasoline additives uses the valuable product isobutane. Under high pressure ($P = 20$ bar) and utilizing practically trace amounts of a liquid catalyst, the reaction is to be carried out adiabatically in the liquid phase. Calculate the length of PFR and volume of CSTR required to treat a feed mixture comprising 90 mol% n-butane and

10 mol% *i*-pentane (2-methyl-butane), which is regarded as an inert, at 163 kmol/h at 70% conversion. The feed enters at 20 bar and 330 K. The reaction occurred as follows [16]:

$$n - C_4H_{10} \leftrightarrow i - C_4H_{10} \tag{8.33}$$

Eq. (8.33) above can be resolved into forward and backward steps:

$$n - C_4H_{10} \overset{k1}{\Longrightarrow} i - C_4H_{10} \tag{8.34}$$

$$-r_{n-C_4} = k_1 C_{n-C_4} \tag{8.35}$$

$$k_1 = 2.94^* 10^7 (s^{-1}) * e^{\left(-\frac{65,300}{8.314^* T}\right)} \tag{8.36}$$

$$i - C_4H_{10} \overset{k2}{\Longrightarrow} n - C_4H_{10} \tag{8.37}$$

$$-r_{i-C_4} = k_2 C_{i-C_4} \tag{8.38}$$

$$k_2 = 1.176 \times 10^8 (s^{-1}) e^{\left(-\frac{72,200}{8.314^* T}\right)} \tag{8.39}$$

Initially set the length of PFR = 3 m, and diameter = 1.1 m.
For CSTR, set the initial volume = 2.6 m^3 activation energy in kJ/kmol, Basis units: kmol/m^3, and Rate unit: kmol/m^3s.

P.8.8 Determine the length of the reactor needed to accomplish a 90% reactor conversion. Assume steady-state adiabatic operation, a single tube reactor with a diameter of 2 in., and a feed stream of 100% cis2-butene flowing at a rate of 1 kgmol/h at a temperature of 25°C and 10 atm [17]. (Initially, assume the reactor length equal to 1 m.)

$$\text{Cis2} - \text{butene} \rightarrow \text{trans} - 2 - \text{butene} \tag{8.40}$$

The reaction of 1st-order kinetics, $r_{cis} = kC_{cis}$

$$k = Ae^{E/RT}\ A = 13.8, E = 0, \text{basis units} = \text{kmol/m}^3,$$

$$\text{Rate units} = \text{kmol/m}^3 \cdot \text{h}$$

Fluid package: *NRTL*

P.8.9 It is desired to evaluate the design of a reactor for the production of **allyl chloride** according to the main reaction.

$$Cl_2 + C_3H_6 \rightarrow CH_2 = CH - CH_2Cl + HCl \tag{8.41}$$

And the secondary reaction produces 1,2-dichloropropane

$$Cl_2 + C_3H_6 \rightarrow CH_2Cl - CHCl - CH_3 \tag{8.42}$$

The reaction mixture enters the reactor at 392°F at a rate of 0.85 lbmol/h and contains 4 mol of propylene for every mole of chlorine. With a value of 29.4 psia, the operating pressure can be assumed to be constant and equal to the feed. In the beginning, it is anticipated to use a jacketed, 2-in. diameter tubular reactor that uses boiling ethylene glycol as a coolant to maintain a constant interior temperature at 392°F. The heat transfer coefficient inside is 5 Btu/h ft$^{2°}$F.

The two equations of speed reactions, in lbmol/h ft^3, are as follows [3]:

$$r_1 = 206,000e^{\frac{-27,200}{RT}} P_{C_3H_6} P_{Cl_2} \tag{8.43}$$

$T(°R)$ Main Reaction

$$r_2 = 11.7e^{\frac{-6860}{RT}} P_{C_3H_6} P_{Cl_2} \tag{8.44}$$

$T(°R)$ Secondary Reaction

The activation energy is in Btu/lbmol and the partial pressure in atm. Fluid package: *PRSV*

a) Run a simulation of the reactor's operation using ethylene glycol as a coolant to see how the conversion changes with reactor length. (Presume the starting length to be 2 ft.)
b) Run a simulation of the reactor's adiabatic operation to see how the conversion changes as the reactor's length.
c) Compare the results from earlier experiments with those that will come from a 0.83 ft^3 CSTR reactor.

P.8.10 Calculate the theoretical flame temperature when ethylene at 25°C and 1 atm is burned with:

a) The theoretical amount of air at 25°C, Conversion 100%.
b) 25% excess air at 25°C, Conversion 100%.
c) 50% excess air at 25°C, Conversion 100%.
d) 100% excess air at 25°C, Conversion 100%.
e) 50% excess air preheated to 500°C, Conversion 100%.
f) The conversion of ethylene is 80% in branch (a).

Where the reaction is [15]:

$$C_2H_4 + 3O_2 \rightarrow 2CO_2 + 2H_2O \tag{8.45}$$

Ans. (a) The results of simulation are shown in Figure 8.67.

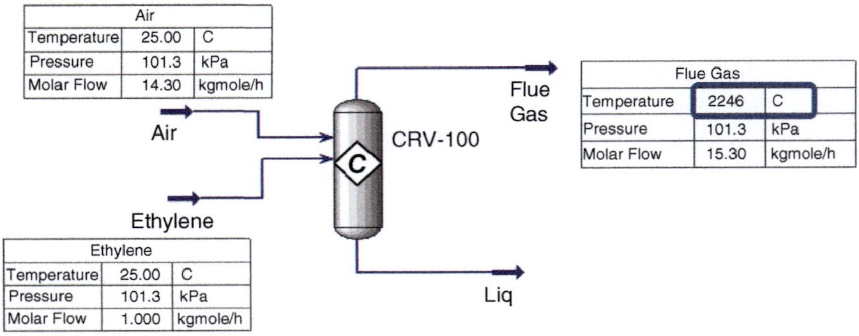

Figure 8.67 Answer of P.8.10 branch (a).

P.8.11 Investigate the effect of temperature (300–600°C) on the conversion of CH_4 in the Gibbs reactor according to the following reactions:

$$CH_4 + H_2O \rightarrow CO + 3H_2 \tag{8.46}$$

$$CO + H_2O \rightarrow CO_2 + H_2 \tag{8.47}$$

where the feed enters the reactor at 300°C, 1 atm and molar flow rate of 5 kmol/h and 17.5 for CH_4 and H_2O, respectively [10].

References

1 (2017). *Aspen HYSYS Unit Operations V10, Reference Guide*. Aspen Technology, Inc.
2 (1981–2005). *HYSYS 2004.2, Tutorial and Application*. New Jersey: Aspen Technology, Inc.
3 Chaves, I.'n.D.'o.G. (2016). *Process Analysis and Simulation in Chemical Engineering*, ©. Springer.
4 House, J.E. (2007). *Principles of Chemical Kinetics*, 2e. San Diego: Elsevier.
5 Hamid, M.K.A. (2007). *HYSYS®: An Introduction to Chemical Engineering Simulation For UTM Degree++ Program*. University Technology Malaysia.
6 (2017). *EHY202 Aspen HYSYS®: Advanced Process Modeling Topics, Course Number EHY202.10.01, AspenTech Customer Education, Training Manual*. Aspen Technology, Inc. 20 Crosby Drive, Bedford, Massachusetts 01730, USA.

7 Froment, G. and Bischoff, K. (1990). *Chemical Reactor Analysis and Design*, 2e. New York: John Wiley & Sons.

8 Harriot, P. (2003). *Chemical Reactor Design*. New York: Marcel Dekker.

9 Rawlings, J.B. and Ekerdt, J.G. (2009). *Chemical Reactor Analysis and Design Fundamentals*, 2e. Madison: Nob Hill Publishing.

10 Haydary, J. (2019). *Chemical Process Design and Simulation Aspen Plus and Aspen HYSYS Applications*. Wiley.

11 (2012). *Ammonia Synthesis with Aspen HYSYS® V8.0, Part 3 Process Economic Analysis*. Aspen Technology, Inc.

12 AL-Malah, K.I.M. (2017). *ASPEN PLUS®, Chemical Engineering Applications*. John Wiley & Sons, Inc. All rights reserved.

13 Felder, R.M., Rousseau, R.W., and Bullard, L.G. *Elementary Principles of Chemical Processes*, 4e. Wiley.

14 Luyben, W.L. (2010). Design and control of the cumene process. *Industrial & Engineering Chemistry Research* 49 (2): 719–734.

15 Fogler, H.S. (2006). *Elements of Chemical Reaction Engineering*, 4e. Pearson Education.

16 Levenspiel, O. (1999). *Chemical Reaction Engineering*, 3e. John Wiley & Sons.

17 Barno, S.K.A., Rashid, S.A., and Abbas, A.S. (2021). Modeling and simulation of an ideal plug flow reactor for synthesis of ethyl oleate using homogeneous acid catalyst. *Chemical and Process Engineering* 42 (1): 53–68.

9

Separation Columns

9.1 Absorption Column

This section introduces the use of HYSYS to model a continuous gas absorption/desorption process in a tray and packed column. The Tray Section is a single-unit activity in the Absorber, and the streams are the bottom liquid and overhead vapor products. The Absorber, which serves as the foundation for all tower layouts, does not have any known specifications. The converged solution that is produced depends on the parameters and make-up of the column feed stream as well as the operating pressure. The vapor and liquid product streams' characteristics are part of the convergent solution [1, 2].

Generally, two types of columns are used: tray columns and packed columns. Packed columns are preferred for absorption/desorption when the column diameter is less than 0.65 m, and the packing height is less than 6 m.

A gas mixture is in contact with a liquid during a mass transfer process known as "gas absorption" with the aim of selectively dissolving one or more gas mixture components in the liquid and creating a solution of the dissolved gas(es) in the liquid. As a result, acetone can be extracted from an air–acetone mixture by immersing the gas in water, where acetone dissolves but air does not. Similar to how ammonia can be recovered from an air–ammonia stream by washing it with water, coke oven gas can be cleaned of benzene and toluene vapor by washing it with oil [3].

The simulator has an icon that is used to calculate the number of trays of column, and, through which, one can calculate the height of the packed column, but there is no icon ready to calculate the height of packed column.

Chemical Process Simulations using Aspen Hysys, First Edition. Khalid W. Hameed.

The equation that is used to find the number of trays/stages in the absorption process is [4]:

$$\text{Fraction of solute } i \text{ absorbed} = \frac{A_i^{N+1} - A_i}{A_i^{N+1} - 1} \tag{9.1}$$

where

A_i is the absorption factor = L_m/mG_m
L_m is the molar flow rate of liquid per unit cross-section area of the column, kmol/m^{2}s.
G_m is the molar flow rate of inert gas per unit cross-section area of the column, kmol/m^{2}s.
m is the equilibrium constant (-).

9.1.1 Problem Statement

A gas absorber is used to remove 99% of the SO_2 from the air–gas stream with pure water. The inlet gas stream molar flow rate is 206 kmol/h. The gas stream contains 3 mol% SO_2. The liquid water mass flow rate is 1.3×10^5 kg/h. The temperature and pressure of the inlet gas and liquid are 20 °C and 1 atm, respectively, and the pressure drop through the column is zero. Determine the number of theoretical stages required to achieve the required exit gas stream specifications. Design a tray column and a packed column for this duty [4].

9.1.2 Simulation

1) Open a new case and enter the following values:
   Components: Water, Air, and SO_2
   Fluid Package: Peng–Robinson
2) Enter the **Simulation** Environment.
3) Add a new Material streams with the following values:

Stream Name	Gas_In	Liquid_In
Temperature	20 °C	20 °C
Pressure	1 atm	1 atm
Flow rate	206 kmol/h	1.3*10^5 kg/h
Composition	3% SO_2, 97% Air by mole	100% water

4) Add the Absorber icon from the object palette/Column Operations group, which looks like this icon .

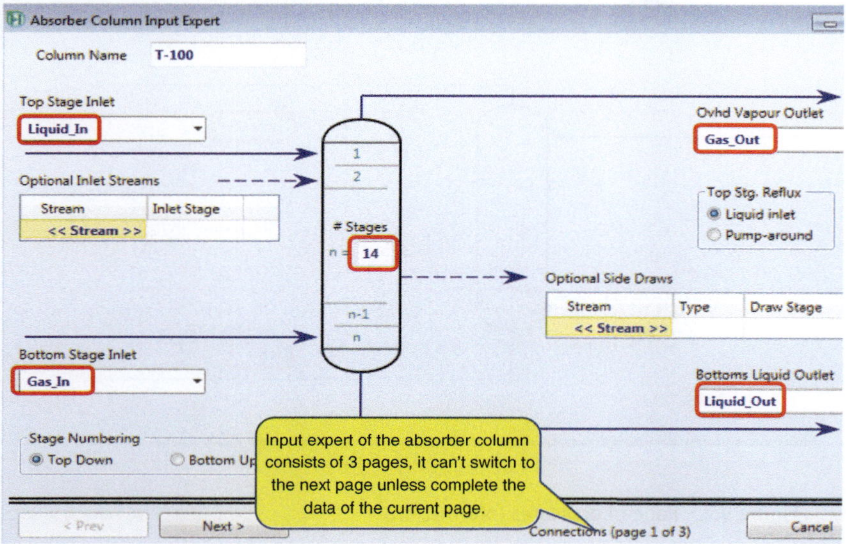

Figure 9.1 Input expert of the absorber column, connecting the inlet and outlet streams.

5) On the Connections page, connect the streams as shown in Figure 9.1. Here, one can assume the number of stages and monitor the composition of outlet gas up to remaining 1% of the gas inlet. So one can assume the number of stages = 14, and one must enter the liquid from the top and gas from the bottom of the column (*i.e. the higher-density stream enters from the top of the column, and the lower-density stream enters from the bottom of the column*).

   So, Top Stage Inlet: Liquid_In, Bottom Stage Inlet: Gas_In.

   Ovhd Vapor Outlet: Gas_Out, Bottoms Liquid Outlet: Liquid_Out.
6) Click **Next**, enter the pressures of the outlets as shown in Figure 9.2.
7) Click **Next**, and leave the temperature information empty, then click the **Done** … button, where HYSYS will bring up a window as shown in Figure 9.3.
8) To run the column: Click the **Run** button near the bottom of the window of Figure 9.3. It should become green for the red Unconverged box. If all the above procedure was followed, the column will converge. However, the result of the gas-out composition should contain ($0.01 \times 0.03 \times 206 = 0.0618$ kmol/h or less). The molar flow rate of SO_2 in the Gas_Out is 0.1407 kmol/h.

So, increase the no. of stages to 20, and re-enter the pressure to 101.3 kPa in the Pn cell, and see the results again, Figure 9.4.

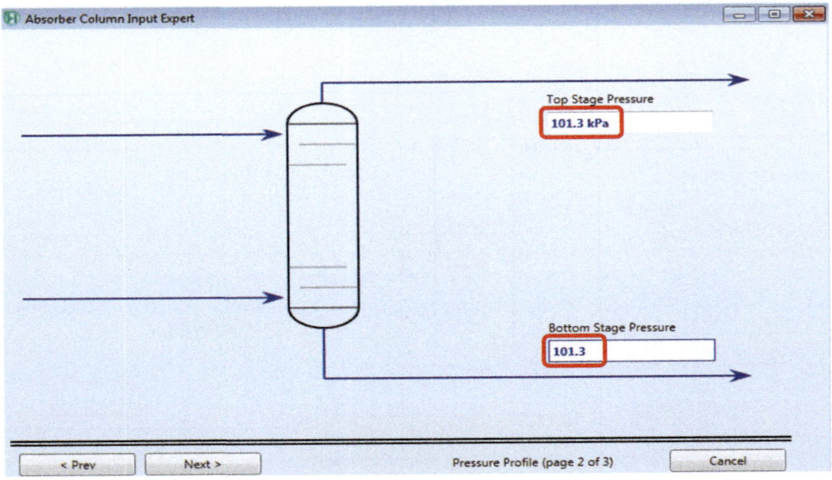

Figure 9.2 Page of pressures information.

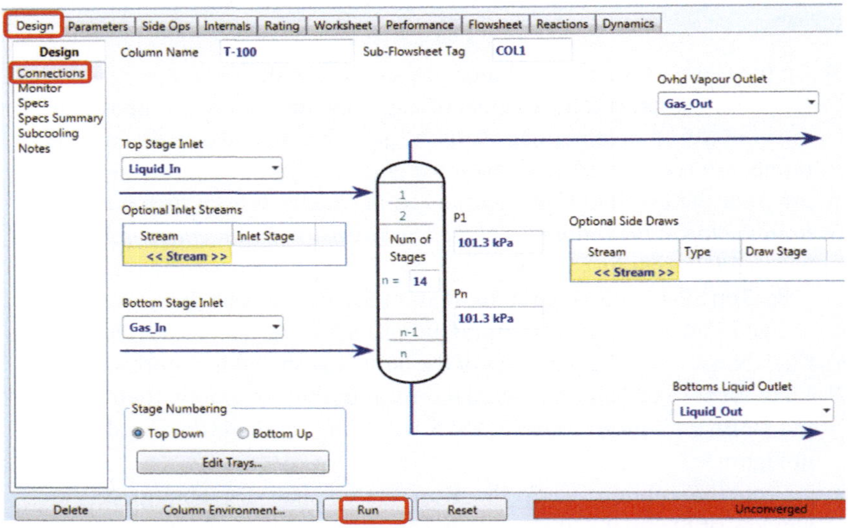

Figure 9.3 Connection page of the absorber column is now ready to run.

***Note**: To increase the separation efficiency, instead of increasing the number of trays, one can increase the flow rate of the liquid water, increase the column pressure, and/or decrease the liquid's temperature.*

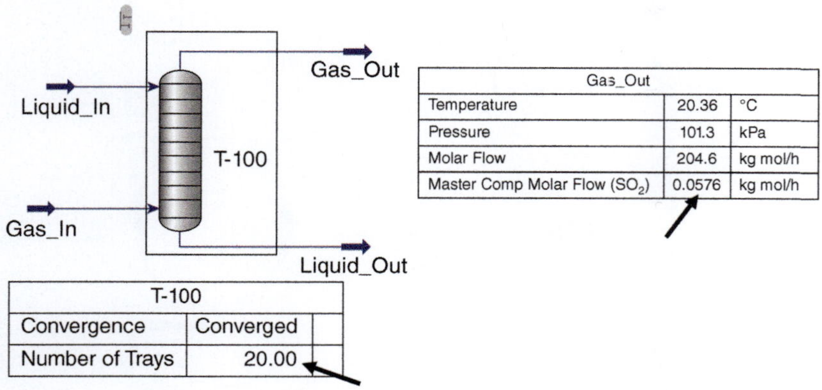

Figure 9.4 Molar flow rate of SO_2 in the exit gas when increasing the number of stages to 20.

9.1.3 Design of Column

9.1.3.1 Tray Column

After increasing the number of stages to 20 and running the column, click the **Internal** tab, Figure 9.5 will appear. Click **Auto Section** button; the Aspen Hysys will insert the Sieve trays as default.

One can change the type of the tray from **Tray/Packing Type** group.

To see the details of results, click the **View** button. The results will appear as shown in Figure 9.6.

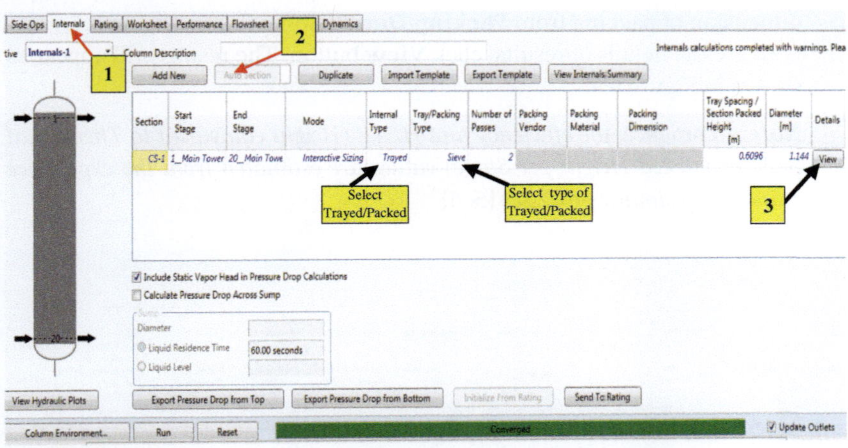

Figure 9.5 Internal tab of the absorber column: How to perform the design of the column.

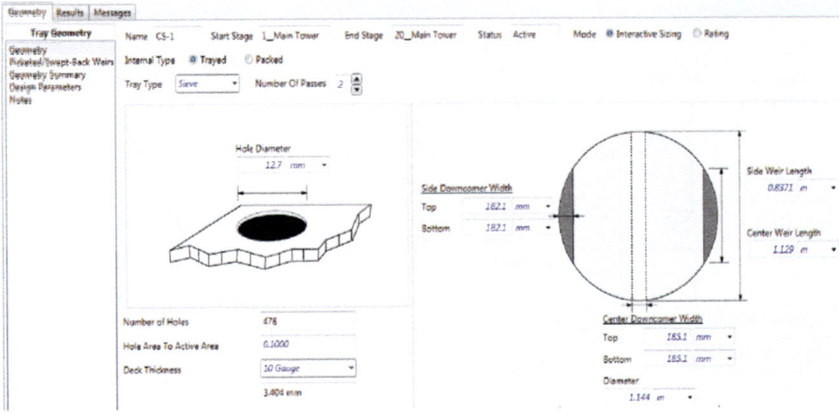

Figure 9.6 Geometry tab of the absorber column/Tray type.

In the **Message tab,** there is a warning error, and one can correct the specification of design as required.

9.1.3.2 Packed Column

To change the tray column to a packed one, from **Internal Type** drop list, change the internal to **Packed** as in Figure 9.7. From the **Tray/Packing Type** drop list, one can change the type of the packing; also one can change the remaining properties of the packing from other properties of the column. One can select the type of packing from **Tray/Packing Type** column, the type of packing material from the **Packing Material** Column, and the dimension of packing from **Packing Dimension** column, Figure 9.7.

To see the details of results, click **View** button. The results will appear as shown in Figure 9.8.

Note: *In practice, the efficiency and HETP (Height equivalent to Theoretical plate or Packed Height per Stage) values are estimated from the experience gained with similar processes* [5, 6].

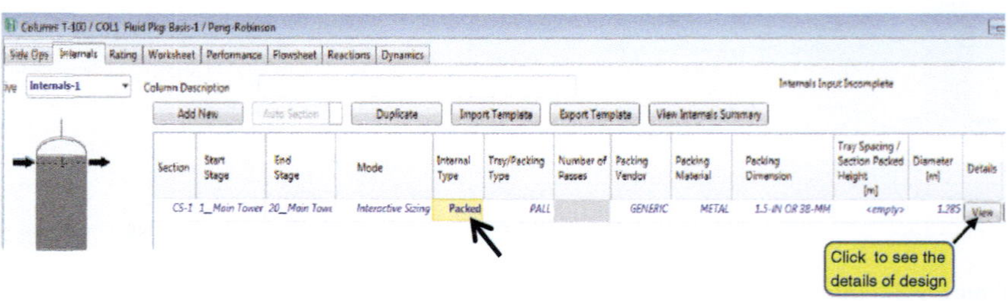

Figure 9.7 Internal tab of the absorber column/packed type.

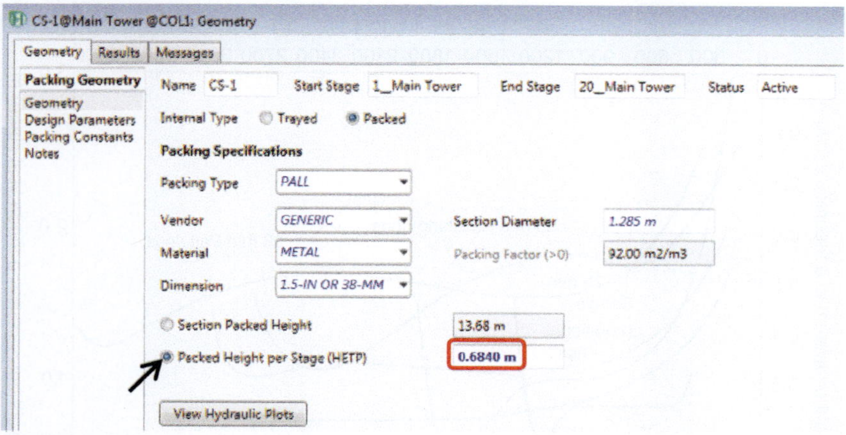

Figure 9.8 Geometry tab of the absorber column/packed type.

The *efficiency* in the absorption is low (< 0.4) but that in the stripping (desorption) is moderate [7].

The efficiency of the absorption column can be entered from **Parameters** tab, **Efficiencies** page.

If **HETP** is not available practically, it can be estimated theoretically using the following equations [3, 5]:

$$\mathrm{HETP} = \frac{G_m}{K_g^{/}a} \frac{\ln\left(\frac{mG_m}{L_m}\right)}{\left(\frac{mG_m}{L_m}\right) - 1} \tag{9.2}$$

$$\mathrm{HETP} = H_{OG} \frac{\ln\left(\frac{mG_m}{L_m}\right)}{\left(\frac{mG_m}{L_m}\right) - 1} = H_{OG} \frac{\ln \varphi}{\varphi - 1} \tag{9.3}$$

where

G_m and L_m are molar flowrates of vapor and liquid per unit area, kmol/m^{2}s.
m is the equilibrium constant, and
$K_g^{/}a$ is the overall mass transfer coefficient, kmol/m^{3}s

HETP can be determined from Figure 9.9 which relates the HETP with the gas/vapor flow rate and the type of packing.

Or simply by using the rules of thumb:

- In the *absence* of data, the HETP = 2 ft for modern random packing and = 1 ft for structured packing [7].
- For small-diameter column (D_T < 2 ft), HETP = D_T (but not less 1 ft), [7].

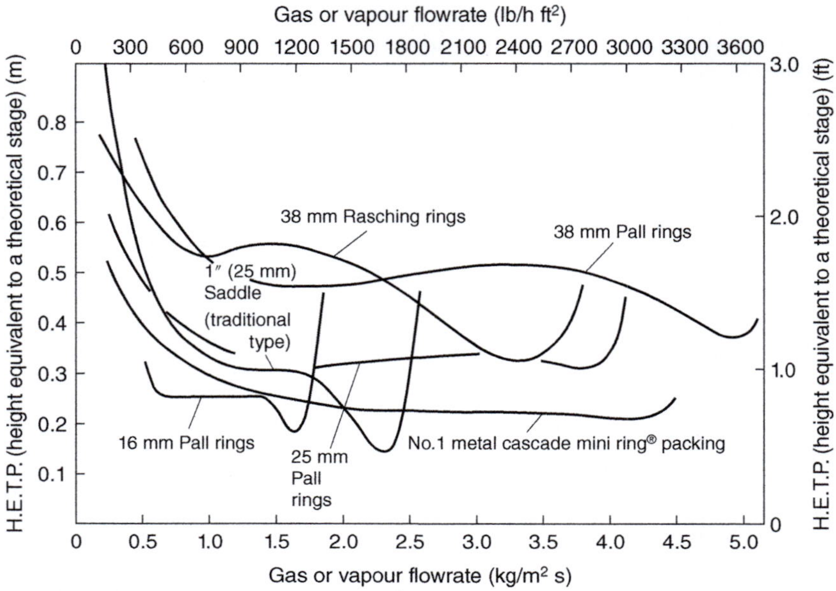

Figure 9.9 Values of HETP for various packing.

Another method to estimate the HETP

$$\text{HETP} = 18d_p \tag{9.4}$$

where d_p is the packing diameter

- For structured packing at low-to-moderate pressures and low-viscosity liquids [8]:

$$\text{HETP} = 100/a + 0.333 \tag{9.5}$$

a = specific surface area of structure packing, ft^2/ft^3.

$$\text{For high pressure: HETP} > 100/a + 0.333 \tag{9.6}$$

- For absorption with a viscous liquid: HETP = 5–6 ft
- For vacuum service:

$$\text{HTEP} = 18d_p + 0.5\,\text{ft} \tag{9.7}$$

In general, lower values of HETP are achieved with smaller-size random packing, particularly in small-diameter columns.

Now, Select **Packed Height per Stage (HETP)** button and change the value of HETP to (0.684 m) according to Eq. (9.4) ($18 \times 38\,\text{mm} = 684\,\text{mm}$)

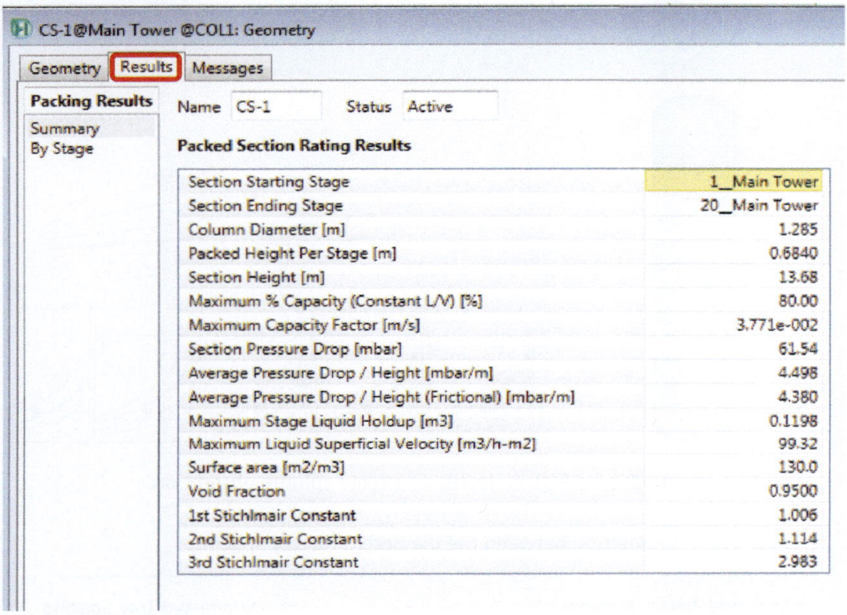

Figure 9.10 Dimensions results of the packed section design of the absorption column.

in order to enable the software to estimate the height of the column, Figure 9.8.

The other results can be seen from **Results** tab as in Figure 9.10.

Write the results required: Section Diameter (m): …………, Section Height (m): ………….,

SO_2 composition in the outlet gas: ……………

To see the Hydraulic Plot in detail, from **Geometry** tab, click **View Hydraulic Plot,** Figure 9.11.

9.1.4 Further Study

Delete the design duty, return the column to 14 stages and change the Liquid_In flowrate from 1.3*10^5 kg/h to 1.4*10^5 kg/h. Run the simulation and see how the column dimensions and exit molar flow rate of SO_2 have changed. Design the column and record the following values.

Section Diameter (m):……, Section Height (m): …, SO_2 flow rate in the outlet gas: …………

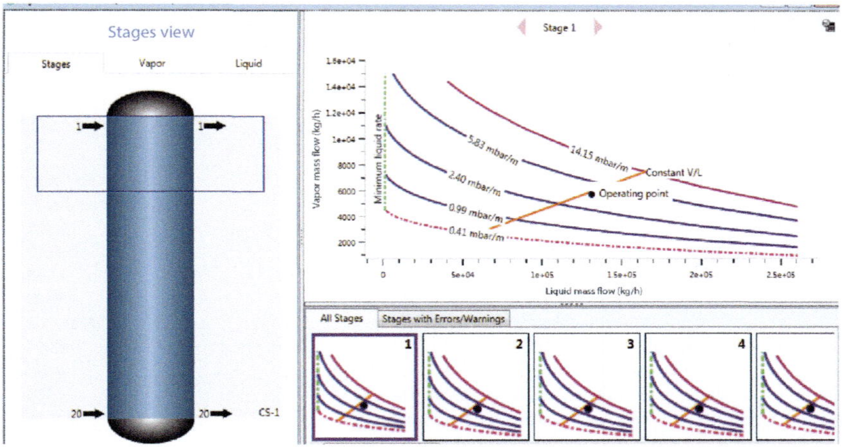

Figure 9.11 Hydraulic Plot in details of packed column absorption.

Table 9.1 The relation between the diameter and tray spacing.

Expected Tower Diameter	Suggested Tray Spacing
—	12 in (30.48 cm) minimum
Up to 4 ft (1.22 m)	18–20 in (45.72–50.8 cm)
4–10 ft (1.22–3.05 m)	24 in (60.96 cm)
10–12 ft (3.05–3.66 m)	30 in (76.2 cm)
12–24 ft (3.66–7.317 m)	36 in (91.44 cm)

Notes

1) *In the case of the tray-type column, the tray spacing is determined by depending on the diameter of the column as shown in Table 9.1.*
2) *In certain types of columns, such as those that separate liquefied air components at cryogenic temperatures, a smaller tray spacing of 4–8 in. may be employed.*
3) *The absorption efficiency is favored when the solvent temperature decreases, the column pressure is high, absorbent flow is high, and the molecular weight of the absorbent is low.*

9.1.5 Summary

In this section, we started with the problem of modeling an absorber that will absorb SO_2 into water in a tray/packed column, where the absorption process is modeled in the absorber column in HYSYS. One can estimate the composition of gas-out and liquid-out, the column height (m), and the column diameter (m). Also, one can reduce the column's height by increasing liquid flow, increasing the column's pressure, and/or reducing liquid temperature.

9.2 Gas Desorption (Stripping)

Gas desorption (stripping): is the reverse of the absorption with transfer of one or more components from the liquid phase to the gas phase. The unit operation that is used in the desorption is the same as that used in the absorption

The equation required to estimate the number of trays/stages in the desorption process is [4]:

$$\text{Fraction of solute is tripped} = 1 - X_N/X_o = \frac{\varphi^{N+1} - \varphi}{\varphi^{N+1} - 1} \qquad (9.8)$$

where

the stripping factor $\phi = mG_m/L_m$

X_N is the concentration or mole fraction of the outlet liquid from the nth stage.

X_o is the concentration or mole fraction of the inlet liquid to the first stage.

9.3 Distillation Column

Distillation is a separation technique based on differences in boiling point, which results in a dissimilarity in vapor pressure. Generally, two types of distillation columns are used: tray columns and packed columns. Typically, tray columns are used when the column diameter is large (more than 0.6 m), and the operating pressure and liquid velocity are high. Packed columns are recommended when the column diameter is less than 0.6 m or when corrosion and foaming is present [6, 8]. Different types of trays, such as sieve, bubble-cap, and valve, can be selected. Sieve trays are most often used.

9.3.1 Vapor–Liquid Equilibrium

It is the basis of all calculations in distillation process; it can be found by several ways according to the type of system and available data [3].

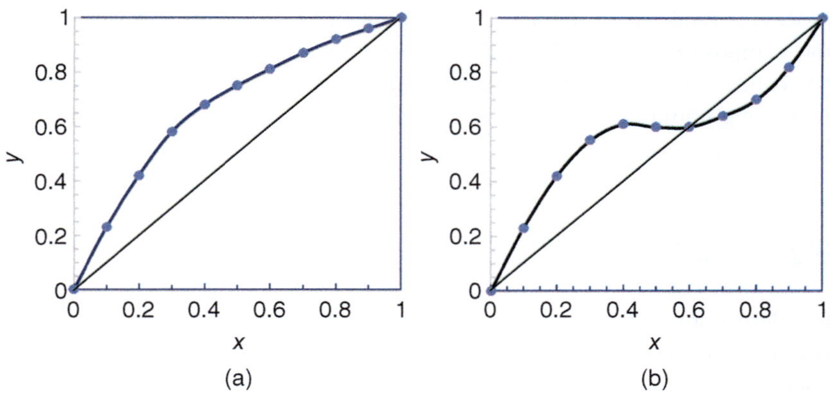

Figure 9.12 Vapor composition versus liquid composition at constant pressure (a) when there is no azeotrope between the two components, (b) when there is azeotrope.

1) From the boiling-point diagram: where many systems are available in the literature survey. Figure 9.12 shows the relation between the composition of the vapor and the liquid at constant pressure, where in figure (a) there is no azeotrope and separation is possible between the two components, while in figure (b) there is an azeotrope between the two components and the separation is impossible.
2) If the boiling-point data are unavailable for a particular system, the system may be:
   a) Ideal solution:
      i)

$$\text{From Dalton's law,} \quad P = P_A + P_B \tag{9.9}$$

and

$$\text{Raoult's law,} \quad P_A = P_A^o \times x_A \tag{9.10}$$

where

P_A, P_B are the partial pressures of components A and B, respectively,

P_A^o, P_B^o are the vapor pressures of components A and B, respectively,

P is the total pressure,

X_A is the mole fraction in the liquid phase.

So, from Eqs. (9.9) and (9.10)

$$P = P_A^o x_A + P_B^o (1 - x_A) \tag{9.11}$$

$$\therefore \boxed{x_A = \frac{P - P_B^o}{P_A^o - P_B^o}} \tag{9.12}$$

$$\text{Partial pressure of an ideal gas is } P_A = y_A \times P \tag{9.13}$$

using Raoult's law, Eq. (9.10)

$$P_A = P_A^o \times x_A = y_A \times P \tag{9.14}$$

$$\therefore \boxed{y_A = \frac{P_A^o}{P} x_A} \tag{9.15}$$

where y_A is the mole fraction in the vapor phase.

To draw the vapor–liquid equilibrium curve, from Eq. (9.12) when the temperature is varied, the vapor pressures P_A^o and P_B^o are varied $\rightarrow x_A$ is varied and then y_A is found from (9.15).

ii) In terms of volatility [3, 4]:

$$\text{Volatility of } A: \frac{P_A}{x_A} = P_A^o \tag{9.16}$$

$$\text{Volatility of } B: \frac{P_B}{x_B} = P_B^o \tag{9.17}$$

$$\text{The relative volatility is } \alpha_{AB} = \frac{P_A^o}{P_B^o} \tag{9.18}$$

$$\rightarrow \alpha_{AB} = \frac{y_A/x_A}{y_B/x_B} = \frac{y_A/x_A}{(1-y_A)/(1-x_A)} \tag{9.19}$$

$$\therefore \boxed{y_A = \frac{\alpha_{AB} x_A}{1 + x_A(\alpha_{AB} - 1)}} \tag{9.20}$$

This is the equilibrium relation for distillation. For separation to be achieved, α_{AB} must be >1.

At a given average α_{AB}, from Eq. (9.20) assume x_A (from 0 to 1) and get y_A.

Equation (9.20) is applied with variance of α less than 15% and takes the average value; if it is varying more, the equilibrium curve must be constructed incrementally by calculating α at several points along the column.

iii)

$$y_A = K_A x_A \quad \text{or} \quad y_i = K_i x_i \tag{9.21}$$

This method is suited for multi-component system, where K is the equilibrium coefficient or distribution coefficient.

b) Non-ideal solution [3, 4]:

i)

$$P_A = \gamma_A P_A^o x_A \tag{9.22}$$

where γ_A is the activity coefficient $= \frac{\textit{chemical activity}}{\textit{molar concentration}}$

And from Eq. (9.13)

$$\rightarrow \boxed{y_A = \gamma_A \frac{P_A^o}{P} x_A} \tag{9.23}$$

And

$$x_A = \frac{P - P_B^o \gamma_B}{P_A^o \gamma_A - P_B^o \gamma_B} \tag{9.24}$$

When the temperature is varied, the vapor pressure and activity coefficient are varied accordingly; from Eq. (9.24) the mole fraction in the liquid phase is obtained, and from Eq. (9.23) the mole fraction in the vapor is also obtained.

ii)

$$y_A = \gamma_A K_A x_A \quad \text{or} \quad y_i = \gamma_i K_i x_i \tag{9.25}$$

9.3.2 Test the Azeotrope

An azeotrope or a constant heating-point mixture is a mixture of two or more liquids whose proportions cannot be altered or changed by simple distillation. This happens because when an azeotrope is boiled, the vapor has the same proportions of constituents as the unboiled mixture [3, 9].

Before initializing the distillation column, one must know if there is an azeotrope between the components that required to be separated since the operation is different in case of azeotrope. To illustrate this phenomenon, one can take an example.

1) Open the Aspen Hysys.
2) Open a new case, then add the components: toluene, benzene, water, and ethanol.
   Fluid Package: NRTL, enter the Simulation environment.
3) From the object palette, Custom Group: select **Equilibrium Unit Operation** 

op-100, Figure 9.13, and drag it to the simulation area.
4) Open the icon (Op-100) by double-clicking on it. From the **Available** group, add ethanol, then water to the **Selected** group (*the less boiling-point component is added firstly*), Figure 9.14.
5) From the **plot Data**, select *XY* Plot.

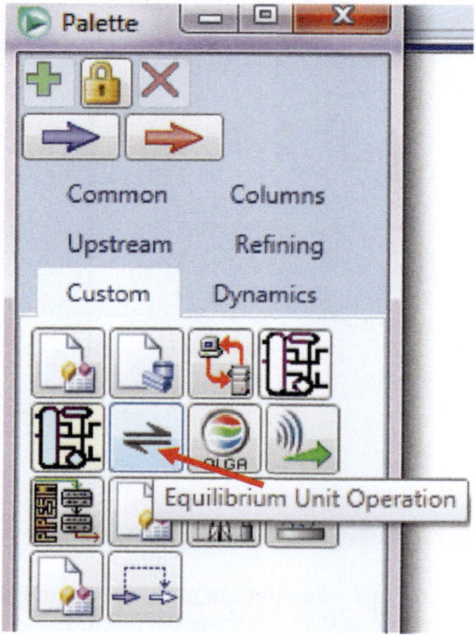

Figure 9.13 Equilibrium Unit Operation in the object palette.

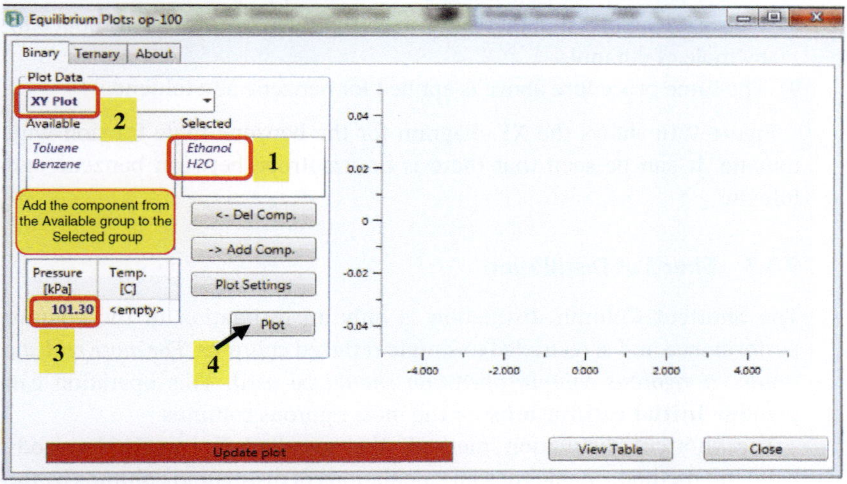

Figure 9.14 Procedure of equilibrium plotting data.

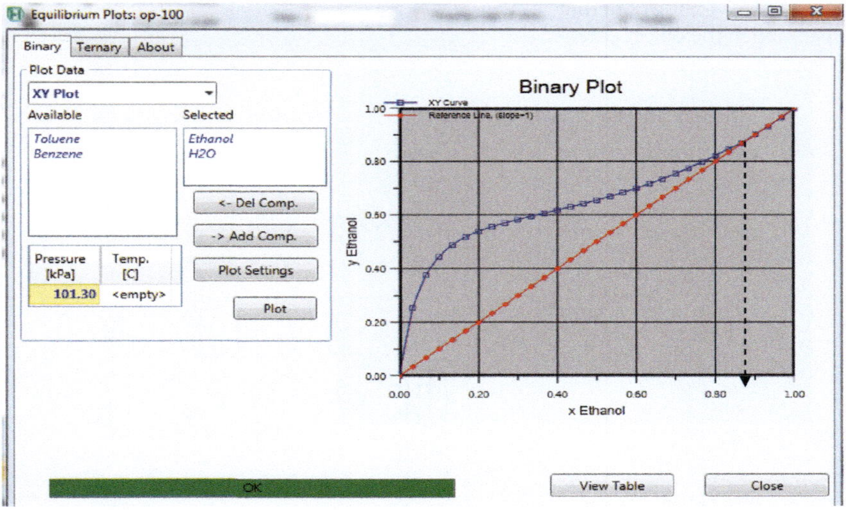

Figure 9.15 Equilibrium plotting data of the ethanol–water that shows the azeotrope at 0.87 of ethanol in mixture.

6) Enter the operating pressure of the column (e.g. 101.3 kPa)
7) Then click the **Plot** button, Figure 9.15 will appear.
8) Figure 9.15 shows that there is an azeotrope at a concentration about 0.87 by mole of ethanol.
9) The same procedure above is applied for benzene and toluene.

Figure 9.16 shows the *XY* diagram for the benzene mole fraction with toluene. It can be seen that there is *no* azeotrope between benzene and toluene.

9.3.3 ShortCut Distillation

The Shortcut Column distillation is only an estimation of the column performance and is restricted to simple refluxed columns. *For more realistic results, a rigorous column operation should be used.* This operation can **provide initial estimations** for the most rigorous columns.

The shortcut distillation method, the so-called Fenske–Underwood–Gilliland method, enables design calculations of distillation columns by the following procedure:

a) selection of key components,
b) estimation of non-key component distribution,
c) estimation of pressures in the column and calculation of condenser and reboiler temperatures,

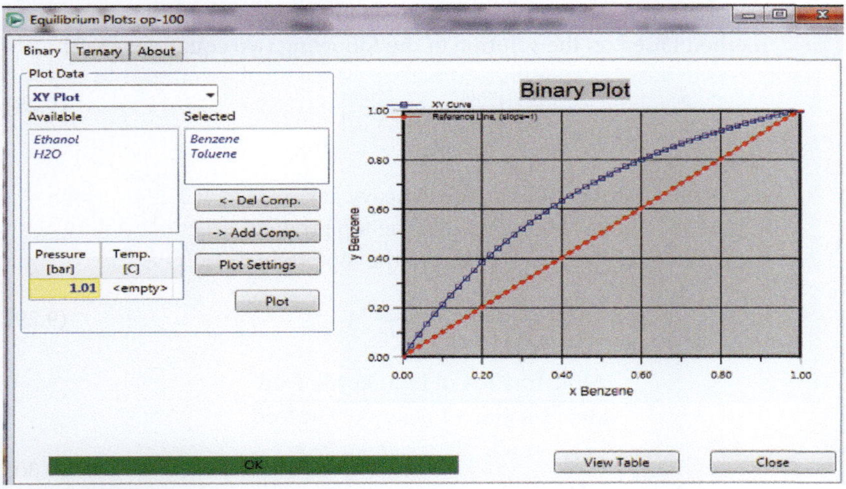

Figure 9.16 Equilibrium plotting data of the benzene-toluene that shows no azeotrope.

d) calculation of the minimum number of theoretical stages, $N_{\min}$, using the Fenske equation [5, 6]:

$$N_{\min} = \frac{\log\left[\left(\frac{x_{LK}}{x_{HK}}\right)_D\left(\frac{x_{HK}}{x_{LK}}\right)_B\right]}{\log \alpha_{LK}} \tag{9.26}$$

where

x_{LK} is the mole fraction of light key (lightest component appearing in the bottom).

x_{HK} is the mole fraction of heavy key (heaviest component appearing in the distillate).

D is the distillate stream.

B is the bottom stream.

α_{LK} is the relative volatility of light key component with respect to heavy key.

e) Distribution of non-key components using the Fenske or Hengstebeck equation [5, 6]:

$$\ln \frac{x_{iD}}{x_{iB}} = N_{\min} \ln \alpha_{iHK} + \ln \frac{x_{HKD}}{x_F} \tag{9.27}$$

where

x_{HKD} is the mole fraction of heavy key in distillate,

x_{iD} is the mole fraction of i component in distillate,

x_{iB} is the mole fraction of i component in bottom,

α_{iHK} is the relative volatility of i component to heavy key.

f) Calculation of the minimum reflux ratio, $R_{\min}$, using the Underwood method based on the solution of the following two equations [5, 7]:

$$R_{\min} = \frac{1}{(\alpha_{AB} - 1)} \left[\frac{x_{dA}}{x_{nA}} - \alpha_{AB} \frac{x_{dB}}{x_{nB}} \right] \tag{9.28}$$

where

α_{AB} is the volatility of light key to heavy key.
x_{dA} is the light key component in the top product.
x_{dB} is the heavy key component in the top product.

$$x_{nA} = \frac{r_f}{(1 + r_f)\left(1 + \sum \alpha x_{fh}\right)} \tag{9.29}$$

$$r_f = \frac{\text{Mole fraction of light key in feed}}{\text{Mole fraction of heavy key in feed}}$$

$$x_{nB} = \frac{x_{nA}}{r_f} \tag{9.30}$$

x_{fh} is the mole fraction of each component in the feed heavier than the heavy key.
α is the volatility of component i relative to the heavy key.

9.3.3.1 Problem Statement

Find the number of trays and reflux ratio of the distillation column required to separate benzene from feed with a flow rate of 850 kg/h at 25 °C and 110 kPa that contains toluene 60% and benzene 40% by mole. The purity of benzene in the distillate must not be less than 98%, and the purity of the toluene in the bottom must not be less than 95%. The condenser and reboiler pressures are 101.3 and 105 kPa, respectively.

9.3.3.2 Simulation

1) Properties Environment: Components: benzene and Toluene. Fluid package: PR.
2) Simulation Environment: Feed stream: $T = 25\,°\text{C}$, $P = 110\,\text{kPa}$, mass flow rate = 850 kg/h,

   And the components mole fraction: benzene = 0.4, and Toluene = 0.6.
3) From the object palette, Columns group; Add the **Short Cut Distillation**  and fill the Connection page, Design tab of the column as in Figure 9.17.
4) Move to the **Parameters** page: Fill the parameters page as in Figure 9.18, where the light key is benzene, and the heavy key is toluene. It is required the purity of benzene in the distillate be 0.98, so the residual is toluene

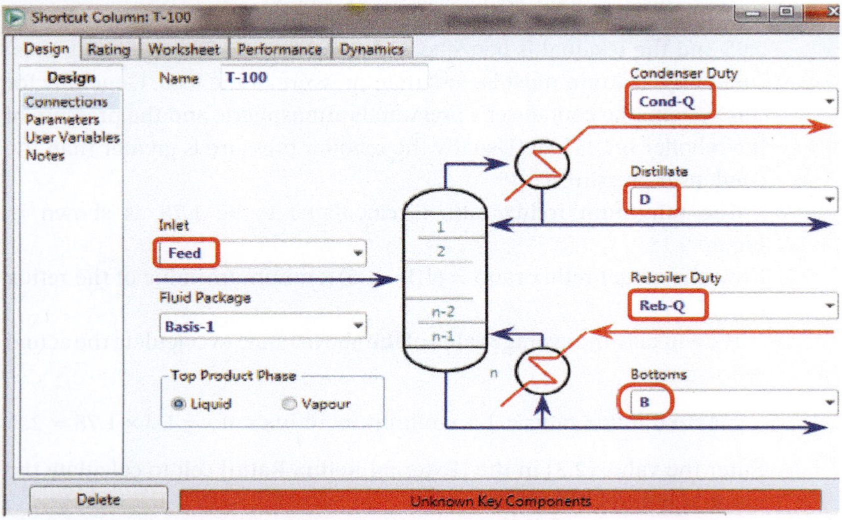

Figure 9.17 Window of the shortcut column, connection page.

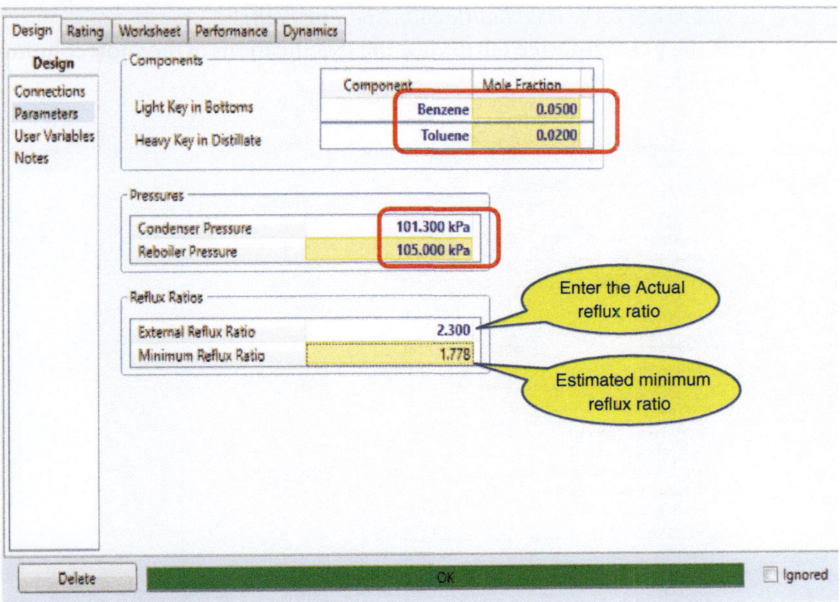

Figure 9.18 Window of the shortcut column/parameters page.

and must be 0.02. While in the bottom, the purity required of toluene is 0.95 and the residual is benzene and equal to 0.05. The operating pressure of the column must be less than pressure of the feed. Generally for this system, the condenser's pressure is atmospheric and the pressure of the reboiler is 105 kPa (Usually the reboiler pressure is greater than the condenser pressure.).

The minimum reflux ratio is calculated to be 1.78 as shown in Figure 9.18.

5) Now, the actual reflux ratio = (1.1 – 1.5) × minimum value of the reflux ratio.

We will take the average value of the above range to calculate the actual reflux ratio:

$$\text{Actual reflux ratio} = 1.3 \times \text{minimum reflux ratio} = 1.3 \times 1.78 = 2.3.$$

6) Enter the value (2.3) in the [External Reflux Ratio] cell to calculate the number of trays.

Note: *For two components (Binary)*
the light key is the more volatile (less boiling point),
the heavy key is the less volatile (more boiling point).

7) Click the **performance** tab to view the specifications of the short column as in Figure 9.19.

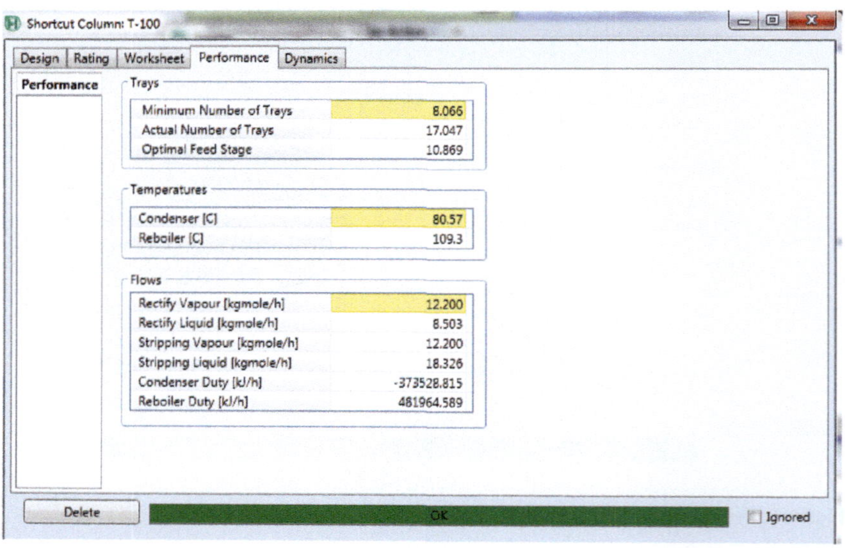

Figure 9.19 Results of the shortcut column in the performance tab.

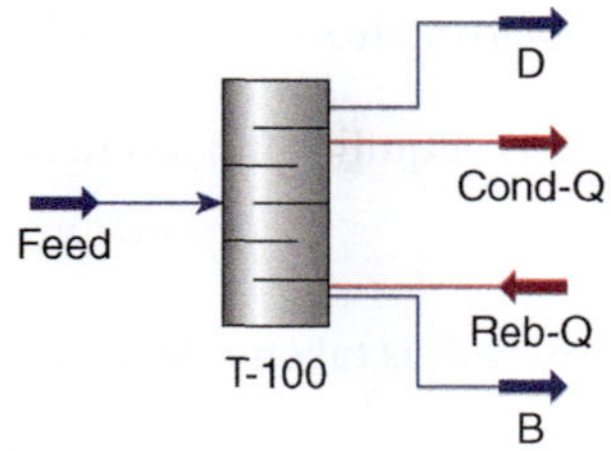

Figure 9.20 PFD of the gshortcut distillation.

The final PFD of shortcut distillation is shown in Figure 9.20.

The number of trays is inversely proportional to the reflux ratio, so one can reduce the number of trays by increasing the reflux ratio and vice versa.

Note: *Increasing the reflux ratio from* R_{min} *affects the capital and operating costs as follows:*

a) *The operating costs rise and are approximately proportional to* $(R + 1)$.
b) *The capital cost initially falls since the number of plates falls off.*
c) *The capital cost rises at high values of R, since there is then only a very small reduction in the number of plates, although the diameter increases because the vapor load becomes greater.*

The associated condenser and reboiler will also be larger and hence more expensive.

9.3.4 Rigorous Distillation Columns

9.3.4.1 Method of Calculation (To Find Number of Plates)

1. Stage-to-stage calculation (Lewis–Sorel method), [5–7].

Overall material balance

$$F = D + W \tag{9.31}$$

where

F is the feed
D is the distillate
W is the waste or bottom streams.

Component material balance:

$$Fx_i = Dx_i + Wx_i \tag{9.32}$$

where x is the mole fraction in the liquid phase, and i represents components A, B, C, etc.

Starting from still (Reboiler) upward

$$\text{Equilibrium relation } y_i = \frac{\alpha_i x_i}{\sum_{i=1}^{i=n} \alpha_i x_i} \tag{9.33}$$

where α_i is taken relative to the *heavy key* component.

$$\alpha_i = K_i/K_{HK}, \quad K_i = y_i/x_i$$

At $x_{i1} = x_{iw}$, find y_{i1} from equilibrium relation (9.33).
Lower operating line

$$y_{i,m} = \frac{L_m}{V_m} x_{i,m+1} - \frac{W}{V_m} x_{i,w} \tag{9.34}$$

Substitute y_{i1} and find x_{i2}
Again, From Eq. (9.33) substitute x_{i2} to find y_{i2}.
Continue up to $x_{i,m} \geq x_{i,F}$.
Now change to upper operating line,

$$y_{i,n} = \frac{L_R}{V_n} x_{n+1} + \frac{D}{V_n} x_{i,D} \tag{9.35}$$

Do not forget equilibrium relation.
where

L_m is the molar flow rate of the liquid in the column under feed point,
V_m is the molar flow rate of the vapor in the column under feed point,
L_R is the molar flow rate of the liquid in the column above feed point and equal to that returns to the column from condenser,
V_n is the molar flow rate of the vapor in the column above feed point,
$x_{i,w}$ is the mole fraction of i component in the waste that leaves the reboiler,
$x_{i,D}$ is the mole fraction of i component in the distillate that leaves the condenser.

Note: *After each calculation, perform the following corrections; thus:*

$$\text{After equilibrium relation } \quad y_i = \frac{y_i}{\sum y_i} \tag{9.36}$$

$$\text{And after operating line } \quad x_i = \frac{x_i}{\sum x_i} \tag{9.37}$$

2. Graphical method (McCabe–Thiele Method) for binary system:
It is a graphical method to determine the number of stages required in the distillation column for binary system. The procedure of calculations is shown in Figure 9.21 and as follows [5–7]:

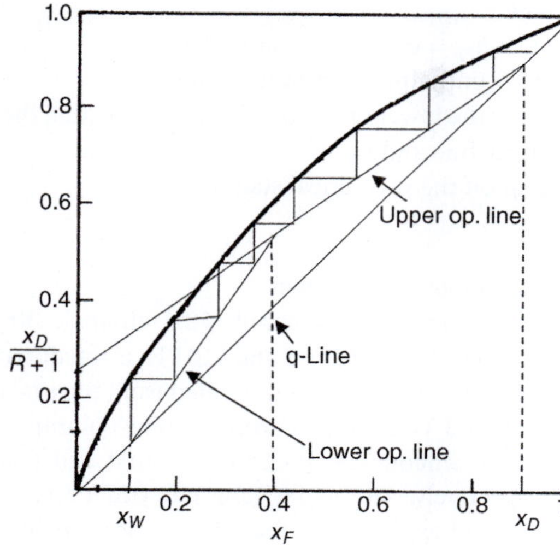

Figure 9.21 Estimating the number of theoretical stages using graphical method.

1) Draw the equilibrium curve from equilibrium data.
2) Draw the upper operating line by connecting between x_D point and $\frac{x_D}{R+1}$ point.
3) Draw the q-line.

   The slope of q-line depends on the condition of feed, where

$$q \text{ is defined as : } q = \frac{C_p \Delta T + \lambda}{\lambda} \tag{9.38}$$

$$\text{And the slope of } q\text{-line} = \frac{q}{q-1} \tag{9.39}$$

where
C_P is the specific heat of feed
ΔT is the difference between the boiling point and the feed's temperature
λ is the latent heat of vaporization of feed.

Table 9.2 shows the slope of q-line according to the condition of feed.

Table 9.2 Slope of q-line according to the feed's condition.

Feed Conditions	Feed Enthalpy, H_F	$q = (C_p\Delta T + \lambda)/\lambda$	Slope $= \frac{q}{q-1}$
Liquid at $T < T_B$	$H_F < H_L$	$q > 1.0$	Slope > 1
Liquid at T_B	$H_F = H_L$	$q = 1.0$	Slope is ∞ (vertical)
Mixture of liq & vap	$H_V > H_F > H_L$	$1 > q > 0$	Negative
Sat. vap. $T = T_B$	$H_F = H_V$	$q = 0$	Slope = 0 horizontal
Super-heated vap $T > T_B$	$H_F > H_V$	$q < 0$	1 > Slope > 0

where
$H_L = h_{Fb}$ = enthalpy of saturated liquid
H_V = enthalpy of saturated vapor.

4) Draw the lower operating line by connecting the cross point of q-line and upper-line and x_W.
5) Step-off the theoretical stages.

9.3.4.2 Problem Statement

A mixture of 10% methanol, 20% ethanol, 30% *n*-propanol, and 40% *n*-butanol by mole at 25 °C and 110 kPa is fractionated using two distillation columns. The methanol is separated from the first column with 0.96 purity, and ethanol is separated from second column with purity of 0.96. The number of theoretical stages of the first and second columns is 18, and 16 respectively. The molar flow rate of feed is 100 kmol/h. Simulate this process and explain how the reflux ratio/number of stages of first column affects the purity of methanol [10].

9.3.4.3 Simulation

1) Open a new case of HYSYS, and enter the components: methanol, ethanol, 1-propanol, and 1-butanol. Select NRTL as a fluid package.
2) Switch to Simulation environment and create the Feed stream with the following conditions:

   T = 25 °C, P = 110 kPa, flow rate = 100 kmol/h, and molar composition: Methanol = 0.1, Ethanol = 0.2, 1-propanol = 0.3, and 1-butanol = 0.4.
3) From the object palette, Columns group, add a rigorous distillation column, Figure 9.22.

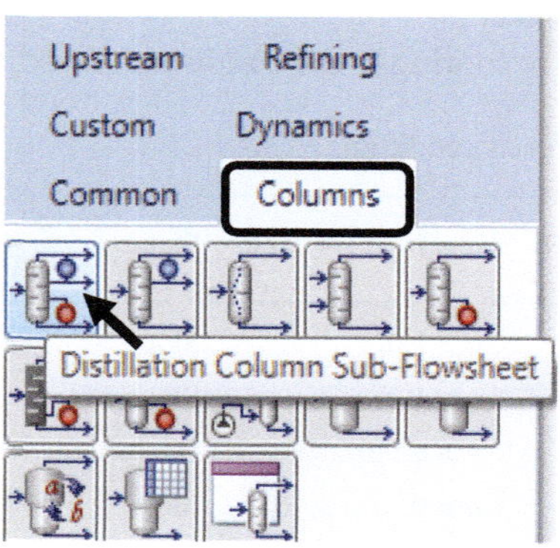

Figure 9.22 Icon of rigorous distillation column in the object palette.

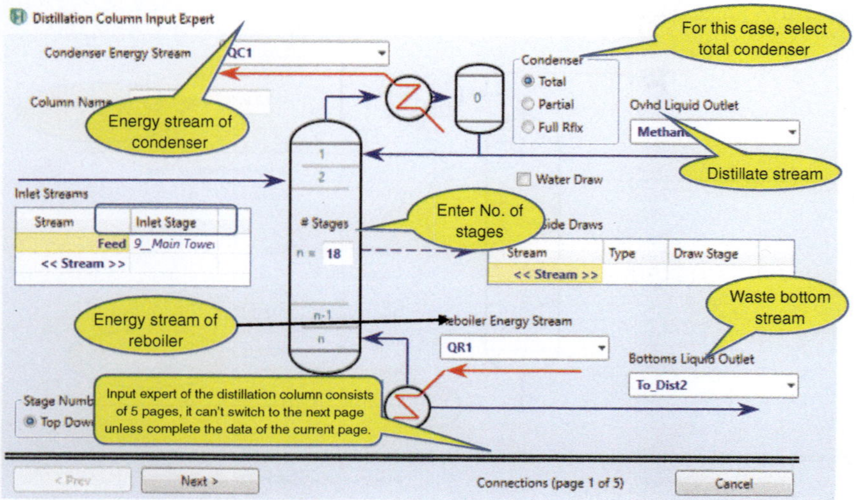

Figure 9.23 Window of Distillation column input expert with connected streams of material and energy.

4) Double-click the distillation column to open it, connect the streams as in Figure 9.23.

   Figure 9.23 shows the Input expert of distillation column, which consists of 5 pages, where it can't switch to the next page unless it completes the streams and data required for the current page.

5) When you complete the first page of the input expert, the "**Next**" button will be active. Click the Next button to move to the second page, where page 2 relates to the Reboiler configuration. Leave this page and click Next to move to page 3.
6) Page 3 is related to the Pressure profile. Enter the pressure of the condenser at 101.3 kPa, and the pressure of reboiler at 110 kPa, and leave the pressure drop of condenser and reboiler at zero, Figure 9.24.
7) After entering the values of pressure, click the Next button to move to the fourth page which is related with temperature estimation. One can specify the temperatures of the condenser, reboiler, and any stage in the column, or leave them to estimate by simulator. Click the Next button to switch to fifth and final page.
8) From page 5 of the Distillation column input expert, which is related to distillate rate and reflux ratio (Specification page). Enter the distillate rate of 10 kmol/h (the amount of methanol in the feed), and leave the reflux ratio cell empty, Figure 9.25.

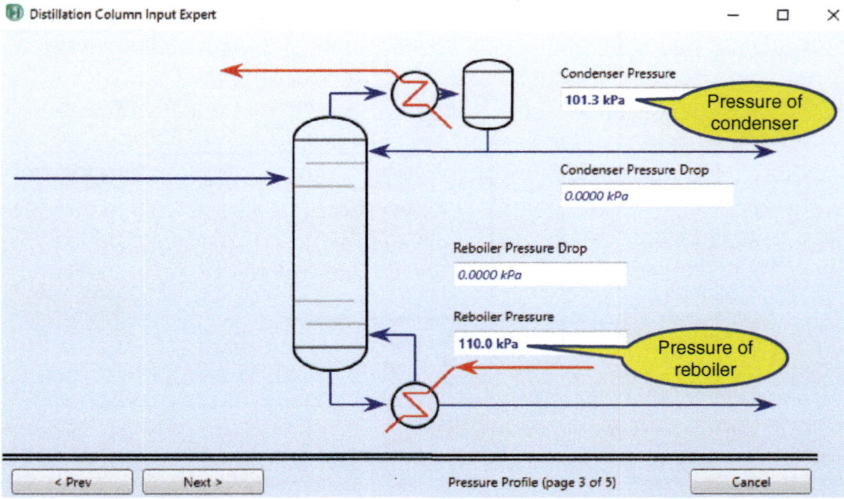

Figure 9.24 Page 3 of 5 (Pressure profile) of Distillation column input expert.

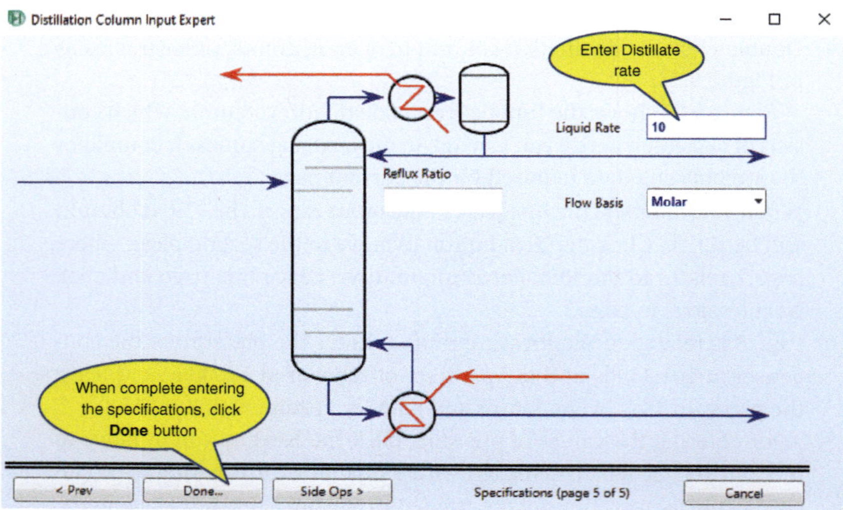

Figure 9.25 Page 5 of 5 (Specifications) of Distillation column input expert, enter distillate rate = 10 kmol/h, and leave cell of Reflux ratio empty.

There are many specifications which can be constrained to converge the distillation column. It can arrive at these specifications from "**Monitor**" or "**Specs**" page of Design table

9) From the fifth page, click the **Done** button; from the new window, click the **Monitor** page of the Design tab, Figure 9.26.
Note, from Figure 9.26, specifications group, the meaning of each specification is as follows:
   - Specified value: In this column, we enter the specified (Constrained) value of the required specifications.
   - Current value: This column will estimate the specified value by simulator.
   - Wt. error: This column represents the difference between the specified value and the current value. For converging the column, the difference must be equal to zero.
   - Active: Activating the box for each value means it must be considered by the simulator, while deactivating the box means the value entered in the specified value column is required to be estimated only.
   - Estimate: The checked boxes indicate to-be-estimated values only.

   For the degree of freedom to be zero, the specified values must be checked boxes under the "**Active**" column. For the current case, the distillate rate has a specified value and checked its box under the Active column, but the Reflux ratio has no specified value but checked

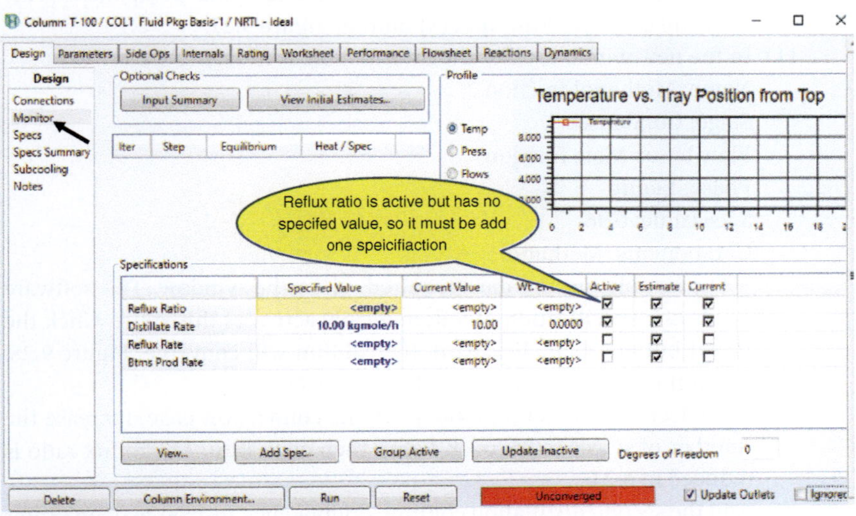

Figure 9.26 Design tab, Monitor page of the column, show the specification group.

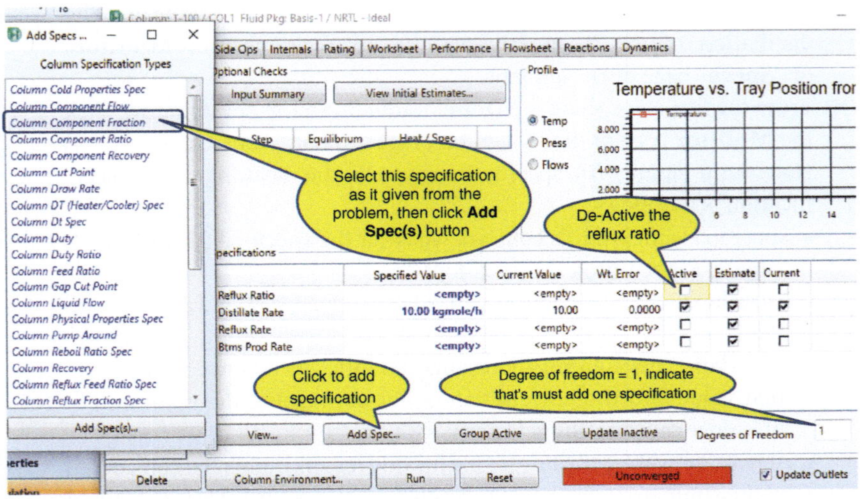

Figure 9.27 Deactivation of Reflux ratio, degree of freedom became 1, and must add one specification.

its box under the Active column. So, deactivate box of the active value of the reflux ratio; the Degree of freedom will equal 1, which indicates adding one specification to the column to be converged.

10) Click the **Add Specification** button, select *Column Component Fraction,* and click the **Add Spec(s)** button, Figure 9.27.
11) In the new window, enter the information as in Figure 9.28:
    Name: Methanol fraction
    Stage: Condenser
    Flow basis: Mole fraction
    Phase: Liquid
    Spec value: 0.96
    Components: Methanol

    On completing the data required, close the window. The software needs one specification; it automatically activates this value. Click the **Reset** button, then click **Run**, the column will converge, Figure 9.28. The reflux ratio is estimated to high value (11.02)

    To decrease the reflux ratio, from the connection page, increase the number of stages to 25, click Reset, then click Run. The reflux ratio is reduced to 6.24.
12) Add the second distillation column, double-click its icon to open. Enter the number of stages = 17, and connect the inlet and outlet streams as in Figure 9.29.

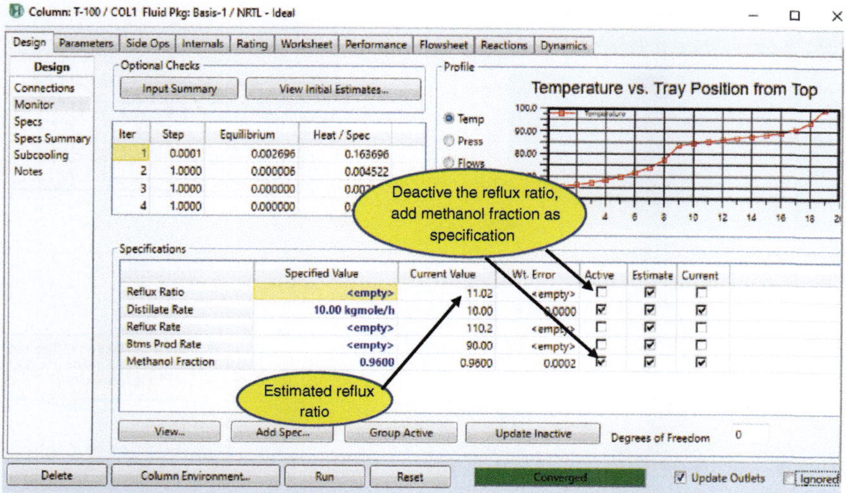

Figure 9.28 Converging the distillation column (T-100), and the reflux ratio is estimated to be 11.02.

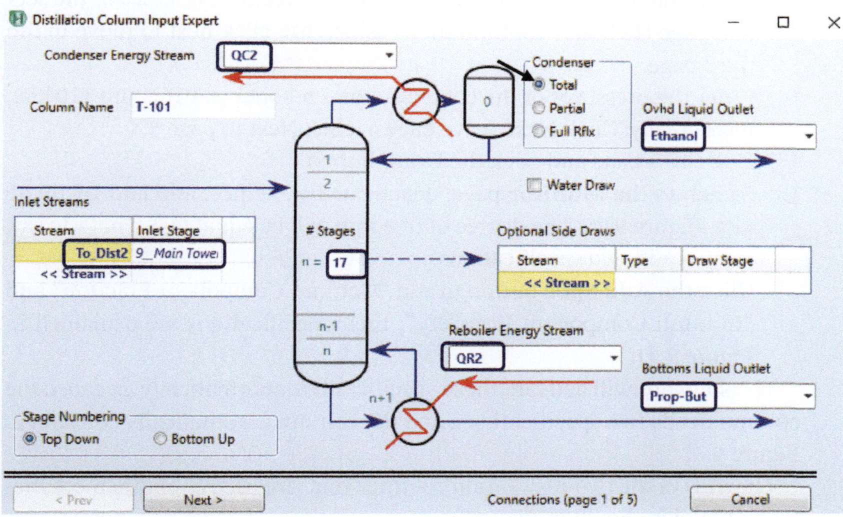

Figure 9.29 Second distillation column with its inlet and outlet streams.

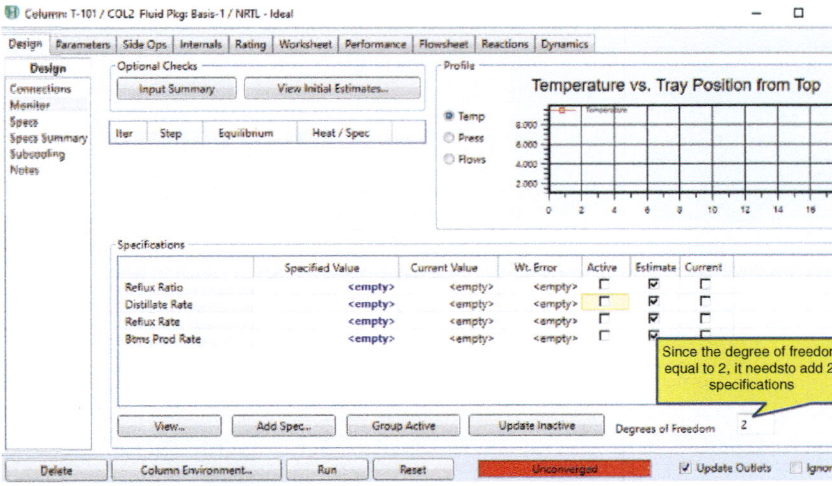

Figure 9.30 Monitor page, Design tab of the second column, Deactivate the Reflux ratio and Distillate rate. Degree of freedom equal to 2, it must add two specifications.

13) When the first page is completed, click the **Next** button, leave the second page (Reboiler configuration) as default, click next to move to the third page.
14) Enter the pressures of the condenser and reboiler at 101.3 and 110 kPa, respectively. Click Next, leave page 4, click Next to page 5.
15) Leave this page and click the **Done** button.
16) Switch to the **Monitor** page, deactivate the Reflux ratio and Distillate rate, Figure 9.30. The degree of freedom will be equal to 2. That's to say, it must add two specifications.
17) Click the **Add Spec** button to add "Column Component Fraction" and "Column Component Recovery"; their specifications are explained in Figure 9.31.

The software will activate these specifications automatically, because the column needs two specifications, and the column automatically converged, Figure 9.32.

The PFD of the two-distillation columns that' are used to obtain methanol and ethanol is shown in Figure 9.33.

9.3.4.4 Design of Distillation Column

The design procedure of the distillation column is similar to the absorption column for tray or packed type. See Section 9.1.3.

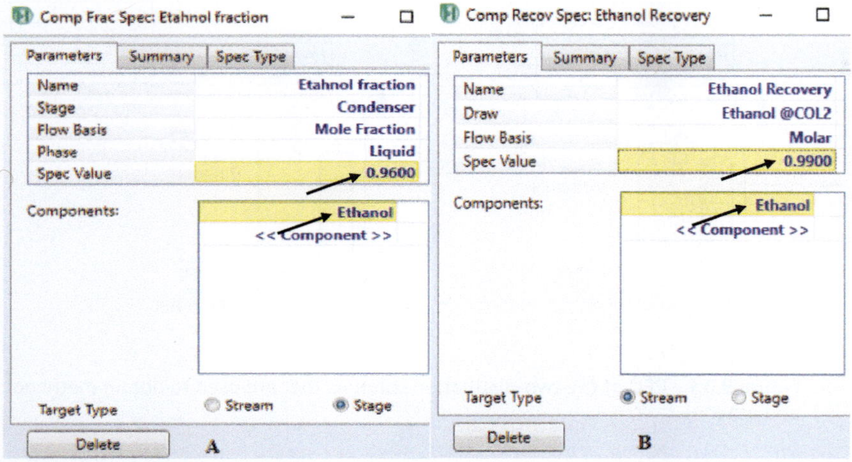

Figure 9.31 Specifications of column that adding; A: Component fraction, and B: Component recovery.

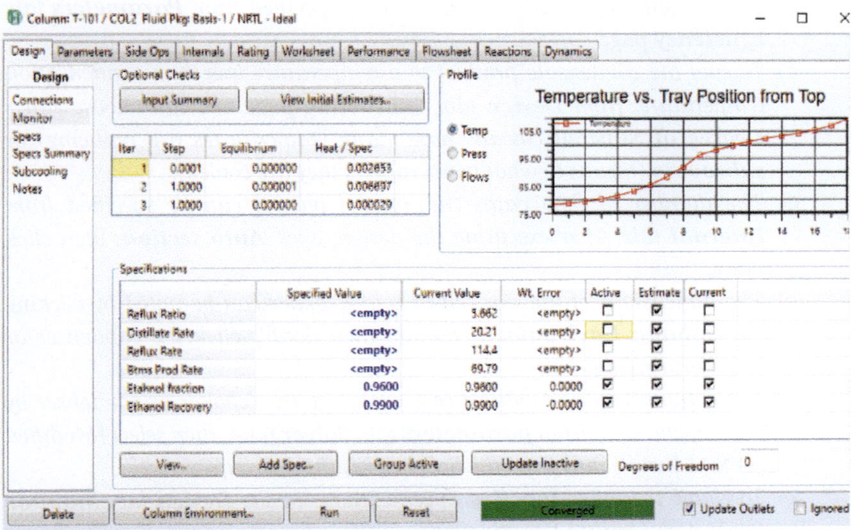

Figure 9.32 Convergence of column after adding the required specifications.

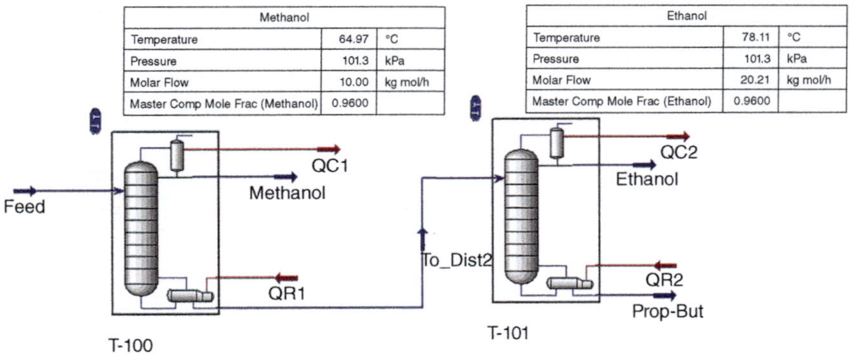

Figure 9.33 PFD of the two-distillation columns that are used to obtain methanol and ethanol.

Notes

a) *In the HYSYS the number of stages = number trays, where the condenser and reboiler are not included.*
b) *The efficiency of the tray column can be specified from* ***Parameters*** *tab,* ***Efficiency*** *page*
c) *To cool the condensate product to a temperature less than condensation temperature, from* ***Design*** *tab,* ***Subcooling*** *page one can choose* <u>*either*</u> ***Degree of Subcool*** *(means how many temperatures are reducing) or* ***subcool to*** *(means to what temperatures they are cooled).*
d) *The number of flow paths (number of passes) can be specified from* ***Internal*** *tab: After executing the design, click* ***Auto section,*** *then click* ***view.***
e) *The dimensions of the Condenser and Reboiler can be noted by clicking the* ***Column Environment*** *button, then double-click the condenser or Reboiler;* ***Rating*** *table*
f) *The column can be modified (e.g. exchange the reboiler or condenser by Heat Exchanger) from* ***parameters*** *tab;* ***Solver*** *page; then select [Modified HYSIM Inside-Out] from the* ***Solving Method*** *Group.*
g) *To Know the other* ***Inactive*** *Specifications: in the* ***Design*** *tab,* ***Monitor*** *Page; click* ***Update Inactive*** *button. The values that appear can be active with any specification to Run the column.*
h) *To Know the Overall Material and Energy balances; in the Home tab of the Ribbon click Flowsheet Summary* Flowsheet Summary.
i) *To Export the Data of the PFD to Excel file: Click* ***Model Summary*** *in the Home tab of the Ribbon.*
j) *There are three cases for condenser operation:*

- ***Total condenser:*** *It means that all rise vapor will condense, some of them will return to the column as reflux and the other will be produced as distillate. A Total condenser has two exit streams:*
  1. *Reflux*
  2. *Distillate*
- ***Partial condenser:*** *It means that some of the rise vapor will condense, and the rest will be produced as Overhead Vapor. Some of the condensed will return to the column as reflux, and the other will be produced as distillate. So, there are two phase products: vapor and liquid. The partial condenser has three exit streams:*
  1. *Overhead vapor*
  2. *Reflux*
  3. *Distillate*
- ***Full-Reflux:*** *It means that some of the rise vapor will condense, and the other will produce as Overhead Vapor. The condensed vapor will fully return to the column as reflux, i.e. the product is only vapor. The Fully Refluxed condenser has two exit streams:*
  1. *Overhead vapor*
  2. *Reflux*

9.3.4.5 Summary

Distillation is a thermal separation process that depends on the difference in boiling point of the liquid components of a mixture. Aspen HYSYS® has a strong calculation module corresponding to distillation columns; this is due to distillation being one of the most important separation operations in process engineering. It is preferable to check the azeotrope of the mixture before initiating separation by distillation. When there is no data about the number of stages or reflux ratio, it is preferable to use shortcut distillation as the initial calculation, then rigorous distillation. It is seen from example 9.3.4.1 that the purity of methanol or ethanol depends on the number of stages and reflux ratio of the column, as well as other parameters such as phase of inlet feed, temperature and pressure of feed, stage of inlet feed, and pressure of the operating column.

9.3.5 Reboiled Absorber column (Stripping Tower)

When the feed is usually a saturated liquid at the boiling point, and the overhead product is the vapor rising from the top plate, the column operates as a stripping tower. A Reboiled Absorber column has a single inlet and two exit streams. When used alone, the Reboiled Absorber has three boundary streams and so requires three Pressure Flow specifications: one pressure and two flow specifications. A pressure specification is always required for the vapor product leaving the column [11].

9.3.5.1 Problem Statement

A liquid feed at the boiling point of 400 kmol/h containing 55 mol% *n*-pentane and 45 mol% *n*-heptane is fed to a stripping tower at 110.0 kPa pressure. It is required to recover 99% of n-pentane as vapor product. Calculate the kmol/h overhead vapor, its composition, and the number of theoretical stages required [11].

9.3.5.2 Simulation

1) Open a new case of Aspen HYSYS, and enter the components: *n*-pentane and *n*-heptane,
   Fluid package: Peng–Robinson
2) Switch to the Simulation environment and add a material stream with the following conditions: Vapor fraction = 0, P = 110 kPa, molar flow rate = 400 kmol/h, and molar composition: 0.55 for *n*-pentane, and 0.45 for *n*-heptane.
3) Add the Reboiled absorber column  from object palette. Double-click on this icon to open it.
4) Initially, it will assume the number of stages = 8. Add the inlet and outlet streams as in Figure 9.34.

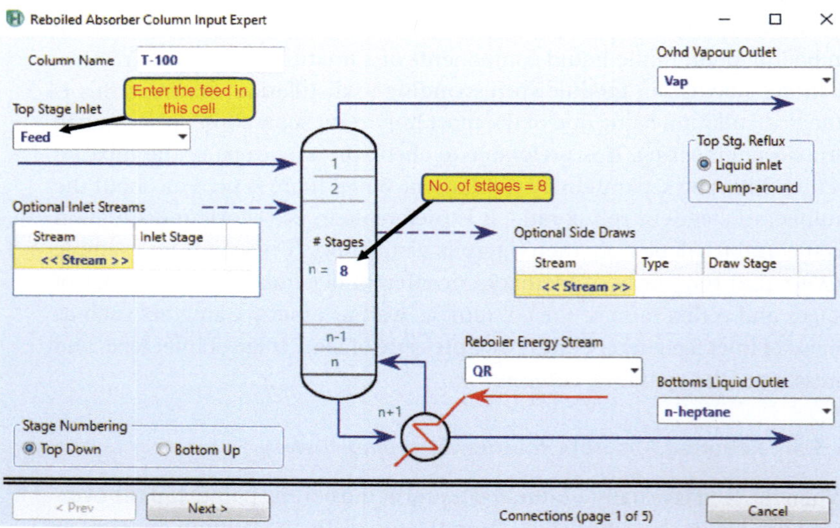

Figure 9.34 Window of Reboiled Absorber Column Input Expert and the data required.

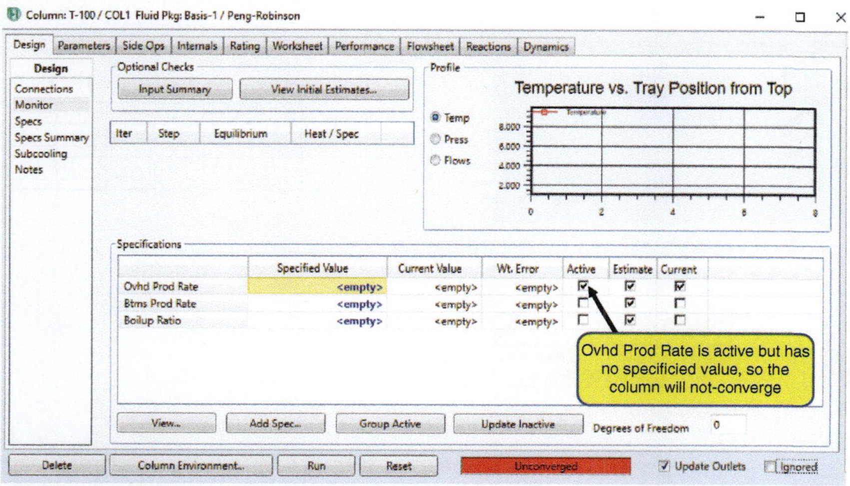

Figure 9.35 Window of Monitor page, Design tab of Reboiled Absorber Column.

5) Click the Next button, then click Next again to move from page 2 to page 3. Enter the top stage pressure and reboiler pressure at 101.3, and 110 kPa, respectively, click Next to move to page 4, again click Next to move to page 5.
6) Click the **Done** button, from the new window, switch to the Monitor page, Figure 9.35.
7) Specify the "Ovhd Prod Rate" equal to 220.0 kmol/h ($0.55 \times 400 = 220$). If the column doesn't run automatically, click **Run** to converge, Figure 9.36. The composition of *n*-pentane in the ovhd stream and *n*-heptane in the bottom stream will equal 0.899 and 0.876, respectively.
8) Change the specification of column as follows: Deactivate the "Ovhd Prod Rate," and the "Degree of freedom" will equal 1. Add a new specification, *Column Component Recovery*, fill the new small window as in Figure 9.37, the column will activate this specification automatically. Click **Run** to converge the column.
9) The PFD of the Reboiled absorber column will look like as in Figure 9.38. After changing the specification, the composition of *n*-pentane in the ovhd stream and *n*-heptane in the bottom stream will now equal 0.899 and 0.986, respectively.

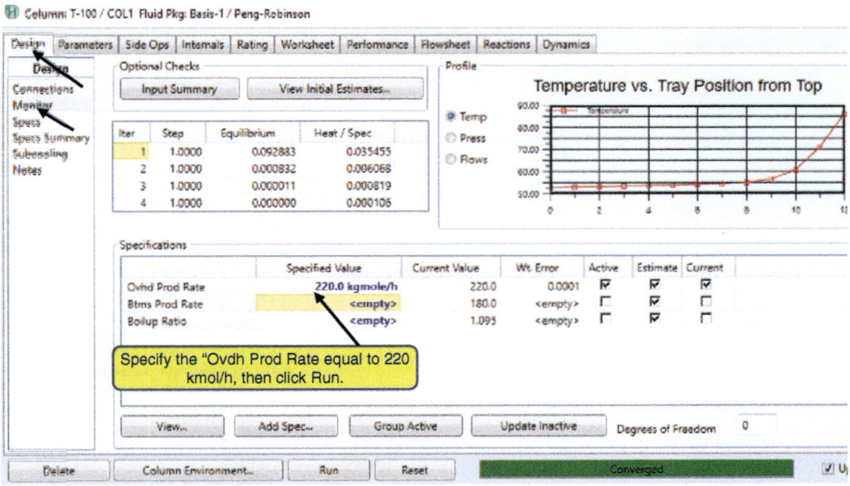

Figure 9.36 Design tab, Monitor page, Specifying the value of "Ovhd Prod Rate" to run the column.

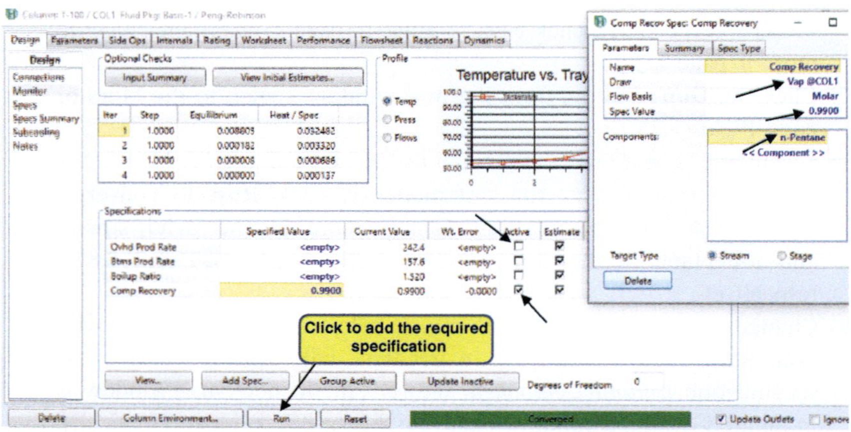

Figure 9.37 Deactivate the "Ovhd Prod Rate" and add a new specification "Component column recovery" for *n*-pentane.

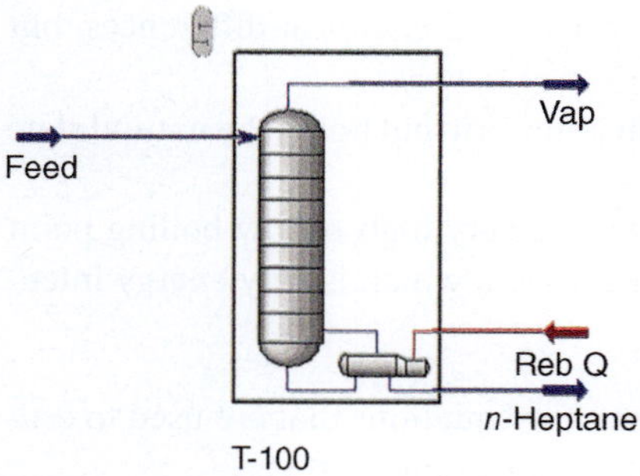

Vap		
Temperature	52.62	°C
Pressure	101.3	kPa
Molar Flow	242.4	kg mol/h
Master Comp Mole Frac (*n*-Pentane)	0.8985	

n-Heptane		
Temperature	99.32	°C
Pressure	110.0	kPa
Molar Flow	157.6	kg mol/h
Master Comp Mole Frac (*n*-Heptane)	0.9860	

Figure 9.38 PFD of the Reboiled absorber column and specification of the Vap stream.

9.4 Extraction Column

Liquid–liquid Extraction process is an operation in which one component (or more) is recovered from a mixture of liquids using a suitable solvent in which the component is soluble. It is employed for recovery and purification of a desired ingredient from the solution in which it was prepared. Extraction may also be used to remove impurities from a feed stream. It is used, for instance, in the pharmaceutical industry, in the processing of coal tar liquids, in the production of fuels in the nuclear industry, and has seen extensive use in the separation of hydrocarbons in the petroleum industry. The liquid-mixture feed and solvent must be at least substantially, if not entirely, miscible to perform this procedure, which essentially involves three stages:

a) intimate contact between the feed mixture and the solvent;
b) separation of the resulting two phases; and
c) extraction and recovery of the solvent from each phase.

Stages (a) and (b) could be combined into a single piece of machinery, like a column, and run constantly. Differential contacting is the term for such an action. The best example of staged equipment for liquid–liquid extraction is a mixer–settler unit, which combines agitation to mix the two liquid phases before settling them by gravity in a different vessel [3, 4].

Extraction is used for one or more of the following reasons [3, 4, 11]:

1) The heat of distillation is undesirable (damage the material or azeotropic mixture).
2) The solute is present in low concentration (most fermentation products).
3) Less energy can be used.

4) Extraction selectivity is favorable because of chemical differences, but relative volatilities overlap.
5) Extraction selectivity is favorable in ionic form but not in the natural state (e.g. citric acid).
6) The components in the mixture have a very high or low boiling point requiring vacuum or cryogenic distillation, which is very energy intensive.

For ***completely immiscible solvents***, the equations that are used to estimate the height of the packed column of extraction are [3]:

$$Z = \frac{L_E^{/}(C_{E2} - C_{E1})}{K_E \cdot a(\Delta C_E)_{lm}} \tag{9.40}$$

$$(\Delta C_E)_{lm} = \frac{\left(C_{E2}^* - C_{E2}\right) - \left(C_{E1}^* - C_{E1}\right)}{\ln \frac{(C_{E2}^* - C_{E2})}{(C_{E1}^* - C_{E1})}} \tag{9.41}$$

$$C_E^* = mC_R \tag{9.42}$$

where

Z is the height of the packed column (m),
$(\Delta C_E)_{lm}$ is the logarithmic mean concentration difference on extract side, kmol/m^3,
$K_E{\cdot}a$ is the overall mass transfer coefficient on extract side, (m/s)(m^2/m^3),
$L_E^{/}$ is the liquid volumetric flow rate of extract phase per unit area, m^3/m^{2}s,
C_E^* is the concentration of solute in the extract phase in equilibrium with C_R, conc. of solute in the raffinate phase, kmol/m^3,
m is the equilibrium constant,

Extract phase: The solvent which has been holding the solute after leaving the equipment.

Raffinate phase: The feed which has lost the solute after leaving the equipment.

Also for complete immiscible solvents, the equations that are used to estimate the number of stages of extraction in case of using mixer–settler are [3, 4]:

For counter-current operation:

$$N = \frac{\log\left[\frac{\lambda - 1}{\frac{X_N}{X_f}} + 1\right]}{\log \lambda} - 1 \tag{9.43}$$

And for cross-current operation:

$$N = \frac{\log\left(\frac{X_N}{X_o}\right)}{\log\left(\frac{\alpha}{\alpha+1}\right)} \tag{9.44}$$

where

$$\lambda = \frac{ms}{a} \tag{9.45}$$

a is the volumetric flow rate of feed solvent A, m^3/s
s is the volumetric flow rate of extract solvent S, m^3/s
X is the concentration or mass ratio of solute B/solvent A
Y is the concentration or mass ratio of solute B/solvent S
N is the number of stages required to extract the solute
X_f is the X_o is the initial concentration of solute in the feed.
$\alpha = 1/\lambda$

For ***partially miscible solvents***, the graphical solution is almost used.

The following example will use a tray column to extract the acetone from the water–acetone mixture.

9.4.1 Problem Statement: Extraction of Acetone–Water with 3-Methylhexane as the Solvent

Water has a high latent heat (heat of vaporization) compared to many other components. For the separation of a water–acetone mixture (50 wt-% each), it may be more energy efficient to use extraction instead of direct distillation. In this example, 3-methylhexane will be utilized as a solvent to extract acetone via liquid–liquid extraction, followed by distillation to remove the solvent from acetone [12].

9.4.2 Simulation

1) Open a new case of Aspen HYSYS and add the components: Acetone, Water, and 3-methylhexane.
   Fluid package: PRSV.
2) Enter the Simulation Environment.
3) Add a **Mixer:** Inlet = Make Up, Outlet = Solvent.
   - **Make Up** stream conditions: $T = 25\,°C$, $P = 1$ bar; composition = 1 for 3-Mhexane.
   - **Solvent** stream conditions: mass flow = 150 kg/h.

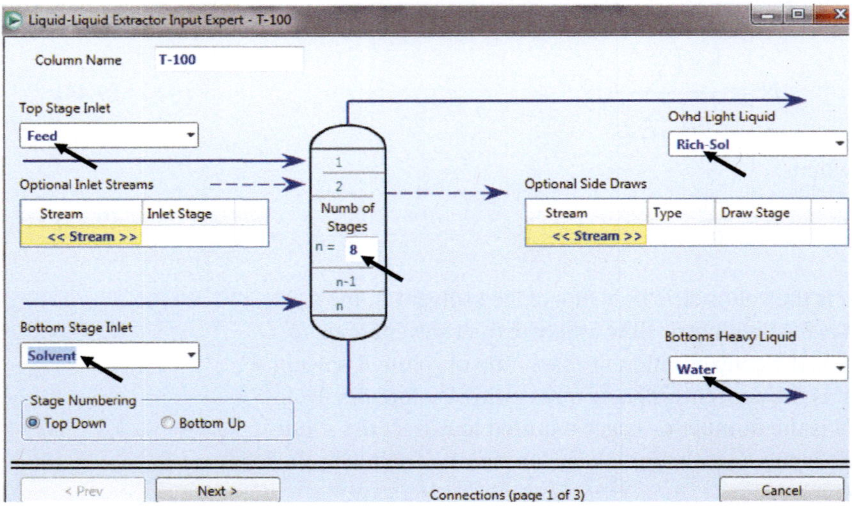

Figure 9.39 Liquid–liquid extractor input expert/connection page.

4) From the object palette, add the Liquid–Liquid Extractor column 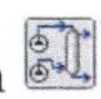, fill the first page as in Figure 9.39: The number of trays = 8, and remember that the *higher liquid density must enter from the top of the column, and lighter liquid density must enter from the bottom*, where the density of each stream can be checked by double-clicking on that stream, Worksheet tab, Properties page.
5) Click Next, enter the pressure for top and bottom equal 1 bar.
6) Next, enter the top temp = 25 °C, click Done.
7) Define the Feed stream: $T = 25\,°C$, $P = 1$ bar, and mass flow = 100 kg/h,
   Composition: Mass fraction = 0.5 for both acetone and water.
8) Click **Run** for the column.
9) Check the composition of the Water stream exiting the bottom of the column. You will see that the mole fraction for water is 1.
10) Insert a Distillation Column; number of stages is 50. On Page 1 enter the information as in Figure 9.40, and click Next when completed.
11) Leave page 2. Next, enter the Condenser and Reboiler Pressures of 1 bar, and click Next.
12) On Page 4, leave all fields blank, click Done on the final page to configure the column.
13) In the Design tab, Monitor page; Enter 1.5 for Reflux Ratio, and De-activate the Ovhd Vap Rate.

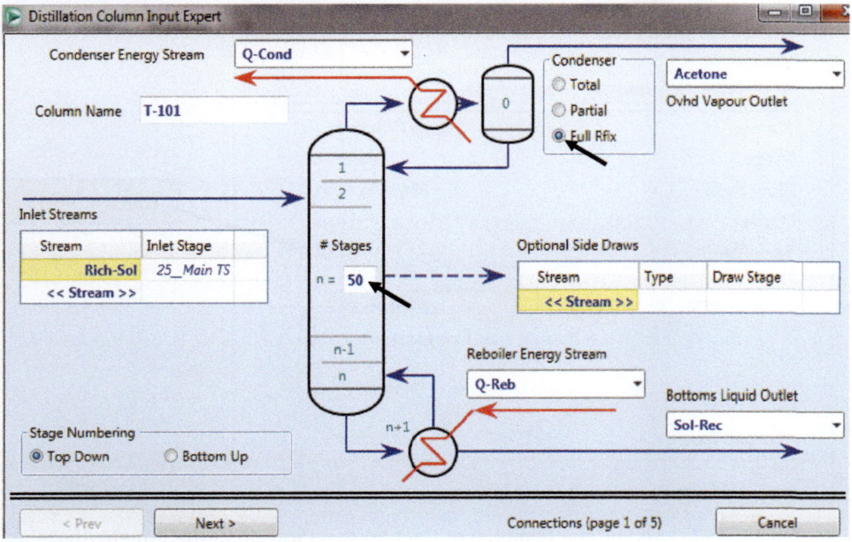

Figure 9.40 Distillation column input expert/ connection page.

14) Add a new specification: *Column Component Fraction* for purity of 3-methylhexane, Figure 9.41.
   The Degrees of Freedom for the column should now be 0. Click the **Run** button to begin column calculations.
15) Add a cooler to cool the bottoms stream in order to recycle it back to the mixer, Figure 9.42.
   - Cooler: Input = Sol-Rec, Outlet = Lean-Sol, Energy = *Q*-Cool
   - Enter the temperature of Lean-Sol = 30 °C, and its pressure = 1 bar
16) Recycle the bottoms streams from the second column in order to prevent throwing away solvent product.
   - Add a Recycle block  to the flowsheet: Inlet = Lean-Sol, Outlet = Rec Sol.
   - Connect the Rec Sol-stream with mixer (MIX-100) as in Figure 9.42.

The final PFD of Acetone extraction followed by distillation should appear as in Figure 9.42.

9.4.3 Summary

In this example, 3-Methylhexane is used as a solvent in a tray column to extract acetone from water. The acetone–solvent mixture is then separated by distillation so that the solvent can be used again in the

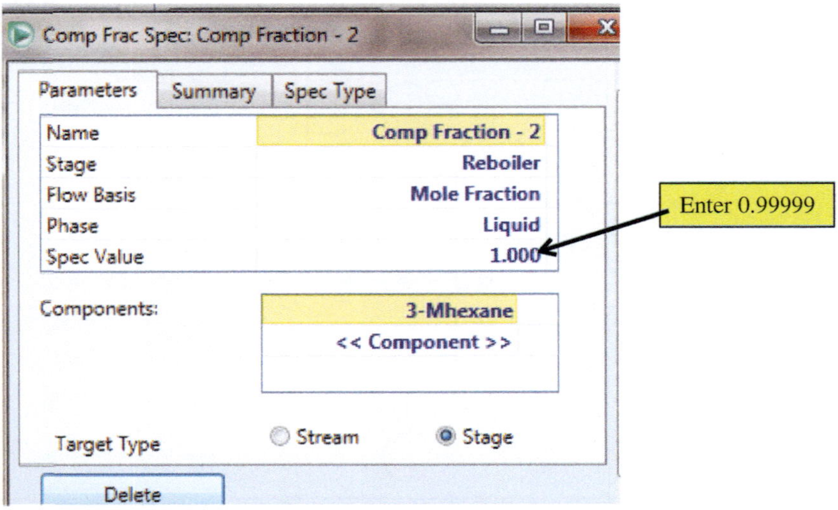

Figure 9.41 Adding the Component fraction specification of the 3-methylhexane.

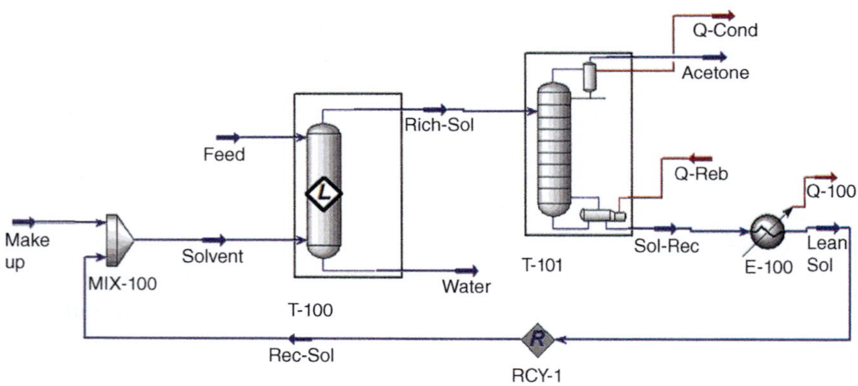

Figure 9.42 PFD of Extraction of Acetone–Water with 3-Methylhexane as solvent followed by distillation.

extraction process. According to the simulation results, liquid–liquid extraction of the water–acetone mixture would require 16270 kcal/h of heating and 9218 kcal/h of cooling. This design has been demonstrated to be workable; however, it might not be the best design. Direct distillation of water and acetone is an additional choice. Although fewer equipment might be needed, direct distillation of water and acetone might use more energy.

9.5 Miscellaneous Examples

9.5.1 Propylene Glycol Production

Propylene glycol is created when water and propylene oxide react. The CSTR reactor receives two feed streams: a water stream at 75 °F, 16.17 psia, and 11,000 lb/h; and a propylene oxide stream at 75 °F, 1.1 atm, and 150 lbmol/h. Regarding propylene oxide, the reaction is of first order. The reaction has the following parameters: $E = 32{,}400$ But/lbmol and $A = 1.7\text{e}13$. Here is the response [13–15]:

$$C_3H_6O + H_2O \rightarrow C_3H_8O_2 \tag{9.46}$$

$$\text{where: } -r_{proxd} = kC_{proxd},$$

where *proxd* is the propylene oxide.

The reaction occurs at a constant temperature and is equal to 140 °F. The reactor has a 280 ft^3 capacity and an 85% liquid content.

The distillation column, which has 10 trays, is used to separate glycol; the reactor products enter the column at a tray number 5. The condenser is a partially condensing type, and the pressures in the condenser and reboiler are 15 and 17 psia, respectively. The water mole fraction in the reboiler is 0.005, the reflux ratio is 1, and the vapor rate is zero.

Figure 9.43 shows the PFD of the Propylene Glycol production process.

9.5.1.1 Simulation

1) Open a new case of Aspen HYSYS, and add the Components: Propylene_Glycol, PropyleneOxide, and water
   - Fluid Package: UNIQUAC
   - Click **Binary Coefficient** button and click **Unknowns only** button

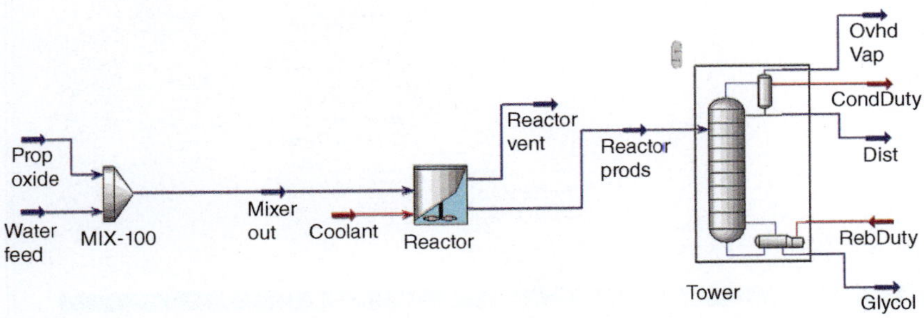

Figure 9.43 PFD of Propylene Glycol production.

2) From the Navigation pane, click the **Reaction** folder, add a **Reaction Set** by clicking the **Add** button, then add the Reaction by clicking the **Add Reaction** button; choose **kinetics**, and click the **Add Reaction** button. Enter the stoichiometry according to reaction (9.46).

   Based on the reaction stoichiometry, HYSYS offers default values for the Forward Order and Reverse Order. The kinetic data used in this tutorial is first order with respect to propylene oxide simply because it is based on an excess of water.
3) In the Fwd Order cell for H_2O, change the value to 0 to reflect the excess of water.
4) Move to the **Basis** group, Figure 9.44.
   - **Basis** = Molar Concn,
   - **Base Component** = PropyleneOxide,
   - **Rxn Phase** = CombinedLiquid,
   - **Basis units** = lbmol/ft^3, Rate units = lbmol/ft^{3}h.

   Forward Reaction group:

   $A = 1.7e13$, E (activation energy) = 3.24e4 Btu/lbmol.

   The Kinetic Reaction window will display as in Figure 9.44.
5) Close the reaction view and add the Reaction Set to the Fluid Package.
6) Enter the **simulation** Environment.
7) Add the propylene oxide Stream with the conditions: $T = 75\,°F$, $P = 1.1$ atm, molar flow rate = 150 lbmol/h, and 100% propylene oxide.

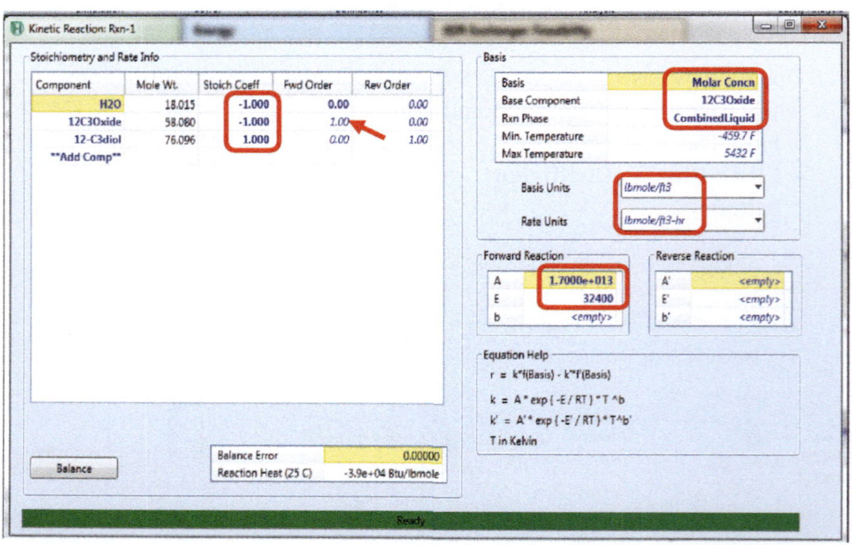

Figure 9.44 Window of Kinetic Reaction of Propylene Glycol production.

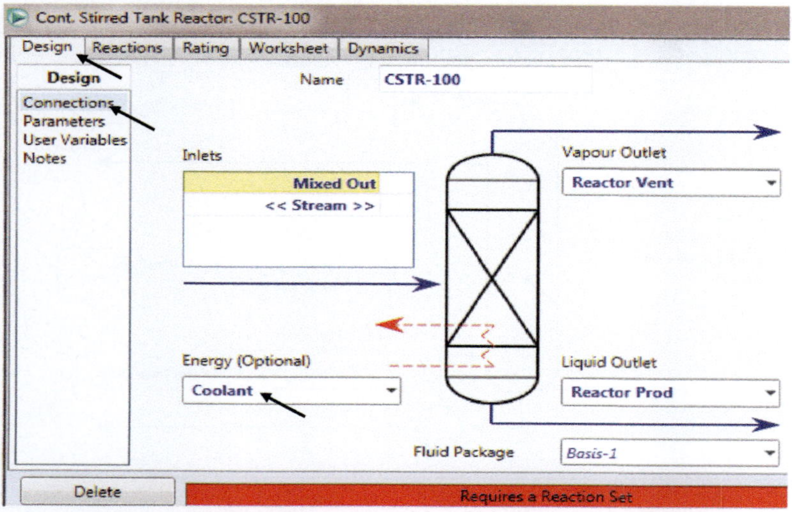

Figure 9.45 CSTR window, Design tab, Connections Page.

8) Add the Water Stream at T = 75 °F, P = 16.17 psia, mass flow rate = 11,000 lb/h, and 100% water.
9) The two streams (PropyOxide and water) are mixed and sent to the kinetic reactor using mixer.
10) Add **CSTR reactor**, fill the **design** tab (Connection page) as in Figure 9.45.
11) Parameters page: Enter Delta P = 0.0, Volume = 280 ft^3 , and the Liquid Volume % = 85%.
12) Reactions tab: Attach the Reaction Set that you created in the Properties environment.

    Initially the Reactor is assumed to be operating at 140 °F; therefore, set the **outlet temperature** of the reactor product to 140 °F. The reactor will converge.
13) Add a Distillation Column to the PFD.
14) Fill the first page as in Figure 9.46.
15) Click the Next button, then click the Next button again to advance to the Pressure Profile page.
16) In the Condenser Pressure field, enter 15 psia, and in the Reboiler Pressure field, enter 17 psia.
17) Leave the condenser and reboiler Pressure Drops at their default value of zero.
18) Click the Next button to advance to the Optional Estimates page. For this example, no estimates are required.

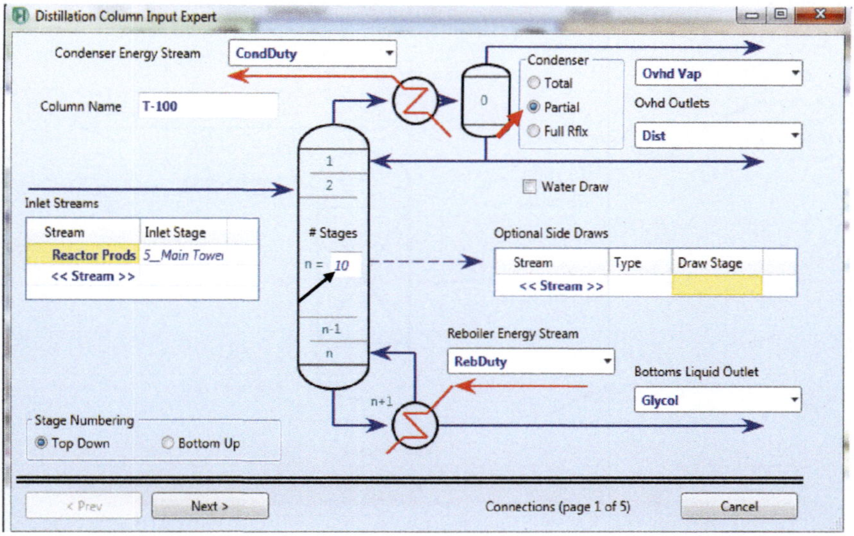

Figure 9.46 Distillation column, input expert.

19) Click the Next button to advance to the fifth and final page of the Input Expert. This page allows you to supply values of the default column specifications that HYSYS has created.

For this Distillation Column, there are three default specifications. The overhead Vapor Rate and Reflux Ratio will be used as active specifications, and later you will create a glycol purity specification to execute the third degree of freedom. The third default specification, overhead Liquid Rate, will not be used.

20) In the Vapor Rate field, enter 0 lbmol/h.

The Flow Basis applies to the **Vapor Rate**, so leave it at the default of Molar.

21) In the Reflux Ratio field, enter 1.0.
22) Click the **Done** button. The Column property view appears.
23) On the **Design** tab, select the **Monitor** page, Figure 9.47.
24) From the Monitor page, add the "Column component Fraction" as a Column Specification.
    - Fill the Component Fraction window as in Figure 9.48.

    Close the property view of Figure 9.48 to return to the Column property view.
25) ***Deactivate*** the Distillate rate specification and Activate the H_2O Fraction Specification where the **Degree of Freedom** will equal **0**.

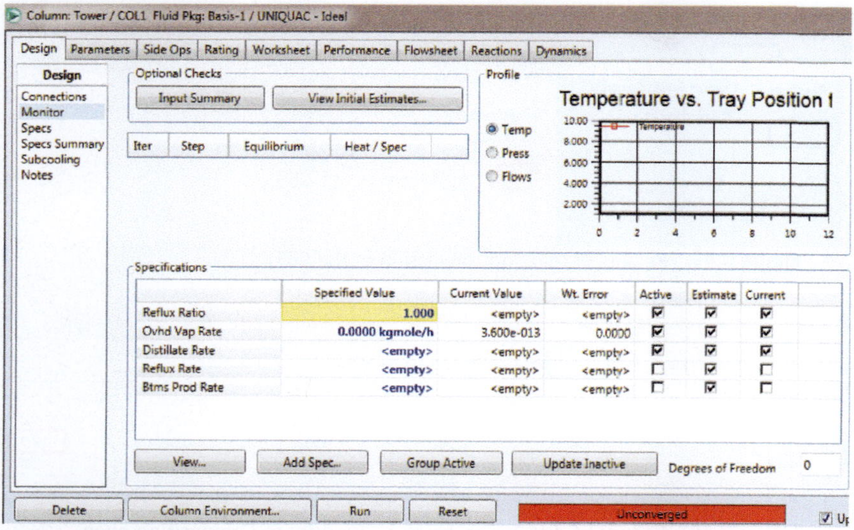

Figure 9.47 Monitor page before converging.

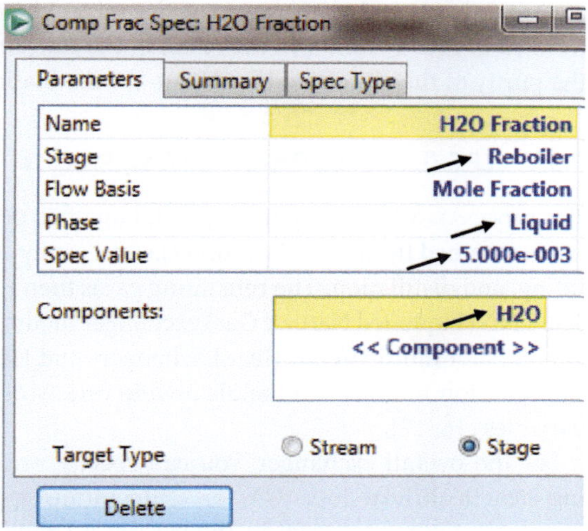

Figure 9.48 Component fraction window: Adding the purity of water as the specification.

26) If the column is not Running, click the **Run** button, the column will Converge.

The results of reactor and distillation column are shown in Figures 9.49 and 9.50, respectively, where the conversion of propylene oxide in the reactor and purity of produced glycol from the bottom of distillation column are 94.7% and 0.995, respectively.

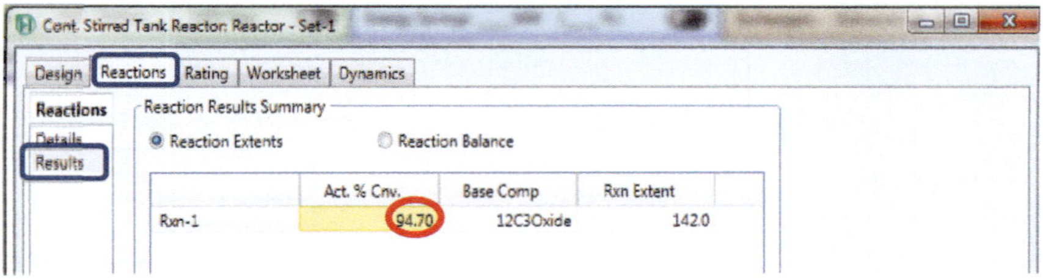

Figure 9.49 Results of reaction conversion of the CSTR.

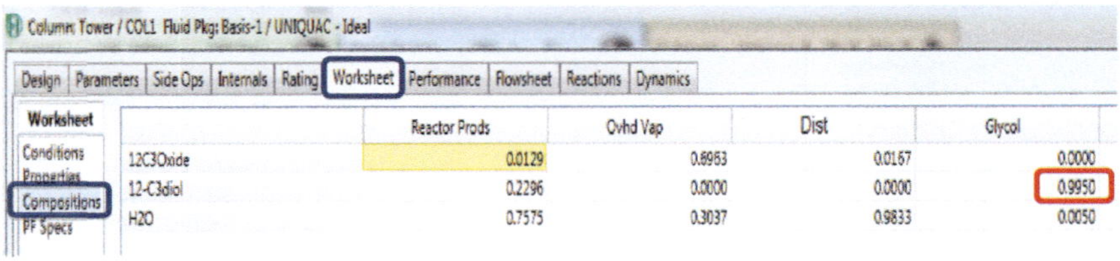

Figure 9.50 Results of distillation column - Worksheet-Composition page.

9.5.1.2 Further Study

Search the effect of temperature of reactor product on the conversion, e.g. from 75 to 160 °F. How will the result of the conversion in the reactor affect the purity of the glycol produced from the distillation column?

9.5.2 LPG (Liquefied Petroleum Gas) Process

The process is a Turbo Expander plant. Liquefied petroleum gas (LPG) product is obtained from a Feed Natural Gas stream by cooling, expanding, separating, and distillation. The remaining gas is then re-compressed for export. The LNG (Liquefied Natural Gas) exchanger model solves heat and material balances for multi-stream Heat Exchangers and Heat-Exchanger networks. The solution method can handle a wide variety of specified and unknown variables [16, 17].

For the overall exchanger, you can specify various parameters, including heat leak/heat loss, UA, or temperature approaches. Two solution approaches are employed; in the case of a single unknown, the solution is calculated directly from an energy balance. In the case of multiple unknowns, an iterative approach is used which attempts to determine the solution which satisfies not only the energy balance, but also any constraints, such as temperature approach or UA.

To install the LNG operation, press F12 and choose LNG or select the LNG Exchanger icon from the object palette .

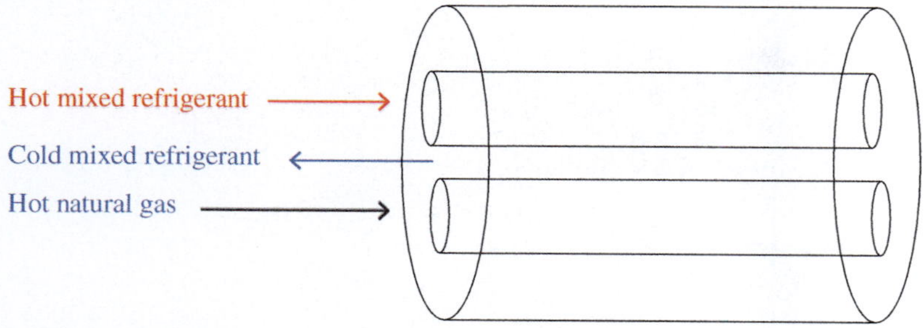

Figure 9.51 Schematic diagram of LNG exchanger.

The dynamic LNG exchanger model performs energy and material balances for a *rating plate-fin-type* Heat-Exchanger model. Figure 9.51 shows a schematic diagram of LNG exchanger. Figure 9.52 shows the PFD of the LPG Process.

9.5.2.1 Simulation

1) After opening a new case, add the following components: N_2, CO_2, C1, C2, C3, *i*-C4, *n*-C4, *i*-C5, *n*-C5, and *n*-C6. Fluid Package: Peng–Robinson
2) Enter the **Simulation** Environment
3) Add the Feed Gas stream with the following conditions:

Name	**Feed Gas**
Temp	30 °C
Pressure	5000 kPa
Flow rate	2988 kmol/h
Component	**Mole Fraction**
N_2	0.0149
CO_2	0.002
Methane	0.9122
Ethane	0.0496
Propane	0.0148
i-Butane	0.0026
n-Butane	0.002
i-Pentane	0.001
n-Pentane	0.0006
n-Hexane	0.0003

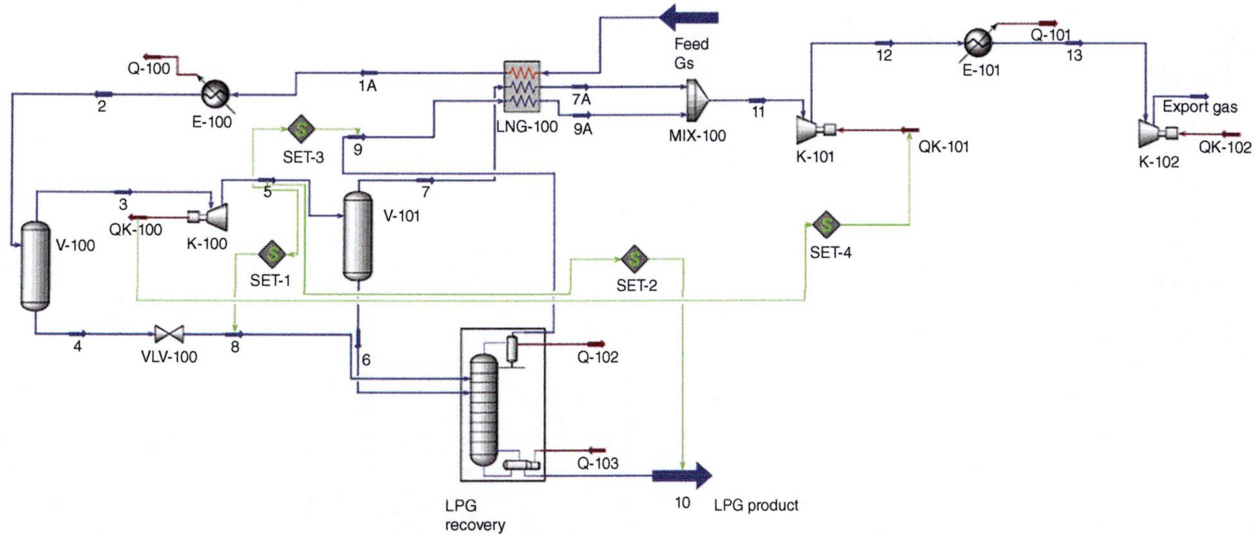

Figure 9.52 PFD of the LPG Process.

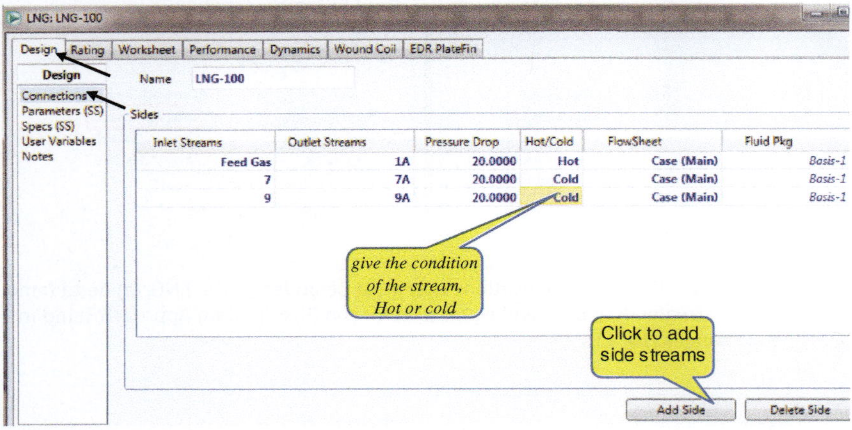

Figure 9.53 Connection page of LNG exchanger.

4) Add a Multi-Pass exchanger, Liquefied Natural Gas (LNG) Exchanger .

5) Select the **Connections** page on the **design** tab and enter the information as in Figure 9.53, (Pressure in kPa)
   - *Add any number of Sides simply by clicking the **Add Side** button.*
   - *In the Hot/Cold column give the condition of the stream. If the designated hot pass is actually cold (or vice versa), the operation will still proceed properly. The actual Hot/Cold designation (as determined by LNG) can be found on the **Performance** tab in the **results** page.*
   - *The **Exchanger Design (Weighted)** method is used to split the heating curve into intervals. The **Step Type** parameter sets whether HYSYS splits the curve by temperature, enthalpy, or auto interval to minimize the errors.*

6) Switch to the **Parameter (SS)** page. These values will be left unchanged.
   - *The **Weighted method** is required for the exchangers with more than **two** sides.*
   - *By selecting the **Dew/Bub pt** checkboxes, HYSYS will point on the heating curve for any phase change that occurs.*

7) Go to the **Spec (SS)** page: Degree of Freedom = 6 = (7 − 1); add two specifications:

   (7-1) The **Cold Out Same** temp (0 °C), where the delta temperature between 7A and 9A streams is zero, Figure 9.54a.

   (7-2) Minimum **Approach Temp** in the exchanger (10 °C), Figure 9.54b.

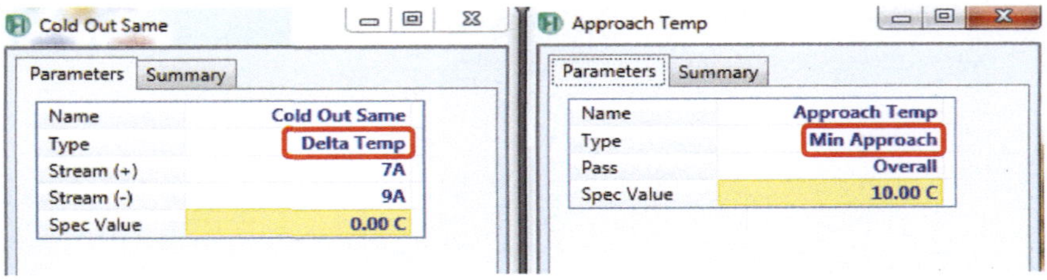

(a) (b)

Figure 9.54 The two specifications that must be added to the LNG, (a) delta temp between stream 7A and 9A and equal to zero, and (b) Minimum Approach Temp in the exchanger (10 °C).

Name	**E-100**
Inlet	1A
Outlet	2
Parameter: Pressure drop	20 kPa
Outlet Temp	−62 °C
Energy Stream	Q-100

8) Add a Cooler with the following conditions:
9) Add a Separator (V-100): Inlet = 2, Liquid Product = 4, Vapor product = 3
10) Add an Expander (K-100): Inlet = 3, Outlet = 5, Outlet pressure = 2800 kPa,
    Adiabatic efficiency = 75%, Energy Stream = QK-100.
11) Add another Separator (V-101) to the Expander outlet:
    Inlet = 5, Vapor product = 7, Liquid product = 6.
12) Add a Valve (VLV-100): Inlet = 4, Outlet = 8.
13) Add a **Set** operation (SET-1) with the following parameters:
    Target variable: click select variable: **Object** = 8, **Variable** = pressure,
    Source: Object = 5.
    Parameters tab: Multiplier = 1, Offset = 0.0
14) Add a material stream and name it: 10.
15) Add a second **Set** operation (SET-2): Target Variable: object = 10, Variable = pressure,
    Source = 5, Parameters: Multiplier = 1, Offset = 0.0.
16) Add a third **Set** operation (SET-3): Target: object = 9, Variable = pressure,
    Source = 5, Parameters: Multiplier = 1, Offset = −5 psi

17) Add a **Distillation** Column (LPG-Recovery): Enter the following information in the **connection** page.

Column Name	LPG Recovery
No. of Stages	5
Feed Stream/Stage	8/ at Stage 1
Condenser Type	Full Reflux
Overhead Vapor Product Stream	9
Bottom Product Stream	10
Condenser Energy Stream	Q-102
Reboiler Energy Stream	Q-103

- The pressure profile page automatically picks these up.
- The temperatures are unknown, so these estimates will be left.

18) Click **Done** button to complete the input expert.
Before the column is ready to run, some more information must be entered.
19) Connect stream 6 up in the **Inlet Streams** box of the column, inlet stage = condenser. (Stream 6 is cold stream that provides the necessary cooling effect).

First Specification	
Type	**Column Duty**
Name	Ovhd Duty
Energy Stream	Q-102 @ Col1
Spec Value	0.001 kJ/h

20) Move to the **Monitor** tab: Add the two following specifications

Second Specification	
Type	**Column Vapor Pressure Spec**
Name	Btms RVP
Stage	Reboiler
Type	Reid Vapor Pressure
Phase	Liquid
Spec Value	200 psia

RVP: Reid Vapor Pressure is a volatility measure commonly used in the refining industry. It is the pressure at which vapor and liquid have a 4:1 volume ratio at 100 °F.

21) Modify the existing Ovhd Vap spec as follows:

Name	Ovhd Vap Rate
Draw	9@Col1
Flow Basis	Molar
Spec Value	400 kmol/h

22) Ensure that the **Ovhd Vap Rate** and **Btms RVP** specs are set to be **Active**, and the **Ovhd Duty** spec is **Estimate**.
    The other unused specifications can be deleted by double-clicking the spec, and then click delete.
23) Click **Run** if the column is not running. When the column has solved, the LNG should then solve as now both streams 7 and 9 are fully defined. What is the duty of the condenser? What is the flow rate of the overhead vapor stream?
    The simulation is now almost complete. The two product gas streams from the LNG must be mixed and re-compressed for export.
24) Mix streams 7A and 9A using a Mixer operation (MIX-100); outlet = 11.
25) Add a **Compressor** (K-101); Inlet = 11, Outlet =12, Energy = QK-101.
26) Add a **Set** operation (SET-104): Target: Object = QK-101, Variable = Heat flow,
    Source: QK-100, Parameters: Multiplier = 1, Offset = 0.
27) Add a **Cooler** (E-101): Inlet = 12, Outlet =13, Energy = Q-101.
    Parameters: pressure drop = 0.2 bar, Temp of stream 13 = 30 °C.
28) Add a **Compressor** (K-102): Inlet = 13, Outlet = Export Gas,
    Energy = QK-102, Outlet pressure = 70 bar.

HYSYS can also solve the flow rate given an otherwise fully specified feed stream, a duty, and an outlet pressure.

Additionally, HYSYS compressors can be supplied with head and efficiency curves.

9.5.3 Reactive Distillation

With the continuous removal of reaction products, reactive distillation has found acceptance as a means of improving the technical and economic operation of processes where it is applicable. Advantages of this process include higher yields, energy savings, and reduced capital costs.

In this example, we will be studying the manufacture of Methyl Acetate, an important oil resin used in the manufacture of artificial leathers. We will be simulating the synthesis of Methyl Acetate from Methanol and Acetic Acid in a catalytic distillation column [16].

9.5.3.1 Simulation

1) Open a new casa and add the following components: Methanol, AceticAcid, M_Acetate, and Water. The fluid package is **Wilson**.
2) Adding the Reaction: The reaction that occurs in this simulation is a **kinetic** reaction type:

$$CH_3OH + CH_3COOH \rightarrow CH_3CH_3COO + H_2O \tag{9.47}$$

3) Add a Set of reactions and then click the **add Reaction** button.
4) Select **kinetic** from the list of reactions and press the **Add Reaction** button.
5) Double-click the reaction (Rxn-1) and enter the stoichiometric coefficients on the **Stoichiometry** tab. Remember that the reactants are negative values and the products are positive values.
   The reaction is second order, $-r = k\, C_M C_A$.
   where M is methanol and A is acetic acid.
6) Go to the **Basis** group. The Base Component is **methanol** and the Rxn Phase is **Liquid**.
   Leave the Basis and Rate Units at their default values (SI units).
7) For the Forward Reaction, enter the Arrhenius parameters as $\boldsymbol{A} = \mathbf{1.0e5}$ and $\boldsymbol{E} = \mathbf{2.3e4\ kJ/kgmol}$. Close the Reaction view.
8) Add the **Rxn Set** to the **Fluid Package** by pressing the **Add FP** button, then click **Add Set to Fluid Package**.
9) Enter the **Simulation** Environment and add the Feed Stream with the following conditions:

Stream Name	Feed
Temperature	75 °C
Pressure	101.3 kPa
Molar Flow	45 kgmol/h
Composition – Mole Fraction	
Methanol	0.4
Acetic Acid	0.4
M_Acetate	0.1
Water	0.1

10) Add a Distillation Column with the following conditions:

Column Name	Reactive Distill
No. of Stages	15
Inlet stream/# Stage	Feed/ 10
Condenser Type	Total
Ovhd Liquid	Distillate
Bottom Liquid	Bottom
Condenser Energy	Cond Q
Reboiler Energy	Reb Q
Condenser Pressure	90 kPa
Delta *P*, Condenser	0
Reboiler Pressure	97 kPa
Specifications	
Reflux Ratio	5
Distillate Rate	20 kgmol/h

11) Go to the **Monitor** page of the Design tab and run the column. Complete the following table with information from the simulation:

Component	Distillate – Mole Fraction	Bottom – Mole Fraction
Methanol	....	....
Acetic Acid	....	....
M_Acetate	....	....
Water	....	....

12) **Adding the Reaction to the Column**
In this case, the reaction occurs on stages 5–10 of the column. Reaction Sets in HYSYS can be attached to any of the stages in the column, including the Condenser, Reboiler, and Side operations.
   - Double-click the column icon: On the **Reactions** tab, press the **New** button, Figure 9.55.

13) Enter the data as in Figure 9.56: From the drop-down list of the **Reaction Set**, select: Set-1.
The reaction will occur in the stages: [5_Main Tower] to [10_Main Tower]

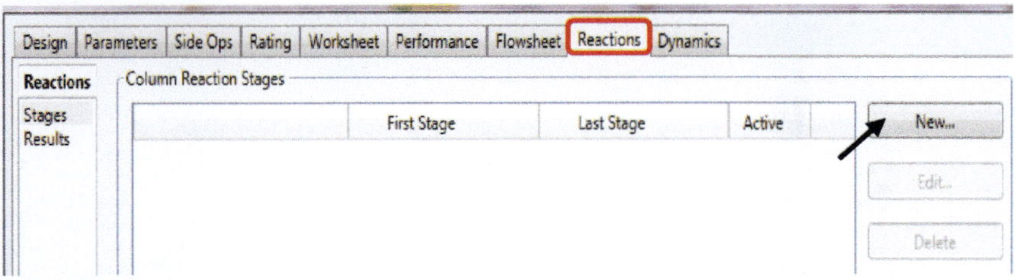

Figure 9.55 Adding the reaction to the Distillation Column.

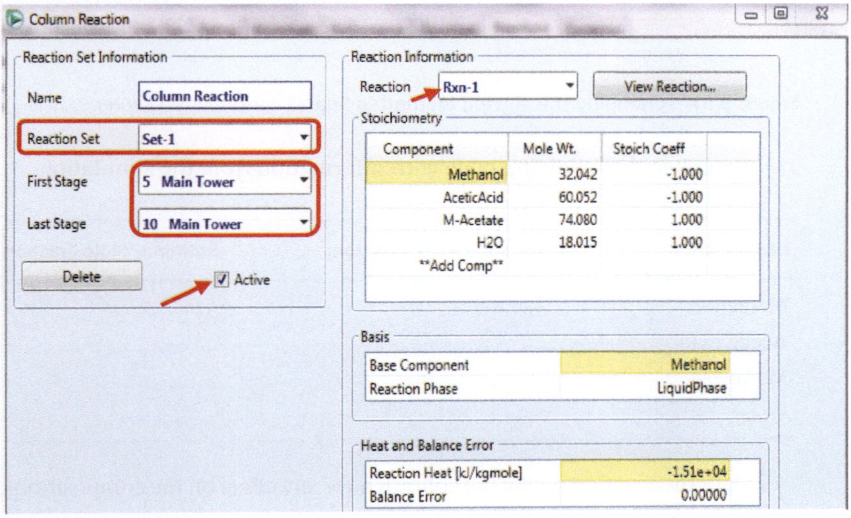

Figure 9.56 window of Reaction in the distillation column.

14) Check the **Active** box on the Column Reaction view. The default column solver in HYSYS "HYSIM Inside-Out" is not capable of handling reactions in the column. Therefore, HYSYS will change the solver to Newton–Raphson Inside-Out.
15) However, for this column, we need to use the **Sparse Continuation Solver**, Figure 9.57. Select this solver on the **Solver** page of the **Parameters** tab.
16) Run the column. (The column may run automatically after the solver is changed.)
17) You can view the results in the **Performance** tab, and the reaction results in the **Reactions** tab, **Results** page.

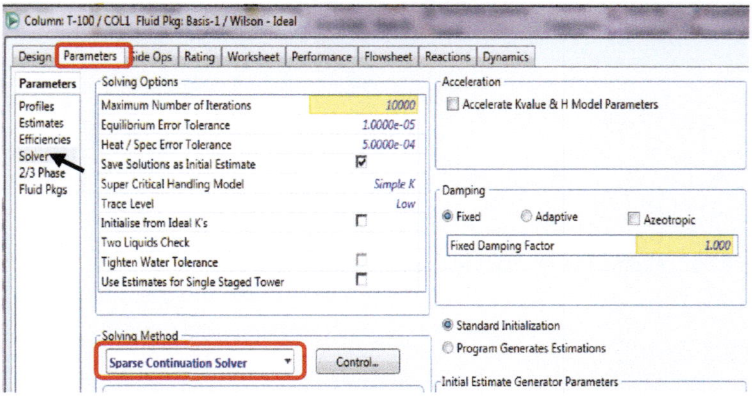

Figure 9.57 Changing the solving method to Sparse continuation solver.

18) Complete the following table with information from the simulation.

Component	Distillate – Mole Fraction	Bottom – Mole Fraction
Methanol	....	....
Acetic Acid	....	....
M_Acetate	....	....
Water	....	....

Does the reaction inside the column have any effect on the compositions of the product streams?

9.5.3.2 Solving Method Group

The Solving Method drop-down list (Parameters tab/Solver page) allows you to select the column solution method. The display field, which appears below the drop-down list, provides explanations for each method, and is restated here in Table 9.3.

9.5.4 Three-Phase Distillation

Due to the addition of a second liquid phase in the column, it is very difficult to model this type of column. For this reason, one must use the **Sparse Continuation Solver** to solve these systems.

With a normal two-phase distillation setup, it would be impossible to distill the stream described previously. Fortunately, HYSYS is able to model

Table 9.3 Explanation of solving method in HYSYS.

Method	Explanation
HYSIM Inside-Out	General purpose method, which is good for most problems.
Modified HYSIM Inside-Out	General purpose method, which allows mixer, tee, and Heat Exchangers inside the column sub-flowsheet. Only a simple Heat-Exchanger Model (Calculated from Column) is available in the Column sub-flowsheet. The Simple Rating, End-Point, and Weighted models are not available.
Newton Raphson Inside-Out	General purpose method, which allows liquid-phase kinetic reactions inside the Column sub-flowsheet.
Sparse Continuation Solver	An equation-based solver. It supports two liquid phases on the trays, and its main use is for solving highly non-ideal chemical systems and reactive distillation.
Simultaneous Correction	Simultaneous method using dogleg methods. Good for chemical systems. This method also supports reactive distillation.
OLI Solver	Only used to calculate the column unit operation in an electrolyte system.

the distillation of streams like this by using a different column setup routine. Different Solvers with two-phase distillation is the most common solver used in the **"HYSIM Inside-Out"** solver. This solver is good for most hydrocarbon systems, but it cannot handle highly nonideal chemical systems, including three-phase systems. The default solver for three-phase columns is the **"Sparse Continuation"** solver. This more advanced solver is able to handle three-phase, non-ideal chemical systems that other solvers cannot calculate.

The second major difference between two- and three-phase distillation is the different input experts that each uses. You have already seen the input expert used by the "normal" two-phase distillation column. The input expert used by the three-phase column is similar though slightly more complex. You will use this interface when we install the three-phase column later in this module.

Finally, the third major difference between the two column types is the additional column specification that are available when using the three-phase column operation. Due to the complications that a second liquid phase adds to the column operation, additional specifications are required to help the column converge [16, 17].

Despite these apparent differences, setting up columns to handle three-phase streams is not that much more complicated than setting them up to handle two-phase streams.

You already have experience with two-phase systems, and you will find that setting up a three-phases system is not that much more complicated

9.5.4.1 Simulation

1) Begin a new HYSYS case and select three components:
   Water, 2-Butanol (or Butanol-2), and *n*-Butyl_Acetate (*n*-B-Acetate)
   Property Package: **NRTL**, On the **Binary Coefficients** tab, press the **Unknowns Only** button.
2) Enter the **Simulation** Environment.
3) Create a new stream and define it with the following information:

Name	Feed
Vapor Fraction	0
Pressure	101.3 kPa
Molar Flow	50 kgmole/h
Mole Fraction	Water = 0.35, 2-Butanol = 0.25, *n*-B-Acetate = 0.4

4) Add a **Three Phase Distillation** Column operation  to the simulation.
5) Select the **Distillation** radio button and press the Next button, Figure 9.58.
6) On the next page, set the "number of stages" = **6**, and set the "Two Liquid Phase Check" on the top five stages (condenser +4 stages), Figure 9.59, Click next.
7) On the third page, enter names for the streams around the condenser. For this column, the Condenser Type is **Partial**, the Outlet Stream is **Both**, and the Reflux Stream will be the **Lights** only; Figure 9.60.
8) Click Next. On the fourth page, set the vapor rate to **0.0**, and the Light Reflux Fraction to **1.0**. The degrees of freedom should read 0.0; once both of these values are entered, Figure 9.61 will appear.

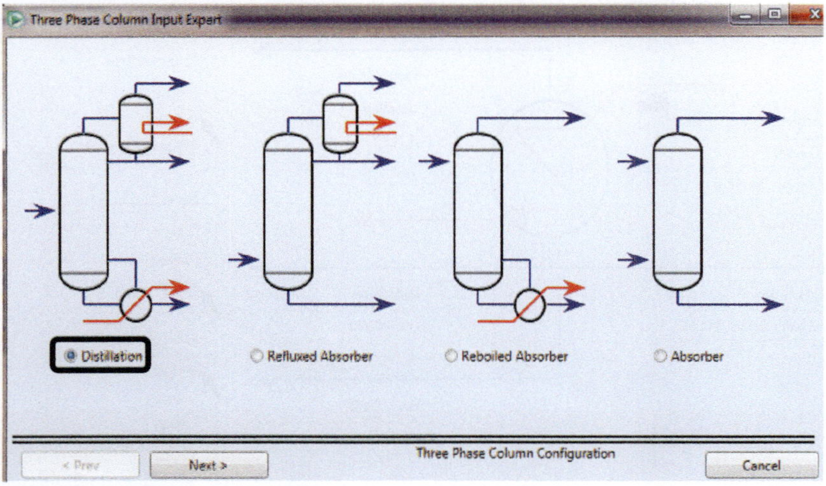

Figure 9.58 Window of the three-phase column configuration.

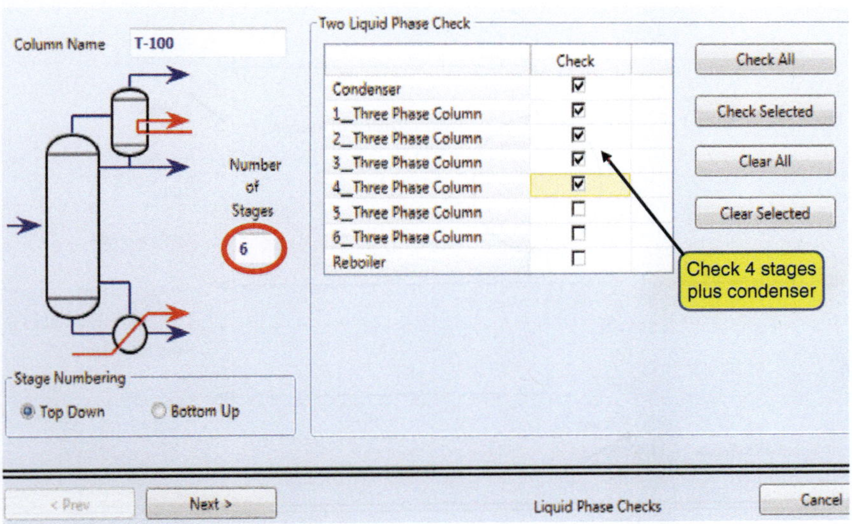

Figure 9.59 Window of two liquid-phase checking.

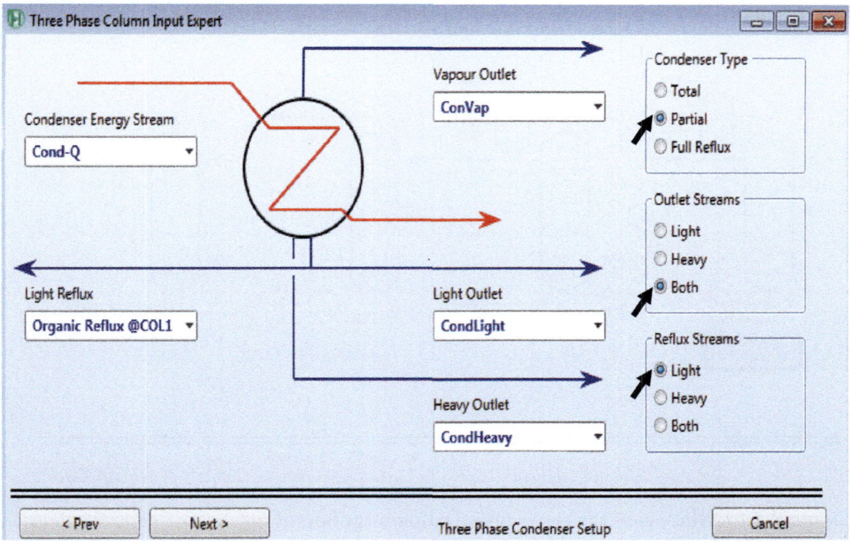

Figure 9.60 Three-phase column input expert/condenser setup.

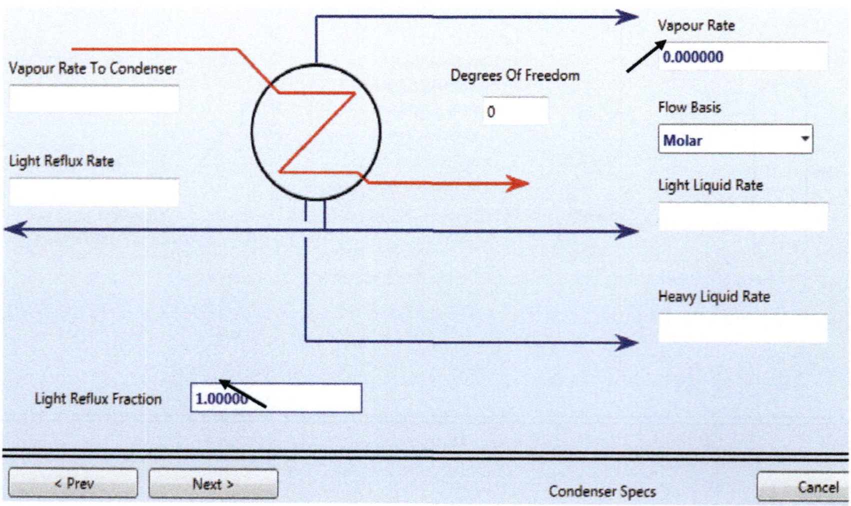

Figure 9.61 Condenser specifications.

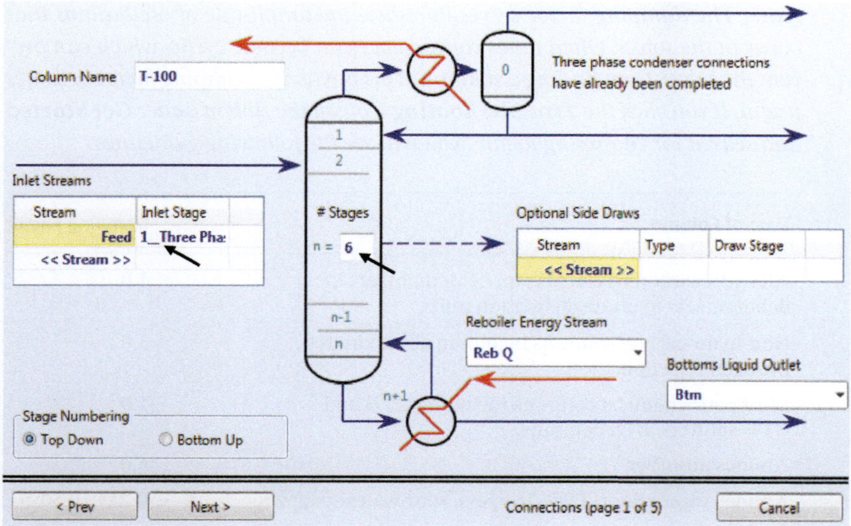

Figure 9.62 Connection page of the three-phases column.

9) Click Next. On the fifth page, attach the **Feed** stream to stage 1. Also, define the reboiler energy stream and attach a liquid product stream to the reboiler. From now on, the screens will look familiar to the **Input Expert** screens that you have seen previously, Figure 9.62.
10) Click Next. On the next page, set the condenser pressure at **101 kPa**, and the reboiler pressure at **105 kPa**.
11) On the next page (page 2), you are asked to supply optional temperature estimates. Enter values of **85, 90,** and **95 °C** for the condenser, top stage, and reboiler temperature estimates, respectively.
12) Press the **Done** button to close the input expert. Note that if the formation of azeotropes were a concern in this column, the ***Azeotropic Initialization*** button would have been pressed, and HYSYS would have checked for possible azeotropes among the selected components.
You are automatically placed on the Column's property view; move to the **Monitor** page of the **Design** table
13) Enter a Bot Prod Rate of **33 kgmol/h**. Press the **Run** button to converge the column; the column should start to solve and will reach a converged solution quickly. If the column does not solve in the first attempt, increase the Bot Rate Spec value to **35 kgmol/h** and try again. Once the column converges, decrease the spec back to its original value. How many Degrees of Freedom are shown here?

***Note:** The damping factor serves to reduce the amplitude of oscillations that occur in the solver. Often times convergence can become cyclic, which can prevent the solver from finding a solution. This is where a damping factor becomes useful. If you click the **Troubleshooting** icon on the ribbon under **Get Started** and search for "damping factor" you will see the following guidelines.*

Type of Column	Damping Factor
All hydrocarbon columns from demethanizers to debutanizers to crude distillation units.	1.0
Non-hydrocarbon columns including air separation, nitrogen rejection.	1.0
Most petrochemical columns including C2 = and C3 = splitters, BTX columns.	1.0
Amines absorber	1.0
Amines regenerator, TEG strippers, sour water strippers	0.25 to 0.50
Highly non-ideal chemical columns without azeotropes.	0.25 to 0.50
Highly non-ideal chemical columns with azeotropes.	0.50 to −1.0

9.5.5 Extractive Distillation for Heptane-Toluene Separation

When the two components in a binary mixture have very close normal boiling points, their relative volatility is likely to be small if they do not form an azeotrope. For such cases, it may be more efficient to use extractive distillation with a solvent than normal distillation. In extractive distillation, a less volatile solvent is used to increase the relative volatilities of the original mixtures, allowing for easier separation [18, 19]. The solvent added to the mixture in extractive distillation differentially affects the activities of the two components, and hence the relative volatility, α_{AB}.

For non-ideal binary mixture:

$$P_A = \gamma_A P_A^* x_A \quad \text{and} \quad P_B = \gamma B x_B \tag{9.48}$$

$$\rightarrow \alpha_{AB} = \frac{P_A/x_A}{P_B/x_B} = \frac{\gamma_A P_A^*}{\gamma_B P_B^*} \tag{9.49}$$

A new solvent creating hydrogen bonds with one component of the original mixture is fed to the column. Boiling point of the new solvent is higher than that of mixture components, and it is usually fed to the top part of the column and obtained in the bottom product.

In this example the n-heptane and toluene are closed in their normal boiling point and will separate firstly without adding a solvent. The feed enters

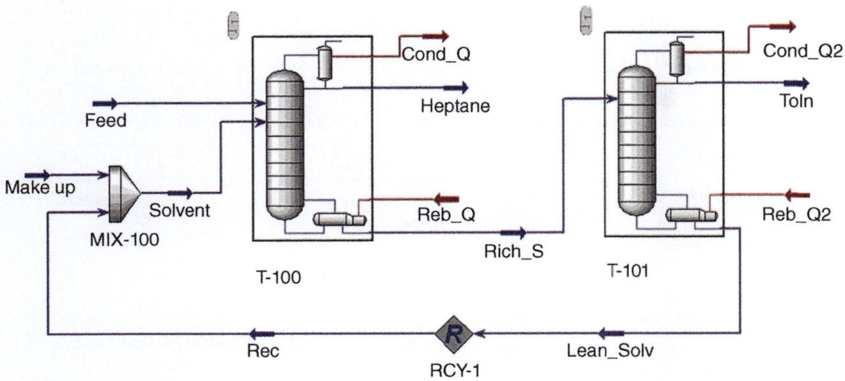

Figure 9.63 PFD for the separation of toluene and n-heptane using extractive distillation.

the column at 1.0 bar, 0.5 vapor fractions, 100 kmol/h and 0.5 mol fractions for each component. In the second case the phenol solvent will be used to see the difference between the number of trays of column and energy operating required.

9.5.5.1 Problem Statement

The n-heptane is separated from toluene by extractive distillation using phenol to alternate the relative volatility. The whole process is explained in Figure 9.63. The feed consists of equimolar n-heptane and toluene with a flow rate of 100 kmol/h, 1 bar, and a vapor fraction = 0.5. The makeup stream of pure phenol at 181 °C, and 1 bar enters the mixer, and outlet solvent has flowrate of 60 kmol/h. The distillation column (T-100) has 50 stages, where the feed enters the column at stage 37 and the solvent enters at stage 4. The stream (Rich_S) from the bottom of column (T-100) is sent to the second distillation (T-101) which consists of 28 stages to separate toluene and solvent, where the stream Rich_S enters at stage 21. The pressures in two columns are 1 bar for condensers and reboilers. The type of condenser for both columns is full reflux. The distillate rate of the (T-100) is 50 kmol/h, and the purity of n-heptane in the condenser is 0.99 by mole. For (T-101) the toluene purity in the condenser is 0.99 by mole, and the purity of phenol in the reboiler is 0.99999. The pressure of Rec-stream is 1 bar. Fluid package = **NRTL**, and vapor model = **RK** [18].

9.5.5.2 Simulation

First, it will begin without using the solvent to see how many stages are required to achieve the separation of *n*-heptane and toluene in the desired purity.

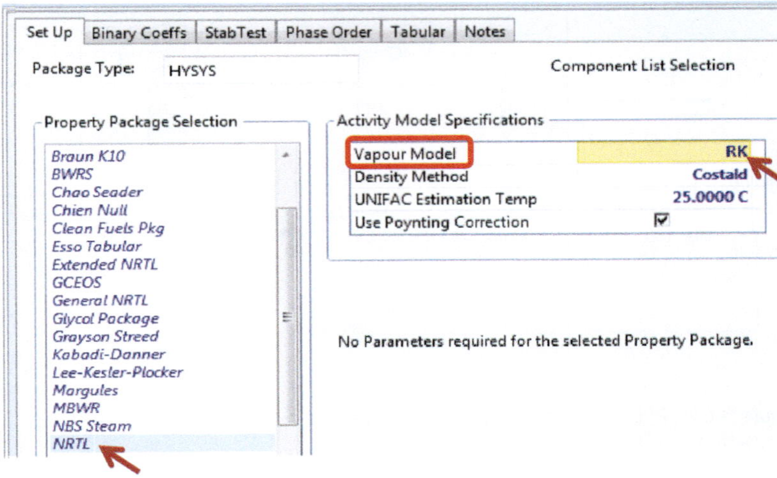

Figure 9.64 Setup of property package.

1) Open a new case and add the Components: *n*-Heptane and Toluene. Property package: NRTL, and select **RK** as the **Vapour Model**, Figure 9.64.
2) Enter the **Simulation** Environment.
3) Add a Distillation column and fill the first page as in Figure 9.65.
4) Click Next, leave the second page, click Next: Enter the pressures of the condenser and reboiler at 1 bar, leave the fourth pages, click Done on the fifth page.
5) For the Feed stream, enter a vapor fraction of 0.5, a pressure of 1 bar, and a molar flow rate of 100 kgmol/h.
   Composition: 0.5 mol fraction for each component of *n*-heptane and toluene.
6) In the **Design** tab; Monitor page; deactivate the **Reflux ratio** and **Ovhd Vap Rate**.
7) Add two specifications:
   (7-1) Column component fraction: Mole fraction of *n*-heptane in the heptane stream = 0.99, Figure 9.66a.
   (7-2) Column component fraction: Mole fraction of toluene in the toluene stream = 0.99, Figure 9.66b.
8) The degree of freedom becomes 0. Click **Run,** the column will converge.
9) Go to the Performance tab, Cond./Reboiler page, make a note of both the Condenser and Reboiler duties. The Condenser Duty is 2.524e7 kJ/h, and the Reboiler Duty is 2.345e7 kJ/h.

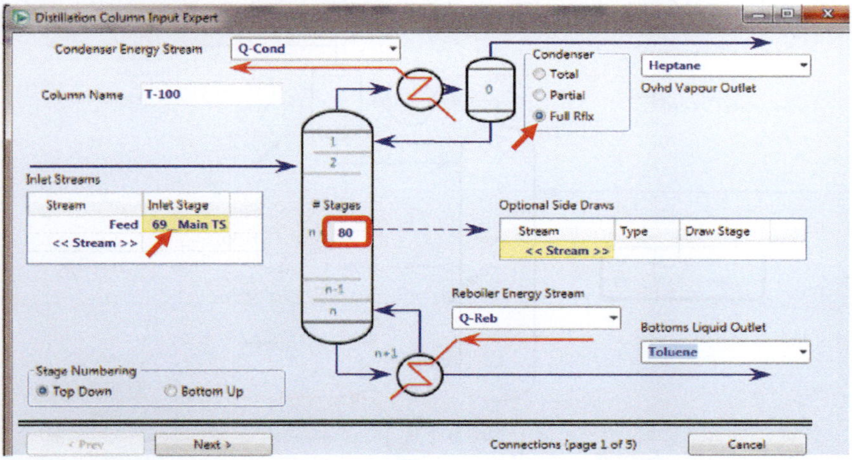

Figure 9.65 Distillation column input expert/connection page without adding solvent.

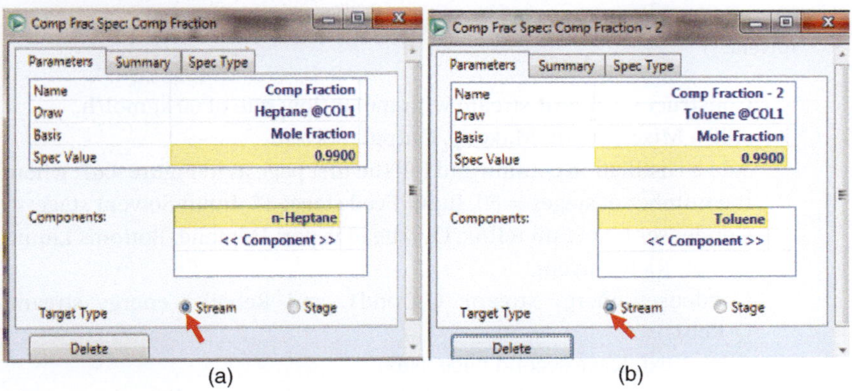

Figure 9.66 Adding two-component fraction specifications: (a) for n-heptane, (b) for toluene.

Save this case and open a **new case**.

1) Create a component list: *n*-Heptane, Toluene, and Phenol.
   Fluid Package: **NRTL**, and select **RK** as the Vapour Model.
2) Enter the **Simulation** environment,
3) Add a **Feed** stream, enter a vapor fraction of 0.5, a pressure of 1 bar, and a molar flow rate of 100 kgmol/h, Composition: 0.5 for *n*-heptane and toluene.
4) Add a **Makeup** stream, enter a temperature of 181 °C, pressure of 1 bar, and a composition of 1 for phenol.

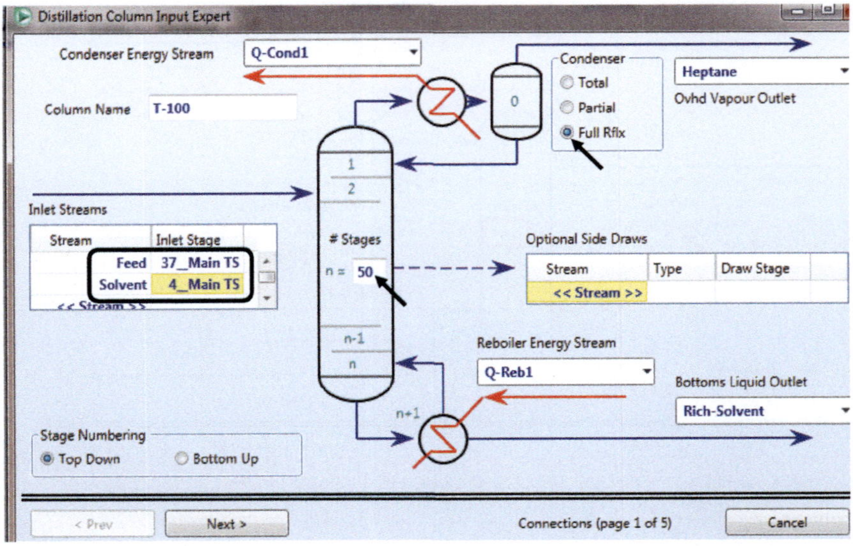

Figure 9.67 Distillation column input expert/connection page when adding solvent.

5) Construct a **Solvent** stream with molar flow rate of 60 kgmol/h,
6) Add a Mixer: Input: Makeup, Output: Solvent.
7) Add a Distillation column and Fill the first page as in Figure 9.67, where the number of stages = 50, Input Feed stage: 37, Input Solvent stage: 4. Condenser type: Full reflux, Overhead vapor: Heptane, Bottoms Liquid Outlet: Rich-Solvent,

   Condenser energy stream: Q-Cond1, and Reboiler energy stream: Q-Reb1.
8) Click Next, leave second page, Next.
9) Enter the pressures of the condenser and reboiler at 1 bar.
10) Leave the fourth page empty. On the fifth page, click Done.
11) In the **Design** tab, **Monitor** page: deactivate the Reflux ratio, in the **Ovhd Vap Rate** cell enter 50 kmol/h.
12) Add a new Specification: Column component fraction as in Figure 9.68.
13) The Degrees of Freedom for the column should now be 0. Click **Run** to begin the calculation. The column should solve.
14) Add a second distillation column to separate the solvent from the toluene in the Rich-Solvent stream. Number of stages = 28, Rich-Solvent enters at stage 28.

    On first Page enter the information as in Figure 9.69 and click Next when complete.

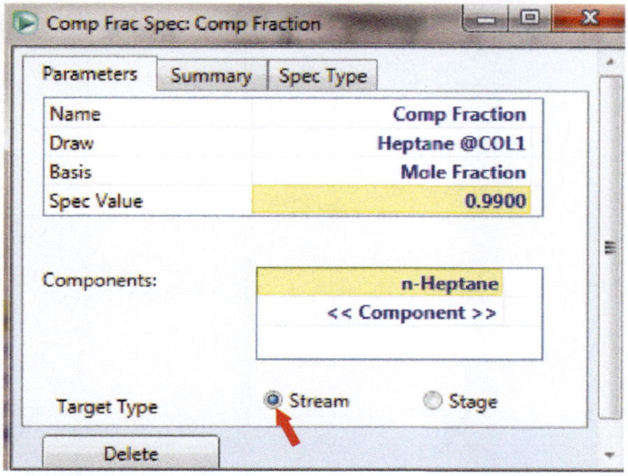

Figure 9.68 Component fraction specification of *n*-heptane.

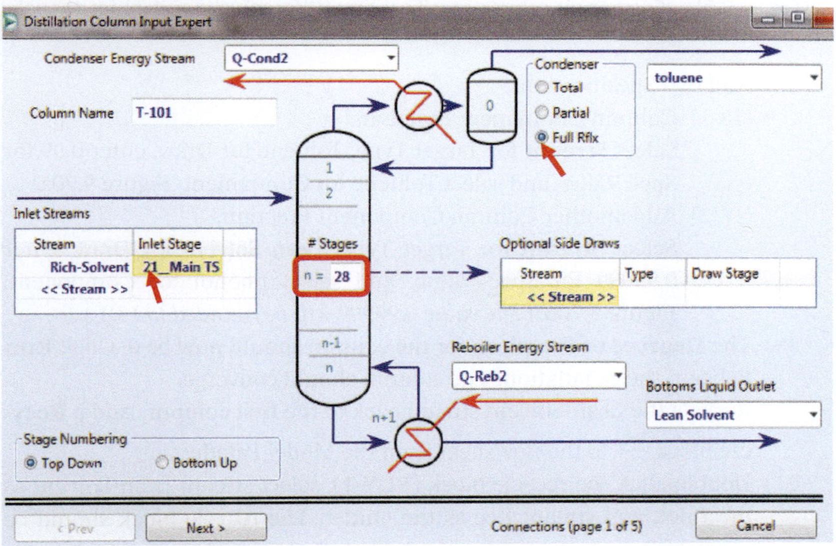

Figure 9.69 Distillation column (T-101), input expert/connection page.

15) Leave the second page, Next: enter the pressures for the condenser and reboiler at 1 bar.
16) Leave the fourth page empty and click Done on page 5.
17) In the **Design** tab, **Monitor** page, Uncheck the active boxes so that there are no active specifications.

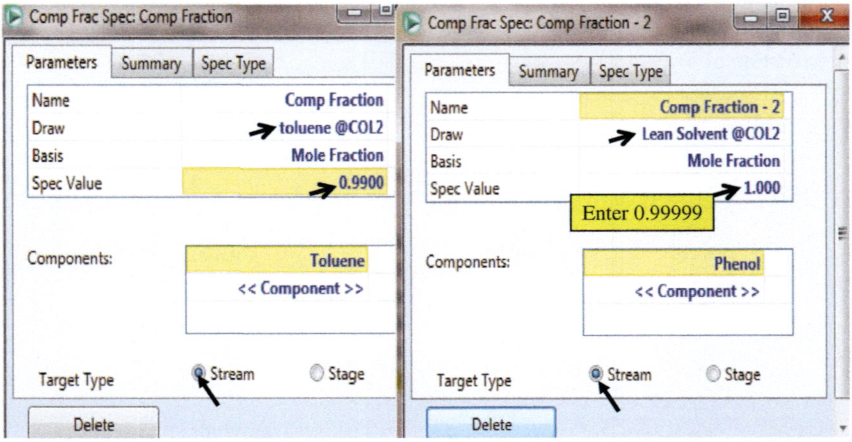

(a) (b)

Figure 9.70 Component column specifications: (a) for toluene and (b) for phenol.

18) Add two Specifications:
    (18-1) Column Component Fraction.
    Select **Stream** for Target Type, Toluene for Draw, enter 0.99 for Spec Value, and select Toluene for Component, Figure 9.70a.
    (18-2) Add another Column Component Fraction.
    Select **Stream** for Target Type, Lean Solvent for Draw, enter 0.99999 for Spec Value, and select phenol for Component, Figure 9.70b. (*The value 0.99999 will be rounded to* 1.0)
19) The **Degrees of Freedom** for the column should now be **0**. Click **Run** to begin the calculation. The column should converge.
20) Recycle the Lean-Solvent stream back to the first column. Add a **Recycle** block to the flowsheet from the Model Palette.
21) Double-click the recycle block (RCY-1), select stream Lean-Solvent as the Inlet, and stream Rec as the Outlet. The recycle block should be solved.
22) Connect the Rec-stream with mixer as a second inlet, see Figure 9.63.
23) Check the results:
    Double-click the first column (T-100) and go to the Cond./Reboiler page, under the Performance tab, make note of the Condenser and Reboiler Duties.
24) Double-click the second column (T-101) and go to the Cond./Reboiler page under the Performance tab, make a note of the Condenser and Reboiler Duties.

The following table will summarize the energy requirements for cases 1 and 2.

	Single-Column Distillation	Extractive Column Distillation
Total Heating Duty kJ/h	1.909e7	7.549e6
Total cooling Duty kJ/h	1.909e7	5.8968e6

9.5.5.3 Summary

For the separation of *n*-heptane and toluene, extractive distillation has a significant advantage in total energy requirements. Adding phenol as a solvent increased the relative volatilities of *n*-heptane and toluene in the mixture and allowed for a much easier separation. However, extractive distillation required more equipment in this case. Therefore, a further analysis on capital versus operational costs would have to be performed in order to make a decision as to which design is the better option.

9.5.6 Azeotropic Distillation: Production of Anhydrous Ethanol Using an Entrainer

An azeotrope is a constant boiling mixture; the vapor and liquid are in equilibrium and have the same composition. A third component is added to the binary mixture which forms a new low-boiling azeotrope with one of the constituents which separates from other original constituent.

The new solvent weakens hydrogen bonds and creates new homogeneous or heterogeneous azeotrope with one of the components of the original mixture. The solvent is fed usually to the bottom part of the column and obtained in the distillate, e.g. the separation of ethanol and water by adding benzene or cyclohexane [9].

The Azeotropic check box on the Solver page of the Parameters tab must be selected for an azeotropic column to converge.

There are two ways to indicate to HYSYS that you are simulating an azeotropic column:

- Enter a negative damping factor, and HYSYS automatically selects the Azeotropic check box.
- Enter a positive value for the damping factor, and select the Azeotropic check box

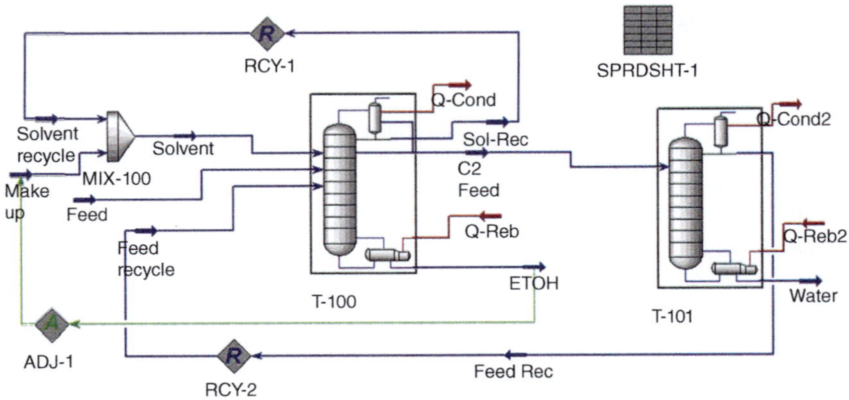

Figure 9.71 PFD of azeotropic distillation process of ethanol–water system.

9.5.6.1 Problem Statement

Ethanol production via fermentation occurs in water, which must later be separated to make anhydrous ethanol (99.95% ethanol). There is an azeotrope in the ethanol–water system at approximately 95 mol% ethanol, which is a barrier to separation. Cyclohexane is one of the solvents used for the production of anhydrous ethanol for food and pharmaceutical usage. It is used as an entrainer: the ternary mixture forms a ternary azeotrope with a different ethanol concentration, which allows ethanol to enrich in the other stream. The azeotropic liquid is separated to recover the entrainer and the ethanol that exits the column in the azeotropic mixture.

The feed to the separation train is a stream at 100 kgmol/h with 87 mol% ethanol and 13 mol% water. Cyclohexane is added to the first column (T-100), and > 99.95 mol% ethanol exits the bottom of the column. The distillate is then separated in a three-phase condenser. The cyclohexane-rich stream is recycled directly to the first column, while the water and ethanol-rich stream are sent to a second column (T-101) from which almost-pure water exits in the bottoms. The distillate of the second column is recycled to the first column. The process flow diagram of azeotropic distillation process of ethanol–water system is shown in Figure 9.71 [9].

9.5.6.2 Simulation

1) Open a new case: Add Ethanol, Water, and Cyclohexane,
   - Fluid package: PRSV.
2) Enter the simulation environment.
3) Add five material streams:
   Feed, Feed Recycle, Solvent, Make Up, and Solvent Recycle.

- **Make Up** Stream: $T = 25\,°C$, $P = 1$ bar, molar flow = 0.01 kmol/h, composition = 1 for cyclohexane.
- **Solvent Recycle** stream: $T = 25\,°C$, $P = 1$ bar, molar flow = 400 kmol/h,
  Composition = 0.5 mol fraction for each of cyclohexane and ethanol.
- **Feed** stream: This is the stream that pumps the ethanol–water mixture into the process.
  Vapor fraction = 0.3, $P = 1$ bar, molar flow = 100 kmol/h,
  Composition: 0.87 for ethanol and 0.13 for water by mole.
- **Feed Recycle** stream: This stream will be the ethanol–water mixture that exits from the condenser of the second column. This stream will be fed back to the first column to prevent losses of ethanol.
  Vap fraction = 0, $P = 1$ bar, mole flow = 25 kmol/h
  Composition: 0.7 for ethanol and 0.3 for water by mole.

4) Add a **Mixer**: Inlet = Make Up and Solvent Recycle, Outlet = Solvent.
5) Add a **Three Phase Distillation** Column.
   In the first window that appears select the Distillation radio button, Click Next.
6) Change the Number of Stages to **62**, Figure 9.72. Make sure that the Condenser is selected to check for two liquid phases. Click Next when complete.

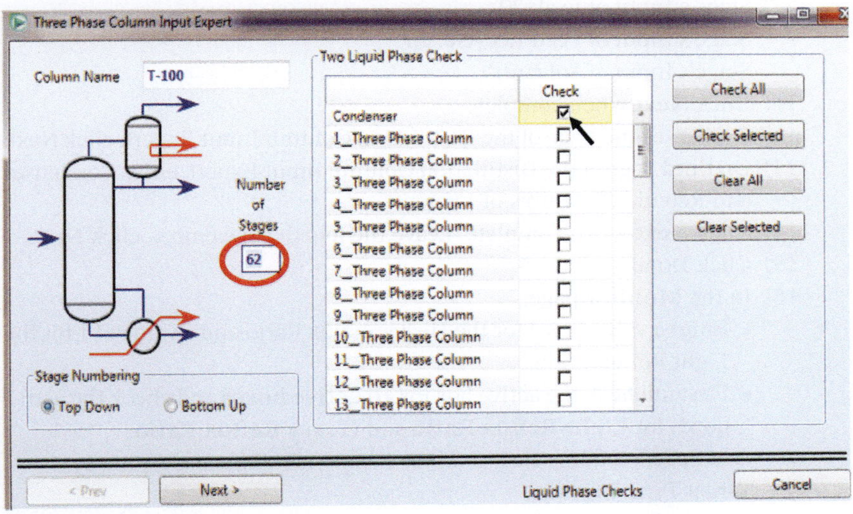

Figure 9.72 Three-phase column input expert/two liquid-phase checking.

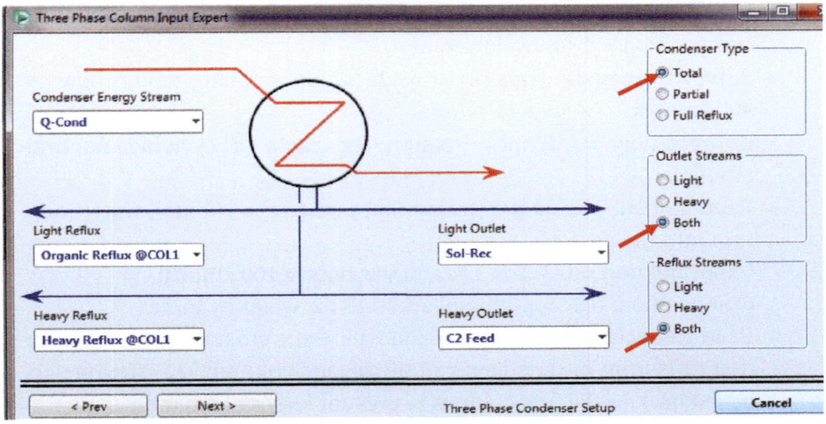

Figure 9.73 Three-phase column input expert/condenser setup.

7) Fill the third window as in Figure 9.73,
   where: Condenser type = Total.
   Outlet Streams = Both.
   Reflux Streams = Both.
8) Click **Next** when complete.
9) In the fourth window, leave all fields blank and click **Next**.
10) The **Distillation Column Input Expert** window will now appear. Fill the first page of the Input Expert as in Figure 9.74, where:
    Stage's Input of feed: 20,
    Stage's Input of Feed Recycle: 20
    Stage's Input of Solvent: 1.
11) Click **Next** when complete.
12) On the second page of the Distillation Column Input Expert, click Next.
13) On third page of the Distillation Column Input Expert, enter Condenser and Reboiler Pressures of 1 bar.
14) Click **Next** when complete. Leave the fourth page empty, click Next.
15) Click Done.
16) In the **Monitor** page:
    - Enter a value of 3.5 for the Heavy Reflux Ratio and a value of 1 for the Light Reflux Ratio.
    - First uncheck the active box for **all active boxes** and check the active boxes for **Light Reflux Ratio** and **Heavy Reflux Ratio**.
17) Add Specification: select *Column Component Fraction*, Figure 9.75.
    Target Type: **Stream**,
    Draw: ETOH@COL1,
    Basis: Mole Fraction,
    Spec Value: 0.9995. Close the window of "Comp Frac Spec."

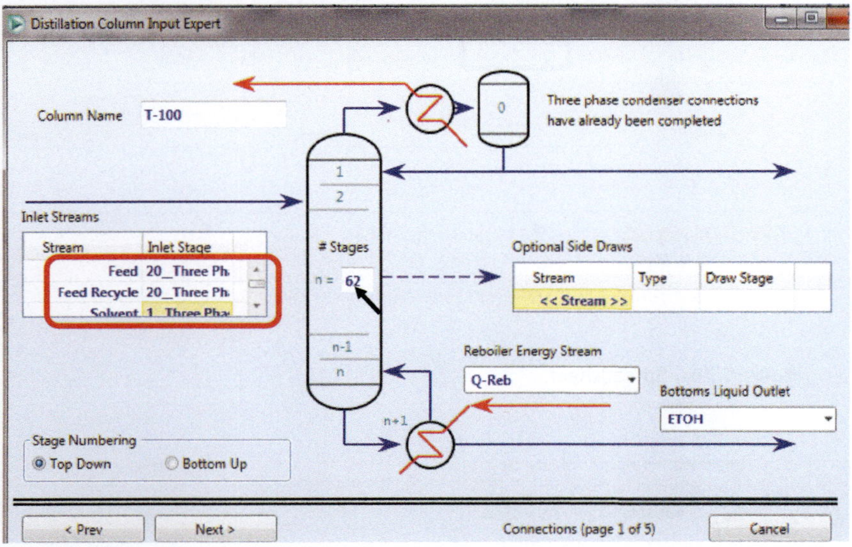

Figure 9.74 Three-phase column input expert/connection page.

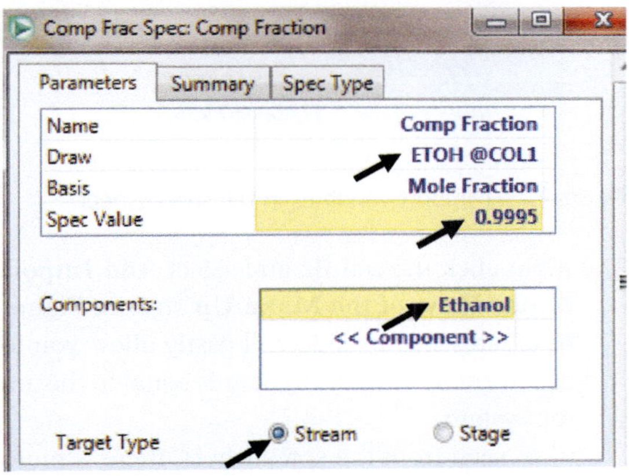

Figure 9.75 Component fraction specification of ethanol.

18) Click Run.
19) Before constructing the second column, an **Adjust** block and a **Spreadsheet** will be added to find the correct flowrate for the **Make Up** stream.
20) Double-click the spreadsheet (**SPRDSHT-1**). Go to the **Spreadsheet** tab, enter the following text in cells **A1** and **A2** as shown in Figure 9.76.
21) Right click the cell **B1** and select **Add Import Variable(s)**, select the **Master Comp Molar Flow** of **Cyclohexane** in stream **ETOH**, Figure 9.77.

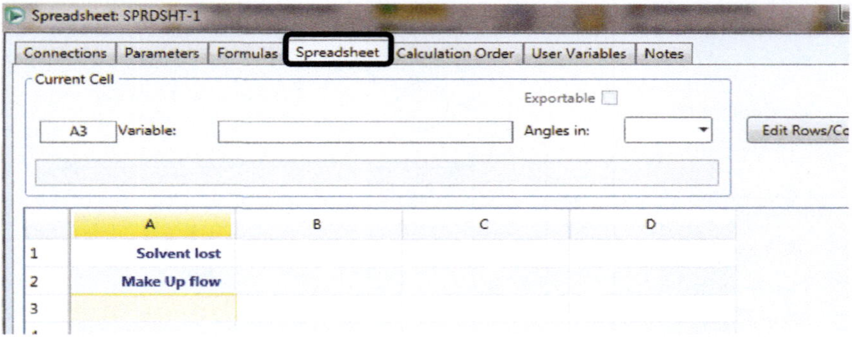

Figure 9.76 Spreadsheet.

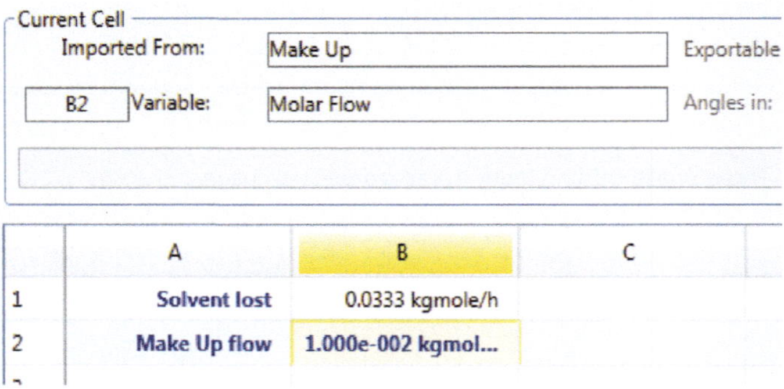

Figure 9.77 Import variables in the spreadsheet.

22) Right click the cell **B2** and select **Add Import Variable(s)**, select the **Molar Flow** of the **Make Up** stream, Figure 9.77. Having these two flow rates side by side will easily allow you to check that the amount of solvent leaving the system is equal to the amount of solvent entering the system.

    As seen from the spreadsheet, there is more solvent leaving the system than entering. This will cause convergence issues when we attempt to close the recycle streams. This is where we will use the adjust block.
23) Add the Adjust block (**ADJ-1**) to the PFD, select the **Adjusted Variable** to be the **Molar Flow** of the **Make Up** stream, select the **Target Variable** to be the **Master Comp Molar Flow (Cyclohexane)** of stream **ETOH**, and set the **Target Value** to cell **B1** in the spreadsheet as in Figure 9.78.
24) The Adjust block will vary the **Make Up** stream flowrate until the amount of solvent leaving the system equals the amount entering

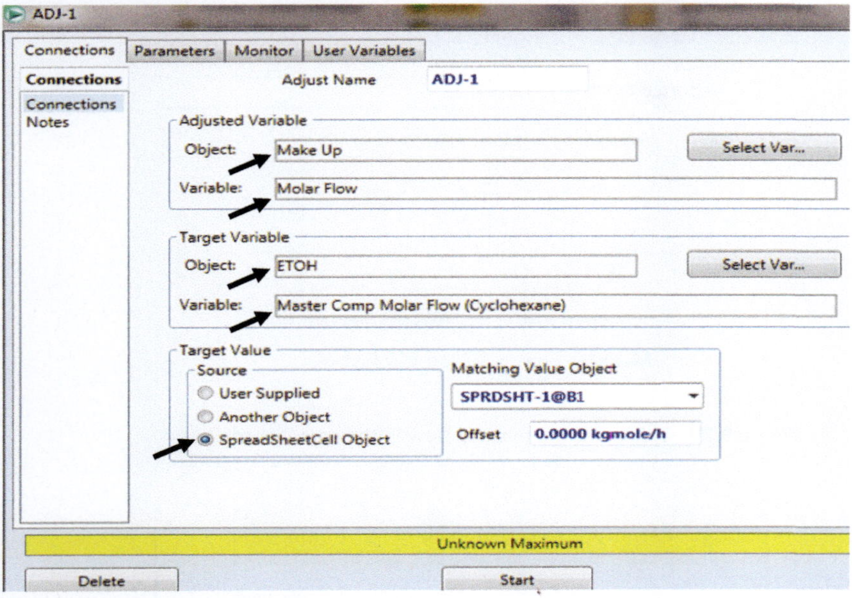

Figure 9.78 Adjusting operation/connection page.

the system. Go to the **Parameters** tab, change the **Tolerance** to **0.001 kgmol/h**, change the **Step Size** to **0.01 kgmol/h**, and enter the iteration to 1000. Click **Start** button to begin calculations. After a few moments the flowsheet will converge.

25) Open the spreadsheet and you will see that the solvent leaving the system is now equal to the solvent entering the system, Figure 9.79.
26) Add a **Recycle** block to close the recycle loop for the solvent.
    Inlet = Sol-Rec, Outlet = Solvent Recycle
    The first column should appear as in Figure 9.80.
27) Now it is ready to add the second column. Add a **Distillation Column** Sub-Flowsheet from the Model Palette.
28) Open the Distillation Column Input Expert window and fill it as in Figure 9.81.
29) Click Next, leave the second page, click Next.
30) Enter Condenser and Reboiler Pressures of 1 bar. Click Next when complete.
31) Leave the fourth page and Click Next.
32) Page 5: Enter Reflux ratio = 0.5, then click the **Done** button.
33) In the **Design** tab, monitor page; deactivate the **Feed Rec Rate.**

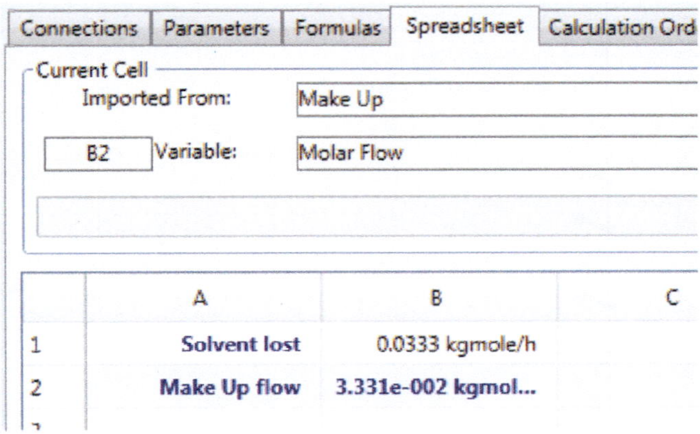

Figure 9.79 Adjusting the makeup flow in the spreadsheet.

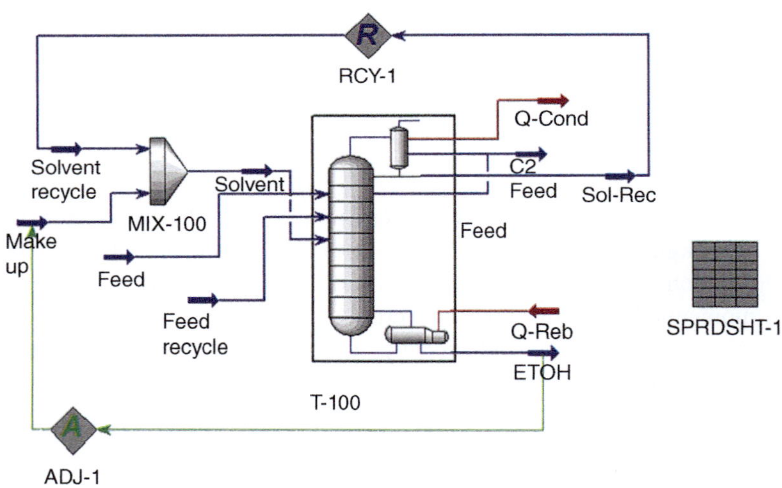

Figure 9.80 PFD of the three-phase column for azeotropic distillation.

34) Adding new specification: Add a Column component fraction, Figure 9.82,
    where the mole fraction of water in the **Water** stream is 0.99.
35) Click **Run** to begin calculations.
36) The last step is to connect the **Feed Recycle** loop. Add a **Recycle** block to the flowsheet.

Inlet = Feed Rec, Outlet = Feed Recycle (*wait some minutes for solving*)

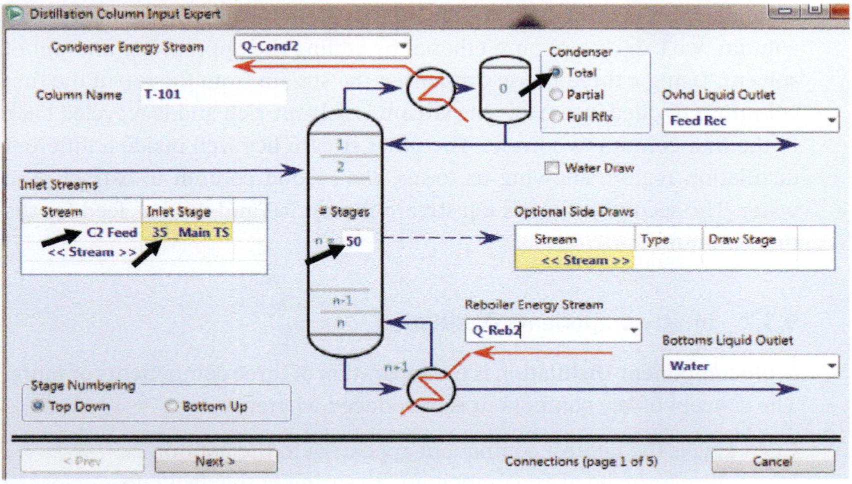

Figure 9.81 The first page of the distillation column shows the separation of water and cyclohexane.

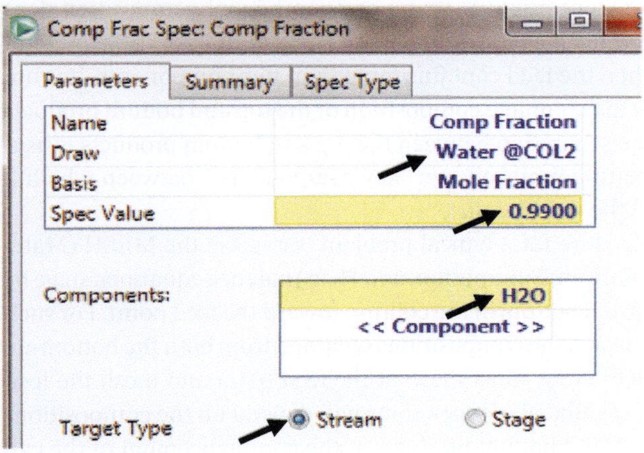

Figure 9.82 Component fraction specification of water.

9.5.6.3 Summary

Azeotropic distillation is a kind of distillation that requires using particular methods to break azeotropes. The use of azeotropic distillation to dehydrate combinations of ethanol and water is a typical historical example. Benzene, pentane, hexane, cyclohexane, heptane, isooctane, acetone, and diethyl ether are just a few of the solvents that can be exhausted during this process. In this instance, water and ethanol are separated to create anhydrous ethanol using cyclohexane as the entrainer. From the bottom of the first column, we can extract pure ethanol by adding the appropriate amount of solvent. Using a three-phase condenser, the stream from the top of the first column is divided into two: One stream is solvent-rich and is recycled back to the first column as solvent. The other stream lies well inside a different distillation region, allowing us to use the second column to extract pure water. The second column's top stream is recycled and used as feed for the first column.

9.5.7 Multi-components Distillation

Multicomponent Distillation is the separation of three components or more. The concept of key component is introduced where:

Light key: is the lightest component appearing in the bottom.
Heavy key: is the heaviest component appearing in the distillate.

The problem of determining the stage and reflux requirements for multicomponent distillations is much more complex than for binary mixtures. With a multicomponent mixture, fixing one component composition does not uniquely determine the other component compositions or the stage temperature. Also, when the feed contains more than two components, it is not possible to specify the complete composition of the top and bottom products independently. The separation between the top and bottom products is usually specified by setting limits on two "key components," between which it is desired to make the separation.

The normal procedure for a typical problem is to solve the MESH (Material balance, Equilibrium, Summation, and Heat) balance equations stage by stage, from the top and bottom of the column toward the feed point. For such a calculation to be exact, the compositions obtained from both the bottom-up and top-down calculations must mesh at the feed point and mesh the feed composition. The calculated compositions will depend on the compositions assumed for the top and bottom products at the commencement of the calculations. Though it is possible to match the key components, the other components will not match unless the designer was particularly fortunate in choosing the trial top and bottom compositions.

The calculations are given in Sections 9.3.3 and 9.3.4, and examples of separating multi-components are also given in Section 9.3.4.2 using rigorous columns. In this example, it will use the shortcut column to determine the **reflux ratio** and the **number of theoretical stages** for multicomponent distillation. The results obtained from the shortcut column will be utilized in the rigorous column [5, 7].

9.5.7.1 Problem Statement

A feed of saturated liquid at 80 °C is composed of 5% propane, 10% *i*-butane, 40% *i*-pentane, 36% *n*-hexane, and 9% *n*-heptane all by mole. It is required to fractionate these components in a distillation column, where in the distillate the *n*-hexane is 1% and, in the bottom, the *i*-pentane is 1%. The feed enters the column at rate of 2500 kmol/h, the pressure of condenser and reboiler is 3.6 and 3.7 bar, respectively. What are the minimum number of stages required? What are the actual number of stages required? What is the feed stage in the column? [15].

9.5.7.2 Simulation

1) Open a new case, and add the components: propane, *i*-butane, *i*-pentane, *n*-hexane, and *n*-heptane.
   Property package: Peng–Robinson
2) Enter the Simulation environment:
   - Add the Feed with the following conditions:
     Vap. Fraction = 0, T = 80 °C, Feed flow rate = 2500 kmol/h, and the mole fraction: 0.05 propane, 0.1 i-butane, 0.4 i-pentane, 0.36 n-hexane, and 0.09 n-heptane.
3) Add the shortcut distillation  from the object palette/Columns group, and fill the first page with the data required as shown in Figure 9.83.

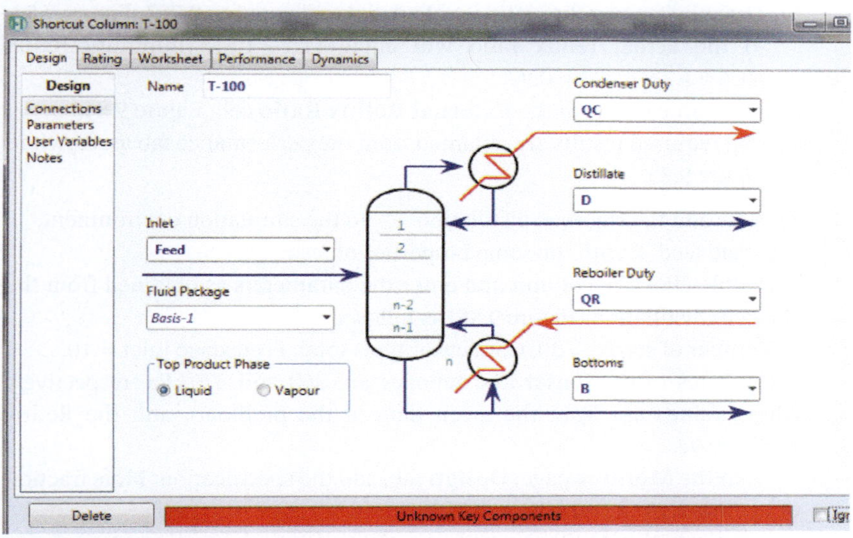

Figure 9.83 First page of the shortcut distillation, Design tab, Connections page.

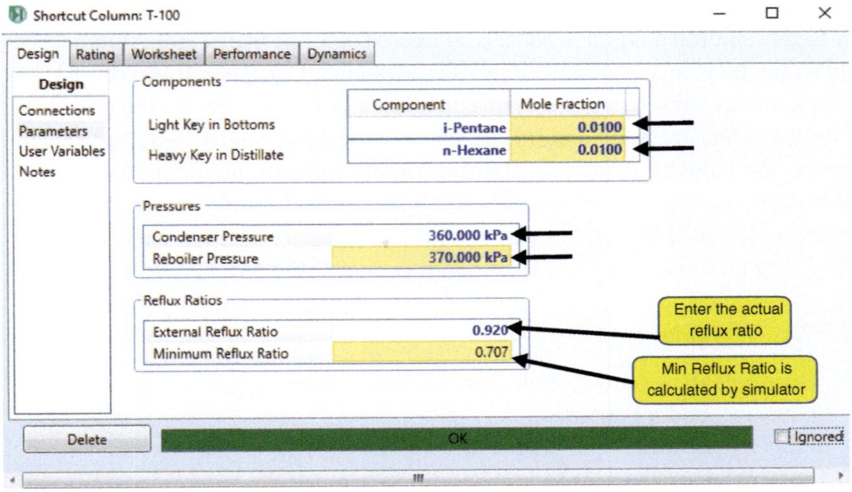

Figure 9.84 Design tab/Parameters page of the shortcut distillation with data required.

4) In the parameters page enter the data as in Figure 9.84.
   According to Section 9.3.3, and as mentioned above, the light key here is i-pentane (has mole fraction of 0.01 in the bottom), and the heavy key is n-hexane (has mole fraction of 0.01 in the distillate).
5) The minimum reflux ratio is calculated and equal to 0.707, Figure 9.84, so the actual reflux ratio will equal $(1.1-1.5) \times$ minimum reflux ratio $\approx 1.3 \times 0.707 = 0.92$.
6) This value enters in the **External Reflux Ratio** cell, Figure 9.84.
7) The required results are obtained from the performance tab as shown in Figure 9.85.
8) Now add the rigorous distillation  to the simulation environment.

Create feed_2 with the same properties of feed.

Double-click the column and enter the parameters as obtained from the shortcut distillation (Figure 9.85) as follows:

Number of stages = 20, Condenser type: total, Feed stage inlet = 10.

Pressures of condenser and reboiler are 360 and 370 kPa, respectively (these values are from the given data of the problem), and the Reflux ratio = 0.92.

From the **Monitor** page, **Design** tab, add the specification: Mole fraction of i-pentane in the reboiler = 0.01 by mole.

Deactivate the Distillate rate and activate the comp fraction, the column will converge, Figure 9.86.

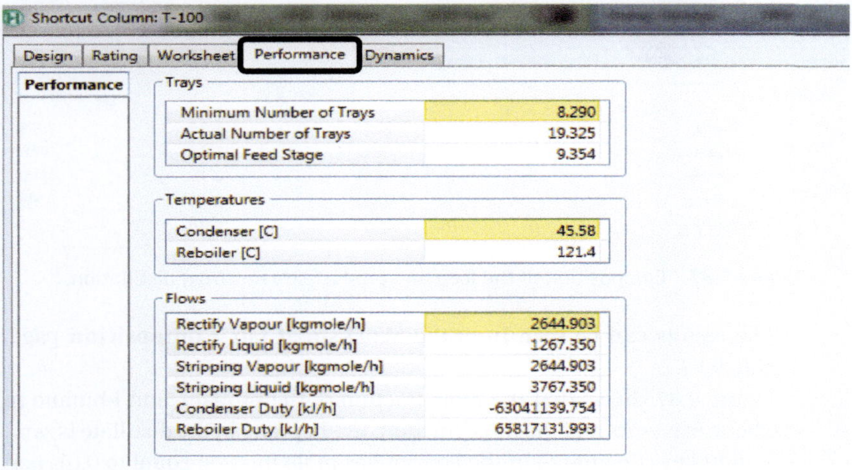

Figure 9.85 Results obtained from the performance tab of shortcut distillation.

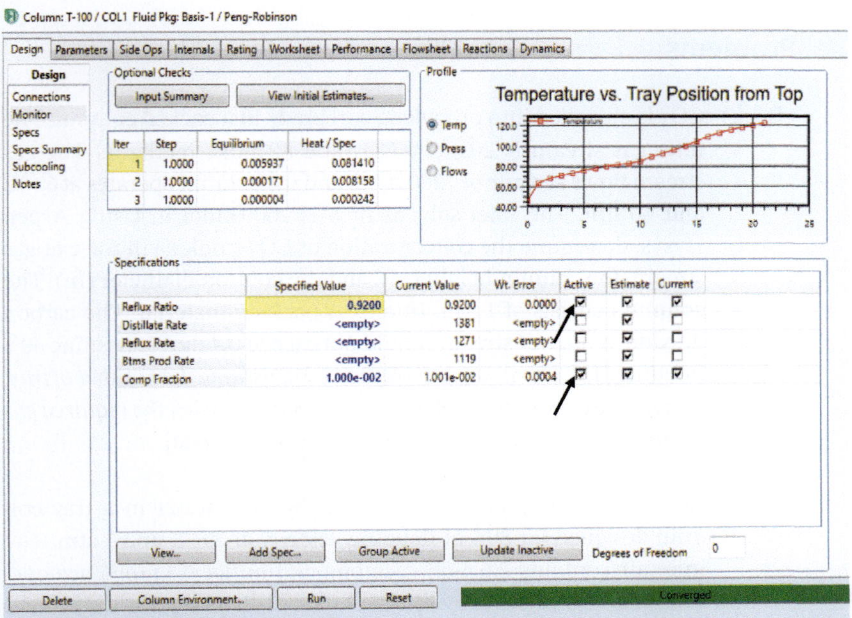

Figure 9.86 Design tab, Monitor page of the rigorous distillation column.

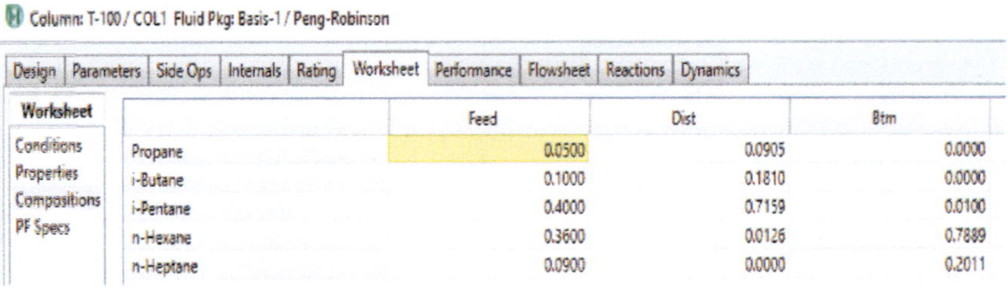
Column: T-100 / COL1 Fluid Pkg: Basis-1 / Peng-Robinson

Design | Parameters | Side Ops | Internals | Rating | Worksheet | Performance | Flowsheet | Reactions | Dynamics

Worksheet: Conditions, Properties, Compositions, PF Specs

	Feed	Dist	Btm
Propane	0.0500	0.0905	0.0000
i-Butane	0.1000	0.1810	0.0000
i-Pentane	0.4000	0.7159	0.0100
n-Hexane	0.3600	0.0126	0.7889
n-Heptane	0.0900	0.0000	0.2011

Figure 9.87 Composition of the feed and product streams from distillation.

The results can be seen from the **Worksheet** tab, **Composition** page, Figure 9.87.

Figure 9.87 shows that the mole fraction of the propane and *i*-butane in the bottom is zero, and the mole fraction of *n*-heptane in the distillate is zero. The light key (*i*-pentane) in the bottom has mole fraction equal to 0.01, and heavy key (*n*-hexane) in the distillate is 0.0126; that means the aim of this column is to separate *i*-pentane from *n*-hexane.

Problems

P.9.1 CO_2 is absorbed into propylene carbonate in a packed column. The inlet gas stream is 20 mol% CO_2 and 80 mol% methane. The gas stream flows at a rate of 7200 m^3/h and the column operates at 60 °C and 60 atm. The inlet solvent flow is 2000 kmol/h. Using Aspen Hysys, determine the concentration of CO_2 (mole%) in the exit gas stream, the column height (m), and the column diameter (m). The column consists of 1.5 in. Intalox-A packing material. The carbon dioxide in the exit stream should not exceed 0.4 mol%. Assume 60% flooding [13]. [*First, solve as tray column, assume the number of tray, and see the composition of outlet gas, when you reach the required gas composition in outlet, change to packed column one*]

P.9.2 Acetone is being absorbed from air by pure water in a tray column designed for 80% of flooding velocity at 293 K and 1 atm. The inlet air contains 2.6 mol% acetone and outlet 0.5 mol% acetone. The total gas inlet flow rate is 14.0 kmol/h. The pure water inlet flow is 45.36 kmol/h. What is the number of trays required? If the column is of packed type, randomly packed with metal Raschig rings, 1.5 in, what is the height of the column? If the efficiency of the column is 0.4, what will the height be? [3, 11].

P.9.3 A gas stream flowing at a rate of 100 kmol/h, temperature of 100 °C, and pressure of 6000 kPa enters a gas absorber. The primary objective of the absorber is to absorb CO_2 contained in the feed stream by contacting counter-currently with methanol solvent in an absorber. The gas stream contains 0.35CO, 0.002H_2O, 0.274CO_2, 0.37H_2, 0.002CH_4, and 0.002N_2. Methanol at 30 °C and 6000 kPa is used as an absorbent solvent. The molar flow rate of methanol liquid is 330 kmol/h. Determine the number of trays required to achieve 0.01 mol fraction of CO_2 in the exit stream, column diameter, and height for efficiency of the column = 0.6 [20].

P.9.4 Ammonia is removed from 10% ammonia – air mixture by scrubbing with water in a tray tower so that 99% of NH_3 is removed, and given that: Liquid flow rate = 1.6 kg/s, Gas flow rate = 0.95 kg/s, calculate

(a) Theoretical number of trays. (b) Number of trays if the column efficiency = 0.7.

The operating conditions of the column are 25 °C and 1 atm [3].

P.9.5 A sulfur burner gas contains 8% SO_2. 98% of SO_2 is to be removed by absorption with water in a packed tower at 15 °C and 1 atm. The liquid flow rate is 4500 kmol/h. Calculate the height of the tower if the flow rate of the gas is 100 kmol/h [11].

P.9.6 An absorption packed tower is fed with a mixture of air and ammonia, the partial pressure of NH_3 is 5 kPa at a rate of gas 4800 kg/h m^2 to recover at least 90% of NH_3, the tower is supplied with water at a rate of 7200 kg/h m^2. The operating conditions of the column are 25 °C and 1 atm. What is the height of the tower for this duty? [4].

P.9.7 A mixture with 33% *n*-hexane, 37% *n*-heptane, and 30% *n*-octane is to be separated in a distillation column. The feed stream is 60% vapor and a temperature is 105 °C. The distillate product should contain 0.01 mol fraction *n*-heptane and a bottom product with 0.01 mol fraction *n*-hexane. The column will operate at 1.2 atm. The feed molar flow rate is 100 kmol/h [11].

a) Calculate the complete product composition and minimum number of ideal plates at infinite reflux.
b) Estimate the number of ideal plates required for the separation if the reflux ratio is 1.5 times the minimum reflux ratio.
c) Find the optimum feed tray.

P.9.8 A mixture of 100 kmol/h saturated liquid contains 30 mol% benzene, 25% toluene, and 45% ethyl benzene is to be separated by distillation column at atmospheric pressure, with 98% of the benzene and only 1% of the toluene is to be recovered in the distillate stream. The reflux ratio is 2. Calculate:

a) The minimum number of ideal plates.
b) The approximate composition of the product.
c) The actual number of trays for 70% efficiency.
d) The minimum number of trays, if the distillation were carried out at 0.2 atm [3].

P.9.9 The feed to a distillation column is at 405 °F and 100 psia enters at 1272 lbmol/h with the following compositions (in mole fraction): Ethane 0.0148, propane 0.7315, *i*-butane 0.0681, *n*-butane 0.1462, *i*-Pentane 0.0173, *n*-pentane 0.015, *n*-hexane 0.0071. Heavy key (HK) in distillate is *i*-butane 0.02 mol fraction; light key (LK) in bottom is propane 0.025 mol fraction for reflux ratios of 1.5. Calculate the minimum reflux ratio and column performance [11].

P.9.10 The feed to a distillation column is at room conditions ($T = 25\,°\text{C}$, $P = 1$ atm). The concentration of the feed stream is 50% ethanol and 50% iso-propanol in mass fractions. The feed is at a rate of 1000 kg/h. Assume number of trays to be 24 and reflux ratio equal 3. Condenser type: Total, all pressures are at atmospheric pressure [11]. Fluid package: NRTL.

a) For the reboiler heat duty equal to 1200,000 Btu/h, find the conditions of the exit streams.
b) If the duty of reboiler is unknown and overhead ethanol concentration is 0.65 by mass, what is the amount of the reboiler load in Btu/h?
c) Calculate the reflux ratio that would give an ethanol concentration of 80% with a reboiler duty of 3000,000 Btu/h is to be calculated.

P.9.11 A mixture of saturated liquid with 4% *n*-pentane, 40% *n*-hexane, 50% *n*-heptane, and 6% *n*-octane is to be distilled at 1 atm with 98% of the hexane and 1% of the heptane recovered in the distillate.

a) What is the minimum reflux ratio for a liquid feed at the boiling point?
b) What are the temperatures and compositions in the upper and lower invariant zones? [3].

***Hint**: First, one must know the light key and heavy key, so for Feed = 100 kmol/h and by hand calculations on mass balance, the following can be concluded:*

		x_F	Fx_F	Moles in D	x_D	Moles in B	x_B
	n-C_5	0.04	4	4	0.092	0	0
LK	n-C_6	0.4	40	39.2	0.897	0.8	0.014
HK	n-C_7	0.5	50	0.5	0.011	49.5	0.879
	n-C_8	0.06	6	0	0	6	0.107
				$D = 43.7$		$B = 56.3$	

Now one can perform the calculations using the HYSYS.

P.9.12 Furfural is extracted from water by toluene as a solvent (S) at 25 °C and 1 atm pressure in a 3 stage counter-current multi-stage extractor as shown in Figure 9.88. If the feed (F) is 500 kg/h with 5% by wt. of furfural, what is the solvent required to recover 99% of the furfural? The vessels V-100, V-101, and V-102 are three-phase separators [4].

P.9.13 A mixture of 15 kg/h of acetic acid and water containing 30 wt.% acid is to be extracted by 140 kg/h of pure diisopropyl ether at 20 °C and 1 atm in tray column. Calculate the number of equilibrium stages required to achieve a concentration of 2 wt.% acetic acid in the product raffinate stream [4].

P.9.14 A countercurrent liquid–liquid extractor is used to remove acetone from a feed that contains 50% acetone (A), 50 wt.% water (B). Pure

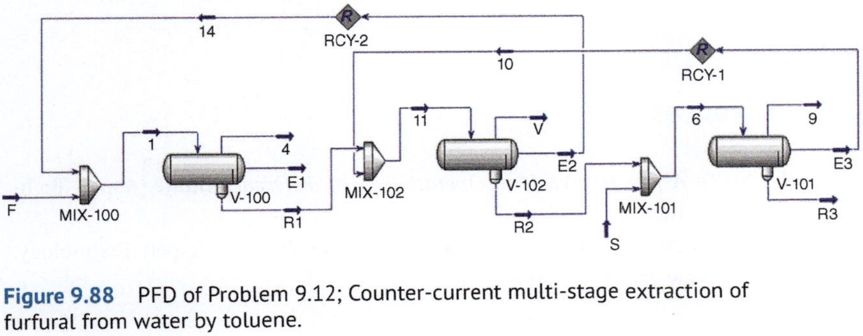

Figure 9.88 PFD of Problem 9.12; Counter-current multi-stage extraction of furfural from water by toluene.

methyl isobutyl ketone (MIBK), at a flow rate of 80 kg/h, is used as the solvent in this separation. A feed flow rate of 100 kg/h is to be treated. It is desired to have a final raffinate of 4 wt.% acetone. The operation takes place at 25 °C and 1 atm. Determine the number of stages necessary for the separation as specified [10]. The system is similar to Problem 9.12.

P.9.15 100 kg/h of solution contains acetic acid 10% by weight in vinyl acetate is extracted by 100 kg/h water as a solvent in cross-current mixer-settler. How many theoretical stages are required to recover 95% of the acid? All the conditions are in 25 °C and 1 atm [11].
Hint: execute the calculations by stage-to-stage calculation.
Repeat the calculations using counter-current operation.

P.9.16 1000 kg/h solution consists of 0.4 water, 0.2 acetone, and 0.4 acetic acid by mass. The acetone and acetic acid to 0.002 mass fraction each are extracted by using chloroform as solvent. All the conditions are 25 °C and 1 atm. Use extraction column with seven stages to execute the duty, property package is UNIQAUC, and use the vapor model: PR in the activity model specification. How much chloroform is required? [14].

P.9.17 Methylcyclohexane (MCH, C_7H_{14}) and toluene are separated by extractive distillation using phenol solvent. The column consists of 21 trays, total condensation, reflux ratio is 8, distillate rate is 200 lbmol/h, and the feed (200 lbmol/h of MCH and 200 lbmol/h of toluene at 20 psia, and 220°F) enters the column in stage 14. First perform the separation without adding solvent (220°F, 20 psia), then use the solvent which enters the column at seventh stage. Condenser pressure = 16 psia, and reboiler pressure = 20.2 psia, distillate rate = 200 lbmol/h [4]. Property package: UNIQAUC.

References

1 (2017). *Aspen HYSYS Unit Operations V10, Reference Guide*. Aspen Technology, Inc.
2 HYSYS 2004.2 (1981–2005). *Tutorial and Application*. Aspen Technology, Inc., New Jersey.

3 Coulson and Richardson's (2001). *Chemical Engineering*, 5e, vol. 2.

4 Seader, J.D., Henley, E.J., and Keith Roper, D. (2011). *Separation Process Principles Chemical and Biochemical Operations*, 3e. John Wiley & Sons, Inc.

5 Coulson and Richardson's (2003). *Chemical Engineering, Chemical Engineering Design*, 3e vol. 6. (ed. R. K. Sinnott).

6 Backhurst, J.R. and Harker, J.H. (1973). *Process Plant Design*. Heinemann Educational Books Ltd.

7 Ludwig, E.E., Rouge, B., and Louisiana (1999). *Applied Process Design for Chemical and Petrochemical Plants*, 3e, vol. 2. Butterworth-Heinemann.

8 Turton, R. and Shaeiwitz, J.A. (2017). *Chemical Process, Equipment Design*. Pearson Education, Inc.

9 AspenTech®, aspenONE, (2012), Azeotropic Distillation with Aspen HYSYS V8.0, production of Anhydrous Ethanol Using an Entrainer.

10 Haydary, J. (2019). *Chemical Process Design and Simulation Aspen Plus and Aspen HYSYS Applications*. John Wiley & Sons, Inc.

11 Binay, K. and Dutta, B. (2007). *Principles of Mass Transfer and Separation Process*. New Delhi: PHI Learning Private Limited.

12 AspenTech®, aspenONE, (2012), Separation of Acetone-Water with Aspen HYSYS V8.0 Liquid-Liquid Extraction with 3-Methylhexane as the solvent.

13 Kamaruddin Abd Hamid, M. (2007). *HYSYS®: An Introduction to Chemical Engineering Simulation For UTM Degree++ Program*. University Teknology Malaysia.

14 AL-Malah, K.I.M. (2017). *ASPEN PLUS®, Chemical Engineering Applications*. John Wiley & Sons, Inc. All rights reserved.

15 Chaves, I.'n.D.'o.G. (2016). *Process Analysis and Simulation in Chemical Engineering*. Springer.

16 Advanced Process Modeling Using, AspenTech Customer Education Training Manual Workbook, Course Number EA1000.06.09, 2007.

17 EHY202 (2017). *Aspen HYSYS®: Advanced Process Modeling Topics, Course Number EHY202.10.01, AspenTech Customer Education, Training Manual*. Aspen Technology, Inc. 20 Crosby Drive, Bedford, Massachusetts 01730, USA.

18 AspenTech®, aspenONE, (2012), Extractive Distillation for Heptane-Toluene Separation using Aspen HYSYS® V8.0.

19 Lastari, F., Pareek, V., Trebble, M. et al. (2012). Extractive distillation for CO_2–ethane azeotrope separation. *Chemical Engineering and Processing* 52: 155–161.

20 Ghasem, N. (2012). *Computer Methods in Chemical Engineering*. Taylor & Francis Group, LLC.

3 Coulson and [illegible] Chemical [illegible] vol. 2 [illegible]
4 Seider, [illegible] Process [illegible] Principles: Computer [illegible] and [illegible] Wiley & Sons, Inc.
5 [illegible] and [illegible] Chemical [illegible]
[illegible] Design [illegible]
6 [illegible]
[illegible] Education [illegible]
7 [illegible]
[illegible] vol. 2 [illegible]
8 [illegible]
Design [illegible]
9 [illegible] ANSYS [illegible]
ANSYS 20.0 [illegible] using [illegible]
10 [illegible]
[illegible]
11 [illegible] and [illegible]
[illegible] Process [illegible]
12 [illegible] Water with [illegible]
[illegible] ANSYS [illegible] as the [illegible]
[illegible]
13 [illegible] for [illegible]
and [illegible] University
[illegible]
14 [illegible] Analysis [illegible]
[illegible]
15 [illegible] Process [illegible]
[illegible]
16 Advanced [illegible]
Training [illegible]
17 [illegible] Modeling [illegible]
[illegible]
[illegible]
[illegible]
18 AspenTech [illegible]
[illegible]
19 [illegible] distillation [illegible]
[illegible] and Processing [illegible]
[illegible]
20 [illegible] Engineering [illegible]
[illegible]

10 Refinery Process

10.1 Refinery of Crude Oil

10.1.1 Assay Setup

Because each oil field has distinct hydrocarbon specifications, crude oil is a complicated mixture of hydrocarbons (liquids and gases) that contain numerous hydrocarbon compounds with diverse looks and compositions [1]. The first step in establishing a refining case in Aspen HYSYS Petroleum Refining is to choose and describe crude tests. Before the incoming crude feeds can be used in the Simulation Environment, their properties must be ascertained. This is done within the HYSYS Properties Environment [2].

Reliable phase behavior predictions and compositional modeling studies require a correct description of the plus fractions' physical properties. Refiners can utilize assay data to help them decide if a feedstock of crude oil is suitable for use in a specific petroleum refinery or if there are any potential issues with yield, quality, production, the environment, or other issues [2].

You can compute and record the physical and petroleum parameters of the crude oil stream using the petroleum assay database found in Aspen HYSYS Refining. A vector called the petroleum assay is used to contain both the physical and assay parameters for a given list of components. All the properties utilized in a standard HYSYS simulation case are considered physical properties [1, 3].

A component list usually contains pseudo-components (hypothetical components) and library components (e.g. methane to *n*-pentane). Since HYSYS Refining is built on a flexible structure, it is not necessary to have a predetermined list of pseudo components. Furthermore, HYSYS Refining can make use of pre-existing lists of fake components produced by the

Chemical Process Simulations using Aspen Hysys, First Edition. Khalid W. Hameed.

HYSYS Oil Environment. A value for a physical and assay attribute is stored in each component [3, 4].

10.1.1.1 Adding a New Assay

To import the petroleum assay, using the "Petroleum Assays" button, in the **Home** tab of the Ribbon as shown in Figure 10.1 [5].

By opening this page, you will get access to the Assay Management ribbon as shown in Figure 10.2.

One can find the "New Assay" button in the far left. Click **dropdown** arrow to select between these options, Figure 10.3.

- Import from Library – Select an assay from the Assay Library.
- Import from File – Import a supported Excel, CSV (comma-separated values file), or AFAM case with assay data.
- Manually Enter – Manually enter data into the Assay Table.

By default, clicking the "New Assay" button will select the first option "Import from Library." Once the Assay Library is clicked, a new window is opened as shown in Figure 10.4 that is required to choose one of the common

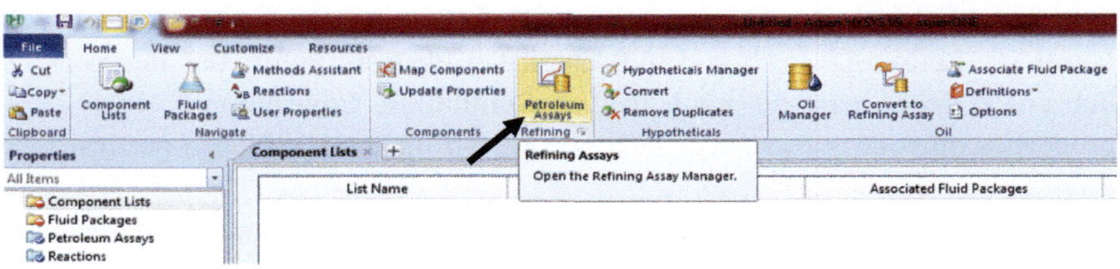

Figure 10.1 Importing the petroleum assay, Properties environment.

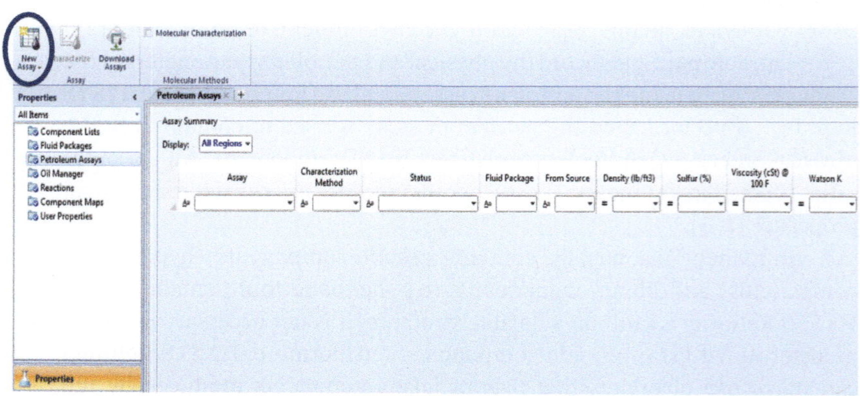

Figure 10.2 Assay management ribbon.

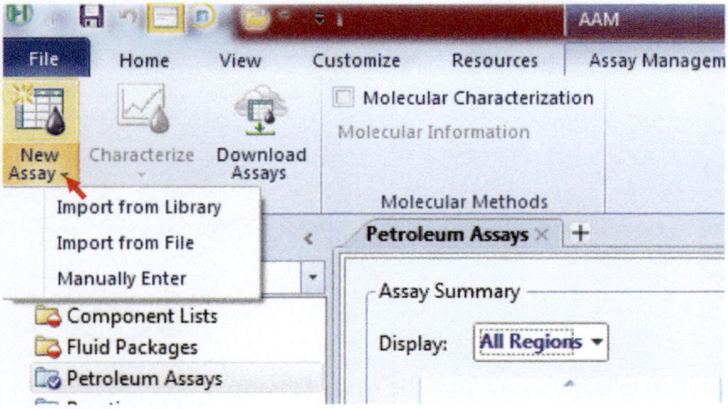

Figure 10.3 How you can add a new assay.

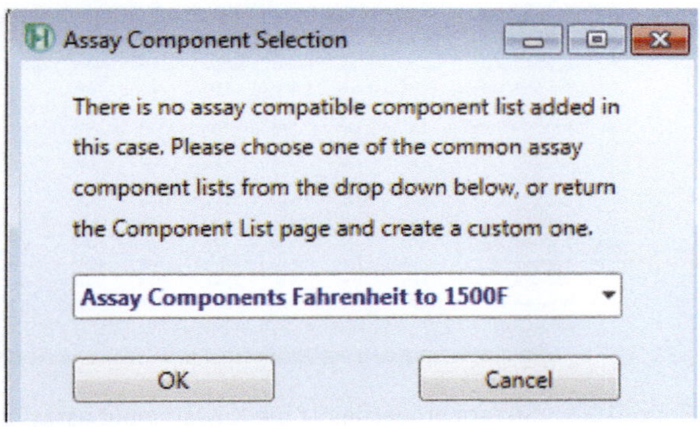

Figure 10.4 Choosing Assay Components Fahrenheit to 1500 °F.

assay component lists from the drop-down list. From drop-down list choose "**Assay Components Fahrenheit to 1500 °F,**" click OK, Figure 10.4.

There will be a large number of assays to select from and multiple ways to filter the results. In this example, one can import an existing assay from the Assay Library called "**Kuwait**" crude. To find the specific crude to use in this case, type the name "**Kuwait**" in the cell "Assay name" and select the correct crude from the list, which will then be highlighted in yellow. To finalize the selection, click "OK" in the bottom-right corner, as shown in Figure 10.5. After the assay has been imported, you can verify stream properties by checking the "**Input Assay**" and "Conventional Results" place in the left-hand Navigation Pane under the Assay name, Figure 10.6.

10.1.1.2 Manually Creating Assays

Since there are a variety of ways where assay information can be entered into a refining model, HYSYS Petroleum Refining provides a way for users

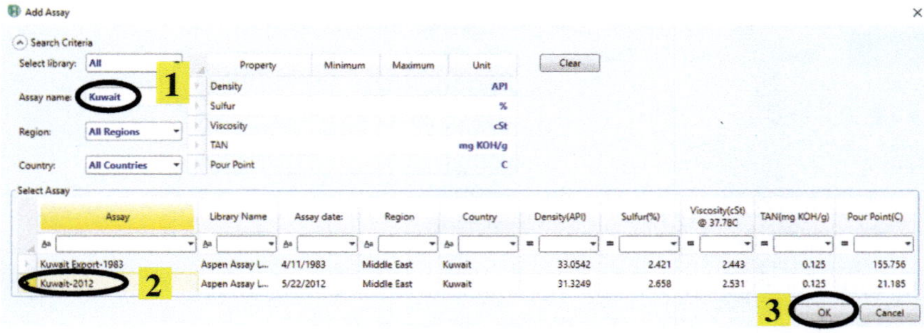

Figure 10.5 Import an existing assay from the Assay Library.

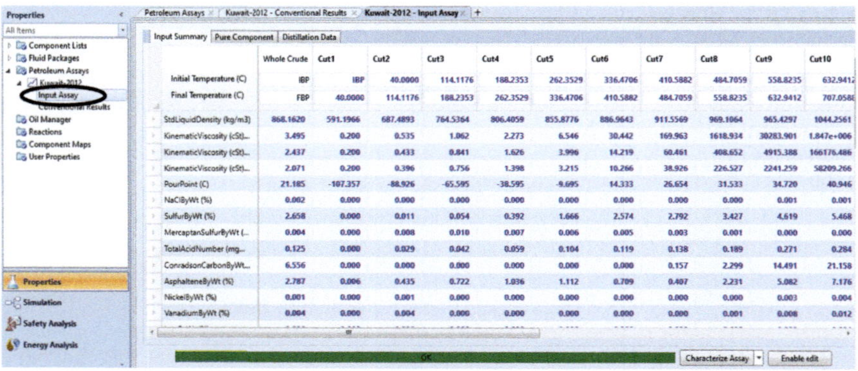

Figure 10.6 Stream properties of the Kuwait input assay.

to manually enter assay data by choosing the **Manually Enter** option when creating a new assay. This will open up the New Assay window, giving a number of custom options to format your assay table, as shown in Figure 10.7 [5, 6].

There are three options to enter assay data:

- Multi-Cut Properties – This is ideal if you have data for specific assay cuts that you wish to enter, such as Cut Yields. Here, you can set the number of cuts from your data and enter the boiling-point ranges in the table provided.
- Single-stream Properties – This option is ideal if you want to evaluate an assay from distillation data and bulk (whole crude) properties to create a single stream for your model.
- Backblending – This will open a custom form designed specifically for users who have data from product streams and wish to simulate a backblended feed by entering pure component, distillation, and property data for specific products [5, 6].

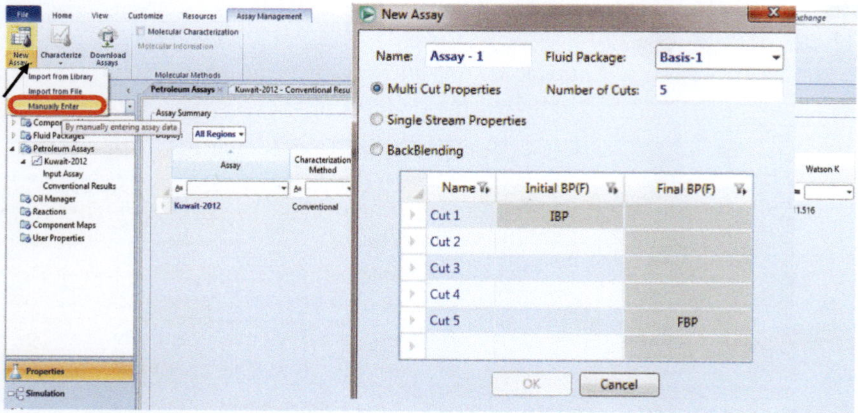

Figure 10.7 Entering assay data manually.

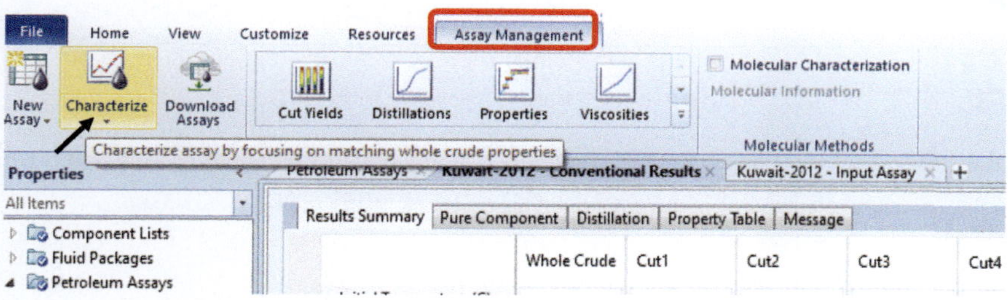

Figure 10.8 Characterizing the whole crude properties.

10.1.1.3 Characterizing Assays

Under the **Assay Management** tab, there are also options for different assay characterizations using the "**Characterize**" button. The options include matching assay cut properties and whole crude properties, and molecular characterization, as shown in Figure 10.8.

The default option to characterize when clicking the **Characterize** button will be to match the **Whole Crude properties** given in the Input Assay table, which means matching the bulk values that have been given by the user will take precedence over matching values for specific cuts [5].

10.1.1.4 Molecular Characterization

When Molecular Characterization is selected, the user is given a list of hydrocarbon molecules present in the crude, Figure 10.9. In addition, viscosity coefficients are also available. After the assay has been characterized using Molecular Characterization, the user can change the characterization

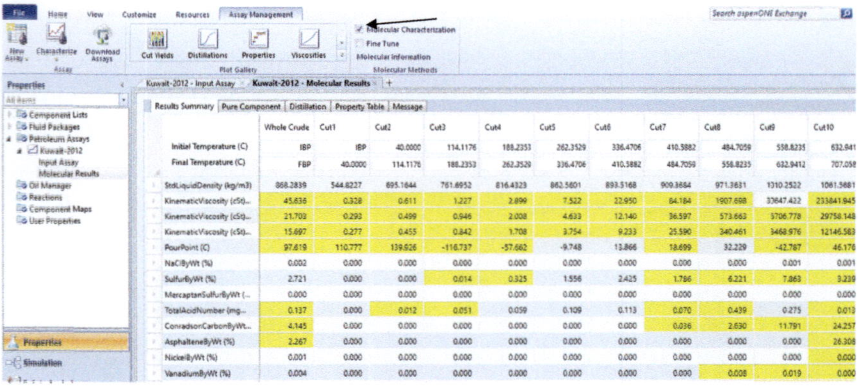

Figure 10.9 Molecular Characterization of the Kuwait assay.

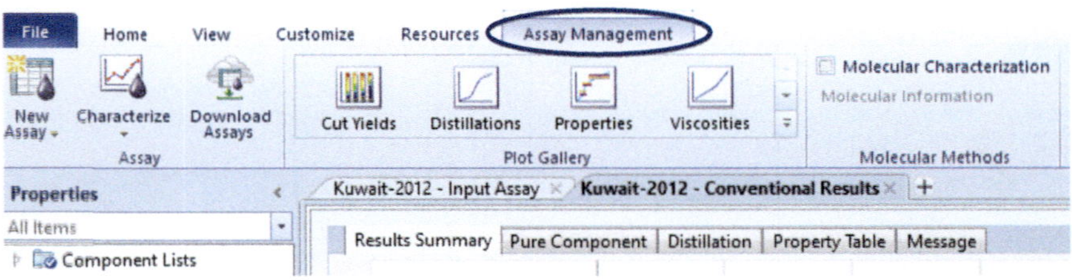

Figure 10.10 Properties that can be plotted for assays.

method by unchecking the "Molecular Characterization" box and using the "Characterize" button.

The properties that can be plotted for assays are shown under the "**Assay management**" tab, in the "Plot Gallery" section, Figure 10.10.

In this example, we have plotted a distillation curve for the Kuwait crude, as shown in Figure 10.11. You can change the graph's coordinates, type, legend, text, etc. from the ribbon [5, 7].

10.1.1.5 Plotting Assay Comparison Data

To compare data from more than one crude assay, the second crude must be added.

Again, from **Home** tab of the Ribbon, click **Petroleum Assay,** click the **Add** button, and in the Assay name, place: Kirkuk, under the Selected Assay; highlight the Kirkuk blend 2015, and click OK [1, 5].

In this example, we will compare the Kuwait and Kirkuk crudes. After Kirkuk has been added to the Petroleum Assays list, you can add it

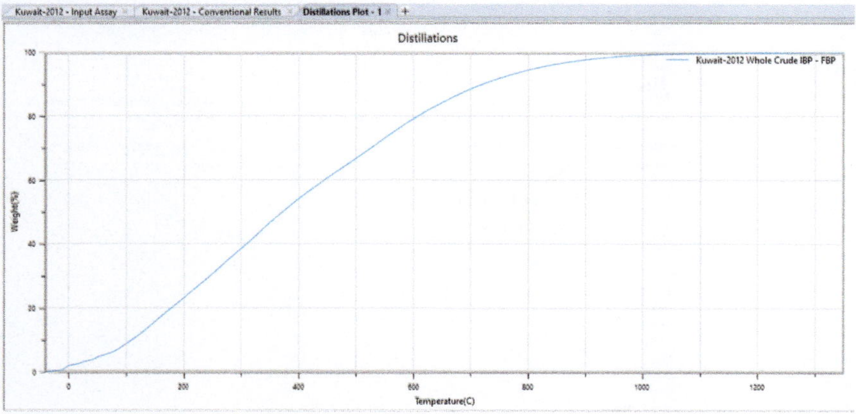

Figure 10.11 Distillation curve plotting for the Kuwait 2012 crude.

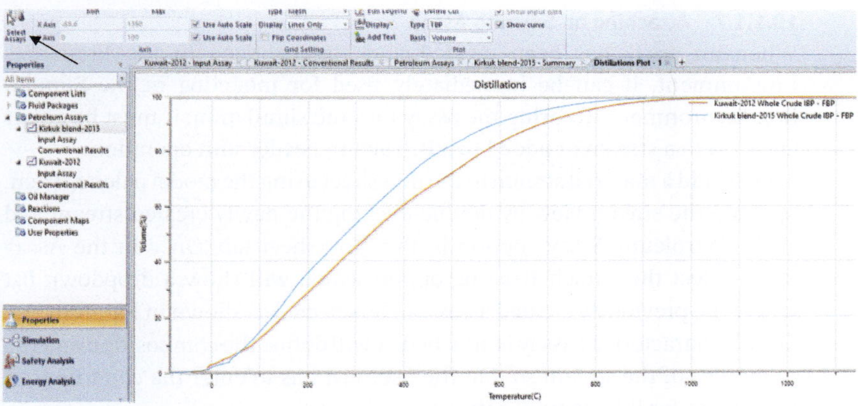

Figure 10.12 Comparison of distillation curve of the Kuwait and Kirkuk assays.

to the plot: from Plot Gallery; click Distillation, and then click "Select Assays" under the Format tab when the Plot is selected, as shown in Figure 10.12.

10.1.1.6 Exporting Assay Data

Assays can be exported from a HYSYS Petroleum Refining case while in the Properties Environment. Go into the Petroleum Assays node to access the Assay Summary, select the assays to export, and click Export button as shown in Figure 10.13.

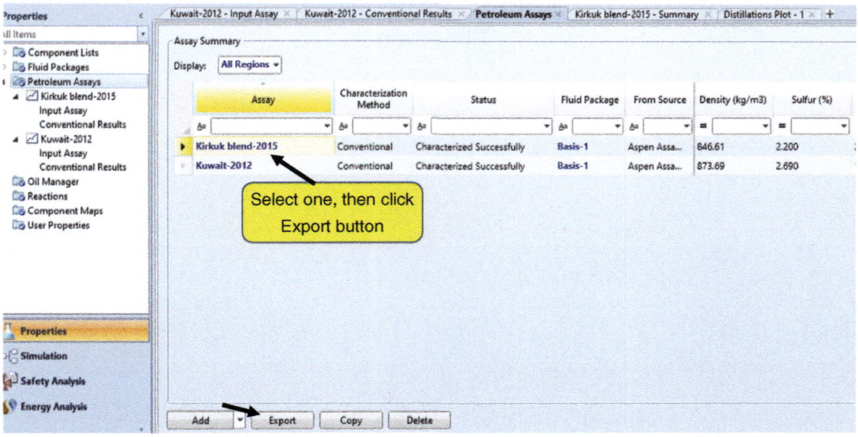

Figure 10.13 How to export assay data.

10.1.1.7 Attaching an Existing Assay

When an assay has been created and characterized in the Properties Environment, it can be immediately used for modeling in the Simulation Environment. To bring the assay into the simulation, it must be either attached to a stream or added to a Petroleum Feeder unit operation.

First, add a material stream to the flowsheet using the model palette. Then, open up the stream view by double-clicking the newly created stream and click "Petroleum Assay" option in the Worksheet tab. Once in the Assay form, select the Attach Existing option which will show a dropdown list where the previously created assay can be selected as shown in Figure 10.14. Once a characterized assay is attached, it will define the composition and the properties of the stream so that the user just has to enter the conditions of the stream for it to be fully defined.

If more than one assay will be used or if the assay will be mixed with another stream, they can be added to a Petroleum Feeder 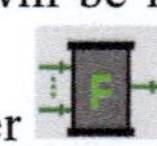, shown in the palette, Refining group. In the Feeder, you can select input feed streams and feed assays (as well as mixing specifications including a fluid package), and HYSYS will calculate a product stream that can be used in the flowsheet [5].

10.1.1.8 Creating a New Assay on a Stream

In some cases, creating a new assay on a stream may be necessary. To do so, add a stream to the simulation and proceed directly to the "Petroleum Assay" option in the stream view. From this form, there are a number of options to add assays, similar to when creating an assay in the Properties

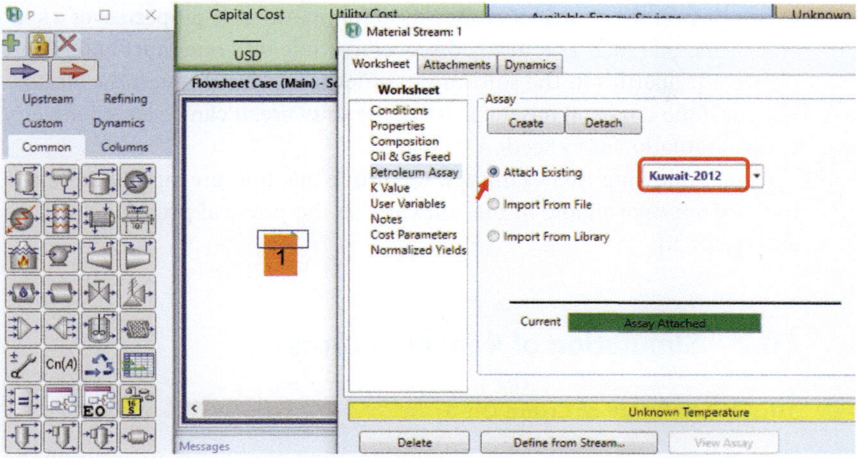

Figure 10.14 Attaching an existing assay, Add material stream, double-click on it, select Petroleum Assay, choose Attach Existing node, then select one of the existing petroleum Assays from drop list.

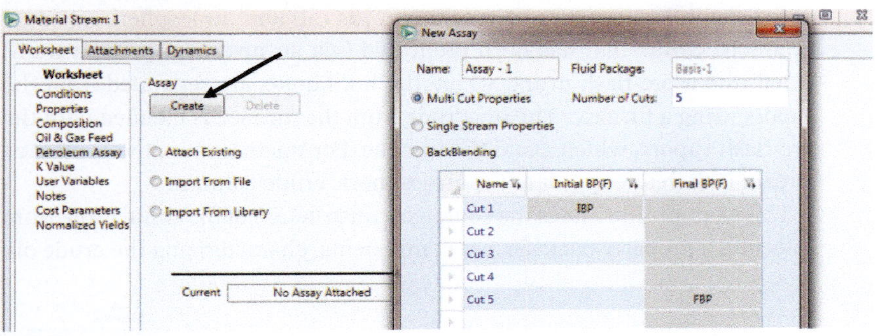

Figure 10.15 View of the assay table while in the simulation environment.

Environment. The user can select Attach Existing, Import from File, Import from Library, or if they click "Create" button, it will open the Manual Entry dialog for the user to begin entering assay data for the stream. Once the assay has been created on the stream, click "View Assay" button at the bottom of the stream view at any time to open up the assay table for that specific stream. Figure 10.15 shows the view of the assay table while in the Simulation Environment.

The cuts of petroleum refinery such as naphtha (light gasoline or light naphtha), Heavy gasoline, Kerosene, atmospheric gas oil AGO, etc. are not found in the HYSYS, and each of them is a mixture of multi-components depending on the source of crude oil, so for unavailable properties of a specific crude oil (such as a newly discovered crude oil), one must enter their physical properties in the software in order to handle through the simulation, or if the crude oil has database in the software, it can be added directly to the simulation as in Section 10.1.

In the following two examples, the crude oils that are simulated in the refinery are unavailable in the software, so the physical properties will be entered [5, 8].

10.2 Simulation of Refinery Process

10.2.1 Refinery of Crude Oil-1

This example represents a crude oil processing plant with an atmospheric crude column to separate the crude into its straight run products and a pre-fractionation train to heat the crude liquids. Figure 10.16 shows the Main Flowsheet for this process [7].

In this illustration, a fractionation facility processes crude oil to create diesel, naphtha, kerosene, atmospheric gas oil, and atmospheric residual products. Crude that has been preheated (via an upstream preheat train) is fed into a pre-flash drum, where the hot liquids are separated from the vapors using a furnace. The hot crude from the furnace is blended with the pre-flash vapors, which avoid the furnace. For fractionation, the combined stream is thereafter put into the atmospheric crude column.

This example illustrates the complete construction of the simulation, from selecting a property package and components, characterizing the crude oil,

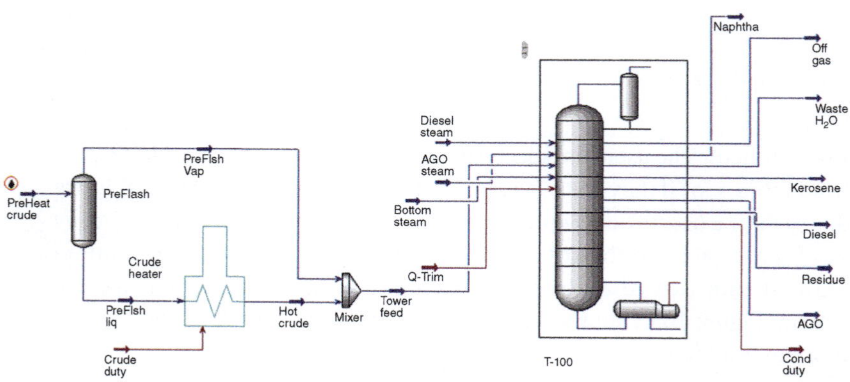

Figure 10.16 PFD of refinery of crude oil_1.

to installing streams and unit operations, through to examining the final results. The tools available in HYSYS are utilized to illustrate the flexibility available to you.

10.2.1.1 Simulation

1) Open a new case and add the components: H_2O, C_3, i-C_4, n-C_4, i-C_5, and n-C_5.
2) Fluid package: Peng Robinson.
3) Change the units by:
   File → select **Options**, from Figure 10.17, select **Units of Measure,** change the units to **Field**. Click the **Copy** button and change the **Mass Density** to **API** and the **Standard Density** to **API_60.**
   Click **OK.**
4) Press **Oil Manager**, Figure 10.18, click on the Arrow beside the **Oil Manager** Folder located in the Navigation Pane then click **Input Assay**, Figure 10.19.

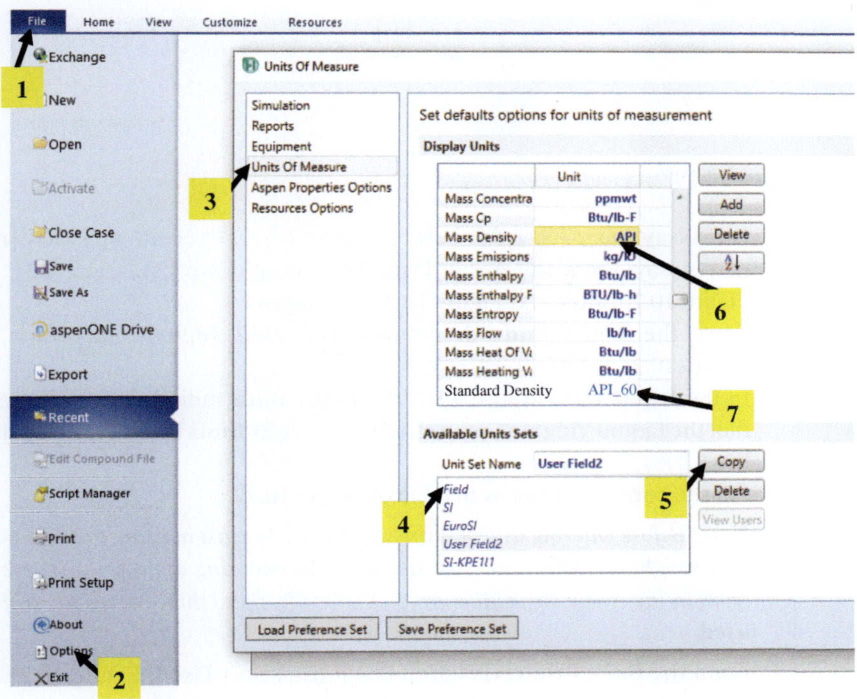

Figure 10.17 How to change the units of measure.

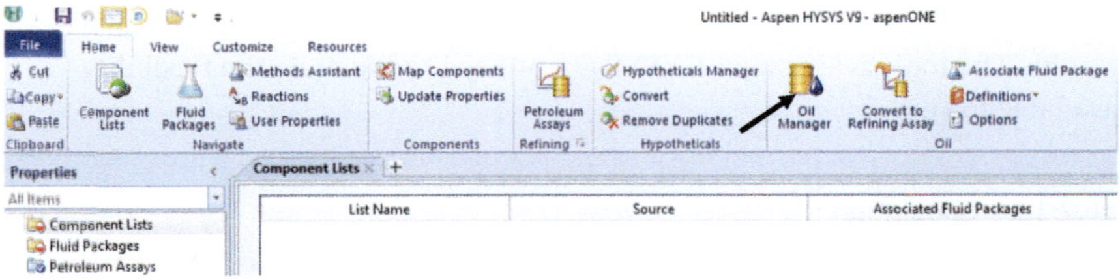

Figure 10.18 Arriving at the oil manager.

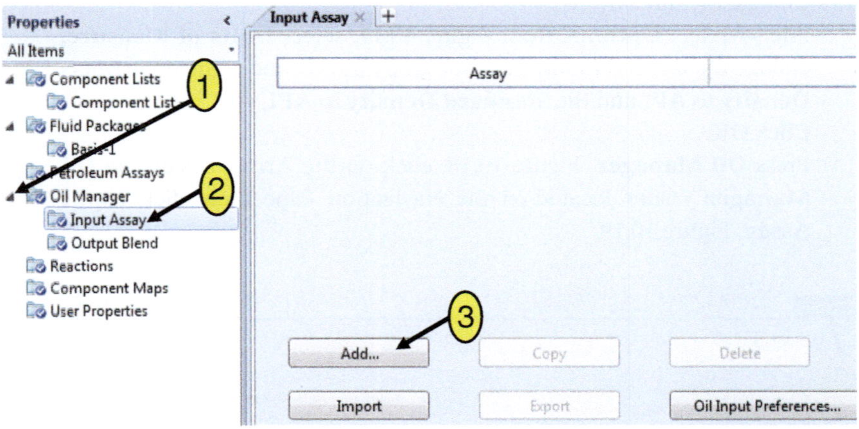

Figure 10.19 Adding a new assay.

5) On **Assay** tab, click **Add.** button, Figure 10.19, to create and view a new Assay. From the **Assay Data Type** drop-down list, select **TBP**, Figure 10.20, after that Figure 10.21 will appear.
6) From the **Light Ends** drop-down list, select Input Composition, Figure 10.22.
7) In the **Input Data** group, select the **Light Ends** radio button. Ensure that the Liquid Volume% is selected in the **Light Ends Basis** drop-down list.
8) Fill the **compositions** as in table of Figure 10.22.

   Note*: Before entering any of the assay data, you must activate the molecular weight, density, and viscosity curves by choosing appropriate curve types in the Assay Definition group. Currently, these three curves are* ***not used****.*

9) From the **Bulk Properties** drop-down list, select **Used**, Figure 10.23. A new radio button labeled **Bulk Props** appears in the Input Data group.

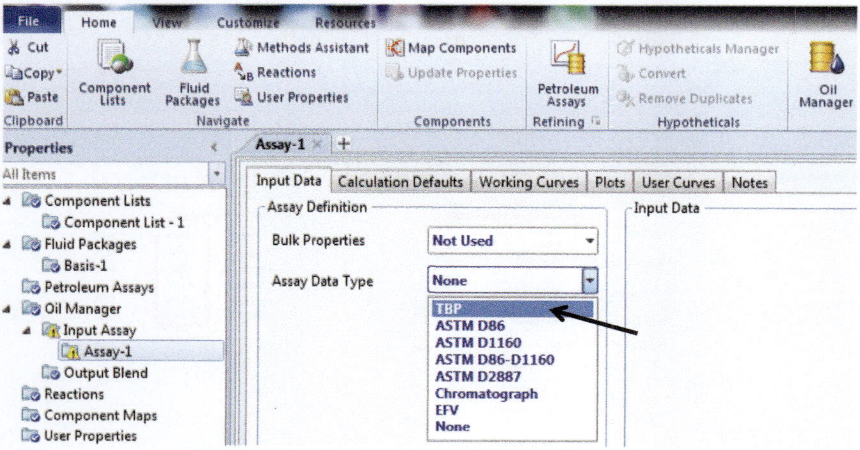

Figure 10.20 How to select TBP from Assay Data Type.

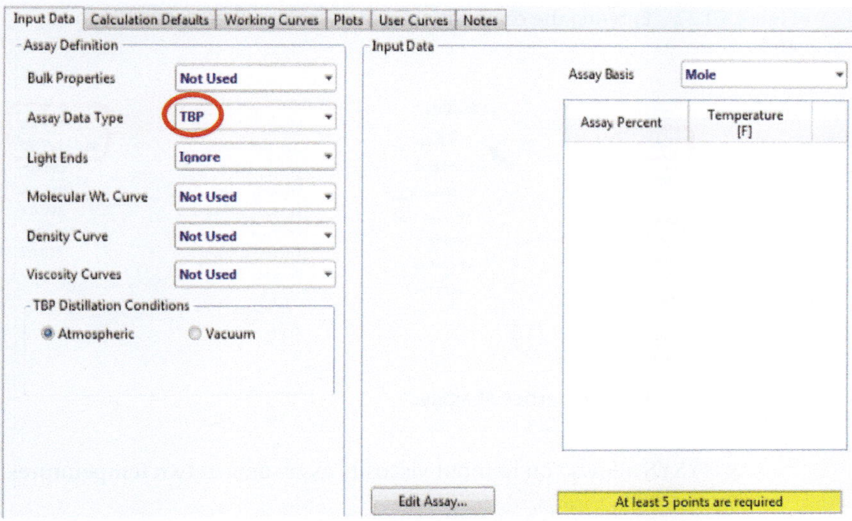

Figure 10.21 Properties of the assay that required defining.

10) From the **Molecular Wt.** Curve drop-down list, select **Dependent**. A new radio button labeled **Molecular Wt.** appears in the Input Data group, Figure 10.23.
11) From the **Density Curve** and **Viscosity Curves** drop-down lists, select **Independent** as the curve type. For Viscosity, two radio buttons appear

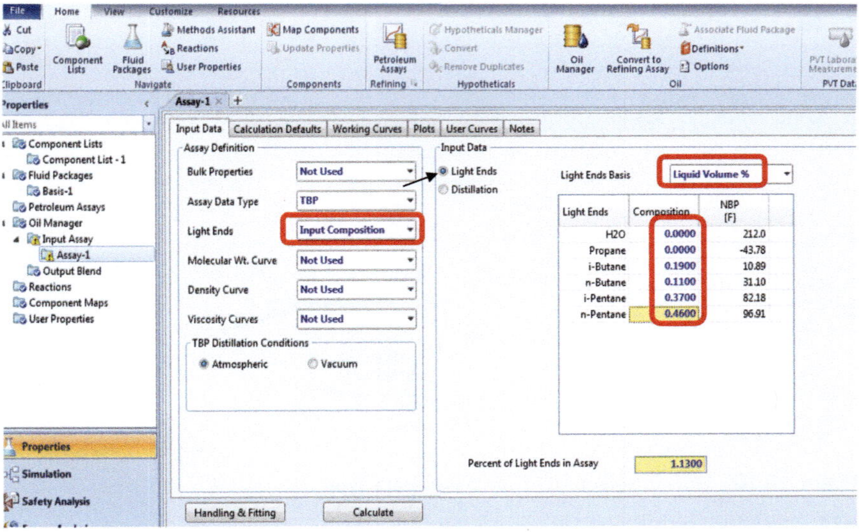

Figure 10.22 Entering the composition in the Liquid volume percent in the light end.

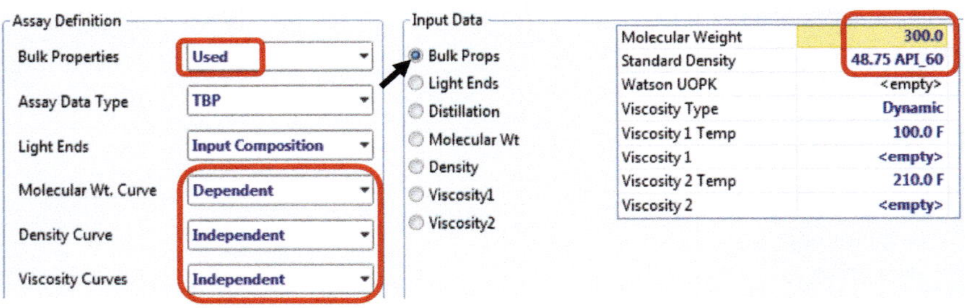

Figure 10.23 Bulk properties of assay.

as HYSYS allows you to input viscosity assay data at two temperatures, Figure 10.23.

12) Entering Bulk Property Data: Select the **Bulk Props** radio button, and the bulk property table appears to the right of the radio buttons. Fill the Molecular weight and Standard Density in the table as in Figure 10.23:
13) Entering Boiling Temperature (TBP) Data:
    - Click the **Calculation Defaults** tab, Figure 10.24. In the Extrapolation Methods group, select Lagrange for *each* method using the drop-down lists.
14) Return to the **Input Data** tab. Select **Distillation** radio button.

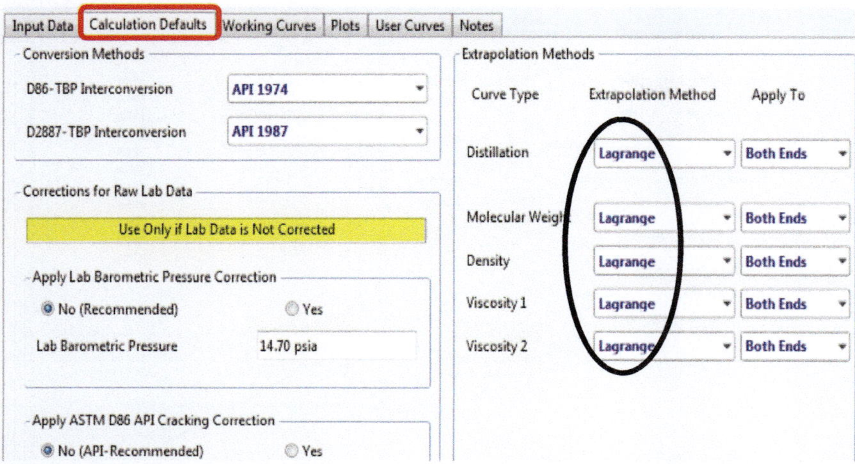

Figure 10.24 Selection of Lagrange method in the calculation of properties.

Note: *HYSYS displays a message under the matrix, stating that "At least 5 points are required" before the assay can be calculated.*

15) From the **Assay Basis** drop-down list, select **Liquid Volume**. Click the **Edit Assay** button. The Assay Input Table property view appears as in Figure 10.25.
   (15-1) Fill the table as in Figure 10.25.
   (15-2) After completing the table, click OK button to return to the Assay property view.

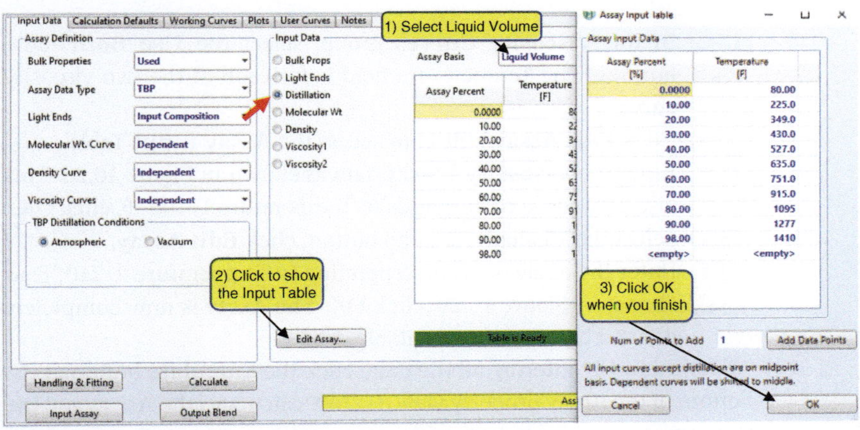

Figure 10.25 Table of assay percent with temperature.

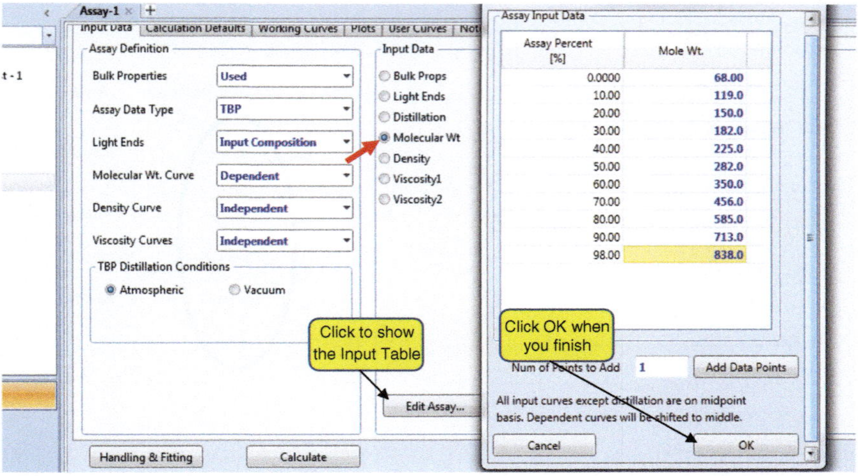

Figure 10.26 Table of assay percent with molecular weight.

16) Enter Molecular Weight Data: Select **Molecular Wt.** radio.
   - Click **Edit Assay**, Fill the table as in Figure 10.26, click **OK**.
17) Enter Density Data: Select **Density** radio
   - Using the same method as for the previous assays, click **Edit Assay**, Enter the API gravity curve data as shown in Figure 10.27.
18) Enter Viscosity Data:
   (18-1) Select the **Viscosity 1** radio button, in the **Viscosity Type** drop-down list above the assay matrix, and ensure Dynamic is selected.
   (18-2) In the **Viscosity Curves** group, select the **Use Both** radio buttons. The Temperature field is for each of the two viscosity curves.
   (18-3) Click **Edit Assay** button to access the **Assay Input** Table. Enter the **Viscosity 1** assay data as shown in Figure 10.28. This viscosity curve corresponds to Temperature 1, 100 °F, click OK.
   (18-4) Select the **Viscosity 2** radio button, click **Edit Assay**.
   (18-5) Enter the assay data corresponding to Temperature 2, 210 °F, as shown in Figure 10.29, click OK. The Assay is now completely defined based on our available data.
19) After you finish entering all the data, click the **Calculate** button at the bottom of the Assay property view. HYSYS calculates the Assay, and the status message at the bottom of the property view changes to **Assay Was Calculated**.

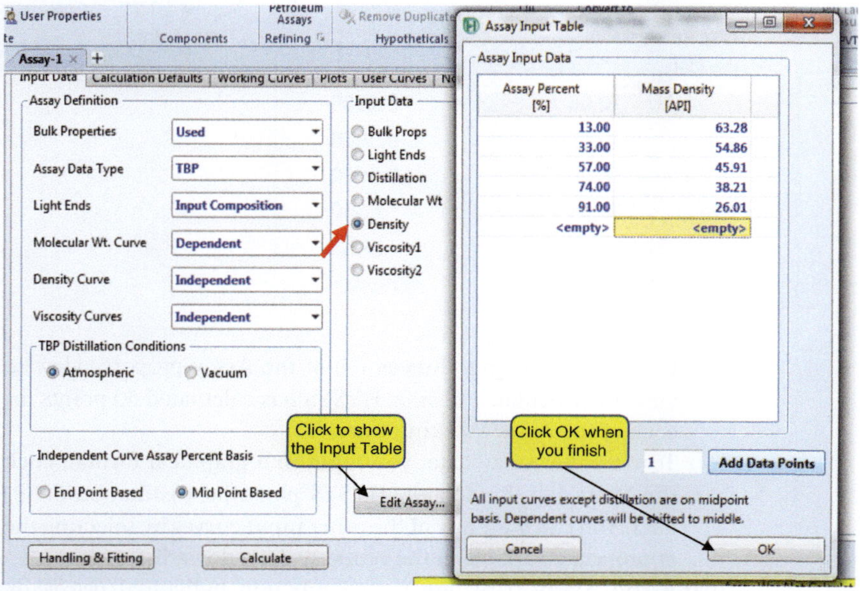

Figure 10.27 API gravity curve data.

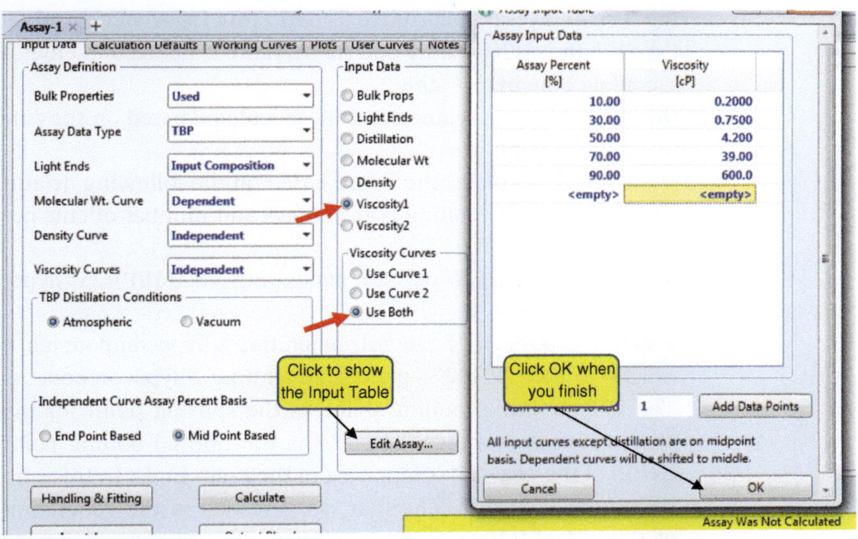

Figure 10.28 Viscosity 1 assay data.

Assay Percent [%]	Viscosity [cP]
10.00	0.1000
30.00	0.3000
50.00	0.8000
70.00	7.500
90.00	122.3
<empty>	<empty>

Figure 10.29 Viscosity 2 assay data.

(19-1) Click the **Working Curves** tab of the Assay property view to view the calculated results. HYSYS has calculated 50 points for each of the Assay Working Curves.

(19-2) To view the Assay data, you input in a graphical format, click the **Plots** tab. By default, HYSYS plots the Distillation (TBP) data. You can view any of the other input curves by selecting the appropriate variable in the Property drop-down list.

20) **Cutting the Assay** (Creating the Blend) into individual *petroleum hypocomponents*.

(20-1) Click the **Output Blend** folder in the Navigation Pane, then click the **Add** button.

(20-2) Click the **Add... >** button, there are two results:

(a) The Assay is transferred to the **Oil Flow Information** table. (When you have only one Assay, there is no need to enter a Flow Rate in this table.)

(b) A Blend (Cut) is automatically calculated based on the current Cut Option.
HYSYS calculates the Blend based on the following default values for the boiling-point ranges and number of cuts per range:

- IBP to 800 °F:25 °F per cut, generating [(800-IBP)/25] hypocomponents
- 800 to 1200 °F:50 °F per cut, generating 8 hypocomponents
- 1200 to 1400 °F:100 °F per cut, generating 2 hypocomponents

The IBP, or initial boiling point, is the starting point for the first temperature range. The IBP is the normal boiling point (NBP) of the heaviest component in the Light Ends. In this case *n*-Pentane at 96.9 °F. The first range results in the generation of $(800-96.9)/25 = 28$ hypocomponents. All the cut ranges together result in a total of $28+8+2=38$ hypocomponents.

(20-3) Click the **Tables** tab to view the calculated properties of these hypocomponents.

(20-4) Return to the **Data** tab. From the **Cut Option Selection** drop-down list, select **User Ranges**, Figure 10.30.

21) In the **Starting Cut Point** field, enter 100 °F. This is the starting point for the first range, Figure 10.31.

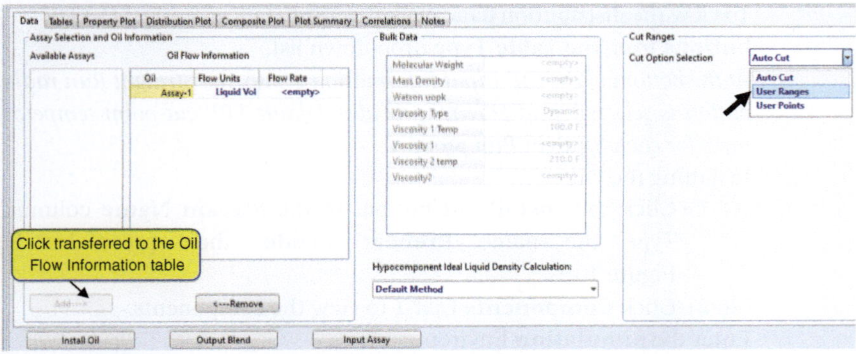

Figure 10.30 Selection of User Ranges from the cut option selection.

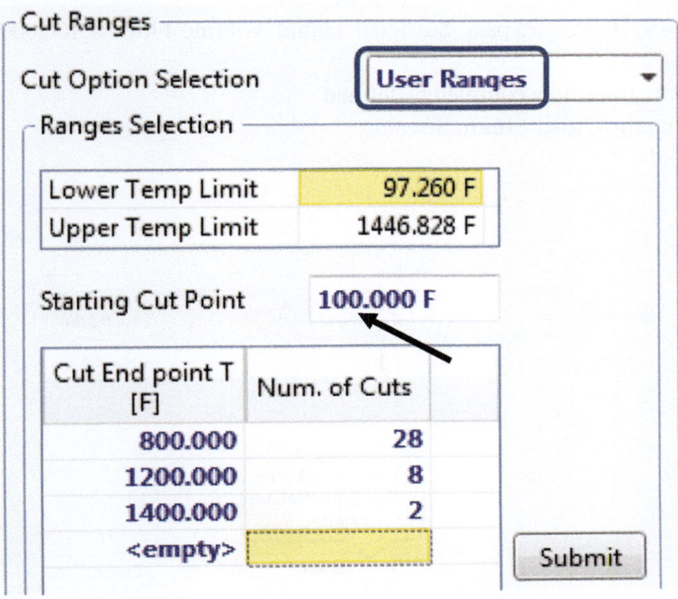

Figure 10.31 Number of cutting in the range of temperatures.

22) Enter the remaining cut-point temperatures and the number of cuts values as shown in the Figure 10.31.
23) Click the **Submit** button to calculate the Blend based on the current initial cut-point and range values.
24) Click Tables tab to view the properties of the petroleum hypo-components.
25) Viewing the Oil Distributions:
    To view the distribution data, click the **Table** tab, and select **Oil Distributions** from the **Table Type** drop-down list.
    At the bottom of the Cut Input Information group, the Straight Run radio button is selected, and HYSYS provides default TBP cut-point temperatures for each Straight Run product.
26) Installing the Oil:
    (26-1) Click the **Install Oil** button. In the **Stream Name** column, Type the name **Preheat Crude**, then press Install, Figure 10.32.
    (26-2) Click **Components List-1** to view the components.
27) Enter the **Simulation** Environment.
    *The stream **Preheat Crude** was already installed during the oil characterization procedure.*
    - Double-click the stream and enter the date:
      T = 450 °F, P = 75 psia, Std ideal Liquid Volume Flow = 100,000 barrel/day.
      The stream is now completely defined.
28) Installing the Utility **Steam** Streams:

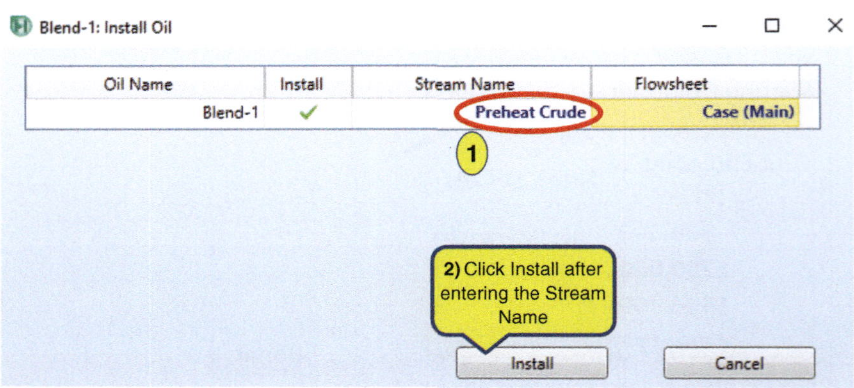

Figure 10.32 Type the stream name of the crude after installing oil.

Steam Name	Bottom Steam	Diesel Steam	AGO Steam
Temperature (°F)	375	300	300
Pressure (psia)	150	50	50
Mass Flow Rate (lb/h)	7500	3000	2500
Composition (%)	100% H_2O	100% H_2O	100% H_2O

29) Installing Unit Operations:
   (29-1) Add a Separator:
   Input = Preheat Crude, Vapor Outlet = PreFlashVap, Liquid Outlet = PreFlashLiq
   (29-2) Add a Furnace (simulated as a heater):
   Inlet: PreFlashLiq, Outlet: Hot Crude, Energy: Crude Duty, **Parameters** Page: **Delta P:**10 psia, from the **worksheet**, enter the temp of the Hot Crude = 650 °F
   (29-3) Add a **Mixer:**
   Input: PreFlashVap and Hot Crude, Output: Tower Feed
30) Add an Energy stream, Name this stream: **Trim Duty**

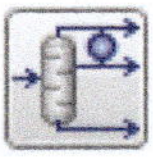

31) Install the Column: Double-click the **Refluxed Absorber** icon on the Object Palette, Columns group.
   (31-1) Fill the first page of the column that should appear as shown in Figure 10.33.
   (31-2) Next, Condenser pressure = 19.7 psia, Condenser pressure drop = 9 psi, Bottom stage pressure = 32.7 psia.
   (31-3) Next, Condenser Temp = 100 °F, Top stage Temp = 250 °F, Bottom Stage Temp = 700 °F.
   (31-4) Next, vapor rate = 0, Reflux ratio = 1. Click **Done**.
32) Adding the Specification values: Click the **Monitor** page, In the Column Specifications group, delete the **Reflux Rate** and **Btms Prod Rate** Specifications.
   - Click the **Add Spec** button, Select **Column Draw Rate,** and click **Add Spec(s)** button.
   In the Name cell, type **WasteH_2O Rate**.
   No further information is required, Close the property view.
33) Install side stripper of **Kerosene** by selecting the **Connection** Page, click **Side Ops**; on the **Side Stripper** page, click the **Add** button, the completed Side Stripper property view **Must** appear as shown in Figure 10.34. Click the **Install** button, close.

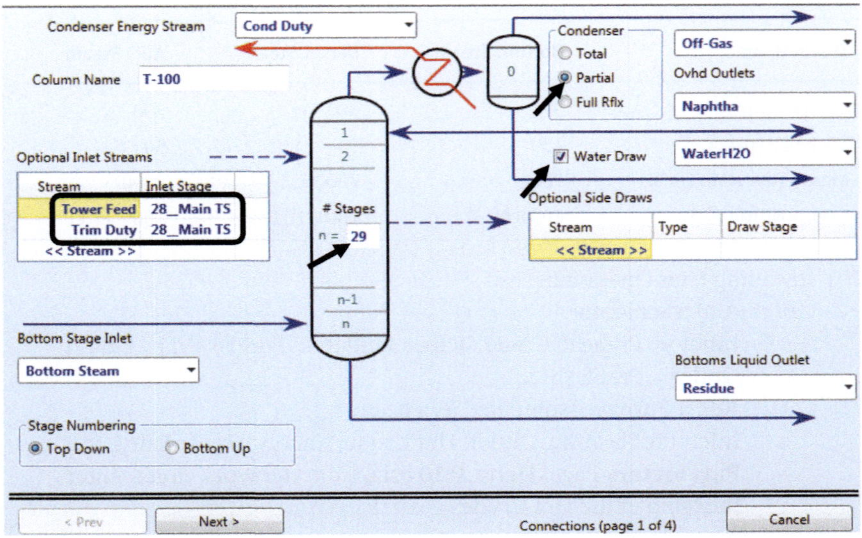

Figure 10.33 Refluxed Absorber input expert/connection page.

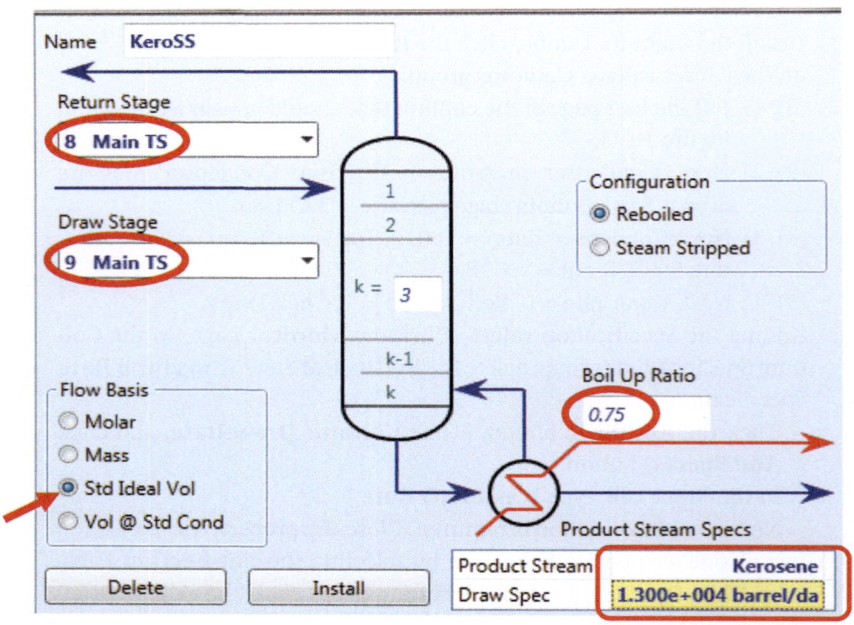

Figure 10.34 Final view of the side stripper column for kerosene.

34) Use the previous step (step 33) to install the two remaining side strippers **DieselSS** (Diesel-side stream) and **AGOSS** (atmospheric gas oil-side stream). The completed DieselSS- and AGOSS-side stripper property views appear as in Figures 10.35 and 10.36, respectively.
The summary of information of side strippers appears in Figure 10.37.
35) Installing Pump Around:
   (35-1) Select the **Pump Arounds** page, click **Add** button.
   (35-2) In the **Return Stage** drop-down list, select stage (**1_Main TS**), in the **Draw Stage** drop-down list, select stage (**2_Main TS**).
36) Click the **Install** button, and a more detailed Pump Around property view appears.
   (36-1) Double-click the [**PA_1_Rate(Pa)**], change **Flow Basis** to **Std Ideal Vol.**
   Enter Spec Value: 50,000 barrel/day, Figure 10.38, close when finishing.
   (36-2) Double-click in the blank space under the **PA_1_Dt(Pa)** specification. In the **Spec Type** drop-down list, select **Duty** and enter the spec value (−55e6 Btu/h).
   Close the icon to return to Pump Around, and close icon of Pump around, Figure 10.39.

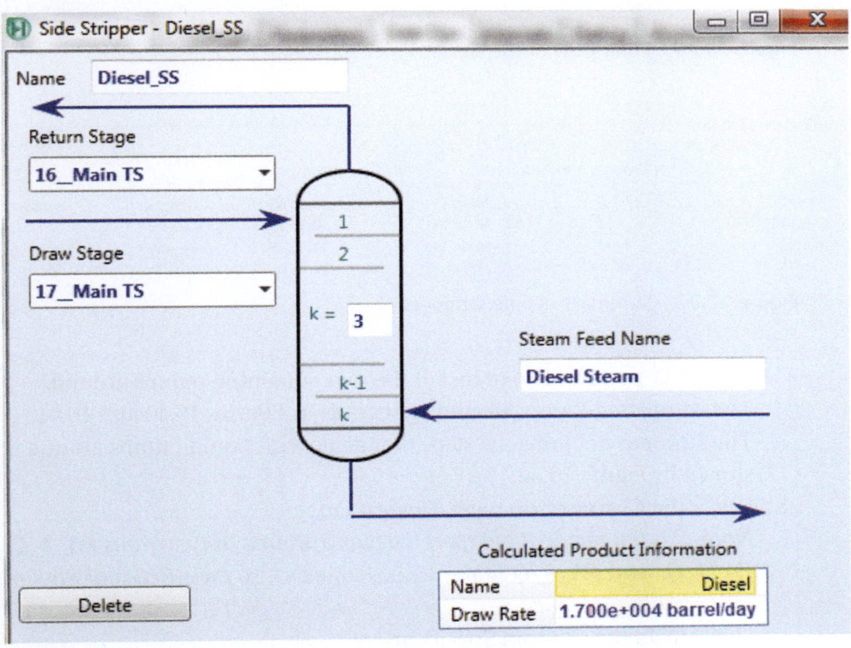

Figure 10.35 Final view of the side stripper column of Diesel.

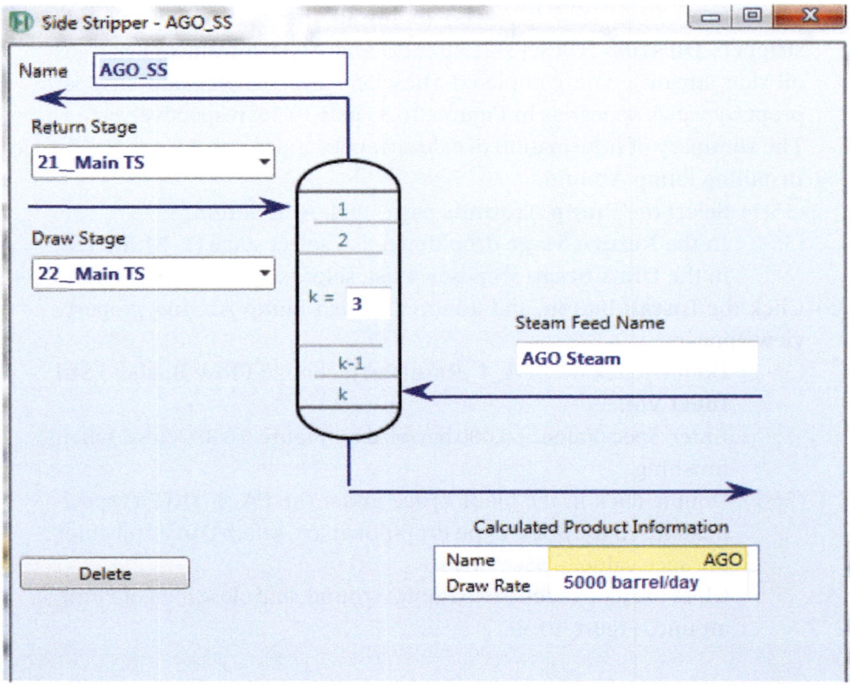

Figure 10.36 Final view of the side stripper column of AGO.

Side Stripper Summary

	# Stages	Liq Draw Stage	Vap Return Stage	Outlet Flow [lbmole/hr]	Reboiler Duty [Btu/hr]
KeroSS	3	9_Main TS	8_Main TS	<empty>	<empty>
DieselSS	3	17_Main TS	16_Main TS	<empty>	
AGOSS	3	22_Main TS	21_Main TS	<empty>	

Figure 10.37 Summary of side strippers.

37) Repeat steps 35 and 36 to install the two remaining pumps around. The completed Pump Around appears as in Figures 10.40 and 10.41. The summary of previous steps that deals with liquid pumps around is shown in Figure 10.42.

38) Select the **Connection** page, **Design** tab.

*Note: It is not required to export the pump around duty streams **PA_1_Q, PA_2_Q, and PA_3_Q** to the Main Flowsheet, so their External Stream cells remain undefined.*

(38-1) Add the Column Specification:

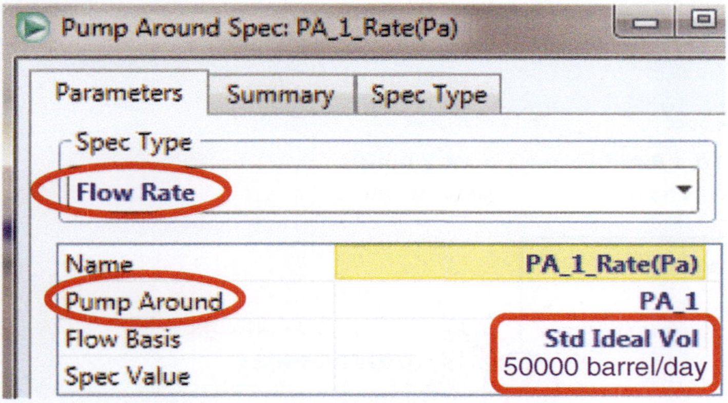

Figure 10.38 specification of pump around.

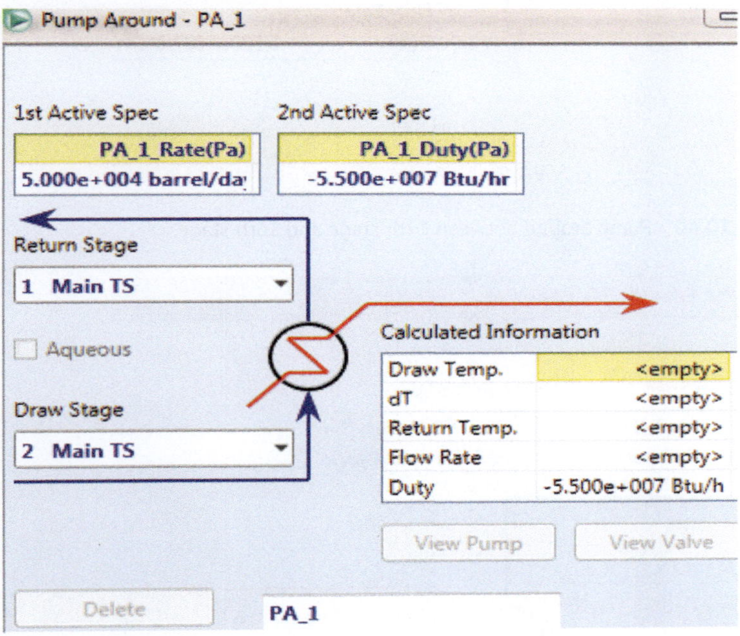

Figure 10.39 Specification of the Pump around between 2nd stage and 1st stage.

(a) In the **Design** tab, **Specs** page: **Column Specification** group, click **Add**, double-click the **Column Liquid Flow**, Change the **name** to **Overflash**, and fill the variables as in Figure 10.43.

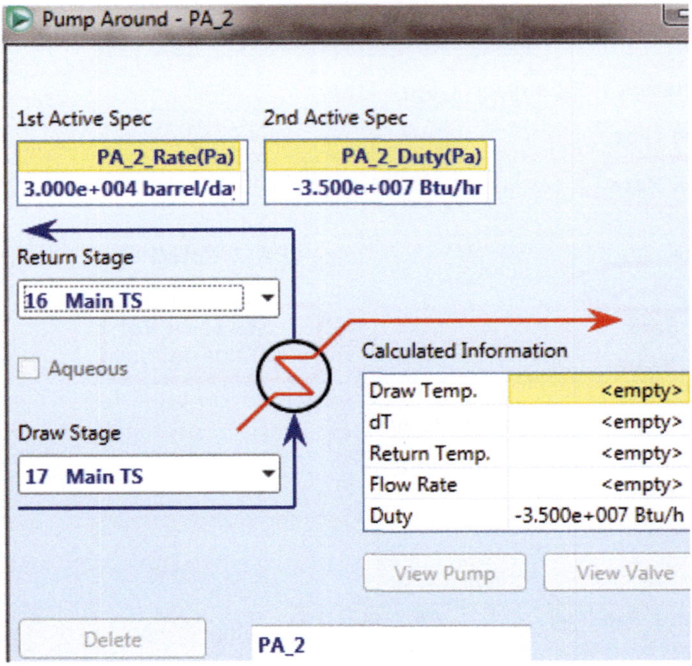

Figure 10.40 Pump around between 17th stage and 16th stage.

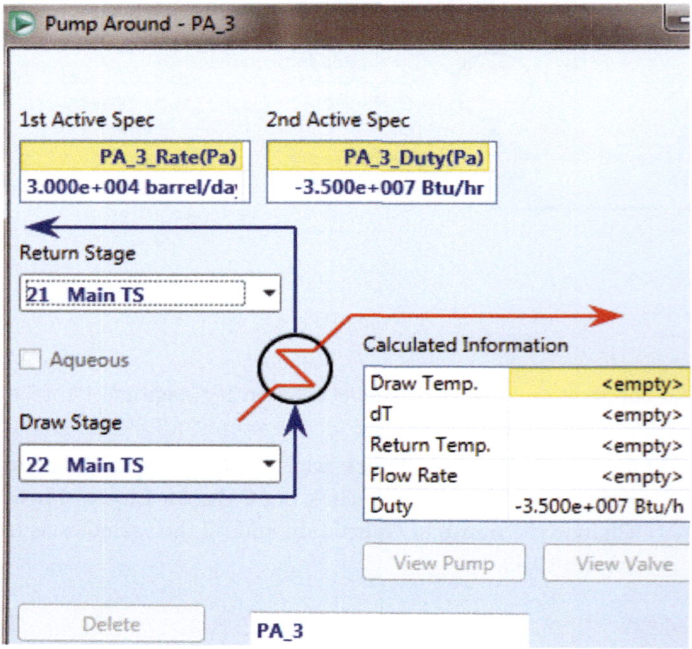

Figure 10.41 Pump around between 22nd stage and 21st stage.

Liquid Pump Around Summary

	Draw Stage	Return Stage	Flow [lbmole/hr]	Duty [Btu/hr]	Draw T [F]	Return T [F]	Expor
PA_1	2__Main TS	1__Main TS	<empty>	-5.500e+007	<empty>	<empty>	☐
PA_2	17__Main TS	16__Main TS	<empty>	-3.500e+007	<empty>	<empty>	☐
PA_3	22__Main TS	21__Main TS	<empty>	-3.500e+007	<empty>	<empty>	☐

Figure 10.42 Summary of pumps around.

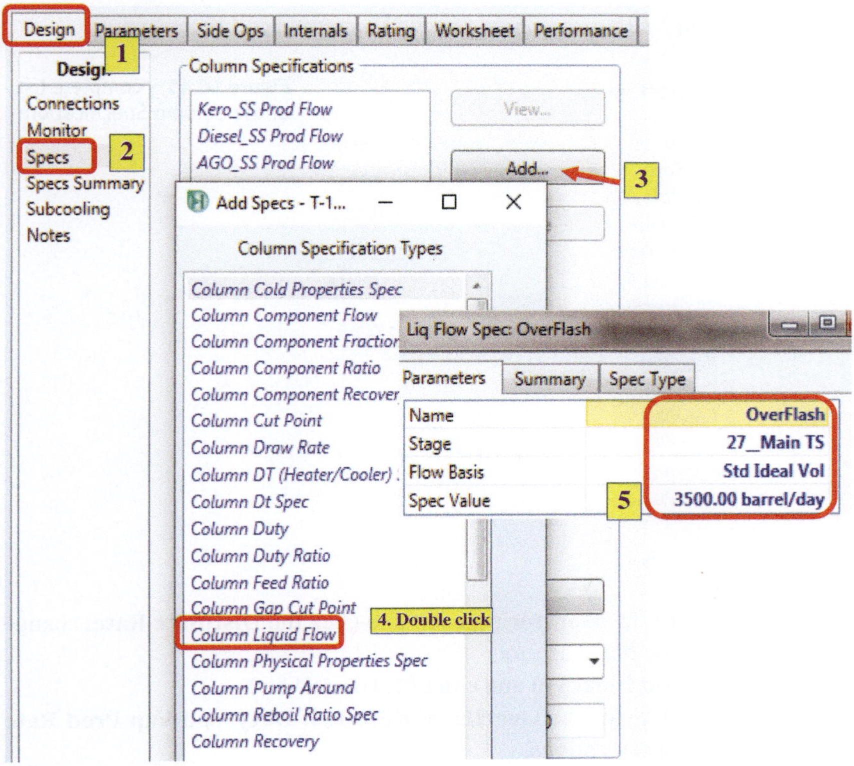

Figure 10.43 Adding the column liquid flow as specification.

A typical range for the Overflash rate = 3–5% (3.5%)
The feed rate = 100,000 barrel/day → Overflash = 3500 barrel/day.

(b) By the same of step a, click **Add** again, double-click the **Column Duty**, and fill the variables as in Figure 10.44.
The completed list of the **Column Specifications** is shown in Figure 10.45.
If the column begins to run on its own before you click **Run** button, click the **Stop** button and continue activating and deactivating specifications.

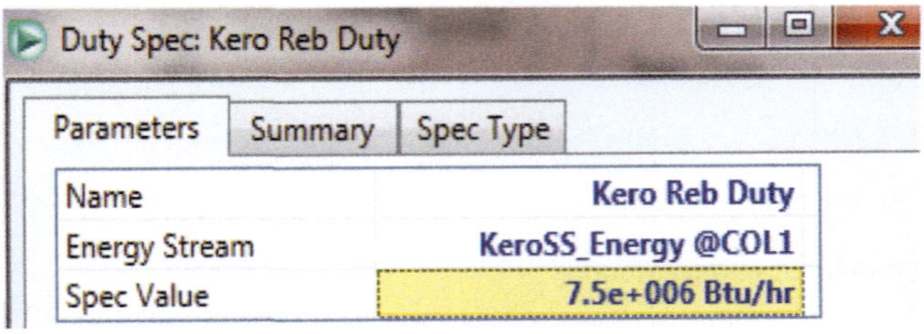

Figure 10.44 Adding the column duty as a specification.

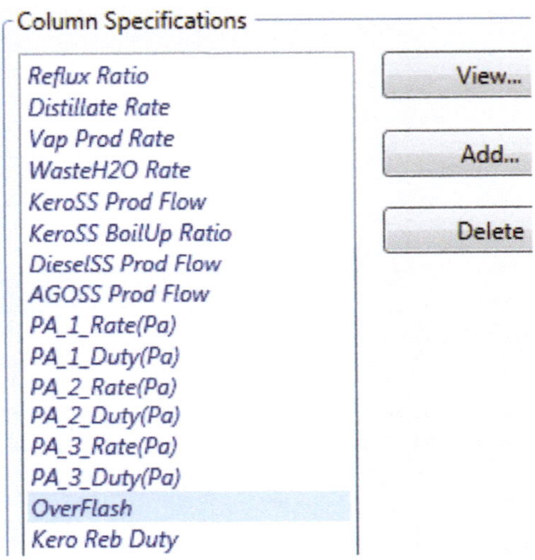

Figure 10.45 Complete list of the Column Specifications.

(38-2) In the **Monitor** page, double-click the **Distillate Rate**, change the flow basis to:
Std Ideal Vol and enter (2e4 barrel/day)

(38-3) Activate the **Overflash, Kero Reb Duty,** and **Vap Prod Rate** Specifications.

(38-4) Deactivate the **Reflux Ratio, Waste H_2O Rate,** and **KeroSS BoilUp Ratio** Specifications.

Click **Run** and see the results.

10.2.2 Refinery of Crude oil-2

In this example, the crude oil is lighter than that in Section 10.2.1, and the column used is from the template columns of the software.

Figure 10.46 shows the process flow diagram of the refinery of crude oil-2. After passing through a preheat train, 100,000 barrel/day of 29.31° API crude

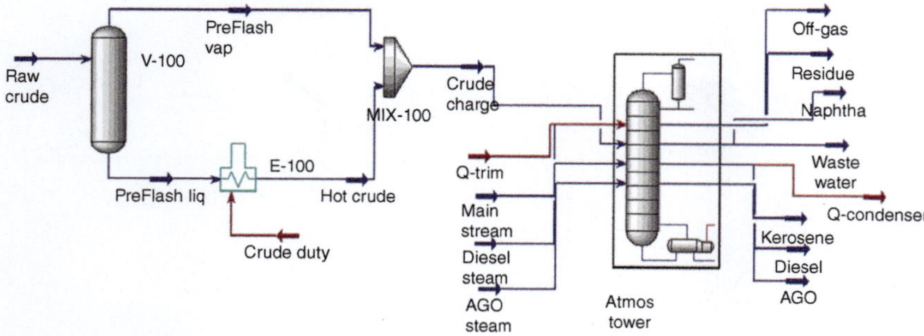

Figure 10.46 PFD of Refinery of Crude oil 2.

is fed into a pre-flash separator operating at 450 °F and 75 psia. The vapor from this separator bypasses the crude furnace and is re-mixed with the hot (650 °F) pre-flash liquids leaving the furnace. The combined stream is then fed to the atmospheric crude column.

The column operates with a total condenser, three coupled side strippers, and three pumps around circuits.

A naphtha product is produced overhead, a kerosene product is produced from the first side stripper, a diesel product is produced from the second side stripper, and an atmospheric gas oil (AGO) is produced from the third side stripper. Both the AGO side stripper and the diesel side stripper are "steam stripped," while the kerosene side stripper has a reboiler.

This example also illustrates the complete construction of the simulation, from selecting a property package and components, characterizing the crude oil, to installing streams and unit operations, through to examining the final results [7].

10.2.2.1 Simulation

1) Components: methane, ethane, propane, *i*-butane, *n*-butane, and water.
2) Property Package: Peng–Robinson
3) From **File** menu: select **Options,** Select **Units of Measure,**
   Change the units to **Field,** click **Copy** button and change
   Mass Density to **API** and **Standard Density** to **API_60**, (see Figure 10.17).
   Click **OK.**
4) **Oil Characterization**
   (4-1) Click **Oil Manager**, Figure 10.47, click arrow beside **Oil Manager** folder located in the Navigation Pane, then click the **Input Assay**, Figure 10.47.

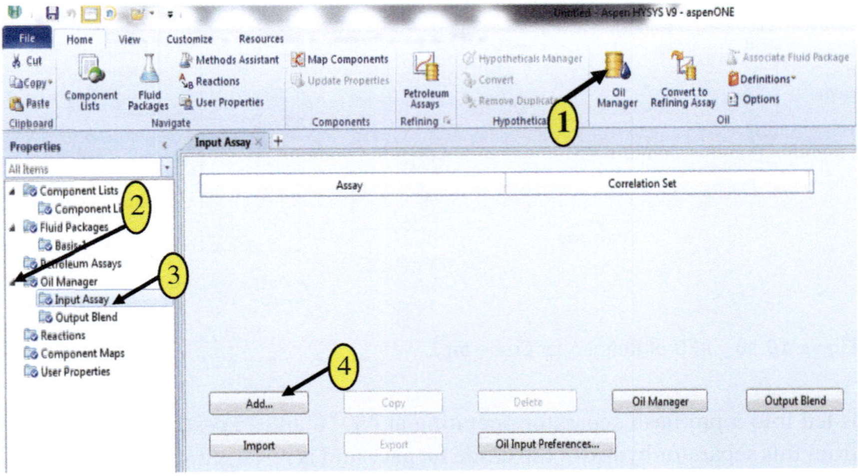

Figure 10.47 Arriving at the Oil manager.

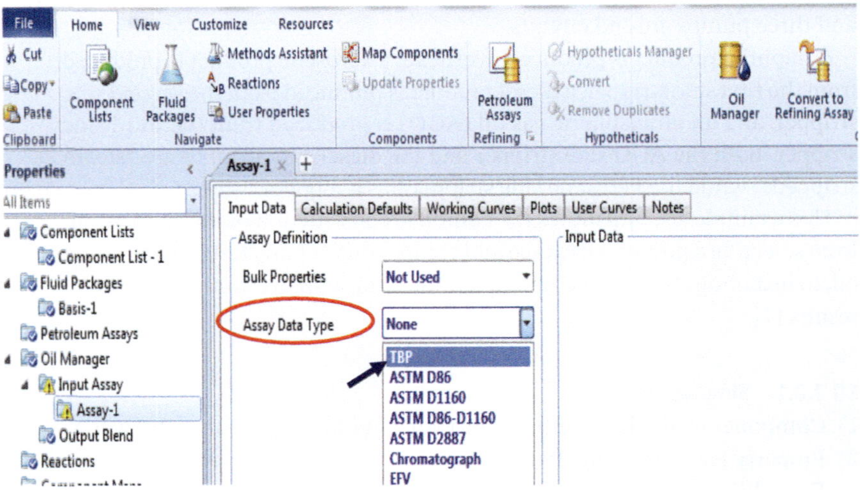

Figure 10.48 Selection of TBP to appear in the properties.

(4-2) On the **Assay** tab of Figure 10.47, click the **Add** button to create and view a new Assay. From the **Assay Data Type** drop-down list, select **TBP**, Figure 10.48, then Figure 10.49 will appear.

(4-3) Change the Bulk Properties setting to Used, Figure 10.50.

(4-4) Select the **Bulk Props** radio, Input the Standard Density = **29.31 API_60**, as shown in Figure 10.50.

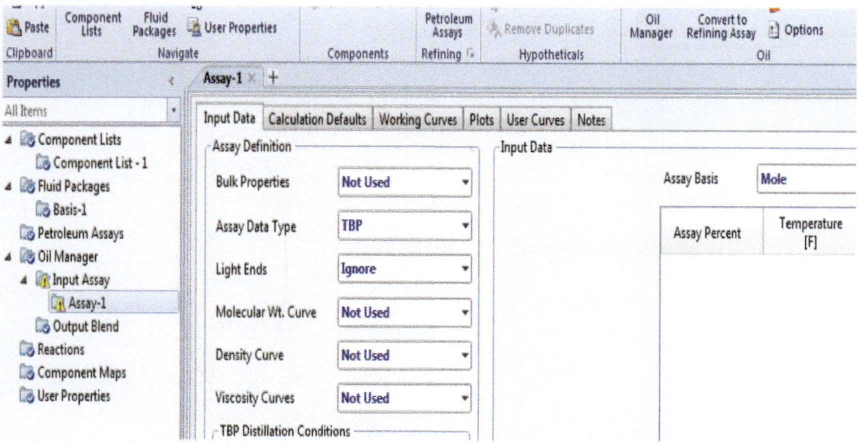

Figure 10.49 The required properties to be entered in the assay.

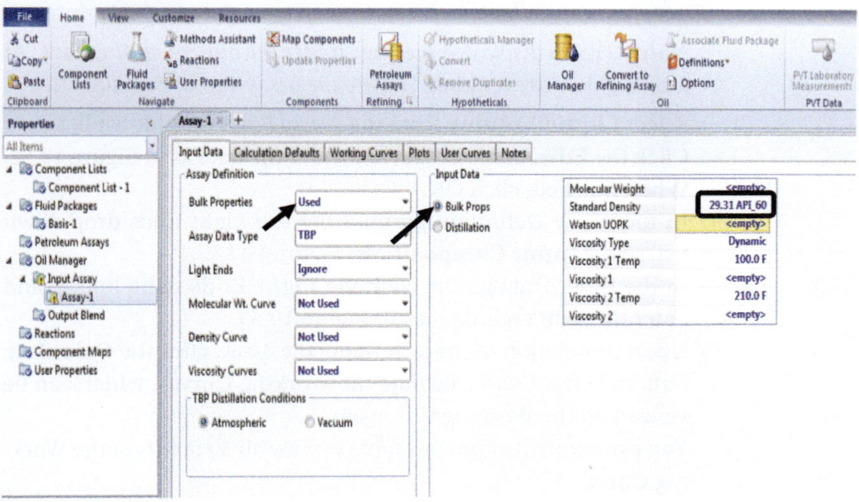

Figure 10.50 Value of Standard Density that must be entered in the Bulk properties.

(4-5) Click the **Calculation Defaults** tab. In the Extrapolation Methods group, select **Lagrange** for each method using the drop-down lists, Figure 10.51.

(4-6) Return to the **Input Data** tab, select the **Distillation** radio button.

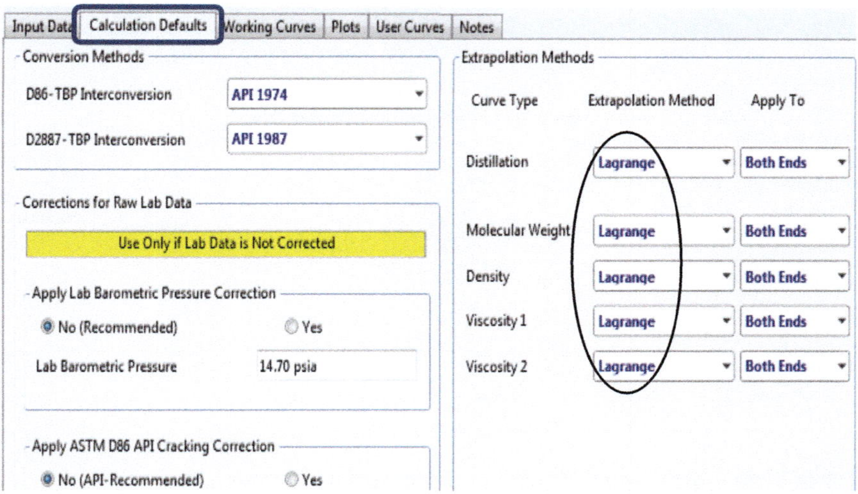

Figure 10.51 Selection of the Lagrange method in the calculation of properties.

Note: *HYSYS displays a message under the matrix, stating that "At least 5 points are required" before the assay can be calculated.*

(4-7) Select **Liquid Volume** from the Assay Basis drop-down list.

(4-8) Click the **Edit Assay** button and enter the data as in Figure 10.52. When finished, click OK.

(4-9) In the Assay Definition group, from the Light Ends drop-down list, select **Input Composition**, Figure 10.53.

(4-10) In the Input Data group, click the **Light Ends** radio button and enter the light ends data as in Figure 10.53.

(4-11) Upon completion of characterizing the assay, click the **Calculate** button. HYSYS will calculate the Working Curves, which can be viewed on the Working Curve tab.
You can scroll through this table to view all 50 points of the Working Curve.

5) **Create the Blend (Cut the Oil)**:

(5-1) Click the **Output Blend** Folder in the Navigation Pane (Oil Characterization property view), and click the **Add** button, Figure 10.54. The Blend: Blend-1 view appears as in Figure 10.54.

(5-2) Click **Add... >** button. HYSYS will transfer that Assay to the Oil Flow Information table. As a guideline, each Outlet stream from the crude column should contain a minimum of five hypocomponents whose composition is greater than 1.0%. Therefore, a total of 30 components should fulfill this requirement, Figure 10.55.

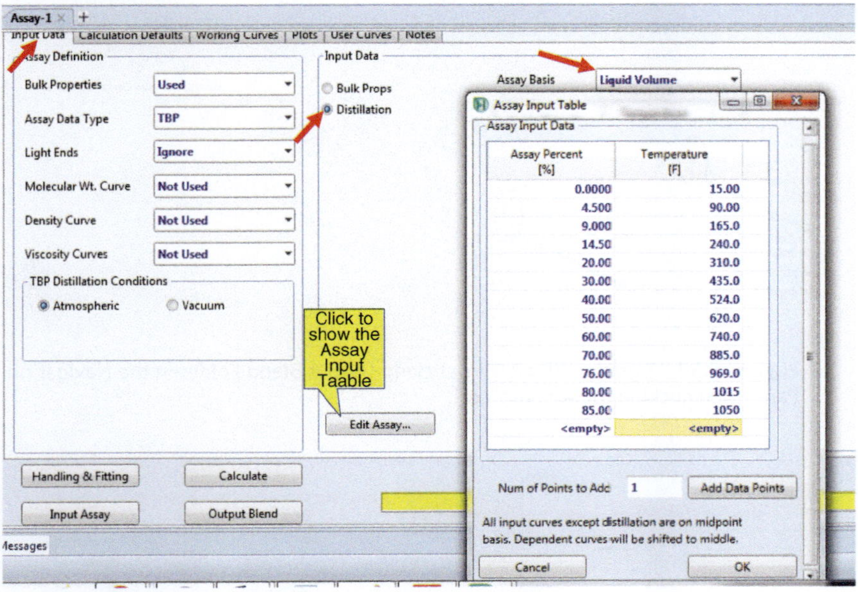

Figure 10.52 Table of assay percent versus temperature.

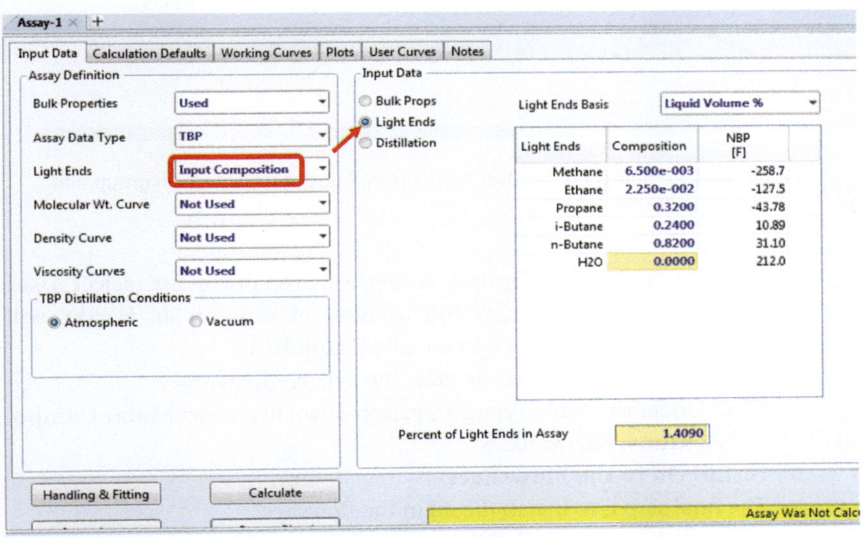

Figure 10.53 Composition of liquid end as liquid volume percent.

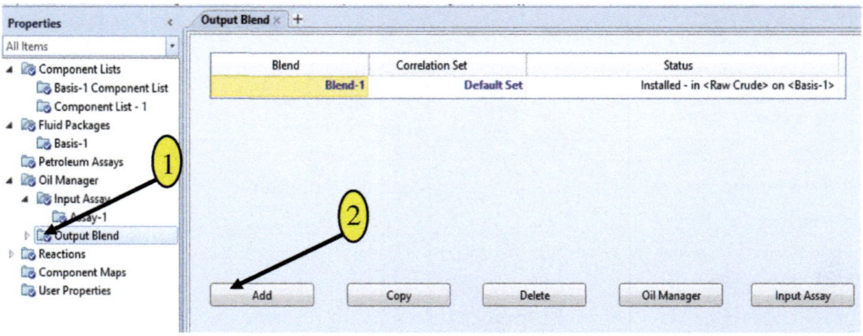

Figure 10.54 Creating Blend-1 by clicking Output Blend Folder in the Navigation Pane, then clicking the Add button.

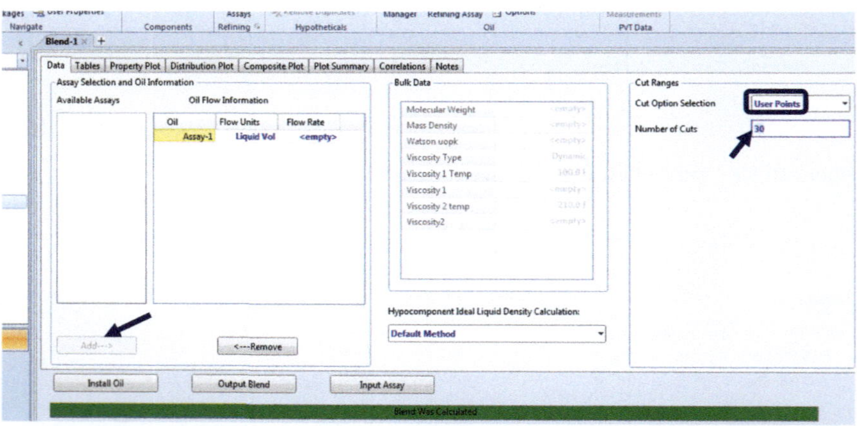

Figure 10.55 Selecting the User Points from Cut Option Selection group and entering the number of cuts.

(5-3) From the **Cut Option Selection** drop-down list, select **User Points**, then specify the Number of Cuts at **30**. HYSYS will calculate the hypocomponents, Figure 10.55.

(5-4) Click the **Tables** tab to view the hypocomponents.

(5-5) Under the Table Type group drop-down list, select **Molar Compositions**, Figure 10.56.

6) **Install Oil in the Flowsheet**

The final step is to install the oil in the flowsheet.

(6-1) Click the **Install Oil** button of the Oil Characterization property view.

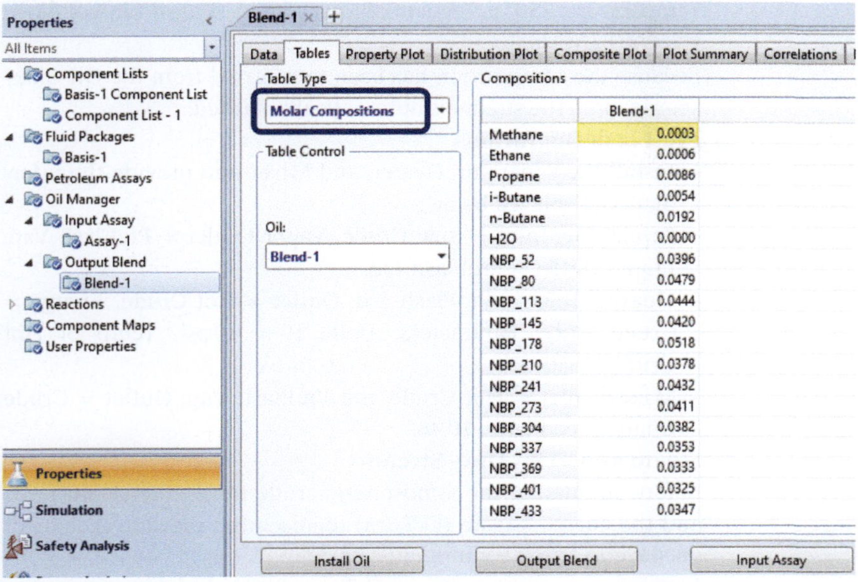

Figure 10.56 Selecting the Molar Compositions under the Table Type.

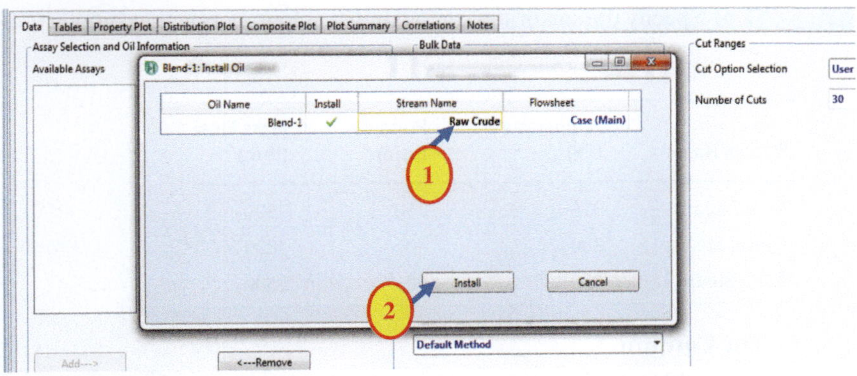

Figure 10.57 Typing the "Raw Crude" under the Stream Name and installing oil.

(6-2) In the Stream Name cell, type **"Raw Crude,"** Figure 10.57. This is the stream name where you would like to "install" the oil. Click the **Install** button.

7) Enter the **Simulation** Environment

(7-1) **Inlet Stream**

(7-2) Specify the conditions of the **Raw Crude** stream as shown below:

Temperature: 450 °F, Pressure: 75 psia, and Std Ideal Liq Vol Flow: 100,000 barrel/day.

Because the composition has been transferred from the Oil Characterization, the stream is automatically flashed.

(7-3) **Pre-Flash Operations**

- Install the Separator, Heater, and Mixer and provide the information displayed below:
- **Separator**: Inlet = Raw Crude, Vapor Outlet = PreFlash Vap, Liquid Outlet = PreFlash Liq.
- **Heater**: Inlet = PreFlash Liq, Outlet = Hot Crude, Energy = Crude Duty, Parameters: Delta P = 10 psi, Temp of Hot Crude = 650 °F.
- **Mixer:** Inlet = Hot Crude and PreFlash Vap, Outlet = Crude Charge, see Figure 10.46.

(7-4) **Steam and Trim Duty Streams**

Before simulating the atmospheric crude tower, the steam feeds and the energy stream (Q-Trim) representing the side exchanger on stage 28 to the column must be defined.

(7-5) Install Energy Stream, Name: **Q-Trim** which does not require any specifications; this will be calculated by the Column.

Three steam streams are fed to various locations in the tower.

(7-6) Specify the steam streams as shown below: Define the composition for each as $\mathbf{H_2O = 1}$

Stream Name	Temperature (°F)	Pressure (psia)	Mass Flow (lb/h)
Main Steam	375	150	7500
Diesel Steam	300	50	3000
AGO Steam	300	50	2500

8) **The Column**

The main column, Atmos Tower, is represented by the following:

- Number of stages is 29 ideal stages (not including the condenser).
- The overhead condenser operates at 19.7 psia and the bottom stage at 32.7 psia.
- The condenser experiences a 9 psi pressure drop.
- The temperature estimates for the condenser, top stage, and bottom stage are 100 °F, 250 °F, and 600 °F, respectively.
- Condensed water is removed via a water draw from the three-phase condenser.

- HYSYS comes with a 3 Stripper Crude Column template. A Refluxed Absorber template could also be used, but this would add the procedure of installing Side Strippers and Pump Arounds.

For this example, install the **3 Stripper Crude Column** custom templates.

(8-1) From the Object Palette, Columns group: Double-click the **Blank Column Sub-flowsheet**  icon, then click the **Read an Existing Column Template** button. The Available Column Templates finder property view appears.

(8-2) In the **Files of type** drop-down list, select **Column Templates (*.col).**

(8-3) From the list, select the **3sscrude** (3 side stream crude) template file, then click the **Open** button. The "**3sscrude**" template installed 40 trays, 29 in the Main Tray section, 3 trays in each of the 3 Side Strippers (1 reboiled and 2 steam stripped), a reboiler, and a condenser.

(8-4) In the Column Property view, connect the Inlet and Outlet streams to the column sub-flowsheet as shown in Figure 10.58 (**Design** tab, **Connections** page).

(8-5) Modify the Draw and Return stages of the **Side Strippers** and **Pump Arounds** on the corresponding page of the **Side Ops** table

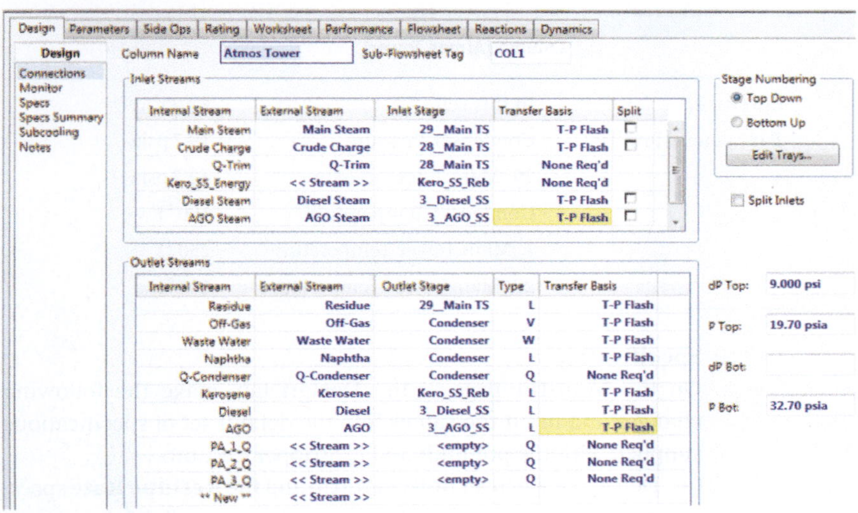

Figure 10.58 Input expert of the crude oil refinery column.

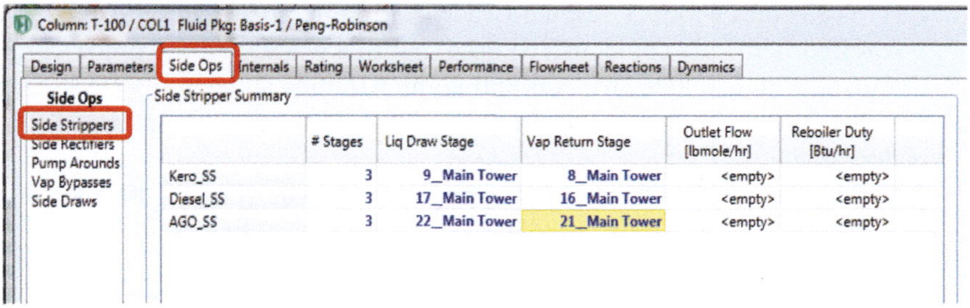

Column: T-100 / COL1 Fluid Pkg: Basis-1 / Peng-Robinson

Design | Parameters | Side Ops | Internals | Rating | Worksheet | Performance | Flowsheet | Reactions | Dynamics

Side Ops: Side Strippers, Side Rectifiers, Pump Arounds, Vap Bypasses, Side Draws

Side Stripper Summary

	# Stages	Liq Draw Stage	Vap Return Stage	Outlet Flow [lbmole/hr]	Reboiler Duty [Btu/hr]
Kero_SS	3	9__Main Tower	8__Main Tower	<empty>	<empty>
Diesel_SS	3	17__Main Tower	16__Main Tower	<empty>	<empty>
AGO_SS	3	22__Main Tower	21__Main Tower	<empty>	<empty>

Figure 10.59 Page's Summary of Side Stripper Summary.

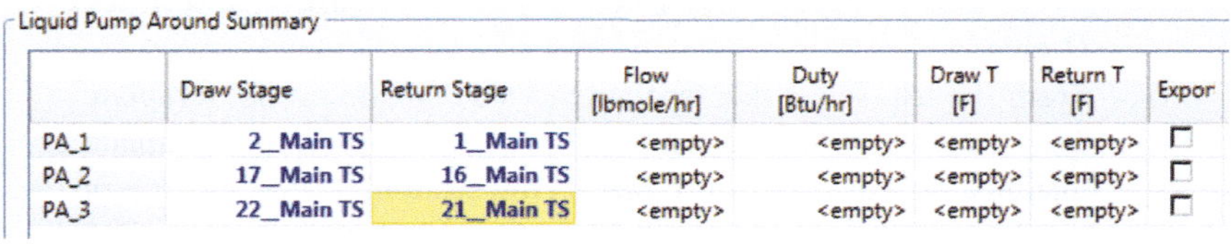

Liquid Pump Around Summary

	Draw Stage	Return Stage	Flow [lbmole/hr]	Duty [Btu/hr]	Draw T [F]	Return T [F]	Expor
PA_1	2__Main TS	1__Main TS	<empty>	<empty>	<empty>	<empty>	☐
PA_2	17__Main TS	16__Main TS	<empty>	<empty>	<empty>	<empty>	☐
PA_3	22__Main TS	21__Main TS	<empty>	<empty>	<empty>	<empty>	☐

Figure 10.60 Page's Summary of the Liquid Pump Around.

(8-6) **Side Stripper** Page: It must be appearing as in Figure 10.59.

(8-7) **Pump Around** Page: It must be appearing as in Figure 10.60.

(8-8) In the Atmos Tower Column property view, specify the column information below.

Column [Atmos Tower]

Tab [Page]	In this cell ...	Enter ...
Parameters [Profiles]	Condenser Pressure	19.7 psia
	29_Main Tower pressure	32.7 psia
	Condenser Temperature	100 °F
	1_Main Tower Temperature	250 °F
	29_Main Tower Temperature	600 °F

(8-9) **Specifications**

On the **Monitor** page of the **Design** tab, make the following changes and input the values into the default set of specifications supplied with the pre-built 3-Side Stripper Column.

(a) Delete the **Kero SS BoilUp Ratio** and the **Residue Rate** specs, where in the specification group, click on each of these specifications, click the **View** button (or double-click on each of

Table 10.1 Specification of the column.

Specification	Flow Basis	Spec Type	Spec Value
Kero_SS Prod Flow	Std Ideal Volume		9300 barrel/day
Diesel_SS Prod Flow	Std Ideal Volume		1.925e+04 barrel/day
AGO_SS Prod Flow	Std Ideal Volume		4500 barrel/day
PA_1_Rate(Pa)	Std Ideal Volume		5.0e+04 barrel/day
PA_1_Duty(Pa)		Duty	−5.50e+07 Btu/h
PA_2_Rate(Pa)	Std Ideal Volume		3.0e+4 barrel/day
PA_2_Duty(Pa)		Duty	−3.50e+07 Btu/h
PA_3_Rate(Pa)	Std Ideal Volume		3.0e+04 barrel/day
PA_3_Duty(Pa)		Duty	−3.50e+07 Btu/h
Naphtha Prod Rate	Std Ideal Volume		2.30e+04 barrel/day

these specification), and click the **Delete** button to delete the specification.

(b) Specify the **Reflux Ratio** spec to have a value of 1. *Deactivate* the **Reflux Ratio.**

(c) Change the default specifications as in Table 10.1 by selecting the specification in the table and clicking the **View** button.

(d) Change the *Flow* Basis to **Std Ideal Volume** before entering values.

(e) Change all the **Pump Around** delta *T* specifications to a **Duty** specification.

(f) On the **Monitor** page, **Design** tab, add the following new specifications as in Table 10.2 by clicking the **Add Spec** button under the Specifications group.

The final specification table will appear as shown in Figure 10.61:

9) Once you have provided all of the specifications, **Run** the column.

When the column is converged, Figure 10.62 will appear.

10.2.3 Sulfur Recovery Using Claus Process Treating Unit

The Claus process is one of the promising technologies for acid gas processing and sulfur recovery. Hydrogen sulfide primarily exists as a byproduct in the gas processing unit. It must be removed from natural gas. The Environmental Protection Agency (EPA) noticed that increasing SO_2 and CO_2 in the air harms the environment. Sulfur generally has an elemental content

Table 10.2 Specification of the column that must be added.

Specification Type	Variable (Field)	Value
Column Liquid Flow	Name	Overflash Spec
	Stage	27_Main Tower
	Flow Basis	Std Ideal Vol
	Spec Value	3500 barrel/day
Column Duty	Name	Kero Reb Duty
	Energy Stream	Kero_SS_Energy @Col1
	Spec Value	7.5e+06 Btu/h
Column Vapor Flow	Name	Vap Prod Flow
	Stage	Condenser
	Flow Basis	Molar
	Spec Value	0.00

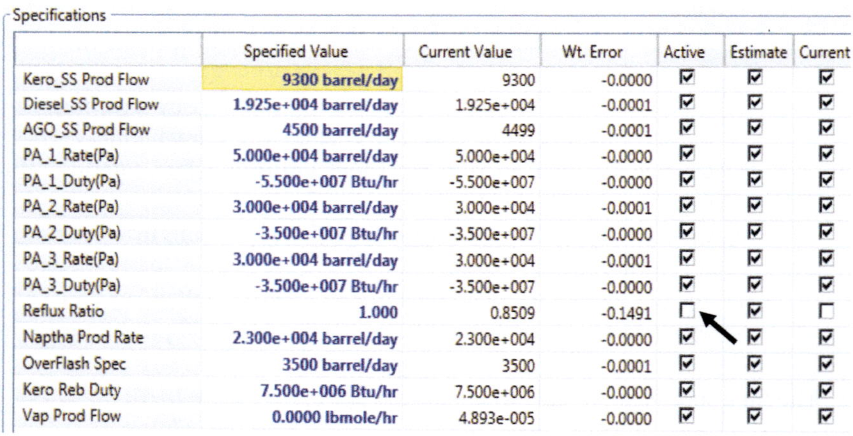
Specifications

	Specified Value	Current Value	Wt. Error	Active	Estimate	Current
Kero_SS Prod Flow	9300 barrel/day	9300	-0.0000	☑	☑	☑
Diesel_SS Prod Flow	1.925e+004 barrel/day	1.925e+004	-0.0001	☑	☑	☑
AGO_SS Prod Flow	4500 barrel/day	4499	-0.0001	☑	☑	☑
PA_1_Rate(Pa)	5.000e+004 barrel/day	5.000e+004	-0.0000	☑	☑	☑
PA_1_Duty(Pa)	-5.500e+007 Btu/hr	-5.500e+007	-0.0000	☑	☑	☑
PA_2_Rate(Pa)	3.000e+004 barrel/day	3.000e+004	-0.0001	☑	☑	☑
PA_2_Duty(Pa)	-3.500e+007 Btu/hr	-3.500e+007	-0.0000	☑	☑	☑
PA_3_Rate(Pa)	3.000e+004 barrel/day	3.000e+004	-0.0001	☑	☑	☑
PA_3_Duty(Pa)	-3.500e+007 Btu/hr	-3.500e+007	-0.0000	☑	☑	☑
Reflux Ratio	1.000	0.8509	-0.1491	☐	☑	☐
Naptha Prod Rate	2.300e+004 barrel/day	2.300e+004	-0.0000	☑	☑	☑
OverFlash Spec	3500 barrel/day	3500	-0.0001	☑	☑	☑
Kero Reb Duty	7.500e+006 Btu/hr	7.500e+006	-0.0000	☑	☑	☑
Vap Prod Flow	0.0000 lbmole/hr	4.893e-005	-0.0000	☑	☑	☑

Figure 10.61 Specifications of the refinery column to be converged.

of 0.1–6 wt % in crude oil, but the value could be higher than 14% for some crude oils and asphalts. It produces SO_2 and CO_2 gases, which damage the environment and atmosphere of the earth, called primary pollutants.

Sulfur simulation (SulSim) simulates the Claus Process from beginning to end by incorporating industry knowledge and data. The process flow diagram is shown in Figure 10.63. This example case utilizes one thermal

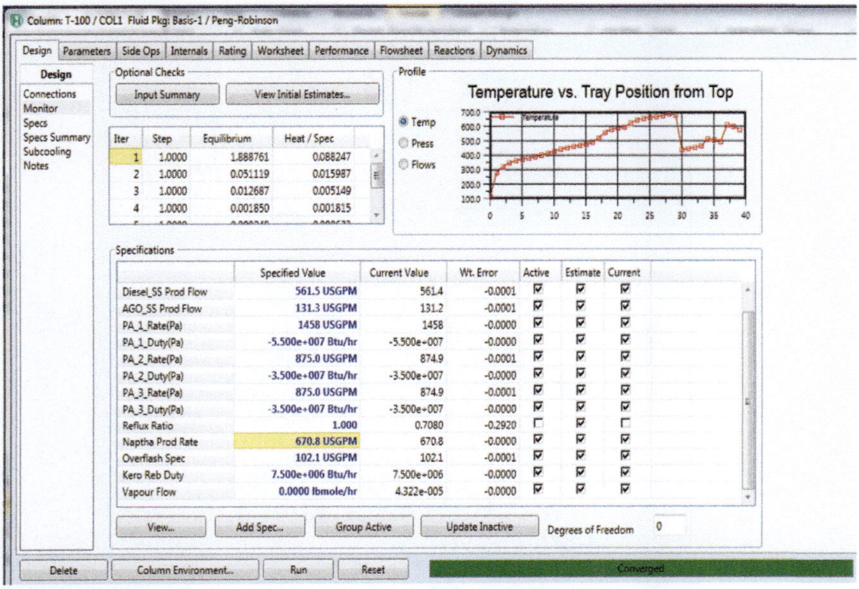

Figure 10.62 Converged column of the refinery crude oil.

stage (furnace, waste Heat Exchanger, and condenser), two catalytic stages (reheater, converter, and condenser), and a tail gas incinerator/stack to remove elemental sulfur from an acid gas containing 75% H_2S, 24.2% CO_2, and trace amounts of light hydrocarbons [9].

The reaction furnace is utilizing the "Rich Feed Acid Gas (Legacy)" model and is operating at 1124 °C with an air supply rate set to achieve 0% air demand (2:1 ratio of H_2S:SO_2 after the last condenser). After the catalytic stages, a cumulative conversion of 96.8% is achieved (96.6% recovered). The incinerator utilized the built-in kinetic correlations (value of "4") which help reduce emissions to the stack [9, 10].

The following features are specifically highlighted in this case:

- An "Air Demand Analyzer" unit op, which is an "adjust" block operation specifically for the (sulfur recovery unit) SRU to determine the amount of furnace air required to meet targeted tail gas specs.
- Incinerator and stack with built-in kinetic correlations.
- Streams have been highlighted using conditional formatting according to H_2S concentration (editable on the ribbon under Flowsheet/Modify| Conditional Formatting.

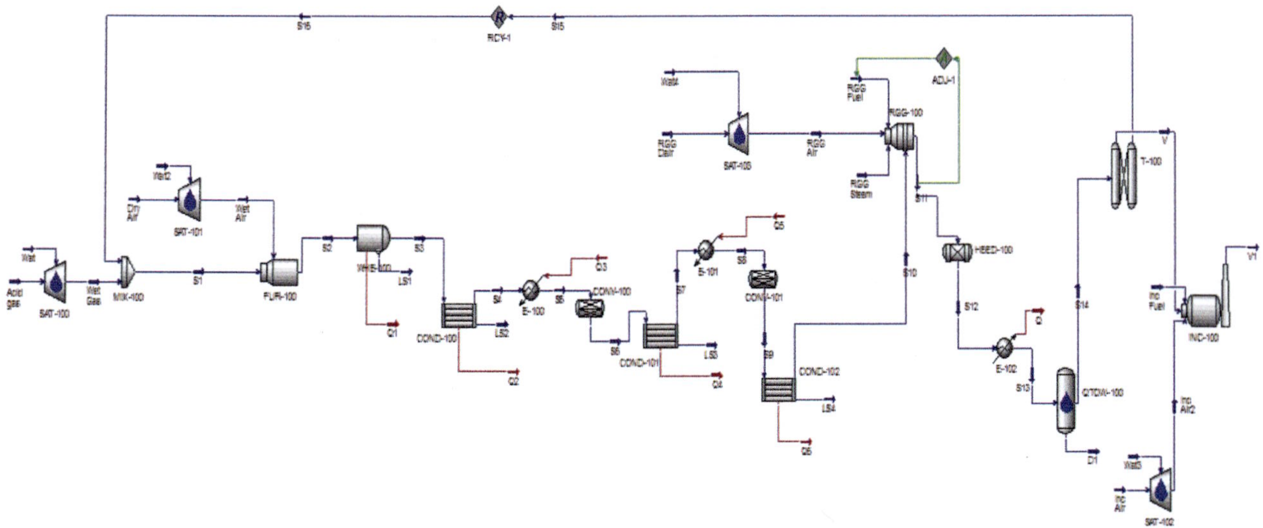

Figure 10.63 PFD of Sulfur recovery using Claus Process Treating Unit.

10.2.4 Simulation

1) Open a new case and add Fluid package: Sulsim (Sulfur Recover)
   It will add all the required components.
2) Enter the Simulation Environment; the Sulfur Recovery Unit (SRU) will be added as shown in Figure 10.64.
3) Click **Start with a Blank Flowsheet**, Figure 10.64, then double-click the icon of SRU-1.
   The Sub-Flowsheet operation of the sulfur recovery will appear as shown in Figure 10.65.
4) From Figure 10.65, click the **Sub-Flowsheet Environment** button.
5) Add the feed of the material stream (Acid gas) with the condition: $T = 35\,°C$, $P = 168.9\,kPa$,
   Molar composition: H_2S = 0.75, CO_2 = 0.242, Methane = 0.005, Ethane = 0.002, Propane = 0.001.
   Its flow will be estimated after adding the mixer.
6) From the object palette add **Saturator** (SAT-100), Figure 10.66.
   Stream Saturator unit ops are used to quickly calculate saturation of acid gas and air streams.
7) Double-click the **Saturator** icon and connect the streams as shown in Figure 10.67.
   Material Streams inside the SRU have a "Sulfur Recovery" page with results of interest.
8) Add a Mixer: Input = Wet gas, Output = S1. Enter the molar flow rate of S1 = 200 kmol/h.

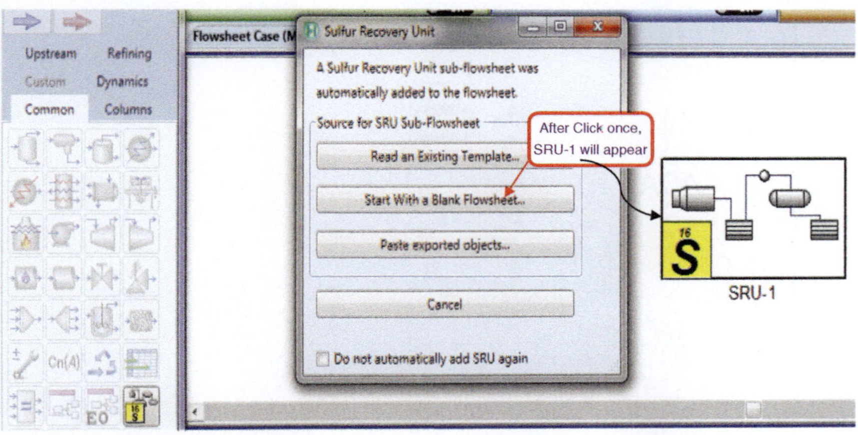

Figure 10.64 Sulfur recovery unit in the simulation environment.

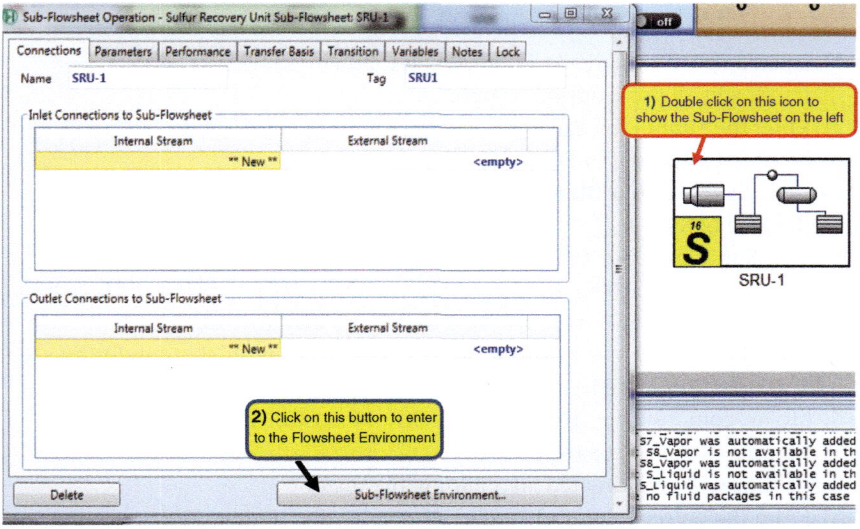

Figure 10.65 Sub-Flowsheet operation of the sulfur recovery.

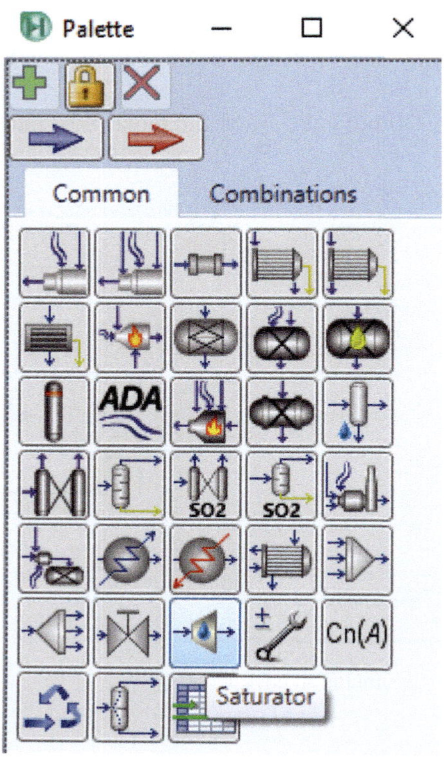

Figure 10.66 Saturator in the Object Palette.

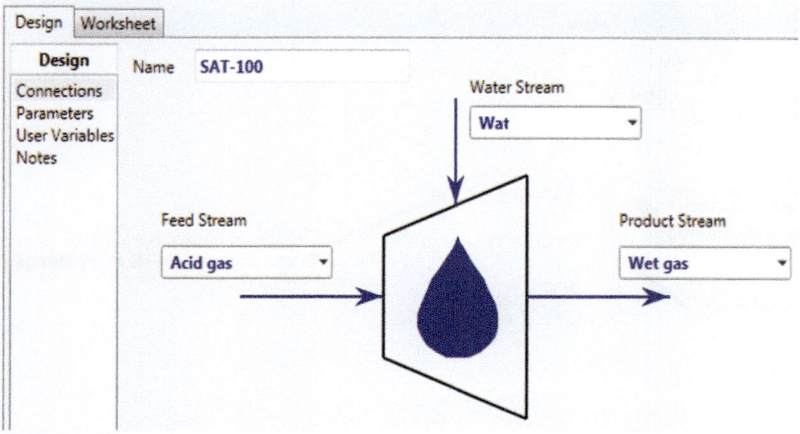

Figure 10.67 Window of Saturator SAT-100, Connection page, and its streams.

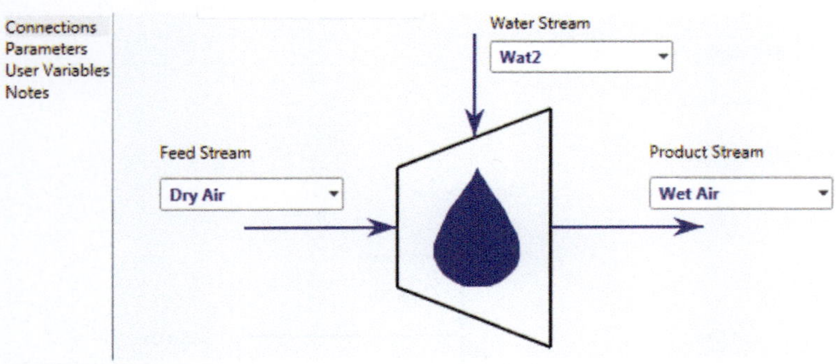

Figure 10.68 Window of Saturator SAT-101, Connection page, and its streams.

9) Add another **Saturator** (SAT-101), and connect it with the streams as in Figure 10.68.
Condition of the Dry Air: $T = 50\,°C$, $P = 168.9\,kPa$, Composition: $N_2 = 0.79$, $O_2 = 0.21$.
Molar flow rate of Wet Air = 200 kmol/h.
10) From the object palette, add a **Single Furnace Reaction Chamber** (FUR-100), Figure 10.69.
Connect the streams with the Single Furnace Reaction Chamber unit as in Figure 10.70.

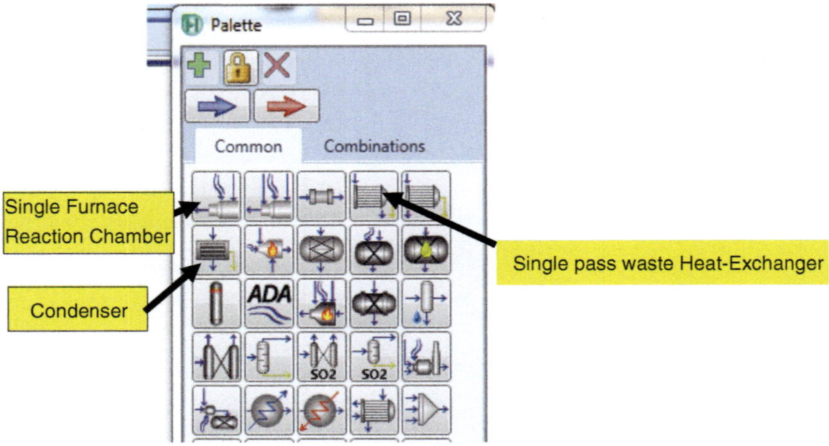

Figure 10.69 Single Furnace Reaction Chamber, Single-Pass Waste Heat Exchanger, and Condenser in the object palette.

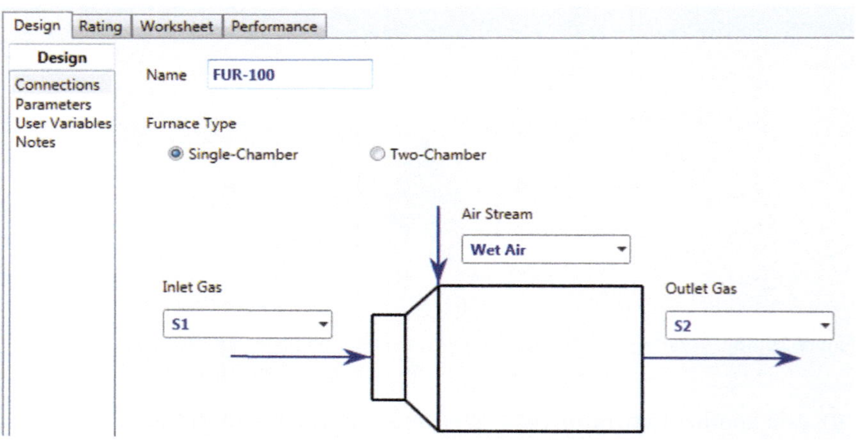

Figure 10.70 Window of Single Furnace Reaction Chamber (FUR-100), Connection page, and its streams.

11) Add a **Single-Pass Waste Heat Exchanger** (WHE-100), from the object palette, Figure 10.69, and connect the streams as in Figure 10.71. Specify the temperature of S3 = 300 °C.
12) Add a **Condenser** (COND-100) from the object palette, Figure 10.69, and connect the streams as in Figure 10.72.
    Specify the temperature of LS2 = 135 °C.
    The LS2 contains liquid sulfur only (S_Liquid).

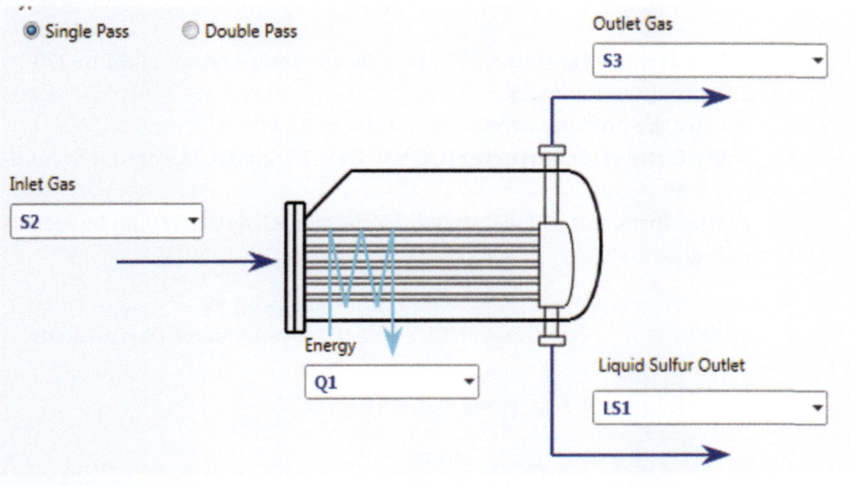

Figure 10.71 Window of Single-Pass Waste Heat Exchanger (WHE-100), Connection page, and its streams.

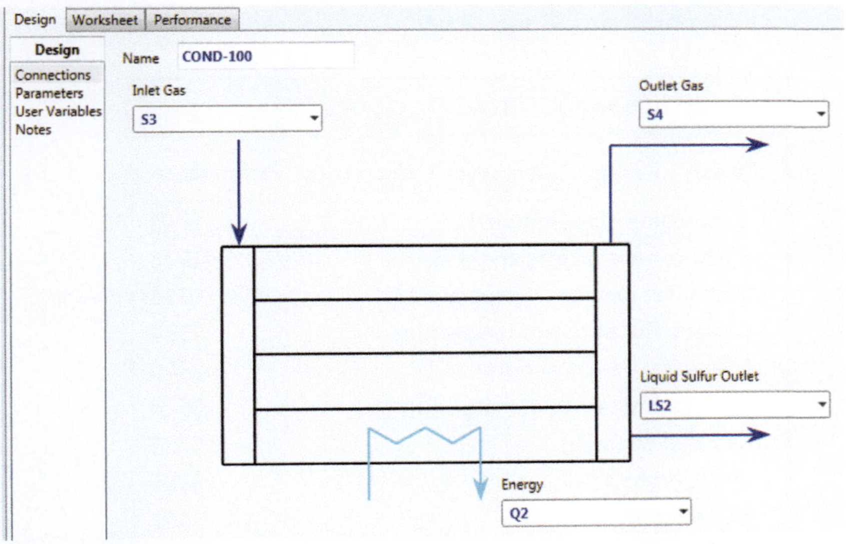

Figure 10.72 Window of Condenser (COND-100), Connection page, and its streams.

13) Add a **Heater (E-100)** 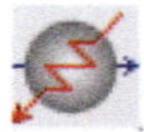 to raise the temperature of S4 to 270 °C and produce stream S5.
    Specify the pressure drop in the heater = 0.
14) Add a **Catalytic Converter** (CONV-100), Figure 10.73: Input = S5, Output = S6.
    The performance of the Catalytic Converter (CONV-100) can be seen as in Figure 10.74.

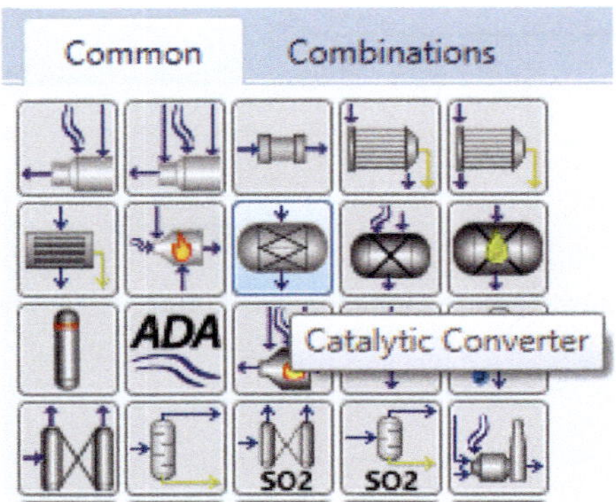

Figure 10.73 Catalytic Converter in the Object Palette.

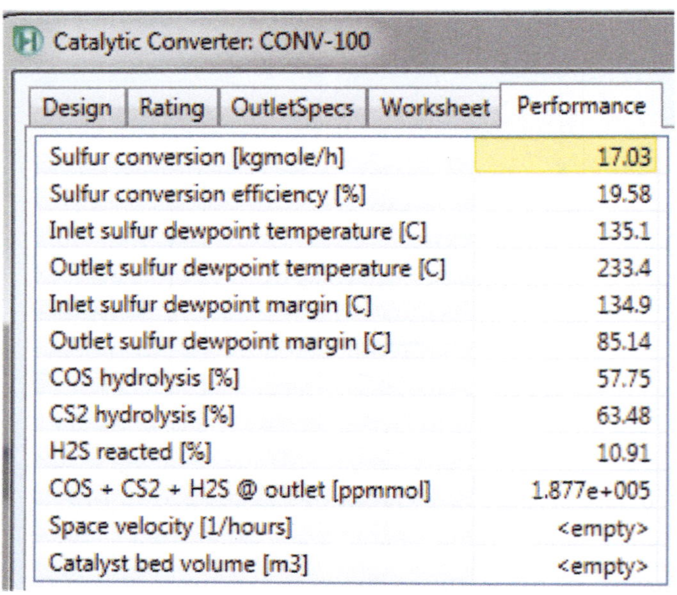

Sulfur conversion [kgmole/h]	17.03
Sulfur conversion efficiency [%]	19.58
Inlet sulfur dewpoint temperature [C]	135.1
Outlet sulfur dewpoint temperature [C]	233.4
Inlet sulfur dewpoint margin [C]	134.9
Outlet sulfur dewpoint margin [C]	85.14
COS hydrolysis [%]	57.75
CS2 hydrolysis [%]	63.48
H2S reacted [%]	10.91
COS + CS2 + H2S @ outlet [ppmmol]	1.877e+005
Space velocity [1/hours]	<empty>
Catalyst bed volume [m3]	<empty>

Figure 10.74 Performance tab of the Catalytic Converter (CONV-100).

15) Add a **Condenser (Cond-101)** 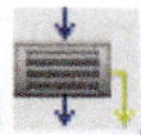
Inlet gas = S6, Outlet gas = S7, Liquid Sulfur Outlet = LS3, Energy = Q4. Specify the temperature of LS3 = 135 °C.
16) Add a **Heater (E-101):** Input = S7, Output = S8, the temperature of S8 = 210 °C, ΔP of heater = 0.
17) Add another **Catalytic Converter (Conv-101)**: Input = S8, Output = S9.
The performance of the Catalytic Converter (Conv-101) can be seen as in Figure 10.75.
18) Add a **Condenser (Cond-102)**: Inlet gas = S9, Outlet gas = S10, Liquid Sulfur Outlet = LS4, Energy = Q6. Set the temperature of LS4 = 135 °C.
19) Add a **Saturator (SAT-103)**, see Figure 10.66,
Feed Stream = RGG Dair, Water stream = Wat4, Product Stream = RGG Air,
Conditions of RGG Dair: $T = 35$ °C, $P = 169.8$ kPa, Composition of RGG Dair: $N_2 = 0.79$, $O_2 = 0.21$.
where:
RGG: reducing gas generator
Dair: Dry air

Design | Rating | OutletSpecs | Worksheet | Performance

Sulfur conversion [kgmole/h]	6.507e-002
Sulfur conversion efficiency [%]	0.09
Inlet sulfur dewpoint temperature [C]	135.0
Outlet sulfur dewpoint temperature [C]	139.4
Inlet sulfur dewpoint margin [C]	74.96
Outlet sulfur dewpoint margin [C]	70.98
COS hydrolysis [%]	-1.27
CS2 hydrolysis [%]	4.34
H2S reacted [%]	-0.04
COS + CS2 + H2S @ outlet [ppmmol]	1.889e+005
Space velocity [1/hours]	<empty>
Catalyst bed volume [m3]	<empty>

Figure 10.75 Performance tab of the Catalytic Converter (CONV-101).

20) Add a **Reducing Gas Generator** (RGG-100) from the object palette, Figure 10.76, and connect the streams as in Figure 10.77.
From the **Parameters** Page, set the Burn Stoichiometry = 85%, Figure 10.78.
Steam-to-Fuel Ratio (mass) = 100%, Figure 10.78.
Conditions of the **RGG Fuel:** $T = 14.95\,°C$, $P = 377.1$ kPa, Assume the molar flow rate = 2 kmole/h (*as an initial value*).
Composition: Methane = 0.95, Ethane = 0.04, Propane = 0.01

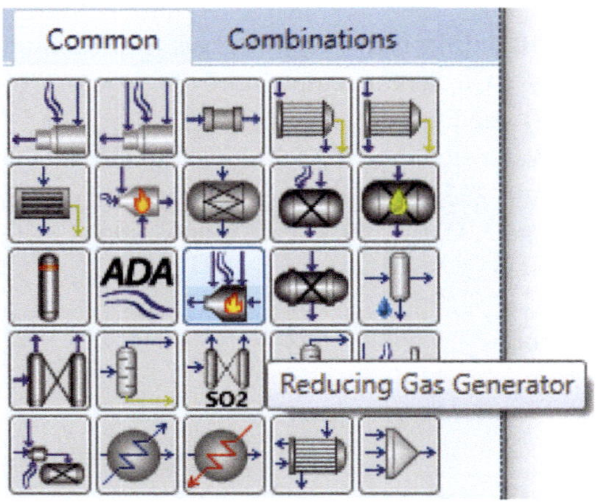

Figure 10.76 Reducing Gas Generator (RGG) unit in the object palette.

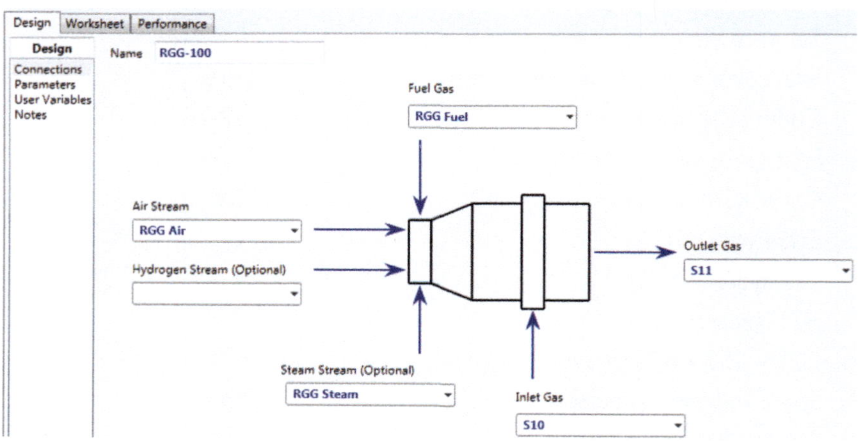

Figure 10.77 Window of Reducing Gas Generator (RGG-100), Connection page, and its streams.

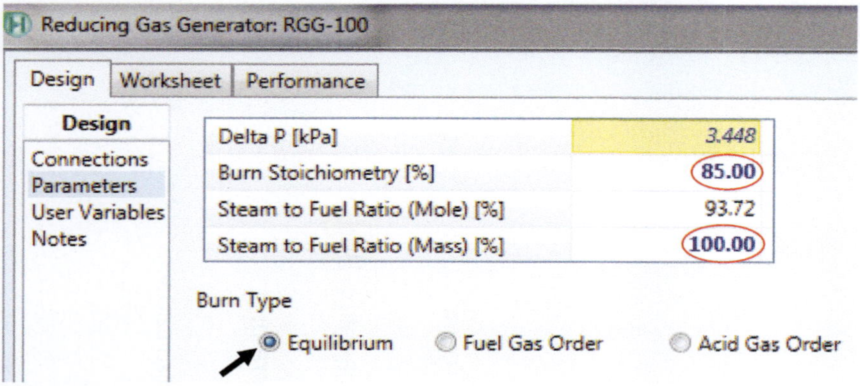

Figure 10.78 Entering the parameters in the RGG-100.

Conditions of the **RGG Steam:** Vapor fraction = 1, P = 451.3 kPa, Composition: H_2O = 1.

21) Add an **Adjust** operation to obtain the temperature of S11 = 285 °C by adjusting the mole flow rate of RGG Fuel.
Where the parameters of the **Adjust** are:
Tolerance = 0.01 °C, Step size = 1 kmol/h, Maximum Iteration = 1000.
22) Add a **Hydrogenation Bed (HBED -100)**, Figure 10.79: Input = S11, Output = S12.
Stream S12 is cooled to 150 °C by Cooler (E-102) and result S13.
Set ΔP of the Cooler (E-102) = 0.
23) Add a **Quench Tower** (QTOW-100), Figure 10.79.
Inlet Gas = S13, Outlet Gas = S14, Water = D1, ΔP = 0, Temp of S14 = 35 °C.
24) Add a **Simple Amine Absorber** (T-100), Figure 10.79.
Inlet Gas = S14, Recycle Gas = S15, Vent Gas = V.
Set the Temperature and pressure of Stream-V = 35 °C, and 100.6 kPa, respectively.
Set the Temperature and pressure of Stream-S15 = 35 °C, and 168.9 kPa, respectively.
From the **Parameters Page**: Concentration of H_2S in Vent = 150 ppm
25) Add an **Incinerator** (INC-100), Figure 10.79.
Inlet Gas = V, Fuel Gas = Inc. Fuel,
Outlet Gas = V1, Air Stream = Inc. Air2.
Conditions of the Inc. Fuel: T = 14.95 °C, P = 377.1 kPa,
Composition: Methane = 0.95, Ethane = 0.03, Propane = 0.02.

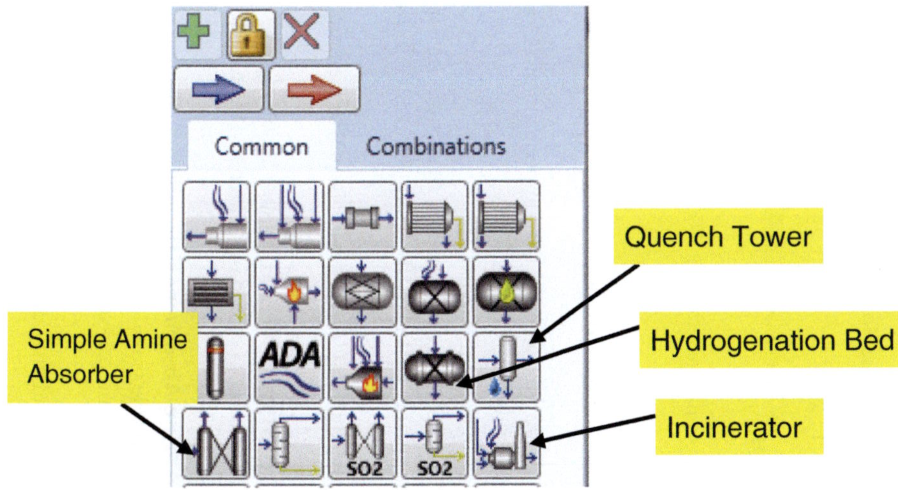

Figure 10.79 Hydrogenation Bed, Quench Tower, Simple Amine Absorber, and Incinerator in the object palette.

26) Add a Saturator (SAT-102)
Feed stream: Inc. Air, Water Stream: Wat3, Product Stream: Inc. Air2
Conditions of the Inc. Air: $T = 35\,°\text{C}$, $P = 170.3\,\text{kPa}$, Composition: $N_2 = 0.79$, $O_2 = 0.21$.
27) Add a Recycle operation: Input = S15, Output = S16.
28) Reconnect the stream S16 with mixer (MIX-100).

Now, one can check the results for efficiency of H_2S removed by comparing between the streams of Acid gas and V1.

10.3 Summary

Crude oil is a complex blend of liquid and gaseous hydrocarbons. The software has library containing many crude oils with their database for many regions around the world, where they can be used as a feedstock, as well as enter assay data manually for unavailable properties of a specific crude oil (such as a newly discovered crude oil). Their physical properties must be entered in the software in order to handle through the simulation. The two examples of Sections 10.2.1 and 10.2.2 illustrated the complete construction of the simulation, from selecting a property package and components, characterizing the crude oil, to installing streams and unit operations, through to examining the final results where a fractionation facility processes crude oil to create diesel, naphtha, kerosene, atmospheric gas oil, and atmospheric residual products. For Section 10.2.1, the distillation column is constructed step by step where refluxed absorber column is added firstly, then the strippers and around pumps were connected with it. While in Section 10.2.2, the distillation column is inserted from the templates that exist in the software.

Section 10.2.3 dealing with Sulsim (sulfur simulation) technology (created by Sulfur Experts) is integrated within HYSYS. Sulsim simulates the Claus Process from beginning to end by incorporating industry knowledge and data. Sulsim incorporates empirical correlations fitted to plant data, letting you:

- Use the rigorous HYSYS simulation to accurately model the reaction furnace, reactors, incinerator, and other related operations of the sulfur recovery plant.
- Meet stringent environmental regulations and standards in flare gases, especially H_2S & SO_2 emissions limits.

References

1 Liu, Y.A., Chang, A.-F., and Pashikanti, K. (2019). *Petroleum Refinery Process Modeling Integrated Optimization Tools and Applications*. Wiley Inc.

2 (2017). *Aspen HYSYS User Guide Version 10.0*. Aspen Technology.

3 (2015). *Aspen HYSYS Petroleum Refining Unit, Operations & Reactor Models, Reference Guide, V8.8*. Aspen Technology, Inc.

4 Haydary, J. (2019). *Chemical Process Design and Simulation Aspen Plus and Aspen HYSYS Applications*. John Wiley & Sons, Inc.

5 Herrmann, L. and Ye, V. (2013). *Assay Management in Aspen HYSYS® Petroleum Refining, Jump Start: A Brief Tutorial (and Supplement to Training and Online Documentation)*. Aspen Technology, Inc. AspenTech®, aspenONE.

6 (2014). *Modeling Heavy Oil & Gas Production and Facilities Using Aspen HYSYS Upstream, AspenTech Customer Education Training Manual, Course Number EHY2351.084.01*. Aspen Technology, Inc.

7 HYSYS® 2004.2 (2005). *Tutorials and Applications*. Aspen Technology, Inc.

8 Aspen HYSYS® (2017). *Process Modeling, Course Number EHY101.10.01, AspenTech Customer Education Training Manual*. Aspen Technology, Inc.

9 SULSIM 6.0 (2003). *Reference Guide*. Aspen Technology, Inc. AspenTech®, aspenONE®.

10 Mondor, C. and Dyment, J. *Using and Optimizing Your Sulfur Recovery Unit in Aspen HYSYS*," Jump Start Guide. Aspen Technology, Inc. V9.

11

Optimizer

HYSYS contains a multi-variable steady-state Optimizer. Once your flowsheet has been built and a converged solution has been obtained, you can use the Optimizer to find the operating conditions which minimize (or maximize) an Objective Function. The optimizer owns its Spreadsheet for defining the Objective functions as well as any constraint expressions to be used. This allows you to construct Objective Functions which maximize profit, minimize utilities, or minimize exchanger UA [1].

There are several optimization terms which will be used [2, 3].

- **Primary Variables** correspond to flowsheet variables whose values are manipulated in order to minimize or maximize the objective function. You set the upper and lower bounds for the primary variables, and these are used to set the search range. Only user-specified process variables (i.e. blue numbers) can be used as Primary Variables.
- **The Objective Function** is the function which is to be minimized or maximized. The function has to be defined within or referenced by the Optimizer Spreadsheet. This allows the user a great deal of flexibility in defining the function.
- **The Constraint Functions** are inequality and equality functions that are defined in the Spreadsheet. In solving the Objective Function, the Optimizer must also meet any constraints that are defined by the user.

11.1 Simple Optimizer

A special tool in HYSYS, the Optimizer, will be used to find the optimum operating conditions, so that the maximum profit can be obtained. This example will use the optimizer to obtain the maximum profit throughout the separation of Tetrahydrofuran (THF) from Toluene using distillation column [4, 5].

Chemical Process Simulations using Aspen Hysys, First Edition. Khalid W. Hameed.

11.1.1 Problem Statement

A simple distillation column to separate Tetrahydrofuran (THF) from Toluene is simulated. The object of this exercise is to select the product specifications such that profit is maximized. The feed at 10°C, 140 kPa, 3700 kg/h, and 44% THF and 56% toluene by mass enters the distillation column, which consists of 10 stages, total condenser, and the pressures of condenser and reboiler are 103, and 107 kPa, respectively. The estimated specifications of reflux ratio = 2 (by mass), distillate mass flow rate = 1500 kg/h, and the active specifications of mass fraction of toluene in the reboiler and THF in the condenser are 0.96 and 0.94, respectively [4, 5].

11.1.2 Simulation

1) Open a new case, and add the following components: Tetrahydrofuran and Toluene.
2) Fluid Package: Wilson, on the **Binary Coeffs** tab, press the **Unknowns** Only button.
3) Enter the **Simulation** Environment.
4) Add a material stream with the following values:
   Name: Feed
   Temp = 10°C
   Pressure = 140 kPa
   Mass Flow rate = 3700 kg/h
   Composition-Mass Fraction: THF = 0.44, Toluene = 0.56.
5) Add a Distillation Column with the following values:

Connections	
Column Name	**T-100**
Number of Stages	10
Inlet stream/ #Stage	Feed/Stage 5
Condenser Type	Total
Bottom's Liquid	Toluene
Ovhd Liquid	THF
Condenser Energy Stream	Cond Q
Reboiler Energy Stream	Reb Q
Pressures	
Delta P for Condenser and Reboiler	0.0 kPa
Condenser	103 kPa
Reboiler	107 kPa

Click Next, and click Next again.

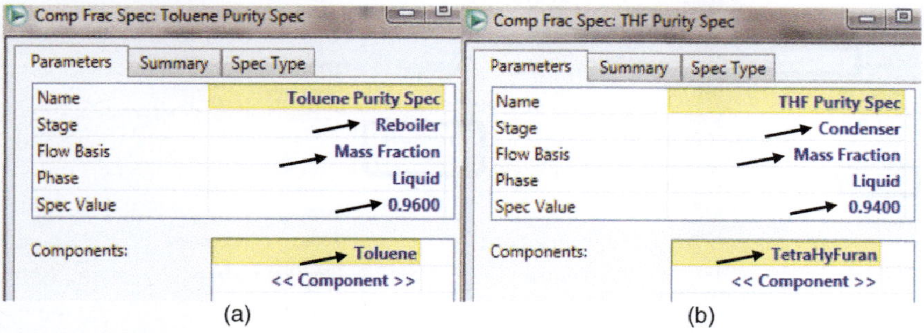

Figure 11.1 Specifying the component fraction specifications for (a) tolugene and (b) THF.

6) Page 5 of 5; Flow Basis: Mass,
   Reflux ratio = 2, Liquid flow rate of distillate = 1500 kg/h
7) Create two new **Column Component Fraction** specifications as shown in Figure 11.1.
8) These two specifications should be Active, and deactivate the Reflux ratio and Distillate rate in the Design tab, Monitor page.
   The **Degree of Freedom** for the column should now be 0.
   The column should now solve automatically. If it does not, press the **Run** button to start the solver.
   Save Your Case and record the following results:
   (THF Mass Flow, THF Mass Frac, Toluene Mass flow, Toluene Mass frac, Condenser Duty, and Reboiler Duty)
9) Changing the Column Tolerances
   In order for the optimizer to work properly, we have to tighten the tolerances of the column solver. There are two tolerances that the column must meet before it can be considered solved.
   - **The Equilibrium Error Tolerance**
   - **The Heat/Spec Error Tolerance**

     We want to set the value for both of these tolerances at 1e−6. Note that this will increase the solving time of the column, but this column solves very fast anyway, so the tighter tolerances are acceptable in this case.

   To change the tolerances, follow these steps:
   - Double-click the column, access the **Solver** page of the **Parameters** tab.
   - Enter 1e−6 for both the tolerance values. These cells are located in the Solving Options group box, Figure 11.2.
10) Adding the Optimizer
    Typically, optimization studies involve an "economic model" composed of **Object Function** or **Profit Function** and operating constraints.

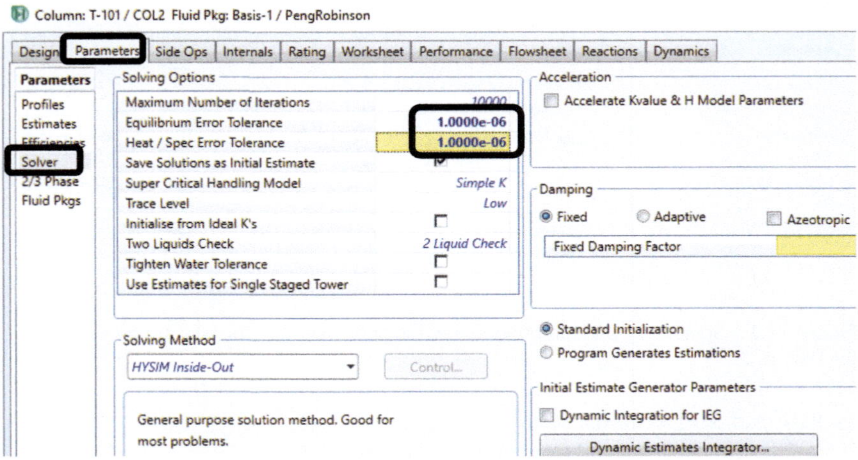

Figure 11.2 Setting the tolerance of the equilibrium error and heat/spec error.

Generally a profit function is given by [4]:

$$PF = \sum_{p} F_p M_p - \sum_{f} F_f M_f - OC \tag{11.1}$$

where

PF = Profit Function/time

$\sum_{p} F_p M_p$ = Product Flows × Product Values

$\sum_{f} F_f M_f$ = Feedstock Flows × Feedstock Costs

OC = Operating Costs.

$$\begin{aligned} \mathit{Profit} = {} & (\text{THF Product} + \text{Toluene Product}) - \text{Heating Cost} \\ & - \text{Cooling Cost} - \text{Feed Cost} \end{aligned} \tag{11.2}$$

- To invoke the **Optimizer**, from the **Home** tab of the Ribbon, click the **Optimizer** icon 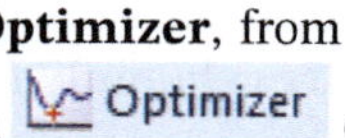

 or press <F5>.
- In the **Variables** tab, Press the **Add** button to add the first variable, Objects = **T-100**, Variables = **Spec Value,** Variable Specification = **THF Purity**, **OK,** Figure 11.3.
- Add the second variable, Objects = **T-100**, Variables = **Spec Value,** Variable Specification = **Toluene Purity**, **OK.**
- Set the Lower Bound at 0.9, and the Upper Bound at 0.99 for each variable, Figure 11.4.

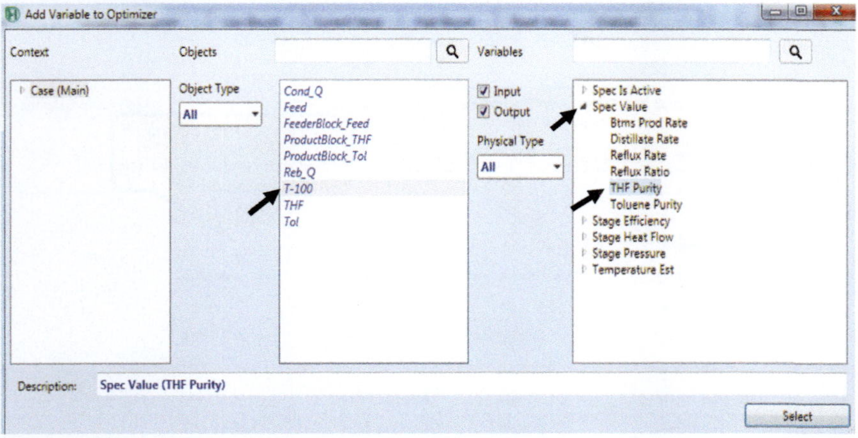

Figure 11.3 Adding the purity specification of THF in the distillation column.

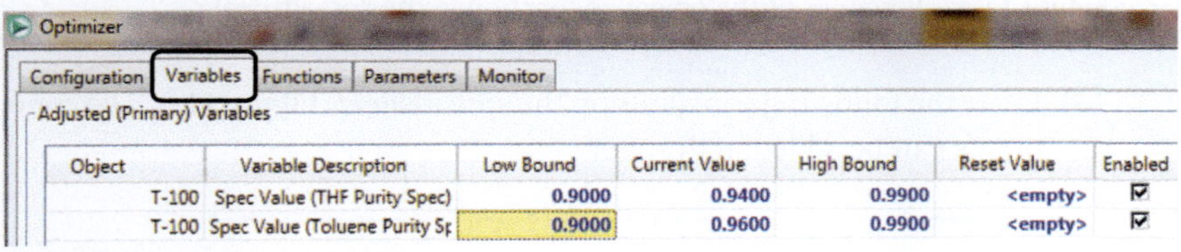

Figure 11.4 Range of adjusted variables in the optimizer.

11) Press the **Spreadsheet** button on the Optimizer view to open the Spreadsheet.
    - Move to the **Spreadsheet** tab.

To Import any Variable, Right click in the cell and select **Add Import variable.**

12) Fill the B cells as follows, and as shown in Figure 11.5:

Cell …	Object …	Variable …
B1	Cond Q	Heat Flow
B2	Reb Q	Heat Flow
B3	THF	Mass Flow
B4	THF	Comp Mass Frac THF
B5	Toluene	Mass Flow
B6	Toluene	Comp Mass Frac Toluene
B7	Feed	Mass Flow

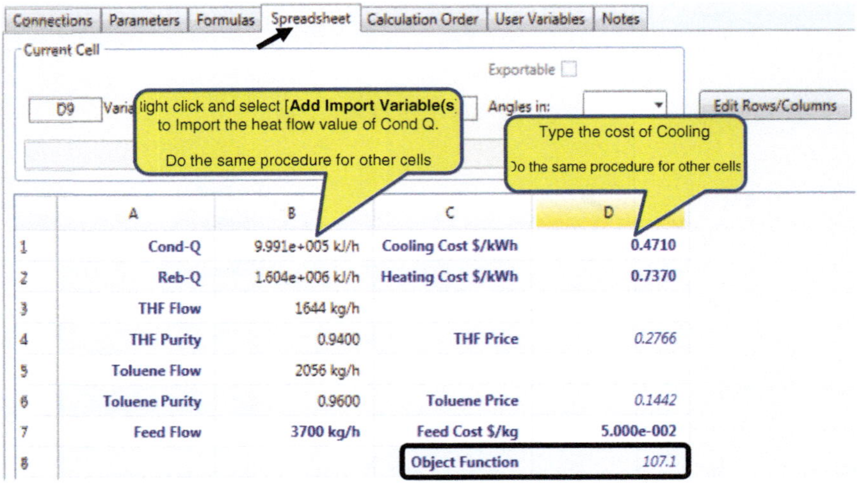

Figure 11.5 Variables of the objective function in the spreadsheet.

13) Enter the following constants on the spreadsheet, Figure 11.5. The comments can be added, if desired.

Cell	Value	Comment
D1	0.471	Cooling Cost $/kWh
D2	0.737	Heating Cost $/kWh
D7	0.05	Feed Cost $/kg

Note: *All common functions must be preceded by a (+ symbol). Special Functions must be preceded by the @ symbol.*
Some of the functions available are:

- *Addition (+): +A1 + A2*
- *Subtraction (−): +A1 − A2*
- *Multiplication (*): +A1*A2*
- *Division (/): +A1/A2*
- *Power (^): +A1^3*
- *Absolute Value (@ABS):@ABS(A1)*
- *Square Root (@SQRT):@SQRT(A1)*
- *Natural Log (@ln):@ln(A1)*
- *Exponential (@exp):@exp(A1)*

14) Enter the following product prices, Figure 11.5:

Cell	Equation	Description
D4	+0.333*b4^3	THF Price (correct for purity)
D6	+0.163*b6^3	Toluene Price (correct for purity)

***Note**: The prices of the products increase as the purity increases.*

15) The Objective Function is placed in Cell D8, Figure 11.5. The equation is:

$$+(b3^*d4 + b5^*d6) - b7^*d7 - (b1^*d1 + b2^*d2)/3600.$$

Close the Optimizer Spreadsheet

16) **Functions** tab:

The Functions tab contains two groups, the Objective Function and the Constraint Functions. However, in this example we do not have constraint functions.

- In the Cell area of the Objective Function group, specify the Spreadsheet cell that defines the Objective Function. Use the drop-down menu in the Edit Bar to select the appropriate cell (D8), Figure 11.6. The Current Value of the Objective Function will be provided.
- Select the Maximize radio button, Figure 11.6.

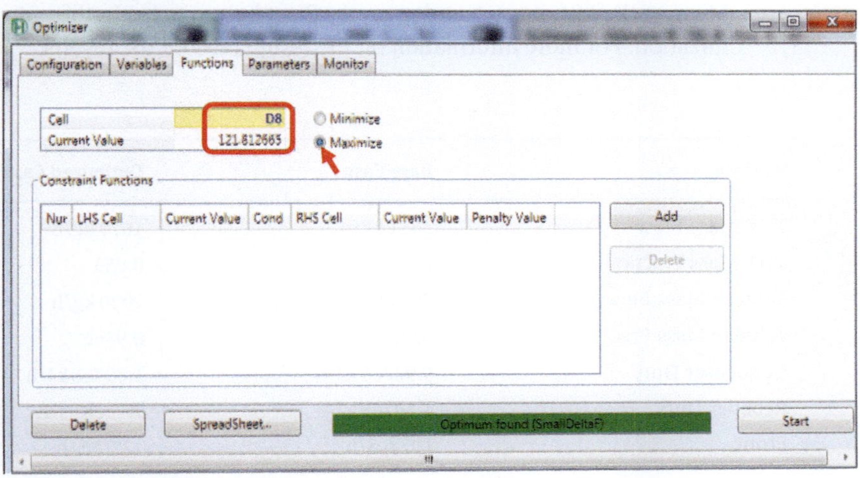

Figure 11.6 Specify whether the objective function is minimize or maximize; the object function is increased from 107.1 to 121.813 $/h.

17) **Parameters** tab:

The Parameters tab is used for selecting the Optimization Scheme [1].

- **Box** – Handles inequality constraints but not equality constraints. It generally requires a large number of iterations to converge on the solution.
- **SQP** – Sequential Quadratic Programming handles inequality and equality constraints. Considered by many to be the most efficient method for minimization.
- **Mixed** – Handles inequality constraints only. It is a combination of the Box and SQP methods. It starts the minimization with the Box method using a very loose convergence tolerance. After convergence, the SQP method is used to locate the final solution.
- **Fletcher Reeves** – Does not handle constraints. Efficient method for general minimization.
- **Quasi-Newton** – Does not handle constraints. Similar method to Fletcher Reeves.
  - Select the **Mixed** method as the Scheme, Use the defaults for Tolerance and Number of Iterations.

18) Analyzing the Results:

Once the Optimization is complete, the Monitor tab displays the values of the Objective Function and Primary Variables during the Optimizer calculations.

- Move to the Monitor tab and press the Start button to begin the optimization. For more information on the results, record the results:

Variable	Base Case	Optimized Case
THF Mass Flow	1651 kg/h	1679 kg/h
THF Mass Fraction	0.94	0.953
Toluene Mass Flow	2049 kg/h	2020 kg/h
Toluene Mass Frac	0.96	0.9866
Condenser Duty	9.922e5 kJ/h	1.097e6 kJ/h
Reboiler Duty	1.597e5 kJ/h	1.71e6 kJ/h
Profit	107.6 $/h	122.2 $/h

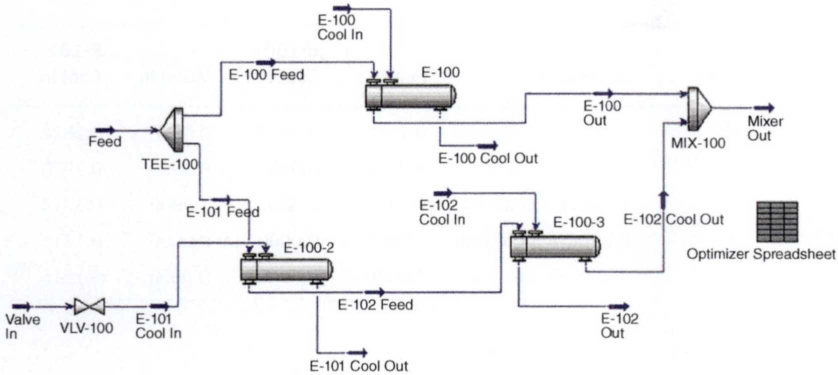

Figure 11.7 PFD of the multiple Heat Exchangers.

11.2 Original Optimizer

11.2.1 Problem Statement

This example will minimize the combined overall heat transfer coefficient times area (UA) for the Heat Exchangers using constraint functions method in the optimizer [4].

Create the following sample case of multiple Heat Exchangers, Figure 11.7, to optimize the overall UA by using the **Original Optimizer.**

11.2.2 Simulation

1) Open a new case, and add the following Components: C_1, C_2, C_3, i-C_4, and n-C_4.
   Property package: Peng Robinson.
2) **Simulation** Environment:
   - Process Streams: Enter the required streams as shown below (Use WorkBook):

Tab [Page]	In This Cell	Feed	E-100 Cool In	Valve In	E-102 Cool In
Worksheet [Conditions]	Temperature (°F)	20	−142	120	<Empty>
	Pressure (psia)	1000	250	350	251
	Molar Flow (lbmol/h)	2745	1542	<Empty>	1640

Tab [Page]	In This Cell	Feed	E-100 Cool In	Valve In	E-102 Cool In
Worksheet [Composition]	Methane mole frac	0.7515	0.9073	0.0000	0.2828
	Ethane mole frac	0.2004	0.0927	0.0000	0.2930
	Propane mole frac	0.0401	0.0000	1.0000	0.1414
	i-Butane mole frac	0.0040	0.0000	0.0000	0.1313
	n-Butane mole frac	0.0040	0.0000	0.0000	0.1515

3) Conditions of Heat Exchangers:
   (3-1) Heat Exchanger **E-100**
   Tube side: Inlet = E-100 Feed, Outlet = E-100 Out,
   Shell Side: Inlet = E-100 Cool In, Outlet = E-100 Cool Out.

Tab [Page]	In This Cell	Enter
Design [Parameters]	Tube side Delta P	10 psi
	Shell side Delta P	10 psi
	Exchanger Model	Weighted
	Intervals (E-100 Feed)	10
	Intervals (E-100 Cool In)	10
	Dew/Bubble Point (Cool In)	Inactive

(3-2) Heat Exchanger **E-100-2**
Tube Side: Inlet = E-101 Feed, Outlet = E-102 Feed,
Shell Side: Inlet = E-101 Cool In, Outlet = E-101 Cool Out.

Tab [Page]	In This Cell	Enter
Design [Parameters]	Tube side Delta P	5 psi
	Shell side Delta P	1 psi
	Exchanger Model	Weighted
	Intervals for two sides	10

(3-3) Heat Exchanger **E-100-3**
Tube Side: Inlet = E-102 Feed, Outlet = E-102 Out,
Shell Side: Inlet = E-102 Cool In, Outlet = E-102 Cool Out.

Tab [Page]	In This Cell	Enter
Design [Parameters]	Tube side Delta P	5 psi
	Shell side Delta P	5 psi
	Exchanger Model	Weighted
	Intervals for two sides	10
	Dew/Bubble Point (E-102 Cool In)	Inactive

4) Streams Specifications:

- Temperature of stream E-102 Out = −40°F
- Vapor Fraction of stream E-101 Cool Out = 1.00
- Temperature of stream E-100 Out = −65°F
- Pressure of E-101 Cool Out = 20 psia
- Temperature of E-102 Cool In = −85°F
- Flowrate of Valve In = 495 lbmol/h.
- Flowrate of E-101 Feed = *Optimized variable* (Initially set to the flow rate of 1670 lbmol/h).

  TEE-100: Inlet: Feed, Outlet: E-100 Feed and E-101-Feed
  Mixer **(MIX-100):** Inlet: E-100 Out, and E-102 Out, Outlet: Mixer Out
  Valve **(VLV-100):** Inlet: Valve In, Outlet: E-101 Cool In

5) Optimizing Overall UA
   The Optimizer determines the optimum ***Tee flow ratio*** *such that the Overall UA is minimized.*
   (5-1) After opening the Optimizer: Click the **Add** button in the **Variables** tab to display the Variable Navigator.
   (5-2) Select the Molar Flow of the stream E-101 Feed.
   (5-3) Specify the Low and High Bounds (lbmol/h) as shown in Figure 11.8.
6) Import the Heat Exchanger UAs,
   To import the three Heat Exchanger UAs to the Optimizer spreadsheet:
   (6-1) After opening the optimizer, click the **SpreadSheet** button (in the bottom of the optimizer) to display the Optimizer Spreadsheet property view, Figure 11.9.
   (6-2) From the Spreadsheet window, click the **SpreadSheet** tab, and in the cell A1, right click and select **Add Import Variable(s)** to display the Variable Navigator property view.
   (6-3) Import the UA value for the Heat Exchanger E-100.

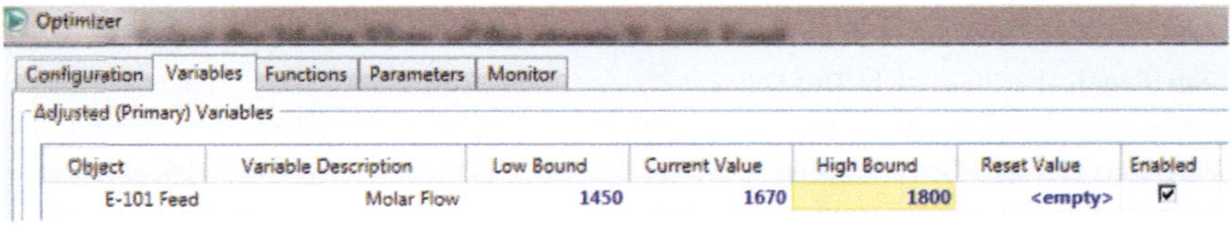

Figure 11.8 Specifying the Low and High Bounds of the adjusted variable (molar flow of E-101 Feed).

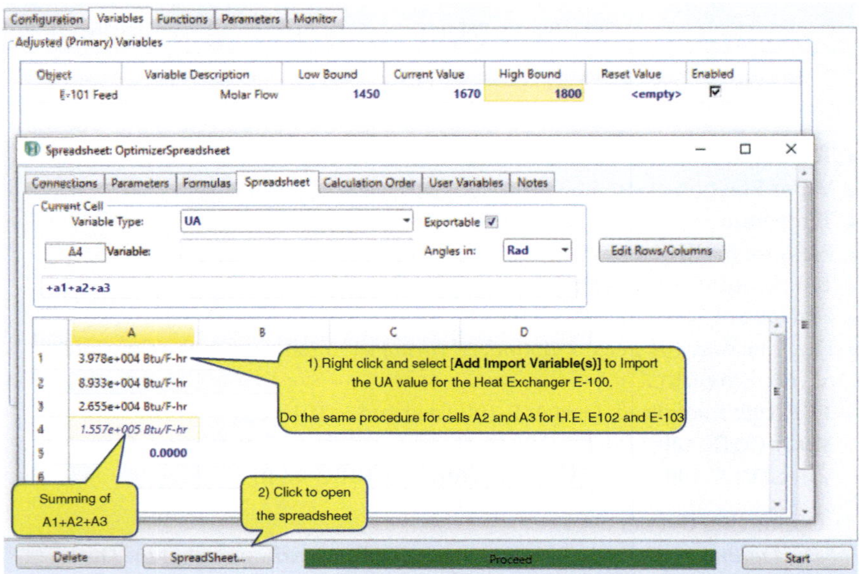

Figure 11.9 Importing the UA value for the Heat Exchangers in the spreadsheet.

(6-4) Repeat steps#2 and#3 for the Heat Exchangers E-102 and E-103 for the cells A2, and A3 respectively, Figure 11.9.

(6-5) In cell A4 enter the formula: +a1 + a2 + a3. This sums the UAs.

(6-6) In cell A5, enter 0.0. This is used in the constraints.

(6-7) Close the Spreadsheet property view.

7) Defining the Objective Function:

You must define the Objective Function and the Constraint Functions. The Objective Function is the expression being minimized, which in this case is the sum of the Heat Exchanger UAs.

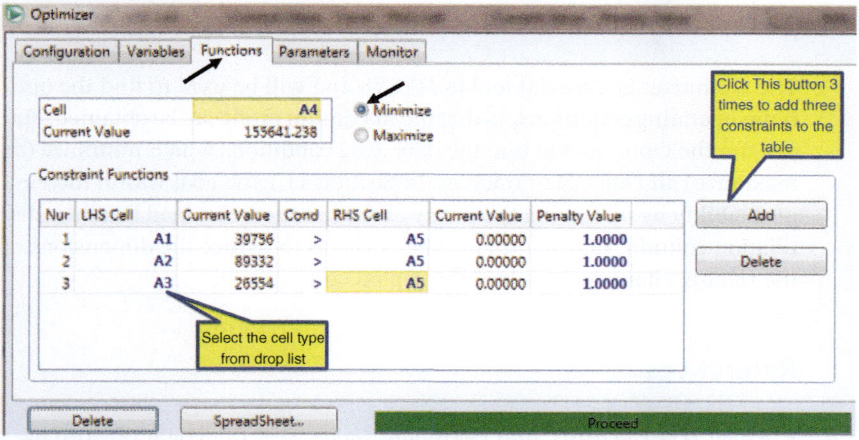

Figure 11.10 Estimating of the objective function in the optimizer.

(7-1) Click the **Functions** tab in the Optimizer property view, Figure 11.10.

(7-2) Click the **Cell** drop-down list and select **A4**. The value of the cell is displayed in the **Current Value** field.

(7-3) Click **Minimize** radio button.

8) Adding Constraint Functions:

Enter constraint functions to ensure the solution is reasonable. Each Heat Exchanger UA *must be greater than zero.*

(8-1) Click the **Add** button three times to add three constraints to the table.

(8-2) In the LHS Cell drop-down list, select the cells A1, A2, and A3 for each of the respective constraints, Figure 11.10.

(8-3) In the RHS Cell drop-down list, select the cell A5 for each of the constraints, Figure 11.10.

(8-4) Click Parameters tab. For this example, use the **Mixed** method, leaving all the parameters at their defaults.

(8-5) Click the **Start** button, and then click the **Monitor** tab to watch the progress of the Optimizer.

An optimum molar flow of **1800** lbmol/h is obtained for the stream "E-101 Feed," corresponding to an overall UA of about **1.43e5** Btu/F-h. This compares to the specified value of 1.5e5 Btu/F-h in the first part of this example.

11.3 Summary

The Optimizer is a special tool in HYSYS and will be used to find the optimum operating conditions, so that the maximum profit can be obtained. You can use the Optimizer to find the operating conditions which minimize (or maximize) an Objective Function. In Section 11.1 the goal was to increase profitability by increasing the purity of the products obtained from the distillation column. In Section 11.2, the goal was to reduce the dimensions of the Heat Exchangers to reduce the capital cost.

References

1 Aspen HYSYS (2010). *Unit Operations Guide V7.2*, Copyright (c) 2010 by. Aspen Technology, Inc.

2 (2000). *Process Modeling Using Hysys With Chemical Industry Focus*. AEA Technology plc.

3 Suhan, M.B., Shuchi, S.B., Abir, M.R., and Hasan, R. (2019). Innovative multi-stage anaerobic modelling of a 10 MW biogas plant using aspen HYSYS. *International Conference on Petroleum Engineering (InICPE 2019)*, Vol. 2019, p. 25.

4 *Advanced Process Modeling Using Aspen HYSYS, AspenTech Customer Education Training Manual Work Book*, Course Number EA1000.06.09, 2007.

5 (2017). *EHY202 Aspen HYSYS®: Advanced Process, Modeling Topics, Course Number EHY202.10.01, AspenTech Customer Education, Training Manual,* Copyright © 2017 by. Aspen Technology, Inc.

12

Solid Operation

12.1 Cyclone

The Cyclone is used to separate solids from a gas stream and is recommended only for particle sizes greater than 5 μm. The Cyclone consists of a vertical cylinder with a conical bottom, a rectangular inlet near the top, and an outlet for solids at the bottom of the cone. It is the centrifugal force developed in the vortex that moves the particles toward the wall. Particles reach the wall, slide down the cone, become separated from the gas stream. The solids being separated must have been previously specified and installed as components in the stream attached to this operation [1, 2].

Cyclone collectors can be designed for many applications, and they are typically categorized as high efficiency, conventional (medium efficiency), or high throughput (low efficiency). High-efficiency cyclones are likely to have the highest-pressure drops of the three cyclone types, while high-throughput cyclones are designed to treat large volumes of gas with a low-pressure drop [3].

Figure 12.1 shows the schematic diagram of the cyclone with its dimensions.

where

D is the diameter of the cyclone's body.
H is the height of duct of inlet dirty gas.
W is the length of duct of inlet dirty gas.
L_b is the length of cyclone body.
L_C is the length of conical section.
D_e is the diameter of outlet clean gas duct.
D_d is the diameter of the solid dirty duct.
S is the section length of the tube of clean gas that insert into the body of cyclone.

Chemical Process Simulations using Aspen Hysys, First Edition. Khalid W. Hameed.

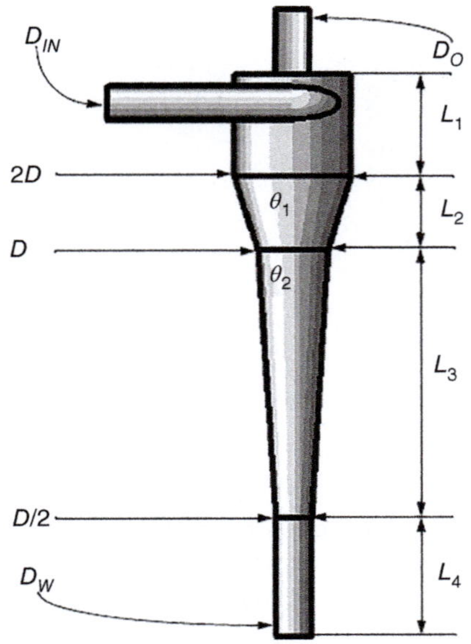

Figure 12.1 Schematic diagram of the cyclone.

Stairmand developed two standard designs for gas–solid cyclones: a high-efficiency cyclone and a high-throughput design, for a given separating efficiency [1, 2]:

$$d_2 = d_1 \left[\left(\frac{D_{C2}}{D_{C1}} \right)^3 \times \frac{Q_1}{Q_2} \times \frac{\Delta\rho_1}{\Delta\rho_2} \times \frac{\mu_2}{\mu_1} \right]^{1/2} \tag{12.1}$$

where

d_1 is the mean diameter of particle separated at the standard conditions, at the chosen separating efficiency.

d_2 is the mean diameter of the particle separated in the proposed design, at the same separating efficiency.

Dc_1 is the diameter of the standard cyclone = 8 in. (203 mm).

Dc_2 is the diameter of proposed cyclone, mm.

Q_1 is the standard flow rate: for high-efficiency design, $Q_1 = 223\ m^3/h$ for high-throughput design, $Q_1 = 669\ m^3/h$.

Q_2 is the proposed flow rate, m^3/h.

$\Delta\rho_1$ is the solid–fluid density difference in standard conditions = 2000 kg/m^3.

$\Delta\rho_2$ is the density difference, proposed design.

μ_1 is the test fluid viscosity (air at 1 atm, 20°C) = 0.018 mN s/m^2.

μ_2 is the viscosity, proposed fluid.

The pressure drop in a cyclone will be due to the entry and exit losses, and friction and kinetic energy losses in the cyclone. ΔP can be calculated according to Shepherd and Lapple [2, 3]:

$$\Delta P = 0.003\rho_f U_t^2 N_k \tag{12.2}$$

where

ΔP is the cyclone pressure drop, millibars
ρ_f is the fluid (gas) density, kg/m^3
U_t is the inlet gas velocity, and

$$N_k = K\frac{HB}{D_C^2} \tag{12.3}$$

B is the width of the inlet cyclone
K is a constant.

The overall efficiency of a cyclone is given by Leith and Licht:

$$\eta_{Dp} = 1 - e^{\left(-M_C D_p^N\right)} \tag{12.4}$$

where

$$N = \frac{1}{n+1} \tag{12.5}$$

$$n = 1 - \left(1 - 0.67D_C^{0.14}\right)\left(\frac{T}{283}\right)^{0.3} \tag{12.6}$$

$$M_C = 2\left[\frac{KQ}{D_C^3}\frac{\rho_P(n+1)}{18\mu}\right]^{N/2} \tag{12.7}$$

K is a geometric configuration parameter that depends only on the relative dimensions of the unit.

12.1.1 Problem Statment

A stream of air at 45°C and 150 kPa with flow rate of 150 kmol/h contains 5% by weight carbon as a pollutant. It must remove 95% of the carbon by cyclone. What are the dimensions of the cyclone required for this purpose, and what is the number of the cyclones required if the maximum diameter of cyclone must not exceed 3 m? Perform the simulation for single-particle diameter and multi-particle diameter.

This case study will focus on removing soot (carbon) from the flue gas of a coal-burning power plant.

12.1.2 Simulation

1) Open a new case with the components (Air and Carbon) and select the Fluid Package: **PRSV.**
2) Enter the **Simulation** environment, and create a new material stream, **Air**, with the following conditions:
   $T = 45°\text{C}$, $P = 150\,\text{kPa}$, molar flow rate = 150 kmol/h,
   Mass fraction: 0.05 carbon and 0.95 air.
3) From the object palette, select and drag the **Cyclone**  into the simulation environment.
4) Double-click the cyclone icon and connect the streams as in Figure 12.2.
5) **Parameters** Page: Select **High Efficiency, Leith/Litch**, Particle Efficiency = 95%, Figure 12.3.
6) **Solids** Page: From **Solids Parameters** group, **Solid Name**, select Carbon, Efficiency Basis: Single-Particle Diameter, Figure 12.4.
7) **Rating** tab, **Constraints** Page: specify the Maximum diameter = 3 m, and Maximum # Cyclones = 40.

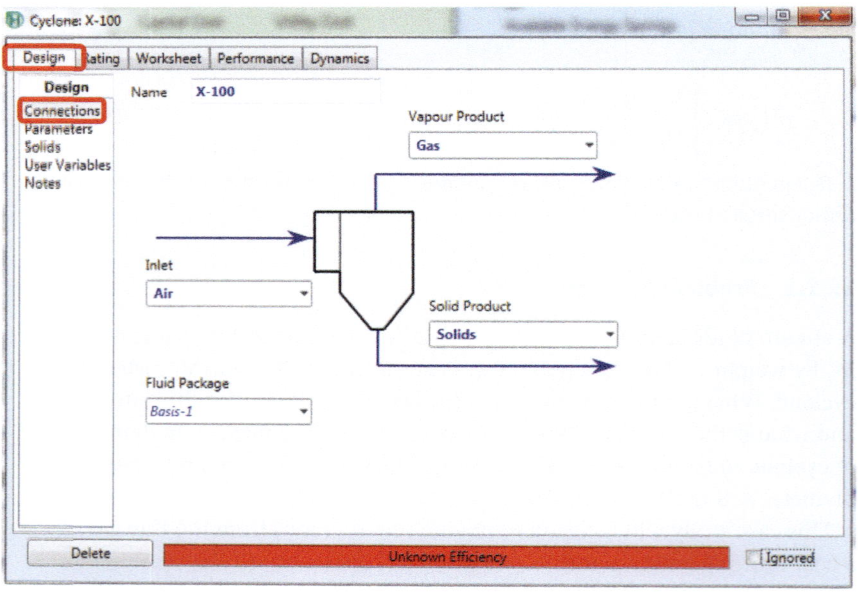

Figure 12.2 Window of cyclone, Connection Page, Design tab.

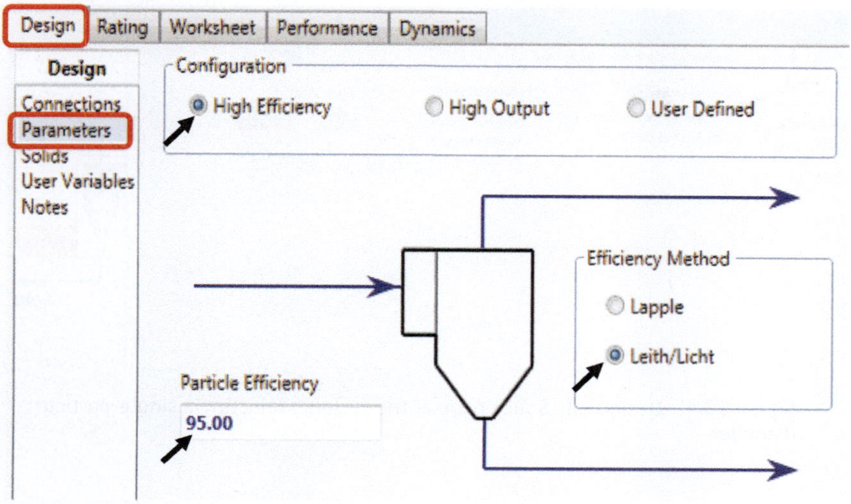

Figure 12.3 Parameters page/Design tab of the cyclone.

The **Results** from the **Rating** tab, Sizing Page, can be seen in Figure 12.5.

Efficiency Method: Select one of the following methods:

Lapple: It is perceived as a standard method and has been considered by some engineers to be acceptable. However, there are several problems associated with this design procedure. The Lapple model greatly underestimated the actual cyclone collection efficiency.

Leith/Licht method: The more rigorous calculation, since it considers radial mixing effects.

Notes

(1) *For high output and less number of cyclones, select* ***High Output*** *radio.*

(2) *When the diameter of the cyclone increased, the number of cyclones decreased.*

8) To see what are the number of cyclones and their dimensions when selecting *High Output* rather than *High Efficiency*, return to the Design tab, Parameters page, and select **High Output** radio.

The number of cyclones and their dimensions can be seen again from the Rating tab, Sizing page, Figure 12.6.

It is clear from Figure 12.5 (when selecting High efficiency), and Figure 12.6 (when selecting High Output), the dimensions of cyclone of Figure 12.5 are less than those of cyclone of Figure 12.6, but the number of cyclones of Figure 12.5 is greater than that of cyclones of Figure 12.6.

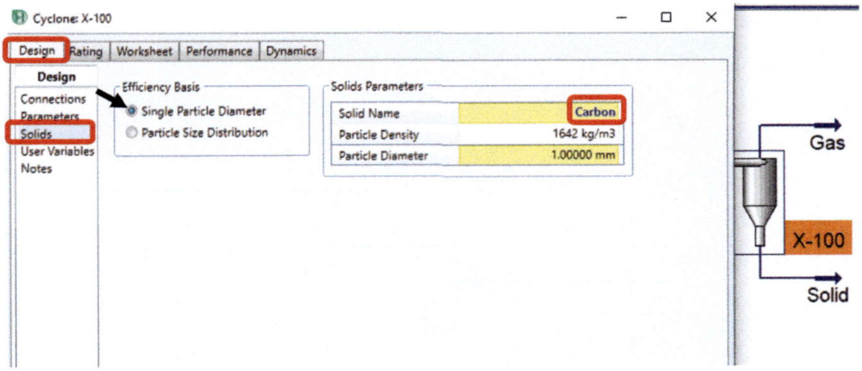

Figure 12.4 Design tab, Solids page of the cyclone, selecting a single-particle diameter.

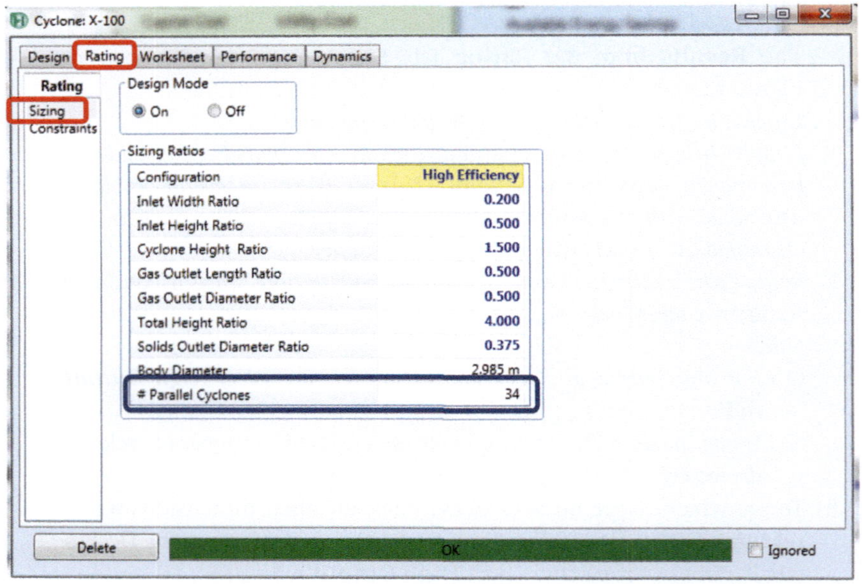

Figure 12.5 Results of the cyclone's design and the number of cyclones when selecting high efficiency.

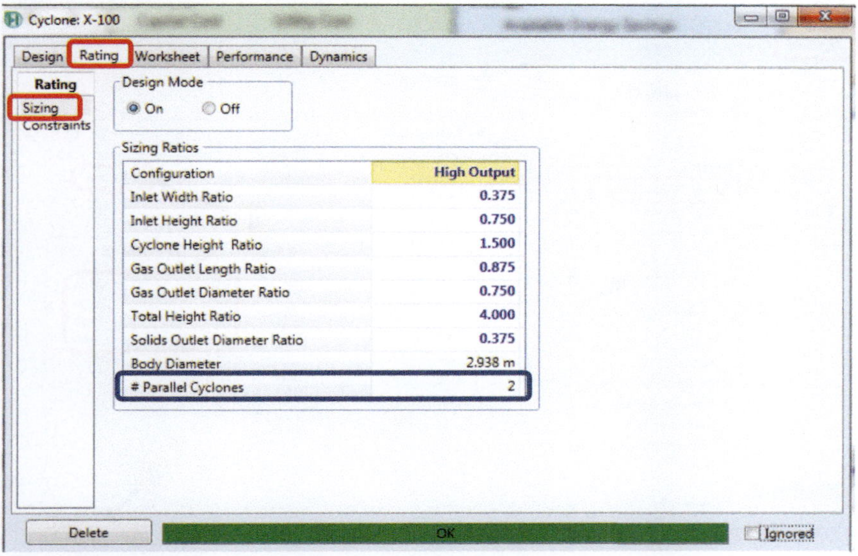

Figure 12.6 Results of cyclone's design and the number of cyclones when selecting high output.

9) Now, one can perform the multi-particle diameter.
Double-click the **Air** stream and from **Worksheet**, select **PSD** Property, click the **Edit** button, and click the **Edit Discrete PSD** button, Figure 12.7. Enter the required data as follows:
Minimum Particle Size = 1.00 mm.

Particle Size (mm)	Mass in Range (%)
1.5	40
2.0	35
2.5	15
3.0	10

When finished, click **OK**, and click the **Calculate PSD** button, then close the Material stream of Air.

10) Return to the Design tab of the Cyclone, Solid page and select **Particle Size Distribution** radio, Figure 12.8.
11) From the Design tab, Parameters page, select **High Output** radio.
12) Again, the dimensions of the cyclone can be seen from **Rating** tab, **Sizing** page, Figure 12.9.

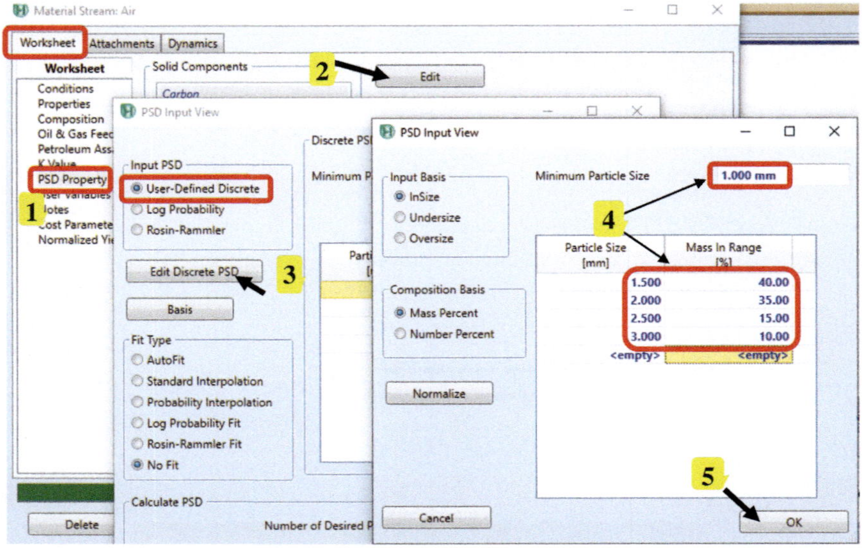

Figure 12.7 Air stream, PSD Property, PSD Input View.

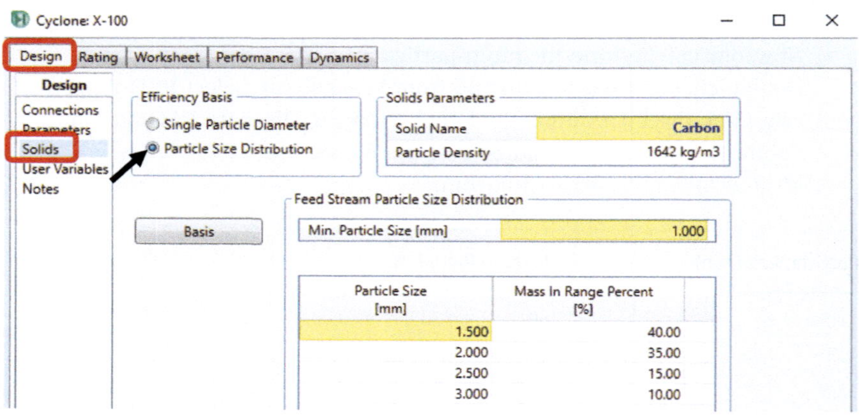

Figure 12.8 Design tab, Solid page, selection of particle size distribution radio.

The meaning of the dimensions of the cyclone unit operation that are presented in the **Rating** tab, **Sizing** page, is shown in Table 12.1.

Exercise 12.1 Simulate Problem Statement in Section 12.1.1 using the **Simple Solid Separator.**

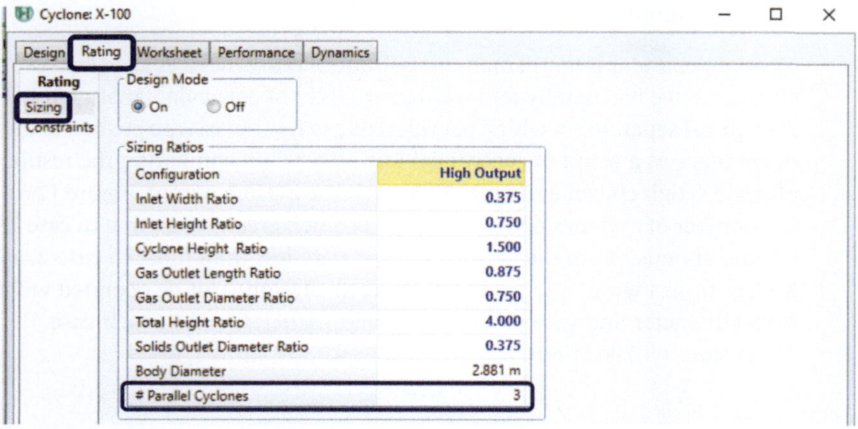

Figure 12.9 Rating tab, Sizing page, dimensions of the cyclone when selecting the particle size distribution, high output.

Table 12.1 Description of the parameters of the cyclone that are present in the Rating tab, Sizing page.

Parameter	Description
Configuration	Select High Output, High Efficiency, or Use Defined. This is also defined on the Parameters page
Inlet Width Ratio	The ratio of the inlet width to the body diameter (must be between 0 and 1). The value must be less than the total Height Ratio.
Inlet Height Ratio	The ratio of the inlet Height to the body diameter
Cyclone Height Ratio	The ratio of the Cyclone Height to the body diameter. The cyclone is the conical section at the bottom of the entire operation.
Gas Outlet Length Ratio	The ratio of the gas outlet length to the body diameter.
Gas Outlet Diameter Ratio	The ratio of the gas outlet diameter to the body diameter (must be between 0 and 1). The value must be less than the total Height Ratio.
Total Height Ratio	The ratio of the total height of the apparatus to the body diameter.
Solid Outlet Diameter Ratio	The ratio of the Solid outlet diameter to the body diameter.
Body Diameter	If Design Mode is on, this is automatically calculated, within the provided constraints. If Design Mode is off, then it can specify this value.
# Parallel Cyclones	If Design Mode is on, this field displays the number of parallel Cyclones (if any) attached to the unit operation. If Design Mode is off and the **Specify Number of Parallel Cyclones** checkbox is selected, it can be specifying this value.

Note: The purification of polluted liquids with solids can be simulated using **Hydrocyclone** similar to Cyclone.

12.1.3 Summary

Cyclone separator is one of many air pollution control devices known as pre-cleaners since it generally removes larger pieces of particulate matter. It is a centrifugal separator in which particles, due to their mass, are pushed to the outer edges as a result of centrifugal force. So, when comparing the results of case 1 (High efficiency, Figure 12.5) and case 2 (High output, Figure 12.6), the number of cyclones needed in case 1 is much greater than that in case 2, but the dimensions of the cyclone in case 1 are less than those in case 2, as is clear from Figures 12.5 and 12.6. Also, the cyclone can be operated with single diameter and various diameter solid particles, and in each case, the dimensions will be different.

12.2 Rotary Vacuum Filter

A Rotary Vacuum Filter is an external filter surface continuous filtration. It is one of the most efficient systems for constant and continuous solid–liquid separation in which liquor is sucked through a moving septum to deposit a cake of solids. It assumes that there is 100% removal of the solid from the solvent stream. This operation determines the retention of solvent in the particle cake, based on the particle diameter and sphericity of your defined solid(s). The diameter and sphericity determine the capillary space in the cake, and thus the solvent retention. The solids being separated must have been previously specified and installed as components in the stream attached to this operation. Figure 12.10 shows the Rotary vacuum filter [1, 4].

The time required for the filtration is obtained by [3, 4]

$$t = \frac{\alpha \mu C}{2A^2 \Delta P} V^2 \tag{12.8}$$

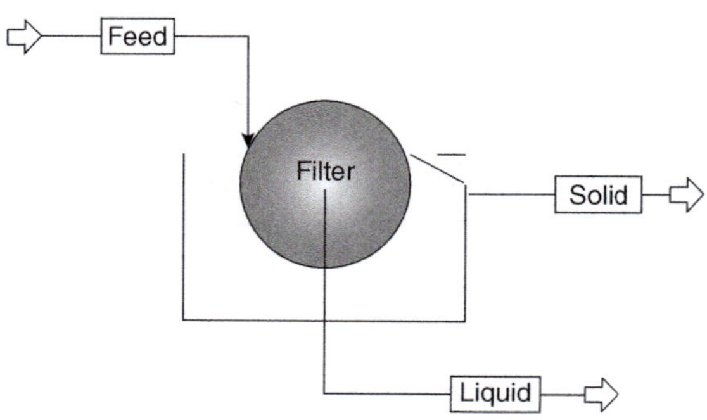

Figure 12.10 Rotary vacuum filter.

where

V is the volume of filtrate,
A is the cross-section area of exposed filter medium,
ΔP is the pressure drop through the bed of solid,
μ is the viscosity of filtrate,
α is the specific cake resistance,
c is the solid concentration.

Noting that the flow rate of filtrate is $Q = V/t$, Eq. (12.8) can be written in the form:

$$Q^2 = \frac{2A^2\Delta P}{\alpha\mu Ct}V^2 \tag{12.9}$$

The time of filtration is lower than the actual time of a cycle t' by a factor ranging from 2.5 to 3.5; hence $t = f \times t'$, where $f = 0.3\text{–}0.4$. Therefore Eq. (12.9) is modified to the form:

$$Q^2 = \frac{2A^2\Delta P}{\alpha\mu Cft'} \tag{12.10}$$

12.2.1 Problem Statement

A stream of slurry at 25°C, 160 kPa, and a flow rate of 4000 kg/h contains 30% calcium by weight in water. It is required to remove the calcium using a Rotary vacuum filter. What is the area of the filter if its diameter is 80 cm?

12.2.2 Simulation

1) Open a new case with the components (water, calcium) and fluid package: **PRSV.**
2) Enter the Simulation Environment and create a new material stream with the following conditions:
   $T = 25°\text{C}$, $P = 160\,\text{kPa}$, mass flow rate = 4000 kg/h, and mass fractions of Water = 0.7, and Calcium = 0.3.
3) From the object palette, add **Rotary Vacuum Filter** to the simulation environment, and enter the information as in Figure 12.11.
4) **Parameters** page: Enter the information as in Figure 12.12, where the pressure drop = 10 kPa, Cycle time = 10 minutes, Dewatering = 45% of the cycle, Submergence = 20% of the cycle.
   The meaning of the dimensions of the Rotary Drum Filter unit operation presented in the **Design** tab, **Parameters** page, is shown in Table 12.2.

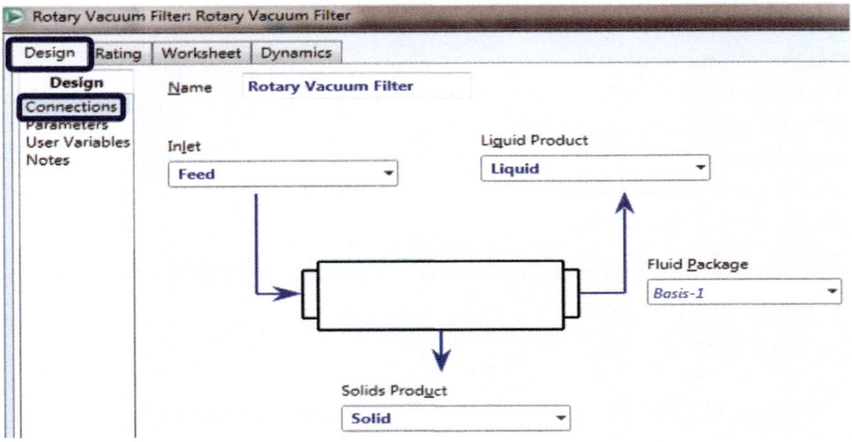

Figure 12.11 Rotary vacuum filter in the HYSYS /connection page.

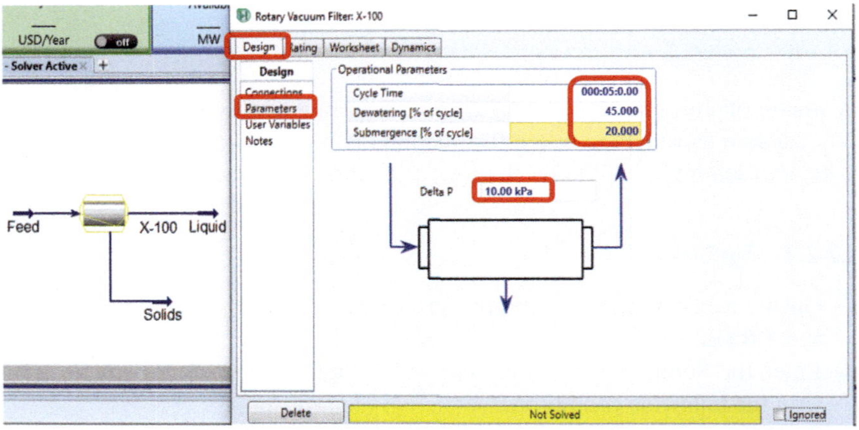

Figure 12.12 Parameters page, Design tab of the rotary vacuum filter.

5) **Rating** tab: Enter the radius of the filter = 0.4 m, Figure 12.13.
So, the area of the filter is 2.44 m^2. The constraints of the cake can be entered in the **Cake** page **Rating** tab. Figure 12.14 shows the PFD of the rotary vacuum filter after converging.
Note: *A single-particle diameter or particle size distribution of calcium can be used as in Example 12.1.*

Table 12.2 Description of the parameters of the cyclone that are present in the Design tab, Parameters page.

Parameter	Description
Cycle Time	The complete time for a cycle (one complete revolution of the cylinder).
Dewatering	The portion of the cycle between time the cake comes out of the liquid to the time it is scraped, expressed as a percentage of the overall cycle time.
Submergence	The percentage of the overall cycle for which the cake is submerged.
Pressure Drop	Pressure drop across the filter.

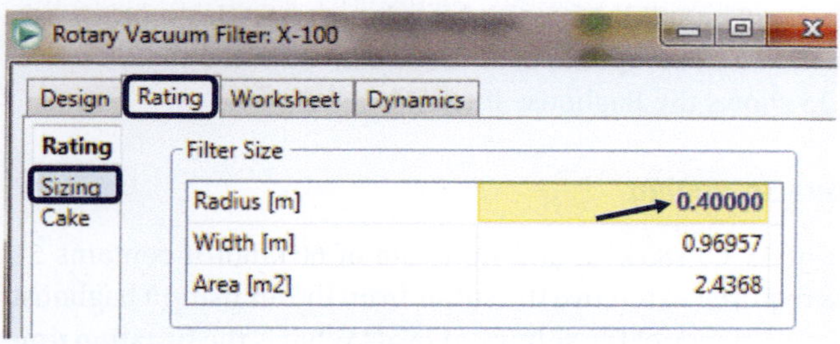

Figure 12.13 Rating tab, Sizing page of the rotary vacuum filter.

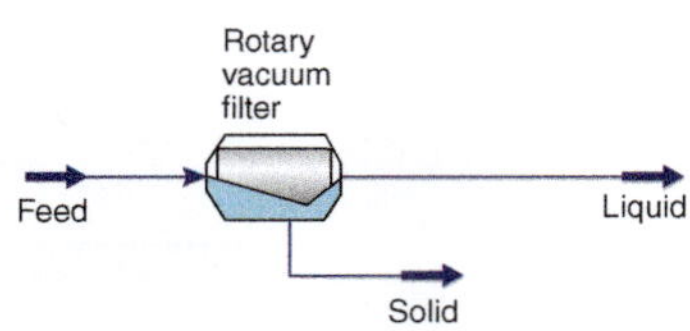

Figure 12.14 PFD of the rotary vacuum filter.

Exercise 12.2 A rotary drum filter with 30% submergence is to be used to filter a concentrated aqueous slurry of $CaCO_3$ (Mw = 100, Density = 2800 kg/m^3) containing 14.7 1b of solids per cubic foot of water (236 kg/m^3). The pressure drop is to be 20 in Hg. If the filter cake contains 50% moisture (wet basis), calculate the filter area required to filter 100 gal/min of slurry when the filter cycle time is 5 minutes and the radius of filter is 0.3 m. The temperature and pressure of the feed are 20°C and 2 bar, respectively.

12.3 Baghouse Filter

One of the most efficient devices for removing suspended particulates is an assembly of fabric-filter bags, commonly called a baghouse. It is an air pollution control device that removes particulates from the air or gas released from commercial processes or combustion for electricity generation. A typical baghouse comprises an array of long, narrow bags that are suspended upside down in a large enclosure. Dust-laden air is blown upward through the bottom of the enclosure by fans. Particulates are trapped inside the filter bags, while clean air passes through the fabric and exits at the top of the baghouse [2, 3].

The Baghouse Filter model is based on empirical equations. It contains an internal curve relating separation efficiency to particle size. Based on your particle diameter, the reported separation efficiency for your solids is determined by this curve. The solids being separated must have been previously specified and installed as components in the stream attached to this operation. Figure 12.15 shows the Baghouse filter [1].

12.3.1 Problem Statement

A stream of air at 35°C, 180 kPa, and flow rate of 60 kmol/h contains 5% mole sulfur. It is required to remove the sulfur from the air using a baghouse filter. What is the efficiency of the sulfur removal? What is the filtration time at which the filter is unable to filter with a 2 kPa pressure drop? What is the relationship between filtration time and pressure drop across the dirty filter?

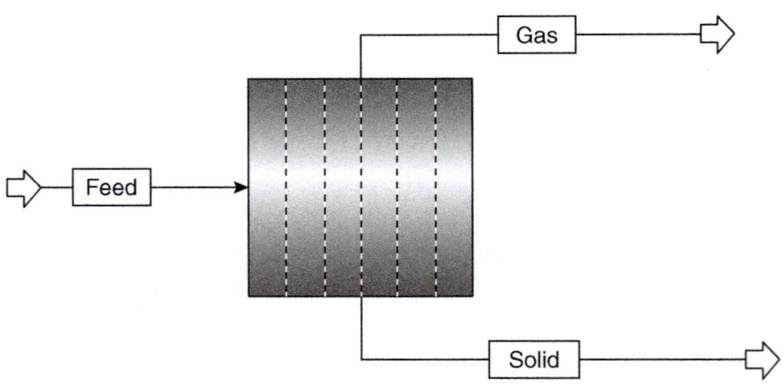

Figure 12.15 Baghouse filter.

12.3.2 Simulation

1) Open a new case and enter the components (Air and Sulfur), and fluid package: Peng Robinson.
2) Enter the **Simulation** Environment and create a new material stream: **Air In** with conditions (T = 35°C, P = 180 kPa, molar flow rate = 60 kmol/h, and molar composition: 0.05 sulfur and 0.95 air)
3) From the object palette add a **Baghouse Filter**  to the simulation area and double-click it to open.
4) Enter the information required as in Figure 12.16. The Baghouse filter will converge automatically because the pressures in the parameters page are setting as default by HYSYS.
5) Parameters: On the Parameters page, you must specify the following information. Set as default values by software, where:

Clean Bag Pressure Drop:	The pressure drop across a clean bag.
Dirty Bag Pressure Drop:	The pressure drop across a dirty bag. This value must be greater than the Clean Bag Pressure Drop.

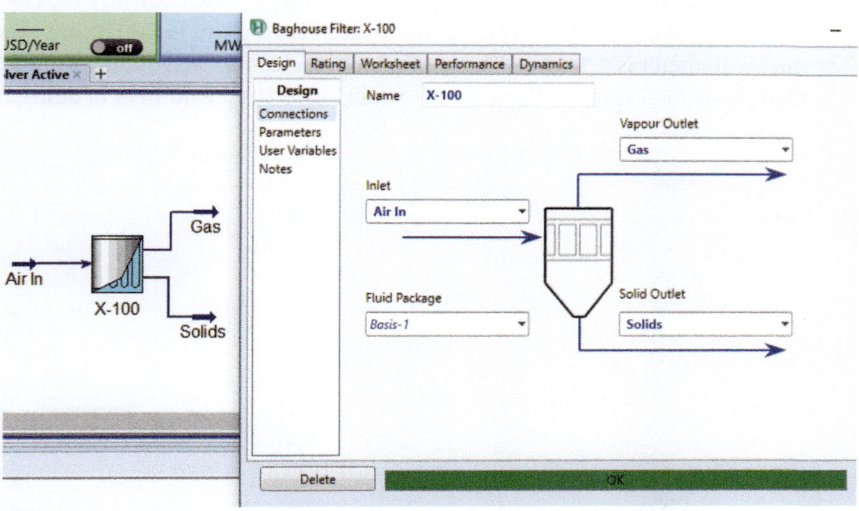

Figure 12.16 Baghouse filter in the HYSYS /connection page.

6. To find the relation between the filtration time and pressure drop across the dirty filter:
   - From **Home** tab of the Ribbon; click Case study 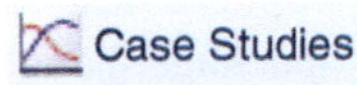

, then click the **Add** button,
   - Independent Variable: Click **Find Variable**, **Object** = Baghouse Filter, **Variable** = Dirty bag pressure drop,
   - Dependent Variable: Click **Find Variable**, **Object** = Baghouse Filter, **Variable** = Filtration Time.
     Figure 12.17 shows the case study with the independent and dependent variables
   - **Case Study Setup:** Start = 1, End = 5, and Step size = 1 (pressure in kPa).
   - Click **Run** then click **Plots**. The results will appear as in Figure 12.18.

It is clear from Figure 12.18 that the filtration time increases with increasing pressure drop across the baghouse filter, but on the other hand the flow rate of air across the filter must be increased to balance the increasing in pressure drop.

Exercise 12.3 ***Cyclone and Baghouse Filter***

A stream of air at 30°C and 5 bar consists of 10 000 kg/h air and 2000 kg/h kaolin. Separate the kaolin using a cyclone with an efficiency of 75%. The standard deviation of the solid particle size distribution is 5.227, and the mean diameter is 8.343 μm. The kaolin has a molecular weight of 258 and a density of 2600 kg/m^3. The number of intervals is 10 (the number of desired

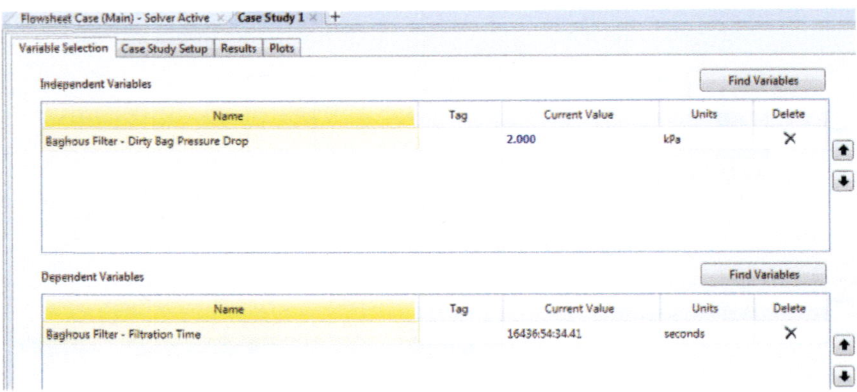

Figure 12.17 Specifying the independent and dependent variables in the case study window to show the effect of pressure drop across the dirty filter on the filtration time.

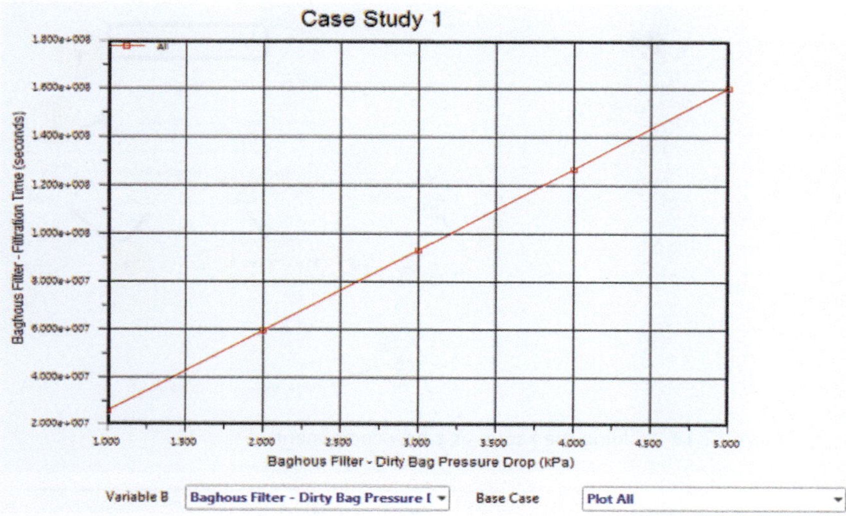

Figure 12.18 Results of filtration time versus dirty pressure drop in the baghouse filter.

points is 10). If the maximum diameter of the cyclone is 2 m, how many cyclones are required, and what are the dimensions of the cyclone?

In problem statement of 12.1, the solid that was removed from the air was a defined material by the simulator, while in this example, the cyclone will be used to remove a *new* solid material from the air.

1) Open a new case and add the components: **Air** and **Kaolin**, where the kaolin is a *new* solid component and it can be added using Hypothetical solids as shown in Figure 12.19.
2) Then click the **New Solid** button, rename to (**Kaolin**) and enter the molecular weight (258 kg/kgmole) and density (2600 kg/m^3). Click the **Estimate Unknown** button, and then click **Add**, Figure 12.19.
3) Fluid package: Peng–'Robinson
4) Enter the Simulation Field.
5) Add a material stream (**Feed**) with conditions (30°C, 5 bar, 10 000 kg/h air, and 2000 kg/h kaolin).
6) In the **Worksheet** tab of the **Feed** stream, **PSD** (Particle size distribution) **Property** page, Figure 12.20, Select ***Kaolin****, click **Edit**, and select **Log Probability**.
7) Enter the mean diameter = 8.343 μm, Standard deviation = 5.227, and Number of Desired Points = 10, Figure 12.21.

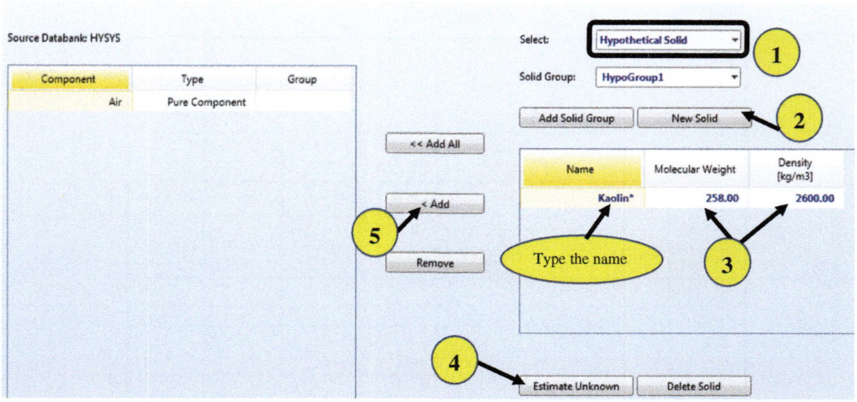

Figure 12.19 Adding the kaolin as a new component.

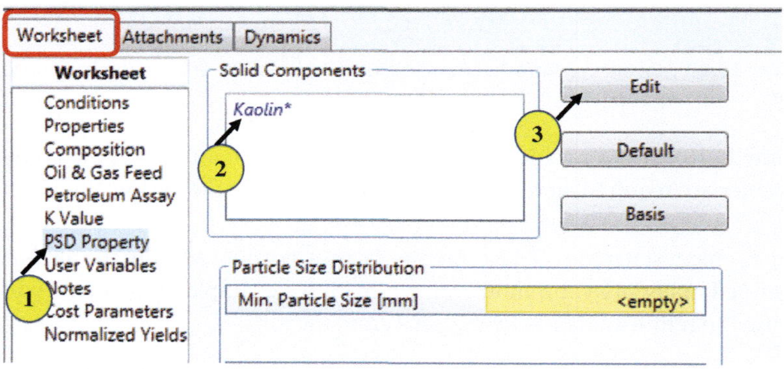

Figure 12.20 Worksheet of the Feed so as to enter the editing of the kaolin component.

8) Click **Calculate PSD**, Figure 12.21.
9) Close the **PSD Input View**. The particle Size Distribution will be estimated, Figure 12.22.
10) Now, add the cyclone icon and connect the feed and outlet streams.
11) From the **Parameters** page, select **High Efficiency**, and enter the efficiency = 75%.
12) **Solids** Page: Under **Efficiency Basis**; select **Particle Size Distribution,** and from Solid Name, select Kaolin*, Figure 12.23.
13) The outlet gas still contains solids of kaolin, so it will be removed using Baghouse Filter as shown in Figure 12.24.

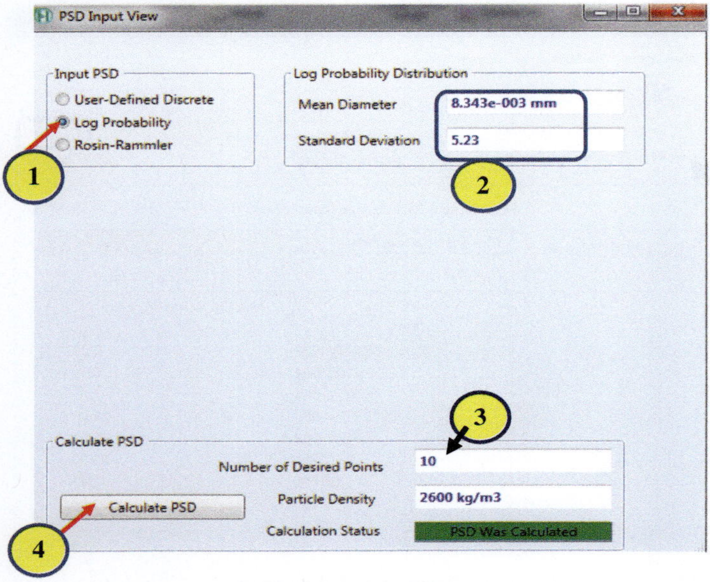

Figure 12.21 Entering the parameters required of the PSD.

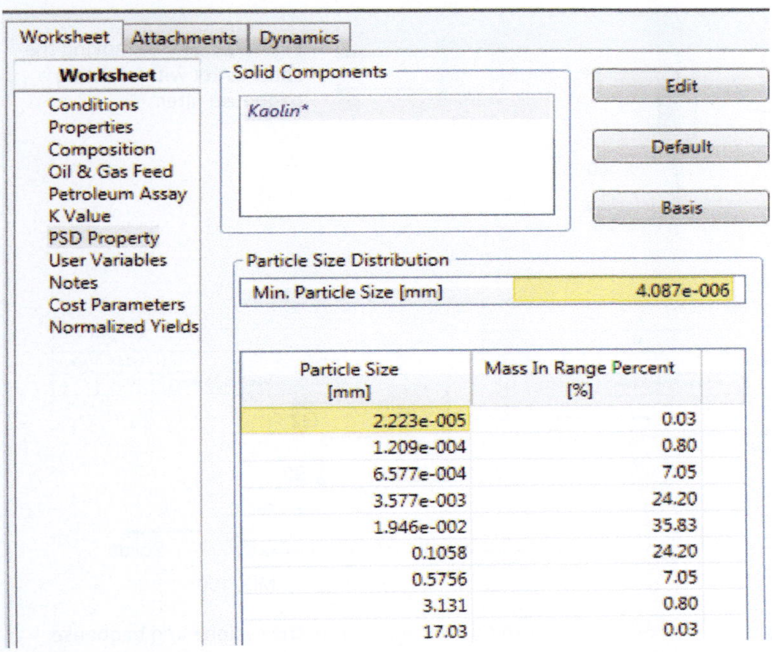

Figure 12.22 Estimated particle size distribution of kaolin in the feed stream.

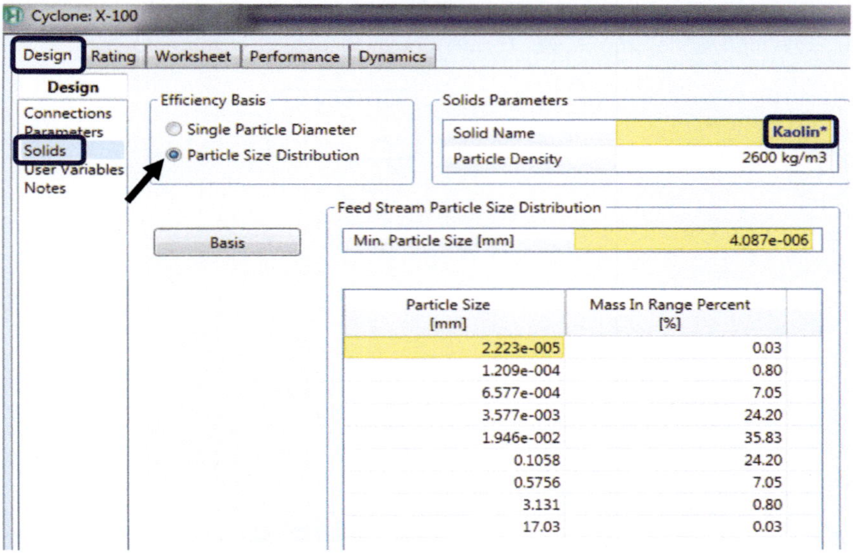

Figure 12.23 Window of cyclone, Design tab, Solids page.

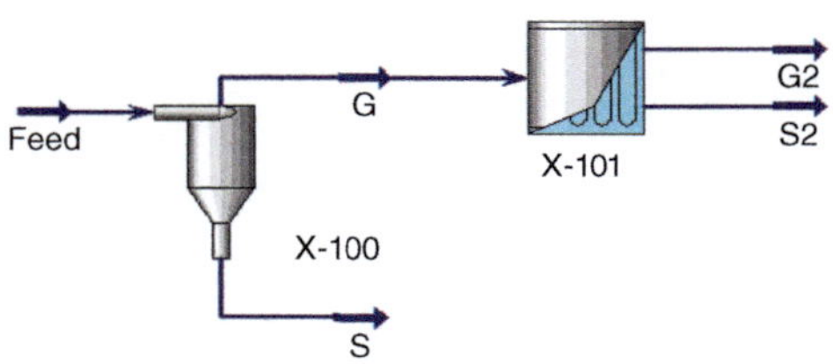

Figure 12.24 Removing the solid outlet with air using baghouse filter.

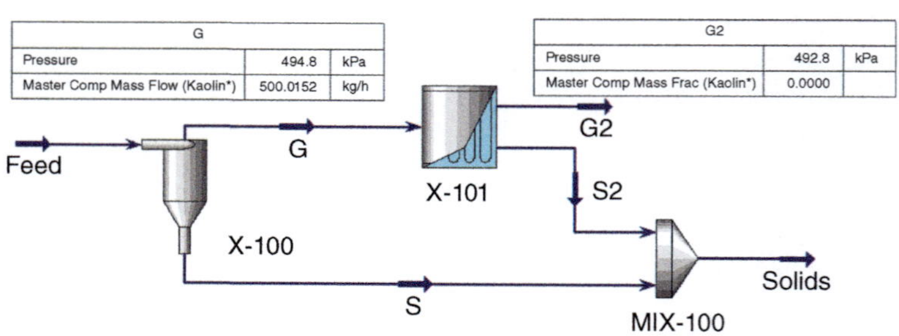

Figure 12.25 Adding mixer to mix the solids from the cyclone and baghouse filter.

14) The streams of solids can be mixed using mixer, Figure 12.25.
It can be seen from Figure 12.25 that the remaining kaolin in the outlet gas stream of the cyclone is 500 kg/h (i.e. removing 75% of the kaolin in the feed), while the outlet gas stream of the baghouse filter is 0.0 (i.e. removing 100% of the kaolin G stream). The cyclone is used first so as to reduce the load on the baghouse filter and hence increase the filter life.

References

1 (2010). *Aspen Hysys Unit Operations V7.2, Reference Guide*. Aspen Technology, Inc.
2 Brauer, H. and Varma, Y.B.G. (2012). *Air Pollution Control Equipment*. Berlin Heidelberg: Springer.
3 (2017). *Aspen Hysys Unit Operations V10, Reference Guide*. Aspen Technology, Inc.
4 Brinkmann, T., Santonja, G.G., Yükseler, H. et al. (2016). *Best Available Techniques (BAT) Reference Document for Common Waste Water and Waste Gas Treatment/Management Systems in the Chemical Sector*. Publications Office of the European Union.

13

Dynamic-State Simulation

Over the past few years, dynamic analysis has grown in significance within the subject of process design. If controllability is taken into account during the detailed engineering stage, novel process designs are more efficient. Dynamic simulation can be used to track how the primary process variables respond to disruptions that are common to the operation of an industrial facility. The main objective of process control is to ensure that a process operates efficiently, safely, and within specified operating conditions [1, 2].

In this chapter, we will learn how to install the control system for the process, where the process *without control* is called an *open system* that may be subjected to outer or inner disturbances that cause the process variable to deviate from the desired value. When installing the control system, the process is called a *closed system* that keeps and insures the process variables under the desired values with changes of conditions.

The following examples explain the procedure for constructing a process control system using Aspen HYSYS. To construct the process control loop, first, it must achieve steady-state simulation before dynamic state.

13.1 Problem Statement 1: Flash Separator

When a stream of saturated liquid mixture at 10 °C, composed of (Ethane 0.5, Propane 0.3, and *i*-Butane 0.2 by mole), is subjected to throttling (e.g. passed through a partially open valve), some of the liquid will evaporate. To separate the vapor from the liquid, a vessel can be used as a separator. Up to now, the process is in a steady state. In order to simulate the real process in which the fluid moves between the unit operation, first and after achieving the steady-state conditions, all streams entering and leaving the system must be terminated with pressure change units (e.g. valve, pipe, compressor, or pump). In this case, the system will be ready to install the

Chemical Process Simulations using Aspen Hysys, First Edition. Khalid W. Hameed.

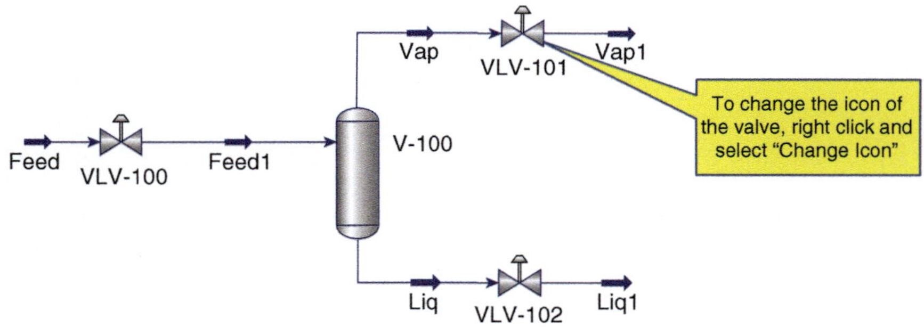

Figure 13.1 PFD of the flash separator.

controller that controls the variables of the streams and unit operation which are required to be stable.

13.1.1 Simulation

1) Open a new case of Aspen HYSYS and add the components (ethane, propane, and *i*-butane).
2) Fluid package: Peng Robinson.
3) Enter the Simulation environment.
4) Construct the process flow diagram as in Figure 13.1.
5) Enter the conditions of the feed stream and unit operation as follows:
   Feed: molar flow rate: basis 100 kmol/h,
   Saturated liquid, i.e. vapor fraction = 0, temperature = 10 °C, and
   Molar composition: $C_2 = 0.5$, $C_3 = 0.3$, and $i\text{-}C_4 = 0.2$.
   Pressure drop of valves VLV-100, VLV-101, and VLV-102 are 500, 50, and 20 kPa, respectively.

13.1.2 Dynamic Simulation

In order to start the dynamic state, ***it must first build a steady state*** like the previous one.

There are some steps to achieve the dynamics simulation [3, 4]:

1) Entering the size of the unit operation: In order to run dynamic simulations, it is important to have all of your equipment sized properly. The response of a vessel to fluctuations in its feed is highly dependent on the size of the vessel relative to the size of the disturbance. If your vessel is inappropriately large, the simulation will result out an unrealistically dampened response. The reverse applies for a vessel that is too small.

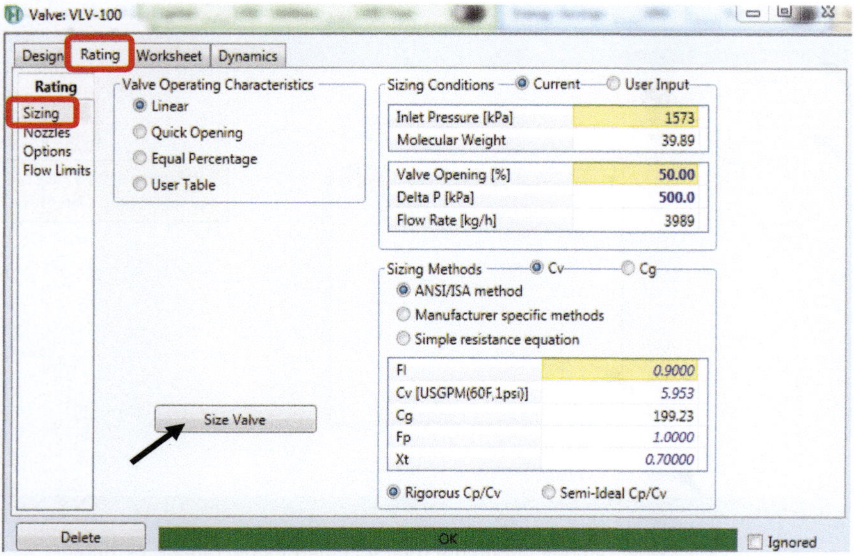

Figure 13.2 Estimating the sizing of valve from Rating tab, Sizing page, then clicking the Size Valve.

The dimensions of unit operation are entered as follows:

- For valves, the sizing of each valve can be entered from the **Rating** tab, Sizing page, and click the **Size Valve** button as shown in Figure 13.2 (*See chapter 7, simulation of problem 7.1, steps 3–4*).
- For the Separator, the volume of the separator is twice the liquid volume that occupied the separator, assuming that the volume of the liquid constitutes 50% of the volume of the separator with a residence time of 5 minutes (*This is a standard value, but it can be altered according to practical experience*). The volume of liquid can be drawn from the **Liquid Volume Flows** that exist in the **Worksheet** of the Feed1 stream, composition page, as shown in Figure 13.3.

  So, the volume of the separator is 2 × liquid volume flow rate × (5/60) = 2 × 7.84 × (5/60) = 1.31 m^3.

  The choice of shape of separator is a vertical cylinder, and the ratio of height/diameter = 3

$$\text{Cylinder volume} = (\pi/4) \times D^2 \times H$$
$$1.31 = (\pi/4) \times 3 \times \text{D}^3$$
$$\rightarrow D = 0.822\ \text{m}^2 \quad \text{and} \quad H = 2.47 \tag{13.1}$$

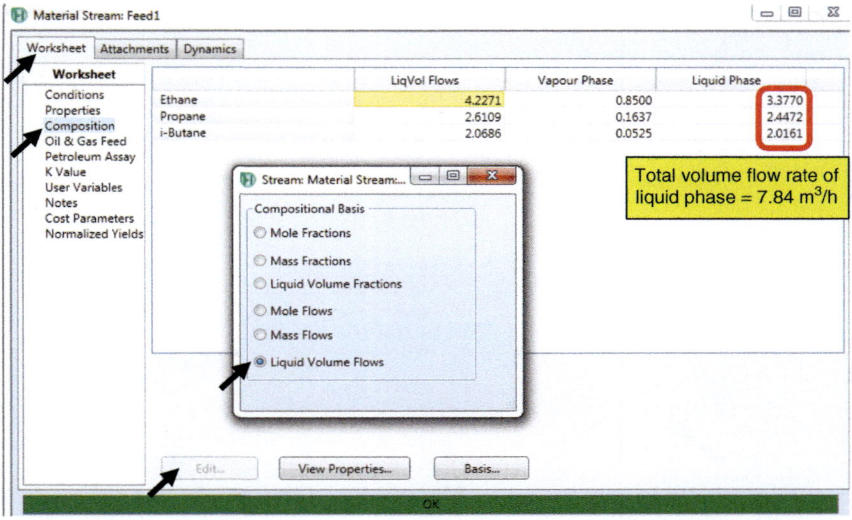

Figure 13.3 How can one obtain the total volumetric flow rate of the liquid phase in the feed stream.

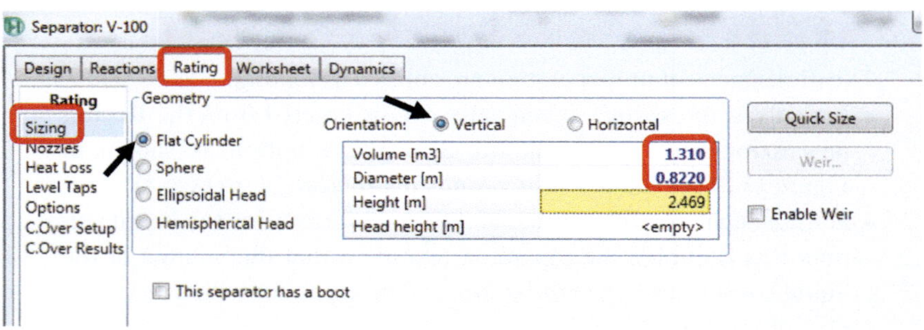

Figure 13.4 Entering the dimensions of the separator.

Now, in the **Rating** tab of the separator, enter the volume 1.31 m^3, and value of diameter = 0.822 m as shown in Figure 13.4.

2) Adding controller operations: The controller is added from the **Dynamic** group in the object palette. The type of controller that will be used in this system is a proportional integral derivative (PID) type, and its icon is 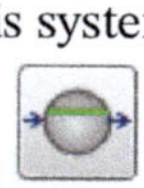. The basic equation of PID is [1, 2, 5]:

$$OP(t) = K_c e(t) + \frac{K_c}{T_i} \int e(t)dt + K_c T_d \frac{de(t)}{dt} \tag{13.2}$$

Table 13.1 Suggestion of controller parameters.

System	K_c	T_i (minutes)	T_d (minutes)
Flow	0.1	0.2	0
Level	2	10	0
Pressure	2	2	0
Temperature	1	20	0
Composition	0.1	0.2	0

This type of controller has parameters (controller gain (K_c), integral time (T_i), and derivative time (T_d)). These parameters are called tuning parameters which are set by either the available transfer function of the process, then estimated by the Ziegler–Nichols or Cohen–Coon methods, or by trial and error until the system shows a stable response. $OP(t)$ is the controller output as a function of time, and $e(t)$ is the error between the set point (SP) and process variable (PV). The error from the set point is the difference between the SP and PV and is defined as $e(t) = SP - PV$. If no transfer function is available for the open-loop system, the trial-and-error technique can be used. The initial guess of the controller's parameters can be obtained from practical experience as shown in Table 13.1 [5].

The process variable (PV) is the actual process value that the controller is trying to regulate. The output target object (OP), sometimes referred to as the manipulated variable, is the parameter that the controller will change to adjust the process variable to the set point value.

Once the controller has been added:

- - select Process Variable Source,
- * set the Minimum and Maximum values,
- select Output Target Object,
- select Controller Tuning Parameters,
- choose the mode of the controller,
- * Off, Manual, Automatic, Indicator, Casc,
- select Controller Action: Reverse or Direct,
  where the Reverse Action: If the PV ***increases*** above the SP, the OP of the controller will ***decrease*** to drive the PV down.
  While the Direct Action: If the PV ***increases*** above the SP, the OP of the controller will ***increase*** to drive the PV down.

3) Add a Controller between Feed1 and Valve (VLV-100); open the controller by double-clicking it.
Name = FC (Flow Controller),
Process Variable Source:
Object; click the **Select PV** button; Object = Feed1, Variable = Actual Volume Flow, OK.
Output Target Object: click the **Select OP** button; Object = VLV-100,
Variable: Actuator Desired Position, OK. See Figure 13.5.
- **Parameters** tab:
The actual volume flow rate of stream F1 at steady state is obtained from the Worksheet tab of F1 stream, Properties page, where its value is 28.99 m^3/h.
So, enter the PV minimum = 5 m^3/h, and PV maximum = 50 m^3/h.
Action = Reverse,
Tuning Parameters: As mentioned in Table 13.1, for flow system,
$K_c = 0.1$, $T_i = 0.2$, Mode = Auto, Figure 13.6.
Control Action: In general, there are two types of control action:
(a) *Reverse control action*

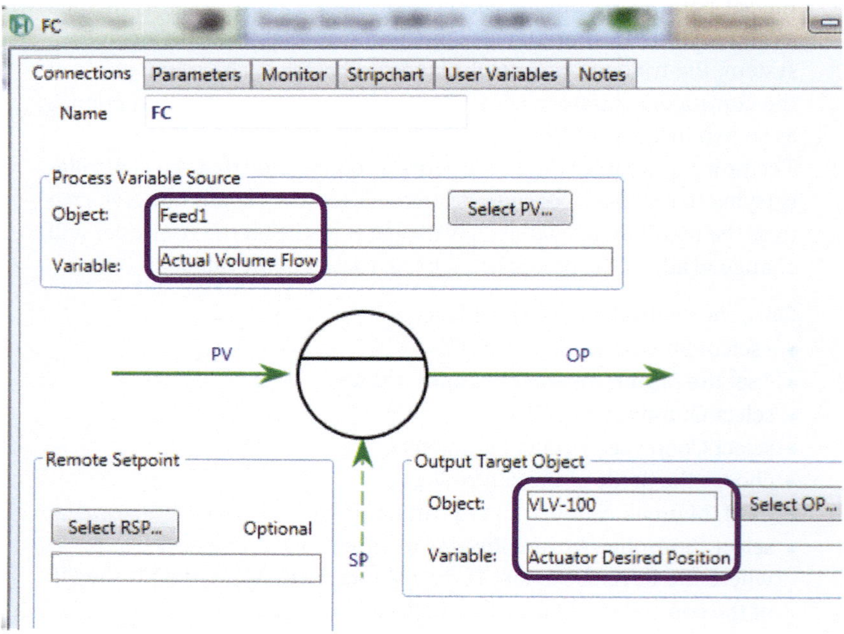

Figure 13.5 Selecting the process variable and output target of the controller.

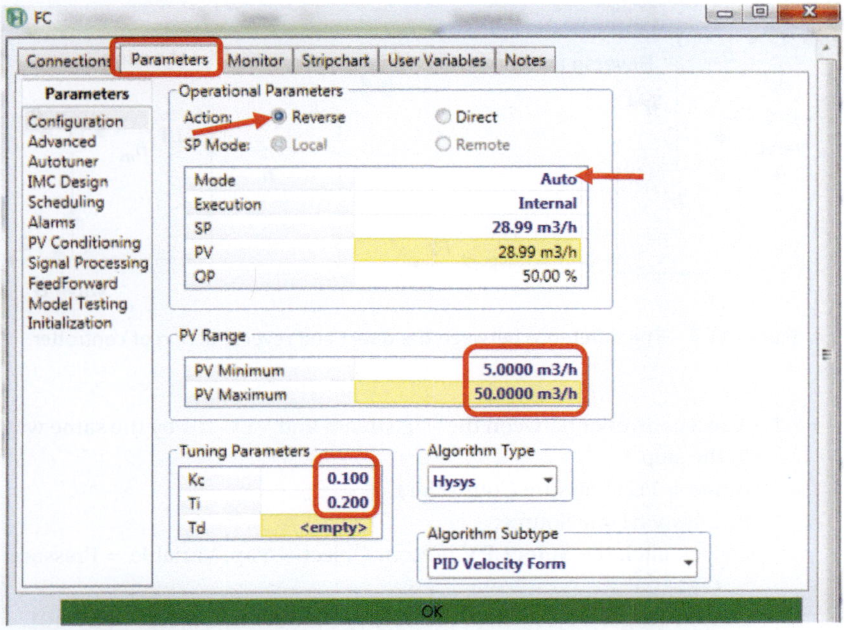

Figure 13.6 Entering the parameters required for the flow controller FC.

(b) *Direct control action*

To understand the difference between direct and reverse action (Figure 13.6, under the group of Operational Parameters), consider Figure 13.7: the process variable is the liquid level in the tank, and the output target object (OP) is the valve **A** opening%, which controls the inlet. When the liquid level falls, the valve opening% must be increased, so the inlet flow will increase and raise the tank level to its original level. This is a **reverse** response. If the output target object is the valve **B** opening %, which controls the outlet flow, the response is **direct**. That way, when the liquid level falls, the valve opening% will decrease, reducing the outlet flow rate and raising the tank level to its original level.

In other words:

Reverse action: Error = *SP 2212 PV; and* $K_c = +Ve$

When the PV or H (liquid level) increases the OP decreases (closing the inlet valve more) to reduce the filling rate.

Direct action: Error = *PV* – *SP; and* $K_c = -Ve$

Whereas the PV increases (tank filling) the OP increases (opening the outlet valve more).

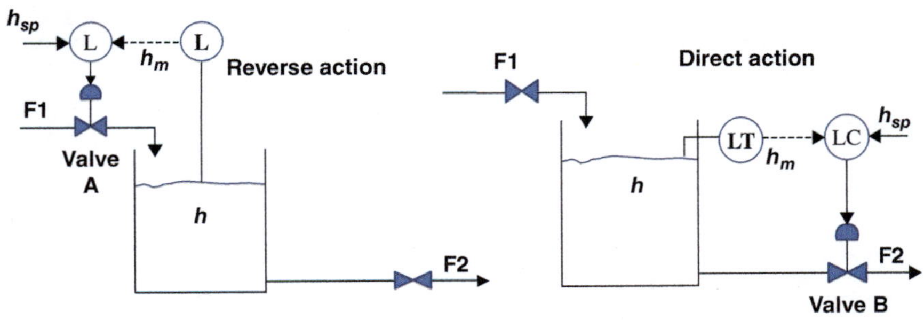

Figure 13.7 The difference between the direct and reverse action of controller.

4) Add a controller between the Vap stream and VLV-101 by the same way in the step 3.
   Name = PC (Pressure Controller),
   Process Variable Source:
   Object; click the **Select PV** button; Object = Vap, Variable = Pressure, OK.
   Output Target Object: click the **Select OP** button; Object = VLV-101, Variable: Actuator Desired position, OK.
   - **Parameters** tab:
     In order to control the flow of the Vap stream, it will control its pressure and, consequently, control its flow since the phase of Vap stream is vapor.
     The pressure value of the Vap stream at steady state is obtained from the Worksheet tab of the Vap stream, Conditions page, where its value is 1073 kPa, or by placing the mouse on the Vap stream, the simulator will show the conditions of the stream.
     So, enter the PV minimum = 200 kPa, and PV maximum = 2000 kPa.
     Action = Direct,
     According to Table 13.1, the Tuning Parameters will be: $K_c = 2$, $T_i = 2$.
5) Adding a level controller for the separator: It can be done by following the same procedure of steps 3 and 4, but in general, for vessels, there is a special option to add the level controller.
   Open the separator, **Dynamics** tab, click the **Add/Configure Controller** button as shown in Figure 13.8. The process flow diagram of the control system can be seen now as shown in Figure 13.9.
   A list of all controllers can be quickly found using the Control Manager button in the Tools group on the Dynamics tab, Figure 13.10.
   Likewise, a list of all face plates can be found using the Face Plate button.

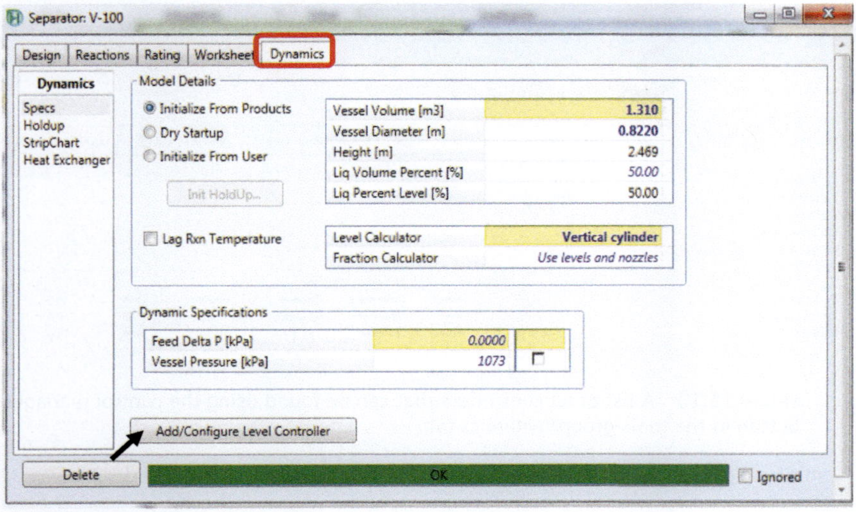

Figure 13.8 Adding a level controller to the separator from dynamics tab.

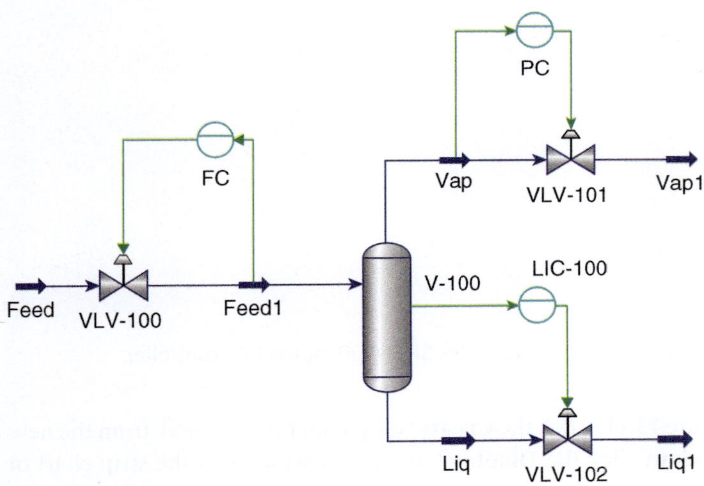

Figure 13.9 PFD of the flash separator after adding the controllers.

6) Test the performance and stability of the system: It can test each controller separately or test all the system.
   - Testing the FC controller: Double-click the FC controller, **Strip chart** tab, and from the **Variable Set** drop down list, select **SP, PV, OP Only** (Set point, Process variable, Output of the controller) as shown in

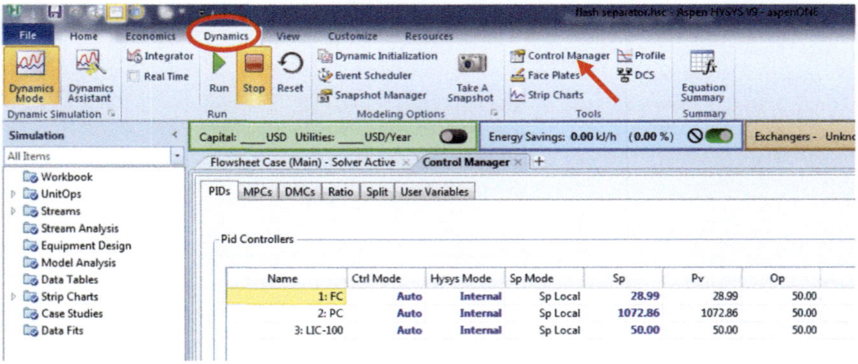

Figure 13.10 A list of all controllers that can be found using the control manager button in the tools group/dynamics tab.

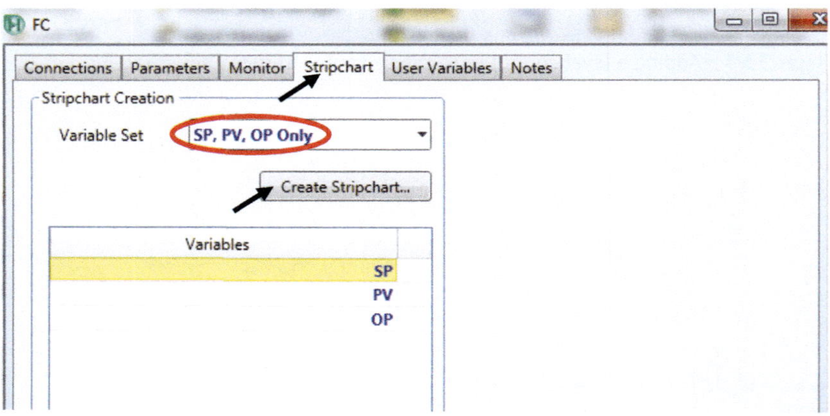

Figure 13.11 Selecting the variables SP, PV, OP for the FC's controller.

Figure 13.11. Click the **Create Strip chart** button, and, from the new window, click the **Display** button. The window of the strip chart of the FC controller will display as shown in Figure 13.12.

In the same way, execute this procedure for the other controllers.

Strip charts *are a way to observe the dynamic fluctuations of the model graphically. They can be produced for any stream, block, or controller.*

7) In order to monitor the signals of the variables slowly, reduce the acceleration of the process variables signals: From the **Dynamics** tab in the Ribbon, click the **Integrator** button, and from new window reduce the acceleration to 0.1 as shown in Figure 13.13. Close the window.

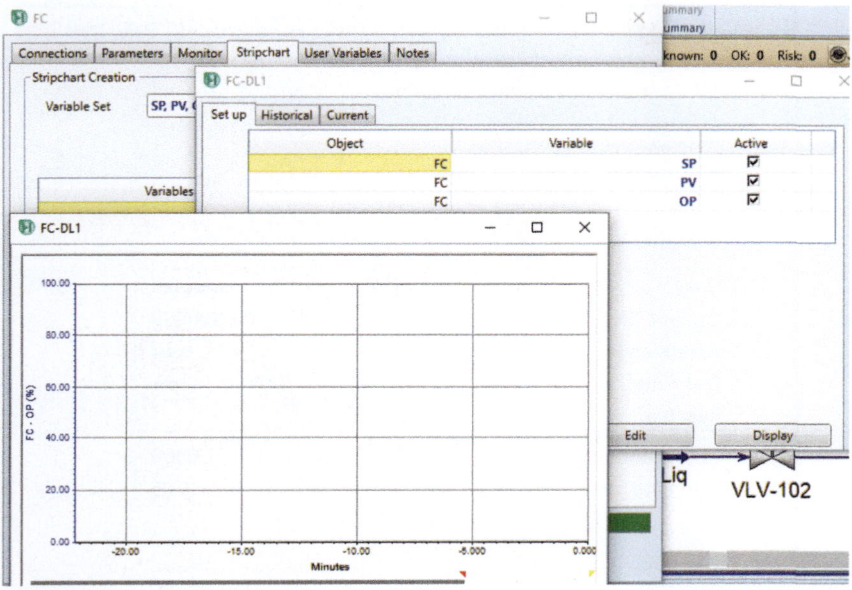

Figure 13.12 Screen of the stripchart of FC controller.

8) Dynamics Assistance: The Dynamics Assistant allows you to see what adjustments you need to make for your model to run dynamically and suggests changes. You should use the assistant as a guide, not a cure-all. To create the best possible models, it is important to use your own engineering judgment.

   It is strongly advised that the first time you convert a steady-state model, you try to make all of the appropriate changes yourself before consulting the assistant.

   Again, from **Dynamics** tab in the Ribbon, click **Dynamics Assistance** button, there is a list of parameters to adjust; all of the items are automatically checked in the "Make changes" column. If you click the Make Changes button, all checked changes will be made at once, no matter what tab you are on. To make only certain changes, uncheck the ones you don't want, Figure 13.14 [3, 6].

   On these tabs, you can see exactly what changes the Dynamics Assistant suggests. You can edit these directly in this window before clicking Make Changes, or you can make them in the flowsheet itself.

   In Aspen HYSYS steady-state models, pressure and flow parameters can be specified independently. For this reason, it is possible to "cheat" by making unrealistic specifications such as having no pressure drop

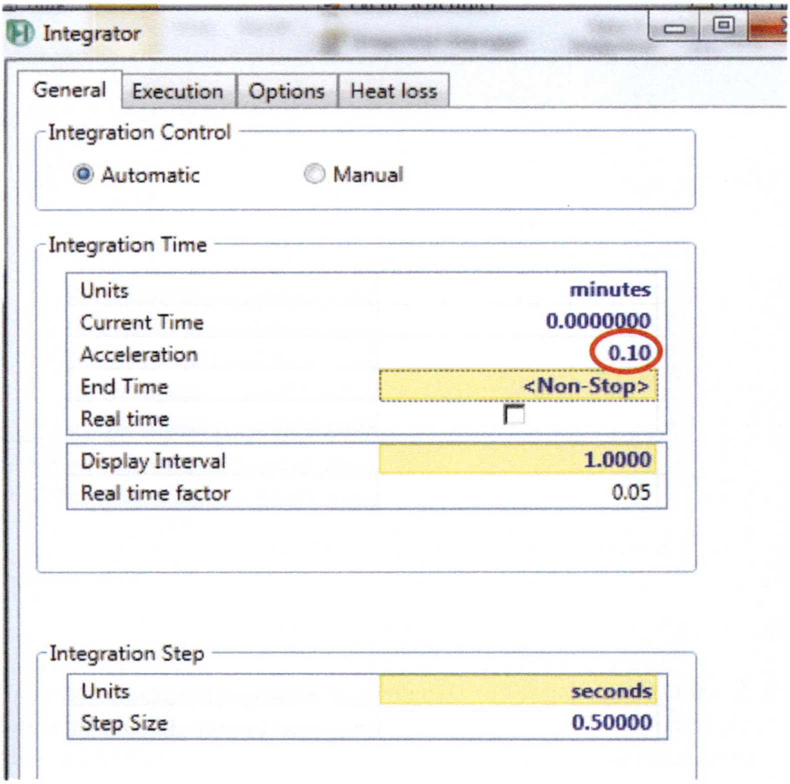

Figure 13.13 Reducing the value of acceleration.

across a unit. *In HYSYS Dynamics, pressure and flow are related such that one cannot be calculated without the other* [6]. If you specify zero pressure drop across a resistance unit in Dynamics, the flow rate through that unit will be zero as well.

Click **Make Change** button, Figure 13.14.

9) Switch to the dynamic state from **Dynamics** tab in the Ribbon, click the 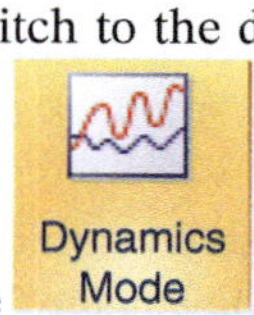

tab, from the new window, click Yes.

10) Now, to see the response curves for each controller, click the **Run** button, for each screen of the controller, right click and select Graph Control, from the Axis tab, activate the **Automatic Auto Scale** and **Show All** radios to display the scales of the all variables, Figure 13.15.

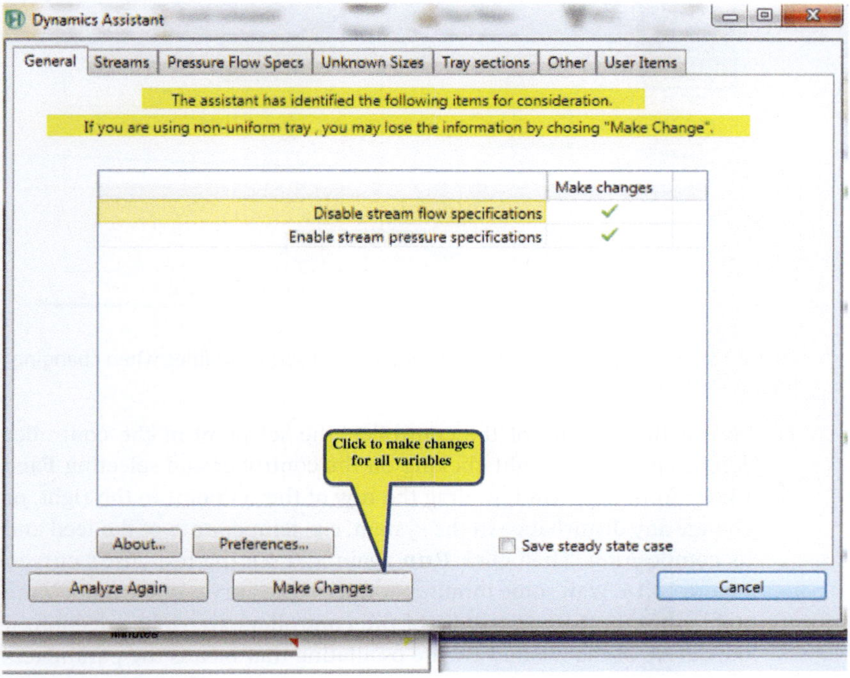

Figure 13.14 Make change of the items from dynamics assistance.

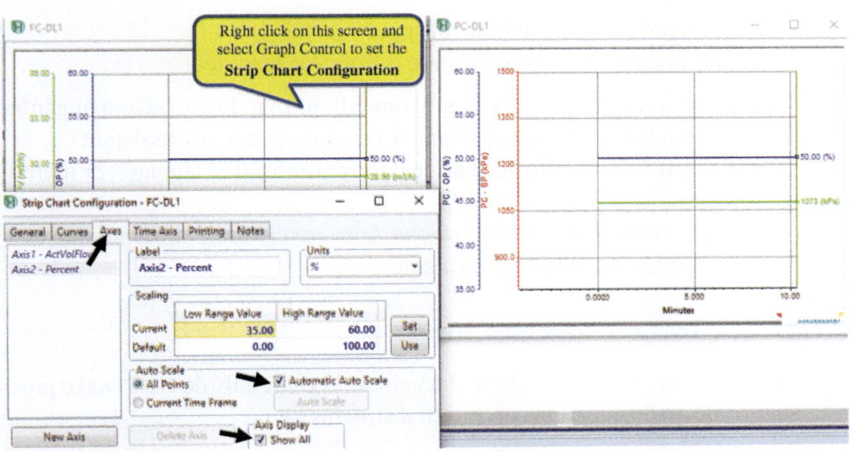

Figure 13.15 Screens of the FC and PC controllers.

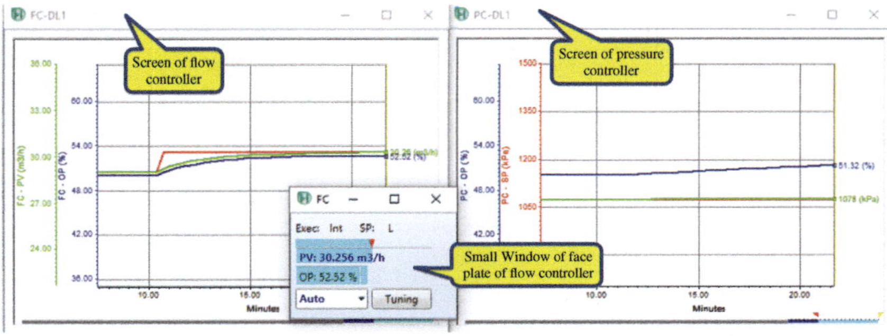

Figure 13.16 Response of controllers and their process variables when changing the set point of the FC.

11) To test the stability of the controller, the set point of the controller can be changed by right clicking on the controller and selecting **Face Plate**; from new window, drag the row of the set point to the right, or change any disturbance in the system, e.g. temperature of the feed and its composition. Then click **Run** again and see the response's curves, Figure 13.16. Wait some minutes and see if the curves of OP and PV are stable; hence, the parameters of the controllers are right selection, if the curves of the OP and PV are oscillating that means the parameters of the controllers need to change, i.e. reduce the value of K_c/increase the value of the T_i.

Dynamics Assistance is a tool that assists with ensuring the proper dynamic specifications are in place

- Recommends a reasonable set of specs
- Recommendations ensure the case is not over specified

The assistant can also size any unit operations that do not have sizing info.

You do not have to make all changes listed In Dynamics Assistant.

Review any messages to determine if they are required changes, or merely suggestions.

Note: *To transition unit operations from steady state to dynamics one,* Table 13.2 *below must be used as a guide.*

13.1.3 Dynamic Transition Summary

Make pressure specs on all boundary streams. Generally do NOT make pressure or flow specifications on internal streams [3, 6].

- Pressure specs are generally preferred over flow specs.
- Add valves to boundary streams attached to separators.
- Size process equipment (valves, separator, heater, Heat Exchanger).
- Enable P/F equations on valves/conductance devices Supply C_V's and k's.

Table 13.2 Guide of transition unit operation from steady state to dynamics.

Dynamics Specifications	
Boundary streams	Insert a valve on all boundary streams (Feed/Product streams) within the flowsheet that are not connected to conductance devices (i.e. Heat Exchanger, cooler, heater)
Pressure specifications	Place a pressure specification on all boundary streams (Feed/Product streams) within the flowsheet.
Distillation columns	Distillation columns with condenser require an extra specification around the condenser. Make a flow specification for the reflux flow.
Valves	Use the "Pressure /flow relationship" as the dynamic specification for the valve.
K-Value	Use the "Overall K-value" as the dynamic specification for coolers, heaters, Heat Exchangers, and LNG exchangers.
Pressure gradients	Be sure to account for pressure gradients throughout the flowsheet. Moreover be sure to specify reasonable pressure drops/raises in the flowsheet. Pressure differential are the driving force for flow through the process flowsheet.
Tray sizing	Use the tray sizing utility to estimate the column geometry and pressure profile.
Mixers	Use the "Equalize All" option as the pressure specification for mixers.
Tees	Remove Use Splits as Dynamic Flow Specs on tees.
Rotating Equipment	Use Efficiency and either Head or Pressure Rise as Dynamic specifications for rotating Equipment. Compressor and Pump Curves, if available, make excellent dynamic specifications.
Hold-ups	Be sure of property size equipment with hold ups.

Dynamic specifications can only be modified when the integrator is stopped. Once the integrator is started, the value of the dynamic specification can be changed (its value appears in blue), but the choice of dynamic specification cannot be changed.

13.2 Problem Statement 2: Gas Processing

A plant takes two different natural gas streams containing carbon dioxide, nitrogen, and (methane through n-butane), combines, and processes them in a simple refrigeration system. A series of separators and coolers removes the heavier hydrocarbon components from the natural gas stream, allowing it to meet a pipeline dew-point specification. The heavier liquid component of the gas stream is processed in a depropanizer column, yielding a liquid product with a specified propane content. Figure 13.17 shows the PFD of the steady-state simulation of the Gas Processing [7].

13.2.1 Steady-State Simulation

1) Open a new case of Aspen HYSYS and add the components: N_2, CO_2, C_1, C_2, C_3, *i*-C_4, and *n*-C_4.
2) Fluid package: Peng Robinson.
3) Enter the **Simulation** environment.
4) Add two streams of material.

Stream 1	**Name**	**Feed 1**					
	Temperature	60 °F					
	Pressure	600 psia (41.37 bar)					
	Molar Flow	6 MMSCFD					
Component	N_2	CO_2	C_1	C_2	C_3	*i*-C_4	*n*-C_4
Mole Fraction	0.01	0.01	0.6	0.2	0.1	0.04	0.04

Stream 2	**Name**	**Feed 2**					
	Temperature	60 °F					
	Pressure	600 psia (41.37 bar)					
	Molar Flow	4 MMSCFD					
Component	N_2	CO_2	C_1	C_2	C_3	*i*-C_4	*n*-C_4
Mass Fraction	0.02	0.0	0.4	0.2	0.2	0.1	0.08

5) Add a **Mixer**

Name	**Mix-100**
Inlet	Feed1, Feed2
Outlet	Mixer out

6) Add a **Separator**

Name	**Inlet sep**
Inlet	Mixer out
Vapor Out	SepVap
Liquid Out	SepLiq

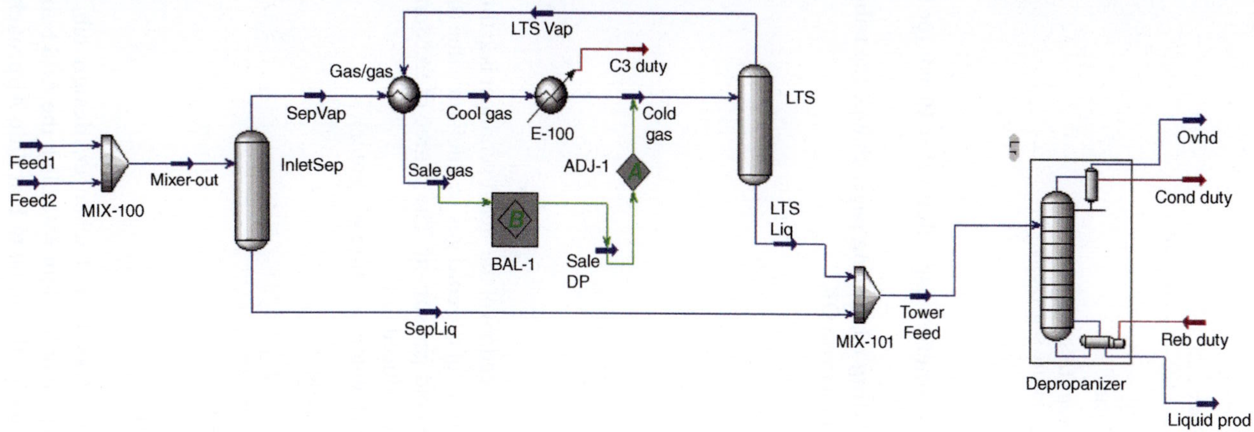

Figure 13.17 PFD of steady-state simulation of the gas processing.

7) Add a **Heat Exchanger**

Name	Gas/Gas
Tube Inlet	SepVap
Tube Outlet	Cool gas
Shell Inlet	LTS vap (LTS: low-temperature separator)
Shell Out	Sale gas

(7-1) From the **Parameters** page, **delta P = 10 psi** (for the shell side and tube side).

(7-2) From the **Rating** tab, **Sizing** page: change the **tube passes per shell** to **1**, then press **OK**.

8) Add a **Cooler**

Name	Chiller
Inlet	Cool gas
Outlet	Cold gas
Energy	*C3 Duty*
Delta P	10 psi.

Note: The degree of freedom of the Heat Exchanger is 2; therefore, determine the temperature of the **cold gas** as a default value (**0 °F**) from the worksheet of the cold gas stream. The degree of freedom of the Heat Exchanger will be reduced to 1.

9) Add a **Separator** (low-temperature separator)

Name	LTS
Inlet	Cold gas
Vapor Out	LTS vap
Liquid Out	LTS liq

10) Return to the **Gas/Gas** Heat Exchanger, **design** tab, **Specs** page, notice that the degree of freedom = 1. Click the **Add** button to display **Exch Spec** and change the name to **Hot Side Approach**; in the **type** cell, enter **Delta Temp**, in **stream (+)** cell, choose **Sep Vap** and in

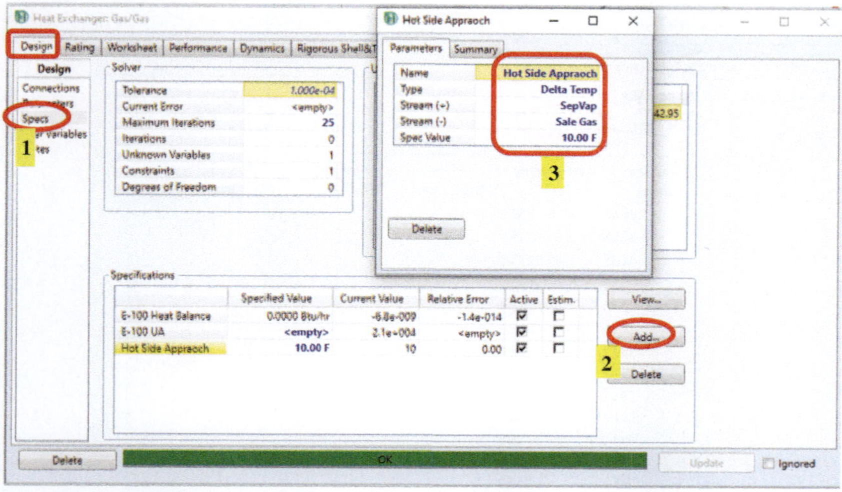

Figure 13.18 Gas/gas Heat Exchanger, design tab, specs page; adding specification of delta temp type.

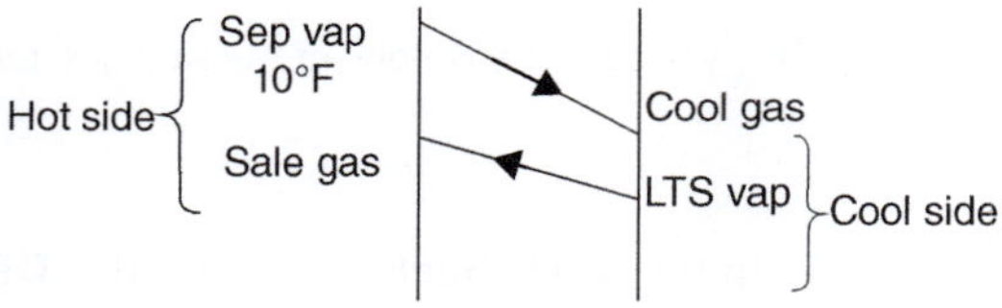

Figure 13.19 How the new specification of (delta temp type) is operating.

stream (−) cell, choose **Sale Gas**, and enter **10 °F** in the **Spec Value**, Figure 13.18.

Figure 13.19 shows how the specification of (Delta temp type) is operating.

To see the results, open the **worksheet** and notice that the temperature of **Cool Gas** = 42.9 °F. Also choose the **Performance** tab to view the results.

11) Checking the Dew Point of Sale Gas:

To ensure that the **Sale Gas** is not converted to liquid during the transferring operation, we will check the **dew point temperature** using the **Balance**, where the transferring operation line specifications are **15 °F** and **800 psia** while the **Sale Gas** specifications are **50 °F** and **570 psia**; therefore, we will create a default line with **15 °F** and **800 psia**.

(11-1) From the object palette, add a **Balance** operation . Choose inlet: **Sale Gas** and outlet: **Sales DP**.

(11-2) In the **parameters**, choose **Component Mole flow**.

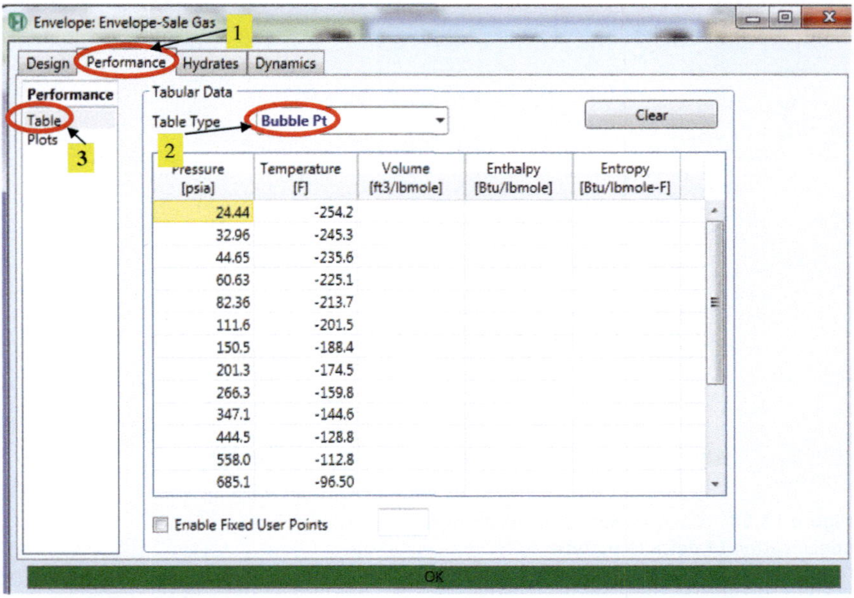

Figure 13.20 Window of envelope of the sale gas, table data.

(11-3) In the **worksheet**, enter the **sales DP** vapor fraction as **1.0** and the pressure as **800 psia**, HYSYS will calculate the temperature for **sales DP** to be **5.27 °F**.
To obtain the phase diagram: from the **Home** tab of the Ribbon, **Stream Analysis → Envelope:** Select **Sales Gas**, click the Performance tab, Table Type: **Bubble point**; the results will be shown in Figure 13.20 in table form. To show the results in graph form, click Plot. Figure 13.21 will appear.

12) Add a **second Mixer**

Name	**Mix-101**
Inlet	LTS liq and Sep liq
Outlet	Tower Feed

13) Add a **De-propanizer** column (distillation column)

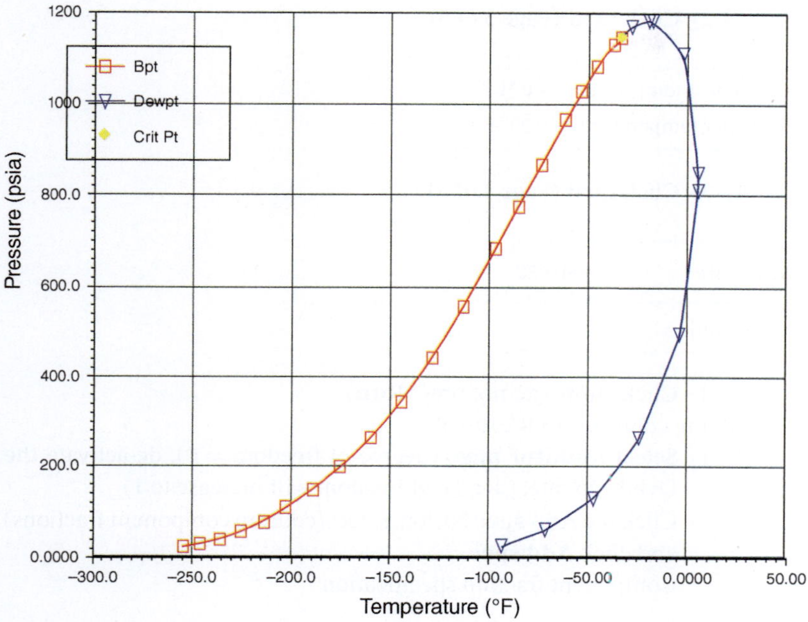

Figure 13.21 Phase diagram of the sale gas.

Name	De-propanizer
#Stages	10
Inlet	Tower Feed
Inlet stage	5
Condenser	Full Reflux
Top Outlet	Ovhd
Bottom Liquid outlet	Liquid prod
Re-boiler Energy Stream	Reb Duty
Condenser Energy Stream	Cond Duty

(13-1) Click next and then click Next (Pressures page)

Condenser pressure	198 psia
Delta *p*	0
Re-boiler pressure	203 psia
Delta *p*	0

(13-2) Click **next** (Page 3 of 5)

Condenser temperature	40 °F
Re-boiler temperature	200 °F

(13-3) Click **next** (Page 4 of 5)

Vapor rate	2 MMSCFD
Reflux ratio	1

(13-4) Click **Done** (do not press **Run**)

14) Adding column specifications:
   (14-1) Select **monitor page** (degree of freedom = 0), de-activate the Ovhd vap rate; (degree of freedom will increase to 1)
   (14-2) Click the **Add spec** button, select (column component fractions) and click **Add spec**.
   Component fraction specification

Name	Propane fraction
Stage	Re-boiler
Flow basis	Mole fraction
Phase	Liq
Spec value	0.02
Component	Propane

   (14-3) Click **group active**
   Propane fraction: active
   Deg. of freedom will be 0.
   (14-4) Click **Run** to converge the column.
   (14-5) Click **Performance** tab and for more details, click the **column profile** page.

15) Installing an **adjust** for calculating the LTS temperature.
   In this logical operation, we will raise the cold gas temp to leave more heavy components in the gas phase in order to raise the heating value for the cold gas and consequently raise the gas price.

(15-1) From the object palette, add the **Adjust** icon.
Adjusted variable: click **Select Var**, object: Cold gas, variable: temp, OK,
Target variable: click **Select Var**, object: Sale DP, variable: temp, OK,
In the **Specified Target Value** enter 10 °F,
Click the **Parameters** tab, in the **Tolerance**: 0.1 °F and in the **step size**: 1 °F, Iteration: 1000,
Click the **Monitor** tab and click **start** button,
→ **Adjusted Var** = 4.4 °F.

(15-2) See the results from the **Workbook**.
save this case "Steady-state Gas processing"

13.2.2 Dynamic Simulation

Figure 13.22 shows the PFD of the Dynamic state.

There are some steps to achieve the dynamics simulation:

1) Double-click the **Feed2** stream in PFD and delete the pressure (it will be calculated by the dynamic mode).
2) Double-click the **Mix-100:** from the parameters page, select **Equalize All** and close **Mix-100** view.
3) Insert a valve (**Sep valve**) between **Mix-101** and **Sep Liq** stream,
   (3-1) Cut the streams (Sep Liq, LTS Liq, and Tower Feed) from the mixer **Mix-101** by double-clicking each and delete them.
   Or use this tool Break Connection found in the **Flowsheet/Modify** tab in the Ribbon to cut the streams from Mix-101.
   (3-2) Now connect the following streams with **Mix-101**:
   Inlet = LTS Exit, and Sep Exit,
   Outlet = Tower Inlet. See Figure 13.22.
4) Open the Valve (Sep Valve).

Name	Sep Valve
Inlet	Sep Liq
Outlet	Sep Exit
Δp	25 psi

Close the valve view.

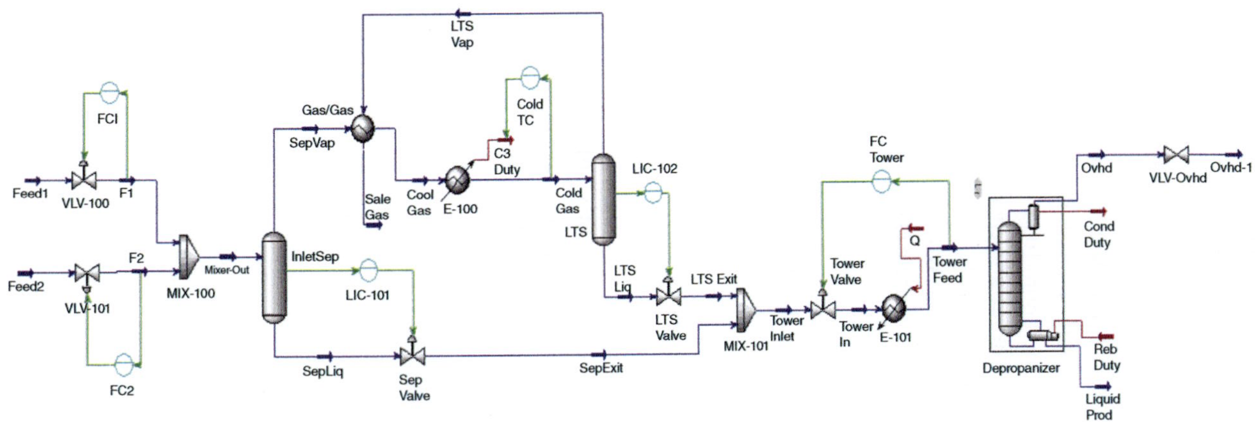

Figure 13.22 PFD of the dynamic state.

5) Insert a **valve** between **LTS Liq** and **Mix-101**.

Name	LTS Valve
Inlet	LTS Liq
Outlet	LTS Exit
Δp	5 psi

Close the valve view.

6) Insert a valve between **Mix-101** and **Tower feed** stream.

Name	Tower Valve
Inlet	Tower Inlet
Outlet	Tower In
Δp	360 psi

Close the valve view.

7) Insert a **heater** between streams of **Tower In** and **Tower feed**.

Name	Heater
Inlet	Tower In
Outlet	Tower feed
Energy	*Q*
Δp	9.7 psi

(7-1) Set the temperature of Tower feed stream to 0 °F. Reset the **de-propanizer column** and re-Run it.

(7-2) Double-click the **de-propanizer column** and click **Column Environment** to view the details, Figure 13.23.

(7-3) Cut the Liquid Prod stream from the Reboiler using the tool Break Connection in the Flowsheet/Modify tab of the Ribbon.

8) Insert a valve, as in Figure 13.23 within the following information:

Name	Reboil Valve
Inlet	Liq Exit
Outlet	Liquid Prod
Δp	25 psi

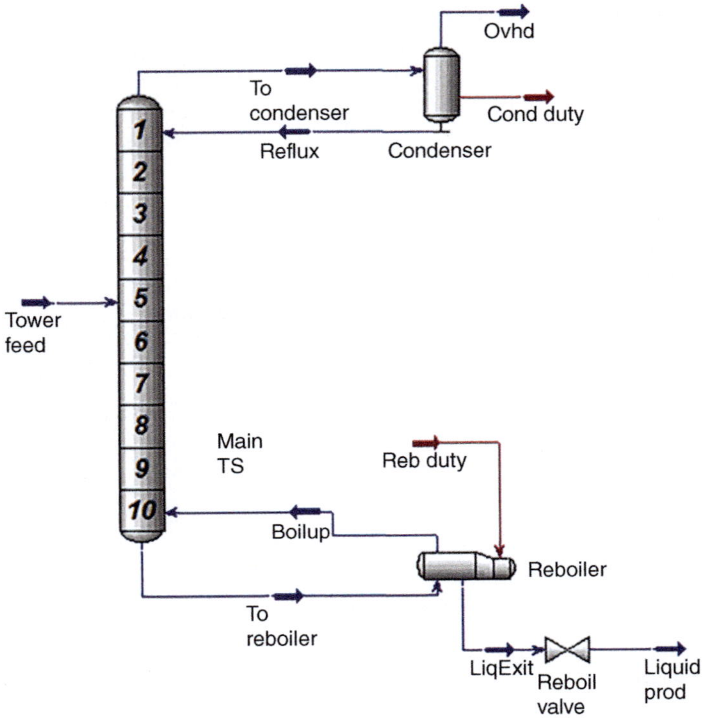

Figure 13.23 De-propanizer column/column environment, inserting a valve between LiqExit and LiqProd.

Double-click the Re-boiler: Bottom Outlet = Liq Exit

9) After adding the valve, click the **Run column solver** icon  in the Home ribbon.

Return to the original PFD by click 

icon in the Flowsheet/Modify ribbon.

10) Removing non-applicable operations:
It will remove some unit operations that are not needed in the dynamic state,
   (10-1) Double-click the **cold gas** icon and record $T = 4.44$ °F (this temperature will be controlled in the dynamic state).
   (10-2) In PFD, delete **ADJ-1**, **Dew point logical operation** (Balance), and **Sales DP** stream.

11) Column sizing:
    (11-1) Double-click the de-propanizer column, From **Internals** tab, click **Add New**, Select the **Trayed** from **Internal Type**, and select **Valve (Ballast-V1)** from **Tray/Packing Type.**
    (11-2) Click the **View** button to open the geometry window, and change the dimensions as follows:

Section diameter	3 ft. (If there are two diameters, type the first 2.5, and the second 3 ft)
Weir height	2 in
Tray spacing	2 ft. (24 in)

Close the window of geometry.
    (11-3) Click the **Rating** tab, and adjust the dimensions as follows:

Diameter	3 ft
Tray spacing	2 ft

12) Add a valve between Feed1 and Mix-100:
    Inlet: Feed1, Outlet: F1, $\Delta P = 5$ psi,
13) Add a valve between Feed2 and Mix-100,
    Inlet: Feed2, Outlet: F2, $\Delta P = 5$ psi,
    The inlet streams of mixer Mix-100 are F1 and F2.
14) Insert a valve after Ovhd Stream, Inlet: Ovhd, Outlet: Ovhd1 ($\Delta P = 5$ psi).
15) Dynamics Assistant
    (15-1) Degree of freedom for the flow sheet must be zero. The following sizing parameters must be specified for these unit operations:

Valves	CV Valve
Vessels	Volume
Cooler/H.E	*k*-value

    (15-2) **For all valves in the PFD**: Double-click the valve; **Rating** tab, click **Size Valve.**
    (15-3) Double-click the **De-Propanizer column,** click column Environment button, Double-click the Reboil Valve, **Rating** tab, and click **Size Valve**. Return to the PFD.

(15-4) Double-click the Inlet Sep vessel, **Rating** tab, enter the **Volume** = 11 m^3.

(15-5) Double-click the LTS separator, **Rating** tab, enter the **Volume** = 10 m^3.

16) Add a Flow Controller type PID 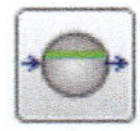 from the object palette, Dynamics group: Insert a controller between F1 and Valve VLV-100,
Name = FC1, Process Variable Source:
Object = F1, Variable = Actual Volume flow, OK,
Output Target Object: Object = VLV-100, Variable: Actuator Desired position, OK.
Parameter tab:
PV minimum = 10,000 barrel/day, PV maximum = 25,000 barrel/day.
Action = Reverse.
Tuning Parameters: $K_c = 1$, $T_i = 5$, (See Table 13.1).
Mode = Auto.

17) Insert another Flow Controller between F2 and VLV-101
Name = FC2, Process Variable Source:
Object = F2, Variable = Actual Volume flow, OK.
Output Target Object = VLV-101, Variable = Actuator Desired position, OK.
Parameters tab:
PV minimum = 8000 barrel/day, PV maximum = 15,000 barrel/day,
Action = Reverse,
Tuning Parameters: $K_c = 1$, $T_i = 5$,
Mode = Auto.

18) Level control:
Double-click each separator vessel, and from the **dynamic** tab, click **Add/Configure Level controller**.

19) Add a flow controller for control the Tower feed mole flow rate.
(19-1) Add a PID controller.
(19-2) Name: FC Tower; process variable: Tower-feed; Molar flow, OK, Output Target Object: Tower Valve.
(19-3) **Parameters** tab;
PV minimum = 200 lbmol/h, PV maximum = 500 lbmol/h, Action: reverse,
Mode: Auto. $K_c = 1$, $T_i = 5$ minutes, OK.

20) Add a temperature controller for control the temperature of cold gas:
(20-1) Add a PID controller.
(20-2) Name: **Cold TC**; process variable: cold gas; temperature, OK. Output target: C_3 duty, OK,

(20-3) **Parameters** tab; Action: direct,
PV minimum = −20 °F, PV maximum = 20 °F,
Mode: Auto. $K_c = 1$, $T_i = 10$ minutes.

(20-4) Click the **Control Valve** button, and **FCV for C_3 Duty** view appears:
From Duty Source: select **Direct Q** radio and in (direct Q) enter the following:
Min. available = 0, Max. available = 2e6 Btu/h, close FCV view.

21) Reboiler level control of distillation column

(21-1) Double-click the De-Propanizer Column: Click **Column Environment,**

(21-2) Add a PID Controller: Name: Reb-LC;
Process Variable (PV): **Case (Main)**; De-Propanizer, **Object**; Reboiler,
Variable: Vessel Liq Percent Level, click Select,
Output Target: **Case (Main);** De-Propanizer, **Object:** Reb duty,
Variable: Control Valve, click Select.

(21-3) **Parameters** tab; PV minimum = 0%, PV maximum= 100%,
Action: direct,
$K_c = 0.1$, $T_i = 3$ minutes.
Mode: Auto.

(21-4) Click the **Control Valve** button, and **FCV for Reb Duty** view appears:

(21-5) Select the **Direct Q** radio, and in (direct Q) enter the following:
Min available = 0, Max available = 6e6 Btu/h, close FCV view.

22) **Top stage TC** Controller: Add a PID controller.

(22-1) Name: **Top stage TC;**
Process variable source, Select PV, **Case (Main):** De-Propanizer,
Object: Main Tower, Variable: Top Stage Temp,
Output Target: **Case (Main):** De-Propanizer, Object: cond duty,
Variable: Control Valve, click Select.

(22-2) Click **parameters** tab; PV min = 50 °F, PV max = 130 °F,
Mode: Auto.
Action: direct, $K_c = 1$, $T_i = 5$ minutes,

(22-3) Click **Control Valve** button and FCV for cond Duty view appears:
Select **Direct Q** radio and in (direct Q) enter the following:
Min. available = 0, Max. available = 3e6 Btu/h, close FCV view.

23) Stage 9 TC controller:

(23-1) Add a PID controller,

Name: **Stage 9 TS**; Process Variable, **Case (Main):** De-Propanizer, Object: main Tower,

Variable: stage temp, 9-main Tower, click Select.

Output Target: **Case (Main):** De-Propanizer, Object: Reboil valve, Variable: Actuator Desired position.

(23-2) **Parameters** tab; Action: Direct, $K_c = 2$, $T_i = 5$ minutes, PV min = 110 °F, PV max = 260 °F,

Mode: Auto.

Figure 13.24 shows the Distillation column with its controllers.

Return to the PFD by clicking 

icon in the **Flowsheet/Modified** ribbon.

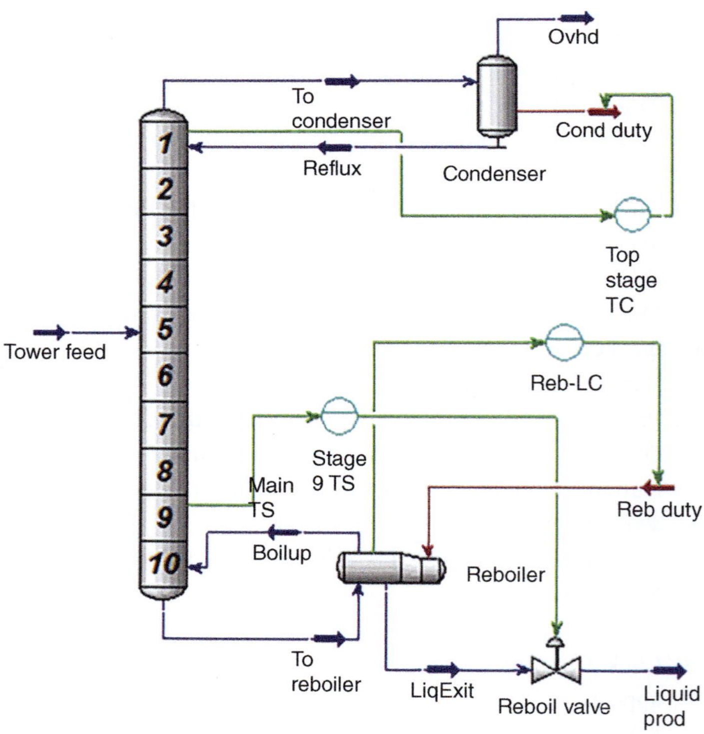

Figure 13.24 Installing control loops in the de-propanizer column.

24) Monitoring in the Dynamics state.
   (24-1) Double-click each Controller: **StripChart** tab; from the Variable Set, select: SP, PV, OP only,
   (24-2) Click the **Create stripchart** button, then click the **Display** button,
   Or it can click the **Monitor** tab.
   Click **Face plate** to see the values of process variable (PV) and Set point (SP).

In any graph, right click and select **Graph control**: to show all the variables; in the Axes tab; Axis Display section; check the **Show All** box.

25) From **Dynamics** tab in the Ribbon, Click **Dynamic Assistant** 

, Figure 13.25.

Now, click the **Make change** button.

26) From the **Dynamics** tab of the Ribbon, Click **Dynamic Mode** 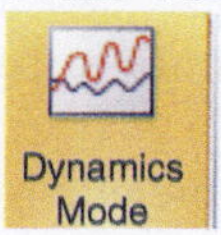

, since the changes are executed in the previous step, to answer about the message, click **No**.

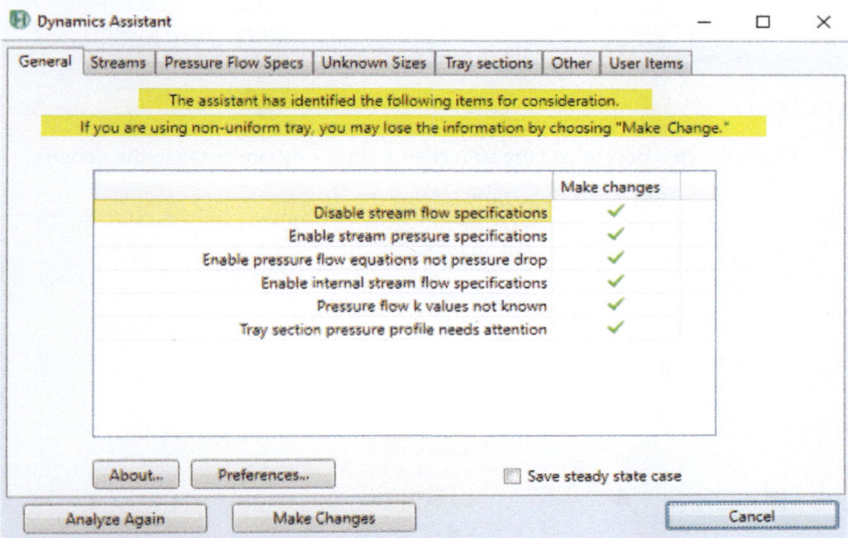

Figure 13.25 Dynamics assistant window/parameters that can be adjusted.

Again, click **Dynamic Mode** 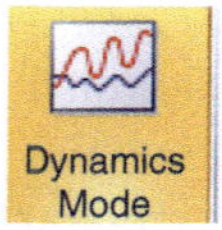

, then click **Yes** to switch to the dynamic mode.

Before clicking Run, present the screen of the controllers then click Run 

.

It can be seen in any set of variables in the Strip Chart, separately from Navigation Pane in the left side of the Simulation environment or from the icon in **Dynamics** tab of the Ribbon, Figure 13.26.

27) Click Strip Chart; from the new window, click **Add,** highlight the DataLogger1, click **Edit**, then click **Add**, Figure 13.27.
Select the variables that are required to be monitored, such as:
   (1) Object: F1, Variable: Actual Volume Flow,
   (2) Object: InletSep, Variable: Liquid Percent Level,
   (3) Object: Depropanizer, Variable: Stage Liq Temperature to below (9_Main Tower).
   One can add more variables to one strip chart, but preferably not more than three variables in one chart.

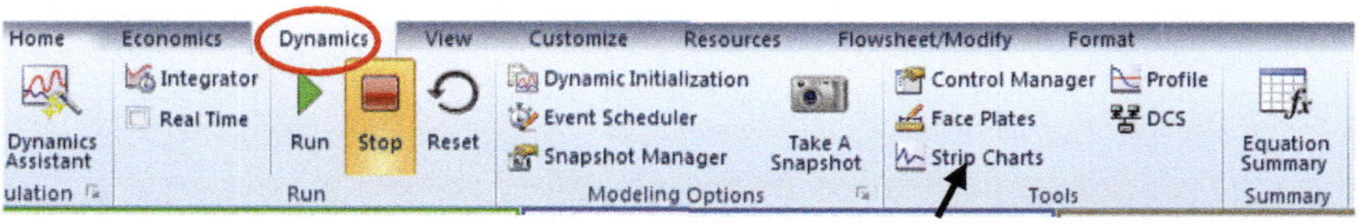

Figure 13.26 How to add the strip chart from the dynamics tab in the ribbon.

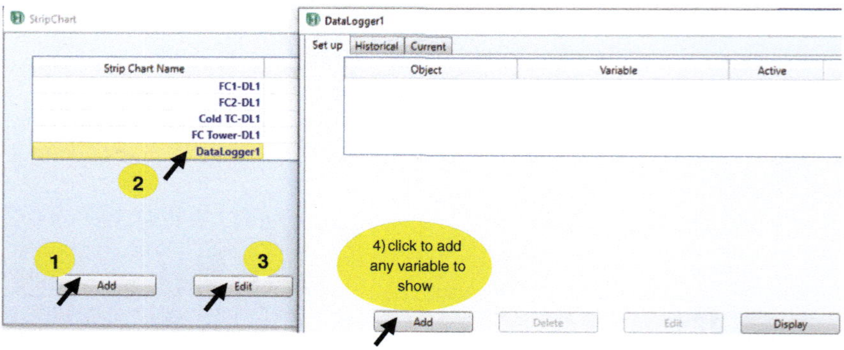

Figure 13.27 Adding variables to the strip chart.

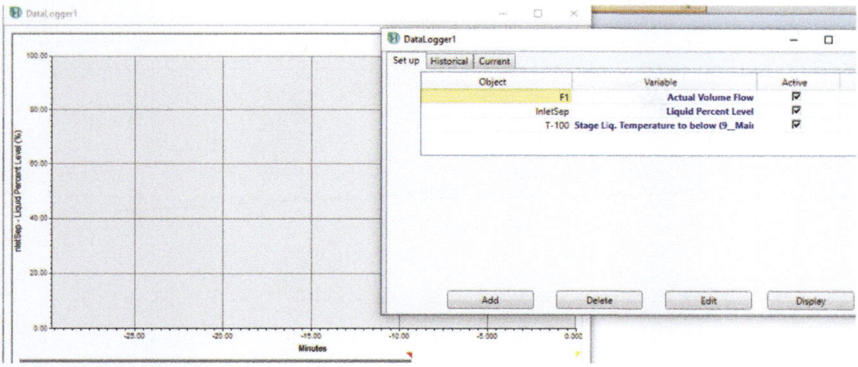

Figure 13.28 Display the screen of the data logger.

28) After adding the variables to the strip chart, click **Display** button to show the screen of the data logger, Figure 13.28.
The time of operating in the dynamic state can be determined from the **Integral** tab, e.g. specifying the current time of 0.0 minute, and the End time of 50 minutes.
One can change the color of the background of screen by right clicking the screen and selecting Graph Control. Right click the screen and select **Graph Control**, from the **Axis** tab, Activate the **Automatic Auto Scale** and **Show All** radios.
In order to follow and monitor the curves easily, it is preferable to reduce the accelerator of the strep chart graph from **Integrator** by the same way of section 13.1.2, step 7, Figure 13.13.
29) Click **Run** and monitor the graphs of the specified variables, Figure 13.29.
Notes of the Dynamic state:
(1) *One must place a resistance operation between every pressure node (e.g. valves, pumps, and compressors) to make a pressure drop between unit operations to run a dynamic simulation.*
(2) *It is recommended to add a valve to all boundary streams, or alternatively the pressures in boundary streams.*
(3) *In the mixer, it is recommended to set to **Equalize all** in the parameters page.*
(4) *To make it easier to determine where you need to make a pressure and a flow specifications, you can change the color scheme of the flowsheet. In the Display Options group on the Flowsheet/Modify tab, select the "Dynamic P/F Specs"*

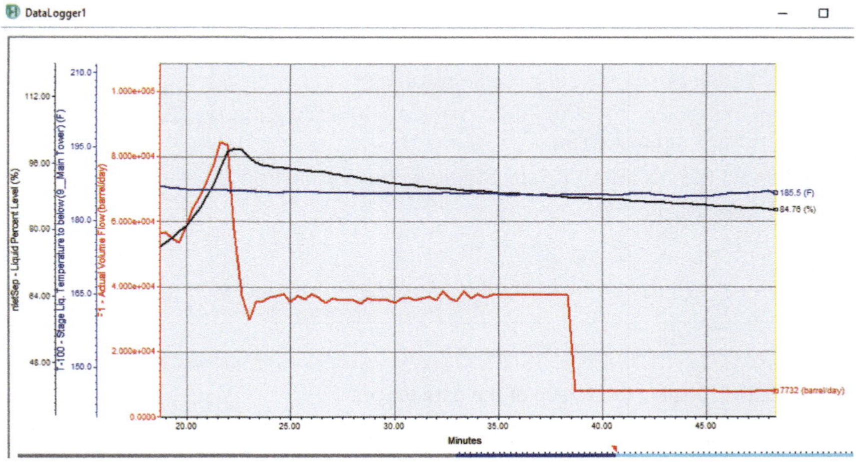

Figure 13.29 Screen of the strip chart (DataLogger1).

Color	Meaning
Green	Presure specified
Yellow	Flow specified
Red	Pressure and flow specified
Blue	No specifications

(5) *In order to run the dynamic simulations, it is important to have all of your equipment sized properly. There are two places where you can change the size of vessels: the Rating tab and the Dynamics tab on the vessel's property form.*

Additionally, you can use the Dynamics Assistant to auto-size your equipment. On the unknown sizes tab, select volume on the left. Change the residence times for the vessels you want to size and click the ***Quick size***... *button.*

(6) *In the Dynamics Assistant window,* Figure 13.25, *you will be given a list of parameters to adjust. Notice that all of the items are automatically checked in the "Make changes" column. If you click the Make Changes button, all checked changes will be made at once no matter what tab you are on. To make only certain changes, uncheck the ones you don't want.*

To review individual changes beforehand, double-click any of the items on the list to redirect to the appropriate tab.

13.3 Advanced Control

13.3.1 Ratio Control

The ratio control architecture is used to maintain the flow rate of one stream in a process at a defined or specified proportion relative to that of another. A common application for ratio control is to combine or blend two feed streams to produce a mixed flow with a desired composition or physical property. So the specifications of ratio control are [2, 5]:

- Ratio control is a special type of feedforward control that has had widespread application in the process industries.
- The objective is to maintain the ratio of two process variables as a specified value.
- The two variables are usually flow rates, a manipulated variable u, and a disturbance variable d.
- Thus, the ratio $R = u/d$ is controlled rather than the individual variables.
- The stream whose flowrate is not under control is usually referred to as *wild* (*Reference*) stream.
- The stream whose flowrate is under control is usually referred to as *controllable* stream.
- The calculation of ratio is performed in terms of the original variables, rather than deviation variables.

Typical applications of ratio control include [1, 8]:

1) Keep a constant ratio between the feed flow and the stream in the reboiler of distillation column.
2) Hold constant the reflux ratio in a distillation column.
3) Control the ratio of two reactants entering a reactor at a desired value.
4) Hold the ratio of a purge stream to the recycle stream constant, in order to maintain the composition of the blend at the desired value.
5) Hold the ratio of a purge stream to the recycle stream constant.
6) Keep the ratio of fuel/air in a burner at its optimum value (most efficient combustion).
7) Maintain the ratio of the liquid flow rate to vapor flow rate in an absorber constant, in order to achieve the desired composition in the exit vapor stream.

13.3.1.1 Problem Statement

The hydrogenation of benzene to produce cyclohexane is a well-known reaction wherein one mole of benzene is reacted with 3 mol of hydrogen in the presence of a suitable hydrogenation catalyst to form one mole of cyclohexane. One must control the ratio of two streams rather than control flow rate of each stream individually. According to the stoichiometry of the reaction [9]:

$$C_6H_6 + 3H_2 \rightarrow C_6H_{12} \tag{13.3}$$

It will take a molar flow rate basis of Bz = 100 kmol/h, so the molar flow rate of H_2 will be 310 kmol/h (where the stream of H_2 consists of 97.5% H_2, 0.5% N_2, and 2% CH_4), i.e. the molar ratio of H_2 stream to Bz stream will be 3.1. Using **Ratio Control,** this value will be kept constant during the operation. In this simulation, two reactant streams that enter the reactor at a specific ratio are considered, with the streams being benzene and hydrogen.

13.3.1.2 Simulation

1) Open a new case of Aspen Hysys, and enter the components: Benzene, Hydrogen, Nitrogen, and Methane.
2) Fluid package: Peng–Robinson.
3) Switch to the **Simulation** environment and construct the following process flow diagram in the steady state as in Figure 13.30.
4) Enter the conditions of the Bz-Feed ($T = 38\,°C$, $P = 2300$ kPa, molar flow rate = 100 kmol/h, and composition = 100% Bz).
5) Enter the conditions of the H_2-Feed ($T = 50\,°C$, $P = 2300$ kPa, molar flow rate = 310 kmol/h, and composition = 97.5% H_2, 0.5% N_2 and 2% CH_4 by mole).
6) The pressure drop on each valve is 100 kPa.
7) Open each valve, and from the rating tab, Sizing page, click **Size Valve** to ensure the sizing of the valve, as in Figure 13.31.
8) From the **object palette,** Dynamics group, insert the Ratio Controller icon  in PFD.

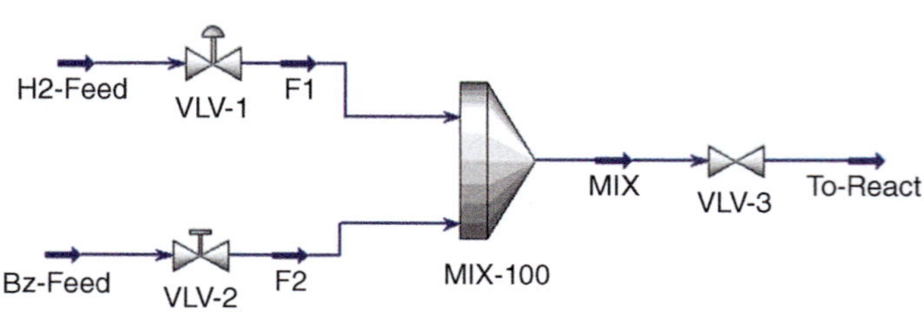

Figure 13.30 Process flow diagram of the mixing of benzene and hydrogen streams in the steady state.

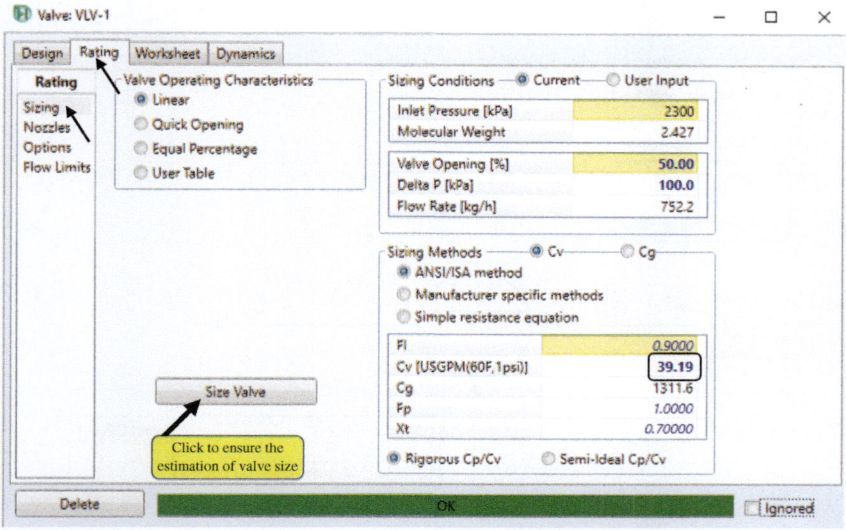

Figure 13.31 Clicking the size valve to ensure the estimation of Cv.

9) Open the Ratio Controller icon 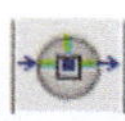 by double-clicking it; the *controllable* variable will be the H_2 stream, and the *wiled* variable will be the Bz stream. So, after opening the Ratio controller, under Process Variable Source, click the **Select PV** button, Object = F1, Variable = molar flow. Under Output Target Object, click the **Select OP** button, choose VLV-1, Figure 13.32.
The software assumes this variable (controlled variable) is PV = 1.
10) Change the PV to 2 (This will be the *wiled* variable = Bz-stream). Under the Process Variable Source, click the **Select PV** button, Object = F2, Variable = molar flow.
11) From the Parameters tab, Range page, enter the range of each variable as in Figure 13.33.
12) From the Configuration page, under Tuning Parameters, enter the values of K_c and T_i, 0.01 and 1, respectively, as in Figure 13.34.
13) Change the Mode to the **Automatic**, and Select **Reverse** as an Action type of the controller, Figure 13.34.
14) The PFD of the ratio control system is illustrated in Figure 13.35.
15) From the StripChart tab, Variable Set, select (SP, PV, OP only), then click **Create StripChart**. Deactivate the OP (OP_1), and lastly, click the **Display** button, Figure 13.36.
16) Close the **RATO-100-DL1** and **RATO-100** windows.

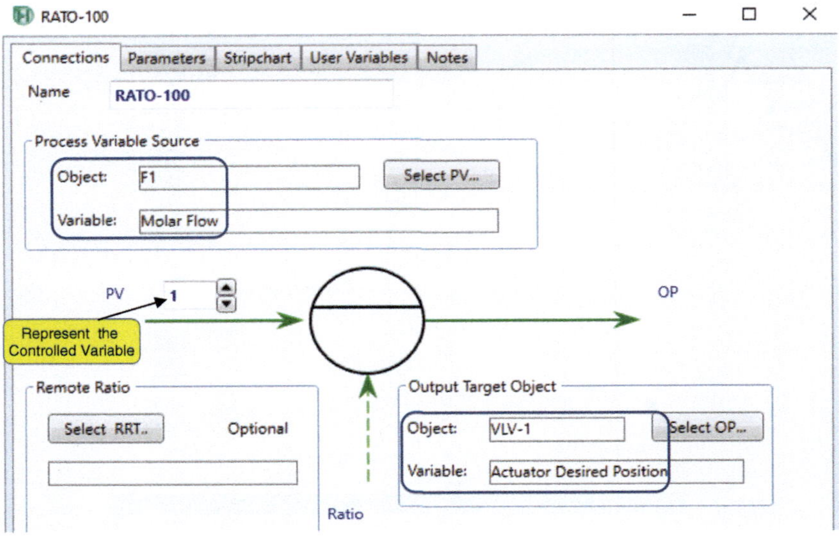

Figure 13.32 Selecting the process variable (PV) and output of controller (OP).

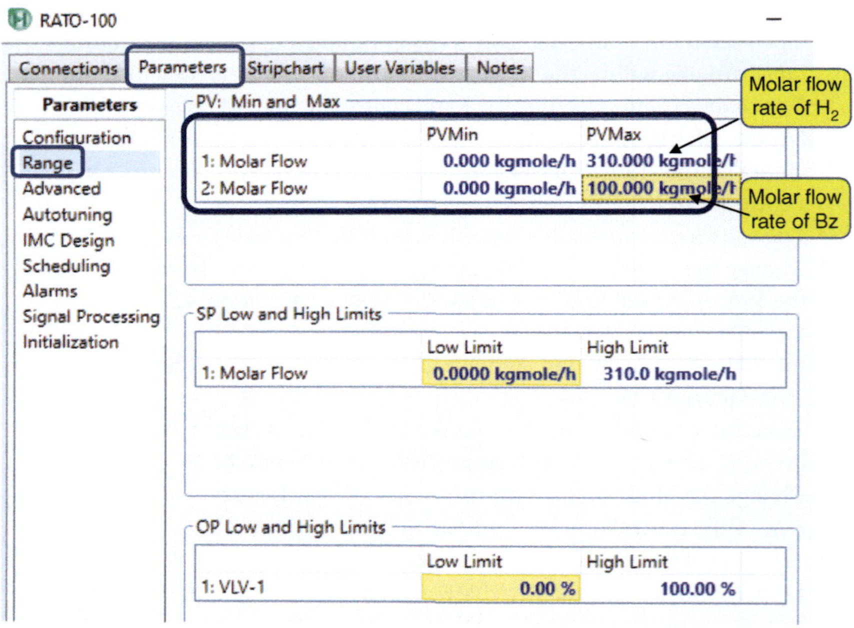

Figure 13.33 Entering the ranges of molar flow rate of H_2-stream (1), and Bz-stream (2).

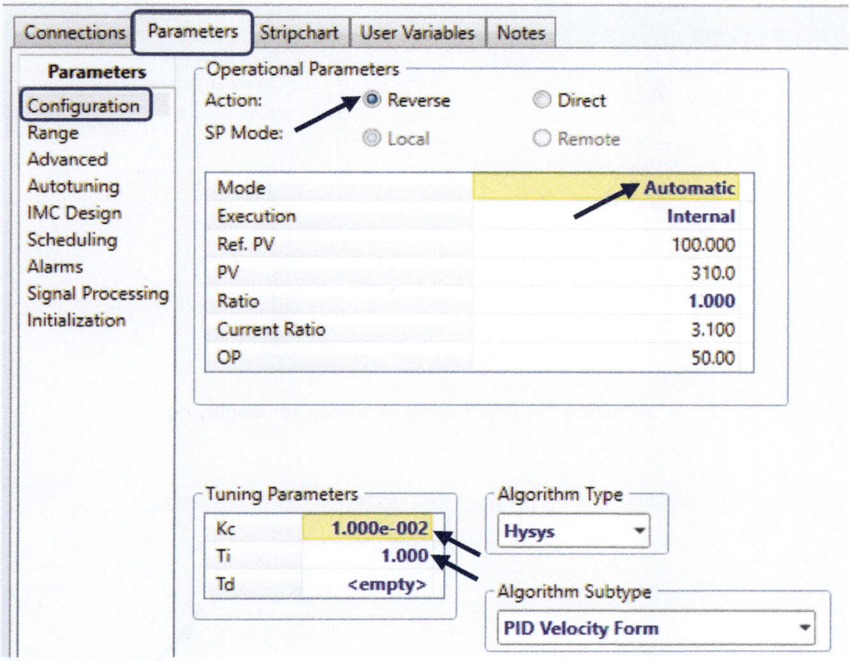

Figure 13.34 Entering the tuning parameters of controller.

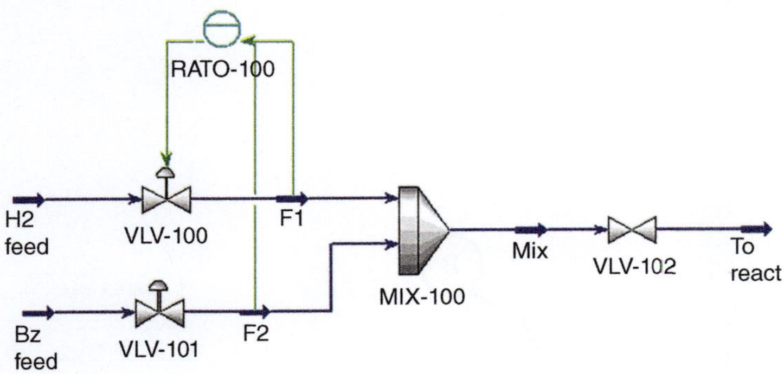

Figure 13.35 PFD of ratio control System.

17) From the **Dynamics** tab of the Ribbon, click **Integrator** tab 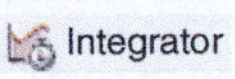

, and reduce the Acceleration to 0.02.

Click the **Dynamic Mode** tab 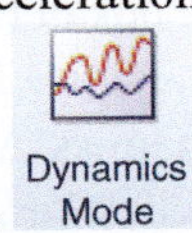

, from the new message, click Yes, then in the new window, click the **Make Change** button. Click the

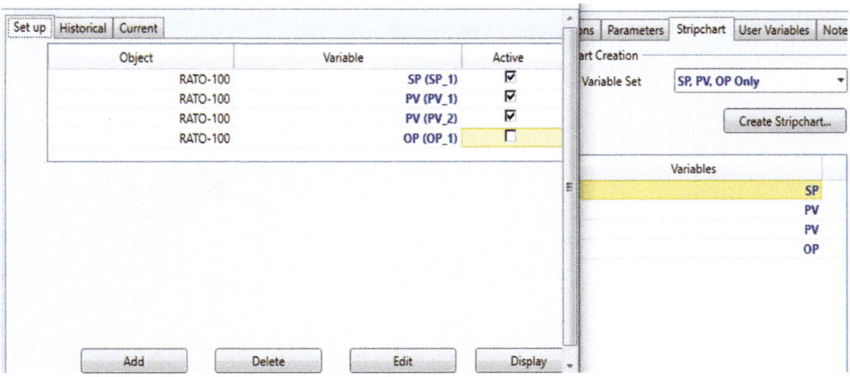

Figure 13.36 Selecting (SP, PV, OP only) as signals for displaying.

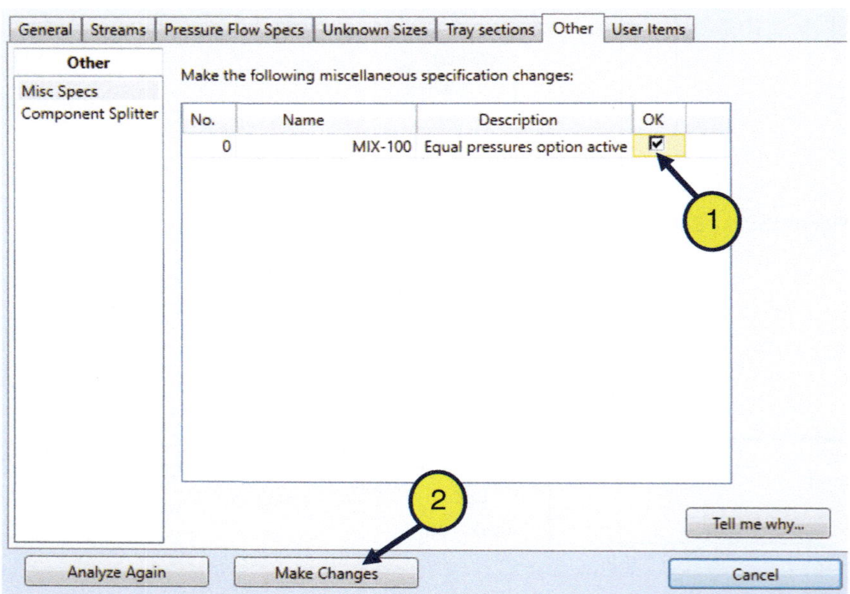

Figure 13.37 Activating the box of (equal pressure option active), then clicking make change again.

Other tab and activate the box of (Equal pressure option active), then click **Make Change** again, Figure 13.37.

18) Close the window of Dynamics Assistance, click **Dynamics Mode** again, from new message, click Yes.

19) Now that we are in the dynamics mode, click **RUN** to display variable signals as in Figure 13.38.
Click the **Stop** button. Double-click the controller, in the Parameters tab, change the Ratio to 3.1, then click **Run.** The process-controlled variable will increase to 310 kmol/h as in Figure 13.39.

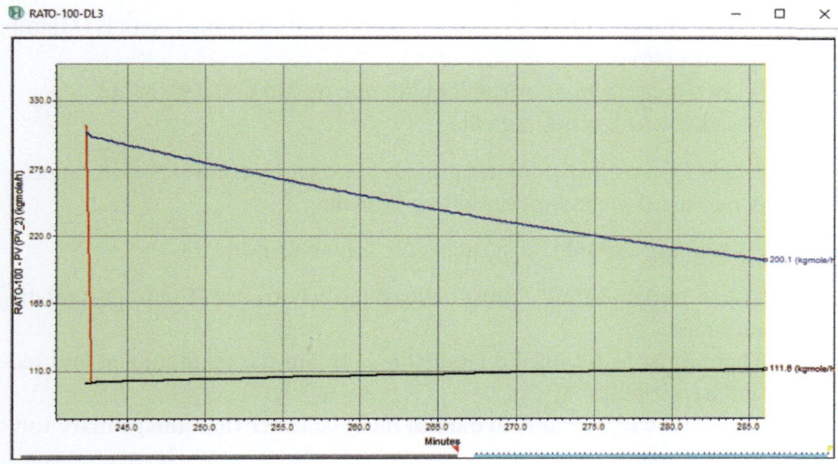

Figure 13.38 Display of the signals on the screen of the controller.

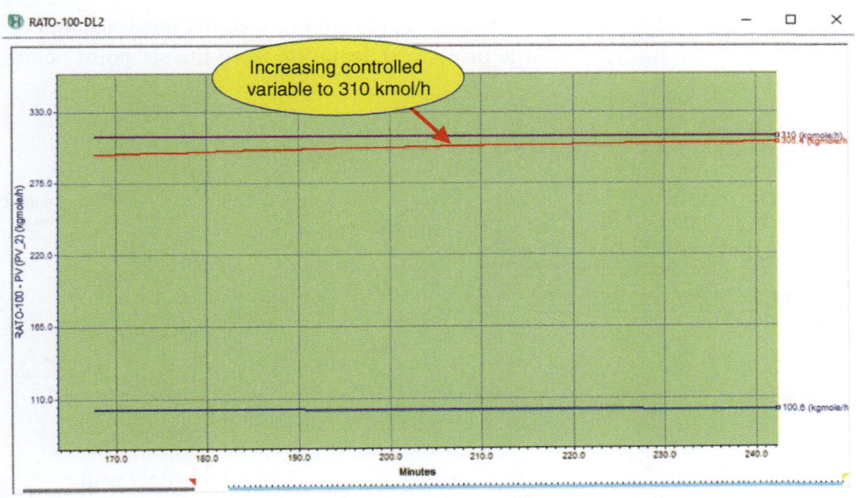

Figure 13.39 Increasing the process controlled variable to 310 kmol/h after increasing the ratio to 3.1.

13.3.2 Cascade Control

It is especially useful in reducing the effect of a load disturbance that moves through the control system slowly. Figure 13.40 shows some examples of cascade control. The inner loop has the effect of reducing the lag in the outer loop, with the result that the cascade system responds more quickly with a higher frequency of oscillation [2, 8].

Number of measurement signals > Number of controlled signals (Manipulated variable)

[i.e. two measurement variables with one manipulated variable]

Cascade control is desired when

1) Single-loop control does not provide satisfactory control performance.
2) A measured secondary variable is available.

A secondary variable must satisfy the following criteria:

1) The secondary variable must indicate the occurrence of an important disturbance.
2) There must be a causal relationship between the manipulated and secondary variables.
3) The secondary variable dynamics must be faster than the primary variable dynamics.

From Figure 13.40b, it is clear that T (reactor temp) will respond much faster to changes in T_i (feed temp) than to changes in T_c (coolant temp). If the temperature of the coolant, T_c, supply should rise, the reactor temperature, T, can undergo a large prolonged deviation from the set point before

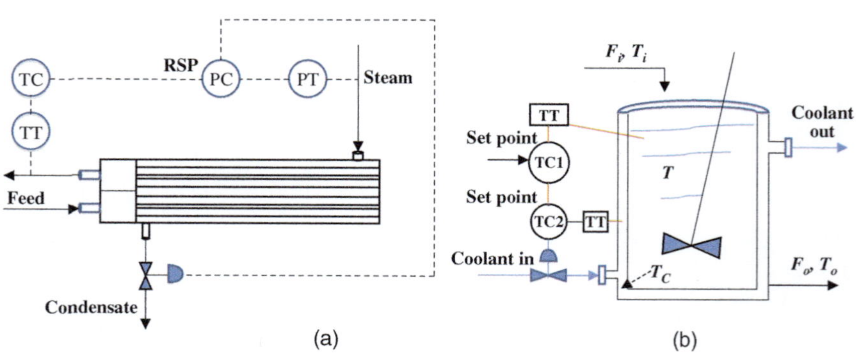

Figure 13.40 Some examples of cascade control: (a) Heat Exchanger with cascade control, (b) CSTR with cascade control [TT = Temp transmeter, TC1 = Primary Temp controller, TC2 = Secondary Temp controller, PT = Pressure transmeter, PC = Pressure controller, RSP = Remote set point], [2].

control is again established. The reason is that the controller does not take corrective action until the effect of the rise in coolant supply temperature has worked itself through the system of several resistances to reach the measuring element.

To prevent the sluggish response of reactor temperature to a disturbance in coolant supply temperature, the control system shown in Figure 13.40b (Cascade control) is proposed.

- The primary controller is also referred to as the *master* controller (Dominant), and
- the secondary controller is referred to as the *slave* controller.

13.3.2.1 Problem Statement

Ethyl acetate is formed by the reaction between ethanol and acetic acid, and the equilibrium is represented as [10]:

$$\text{Ethanol} + \text{Acetic acid} \leftrightarrow \text{Ethylacetate} + \text{Water [liquid phase]} \quad (13.4)$$

Simulate the above isothermal reaction in CSTR, where:

Kinetics: Forward reaction: $-r = kC_{Eth}C_{Ac}$, $k = Ae^{-E/RT}$, $A = 1.9 \times 10^8$, $E = 5.95 \times 10^7$ J/kg mol

Reverse reaction: $-r = k^{/}C_{Eact}C_{wt}$, $k^{/} = A^{/}e^{-E/RT}$, $A^{/} = 5 \times 10^7$, $E^{/} = 5.95 \times 10^7$ J/kg mol

Basis Units: kg mol/m^3, Rate Units: kg mol/m^3 s

CSTR: Volume = 0.2 m^3

Feed: $T = 70\,°C$, $P = 1.1$ atm, Molar flow rate (kmol/h): Ethanol = 186, Acetic acid = 192, Water = 9.

Fluid package: **NRTL.**

Create the cascade control loop to keep the temperature of the reactor at 70 °C.

13.3.2.2 Simulation

1) Open a new case of Aspen HYSYS and enter the following components: Ethanol, Acetic acid, Ethyl acetate, and Water.
   Fluid package: NRTL
2) Add the set reaction with kinetics reaction type. Open the reaction (Rxn-1) and enter the stoichiometry and kinetics data as in Figure 13.41.
3) Add the Set-1 to the fluid package.
4) Switch to the simulation environment and create the feed stream with the following conditions: $T = 70\,°C$, $P = 1.1$ atm, and component molar flow rate (kmol/h): Ethanol = 186, Acetic acid = 192, Water = 9.

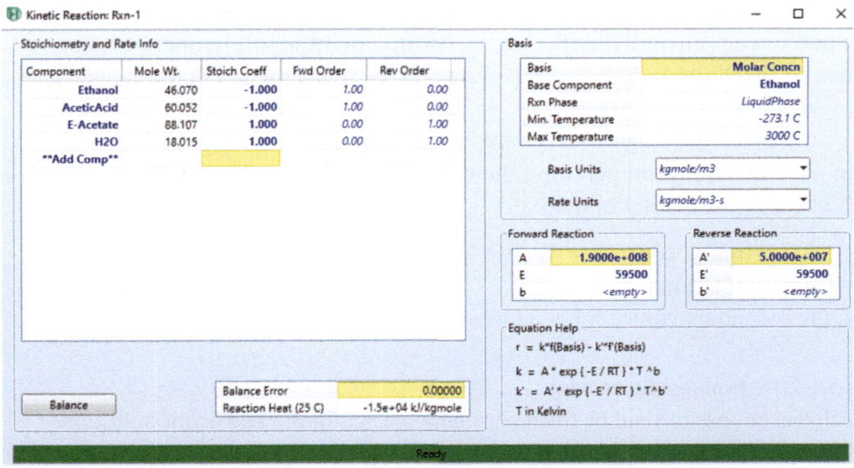

Figure 13.41 Window of the Kinetic Reaction Rxn-1.

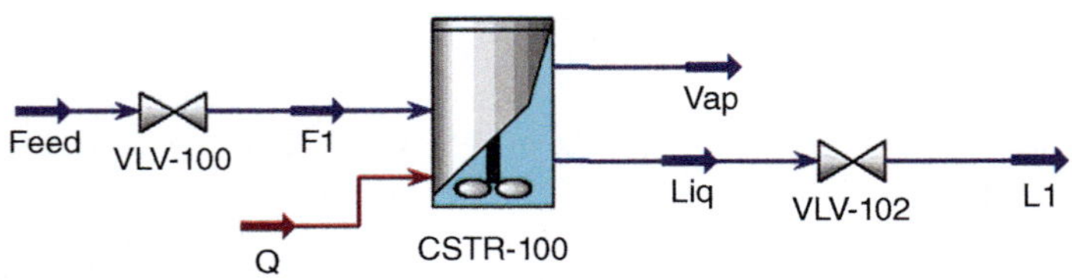

Figure 13.42 PFD of the CSTR.

5) Add a CSTR reactor and connect the inlet and outlet streams as in Figure 13.42. From **Dynamics** tab, enter the volume of reactor = 0.2 m^3, and specify the liquid percent = 100%.
6) Specify the pressure drop of each valve = 5 kPa.
7) From the Reaction tab, select Set-1. Set the temperature of Liq stream = 70 °C.
8) Open each valve, from **Rating** tab, click the **Size Valve** button
9) Specify Cooling water as a Utility Type of Q-Energy, Figure 13.43.
10) Add a PID controller, Name = Slave, and connect the Process Variable Source: Object = Q, Variable = Mass flow.
    Output Target Object: Object = Q, Variable = Control Valve. See Figure 13.44.
11) Switch to the **Parameters** tab, and enter the minimum and maximum values of the cooling water flow rate as in Figure 13.45.

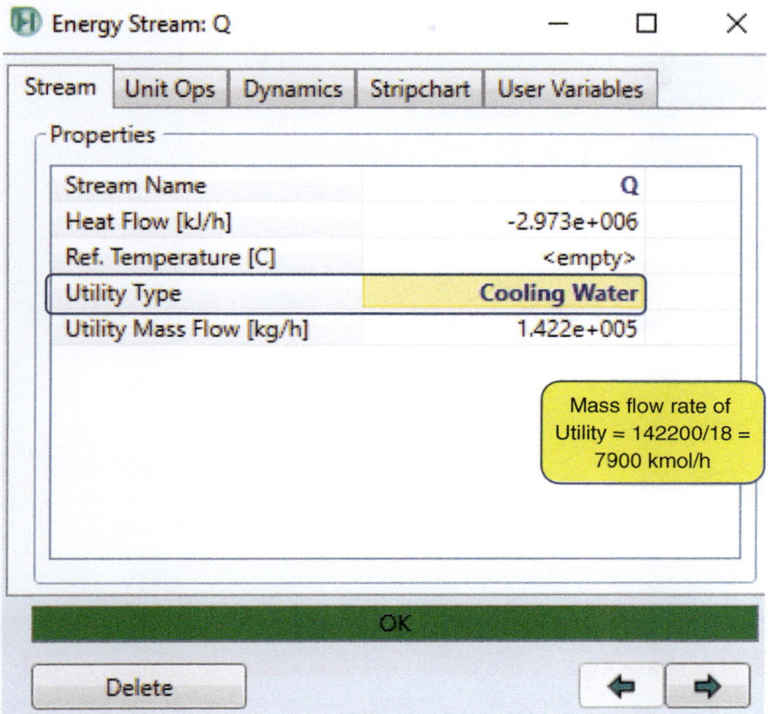

Figure 13.43 Specifying the utility type of Q-energy.

12) Click the **Control Valve** button, and enter the minimum and maximum values of the cooling water flow rate as in Figure 13.46. See the comment in Figure 13.43.
13) Add another controller (Type PID), Name: Master, and enter the data required as in Figure 13.47.
14) Switch to the Parameters tab of the Master controller and enter data as in Figure 13.48.
15) Return to the Slave controller, Parameters tab, and change the Mode to "Casc," Figure 13.49.
16) From Dynamics Ribbon, click Dynamics Assistance, Click the Make Change button, Finish.
17) Click the Dynamics Mode, then click Yes.
18) From the Integral tab, reduce the Acceleration to 0.1.
19) Double-click each controller, create stripchart (SP, PV, OP), and click the Display button.

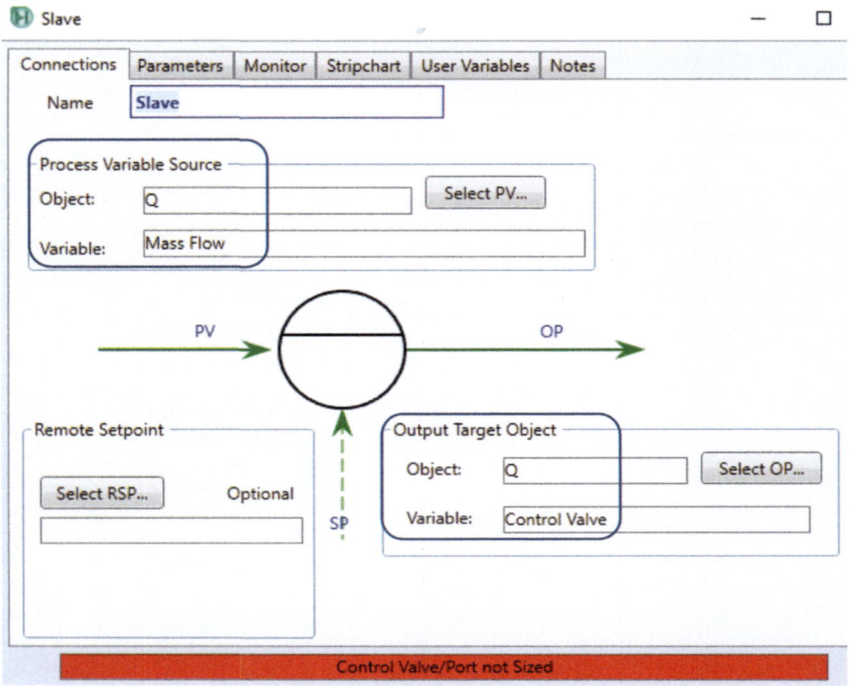

Figure 13.44 Slave controller and its signals.

20) Click the **Run** button to see the results. Figure 13.50 shows the curves of the Slave controller.
21) To obtain the best results, one can change the values of K_c, Ti of the controllers.

13.3.3 Split Range Control

This type of control is used, where there are several manipulated variables, but a single output variable. Figure 13.51 shows the mechanism of split range control, where the number of control signals is greater than the number of measurement signals [2, 11].

From Figure 13.51, as controller output increases from 6 to 9 psi, V_2 opens continuously while V_1 remains fully open; both actions lead to a reduction in the pressure. Above 9 psi, V_1 starts closing with V_2 fully open, both actions lead to a reduction in the pressure until the reactor has returned to the desired operation.

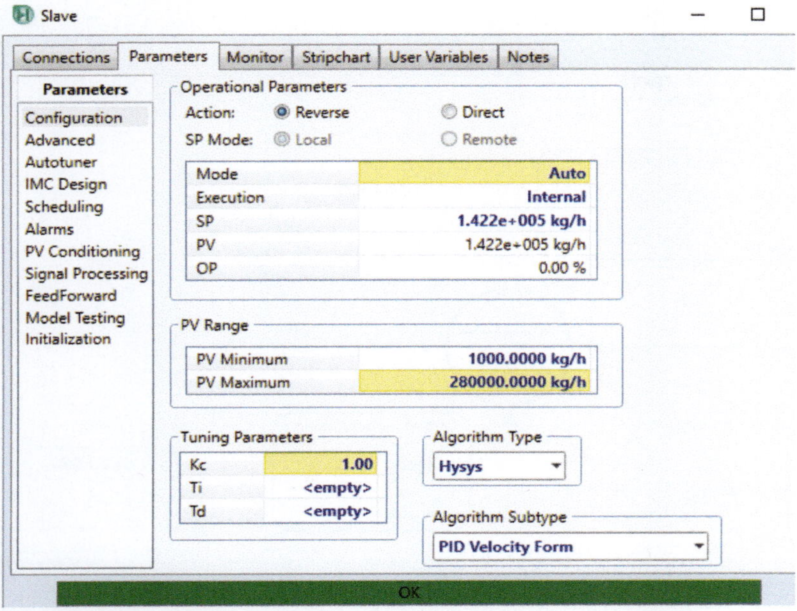

Figure 13.45 Parameters tab of the slave controller.

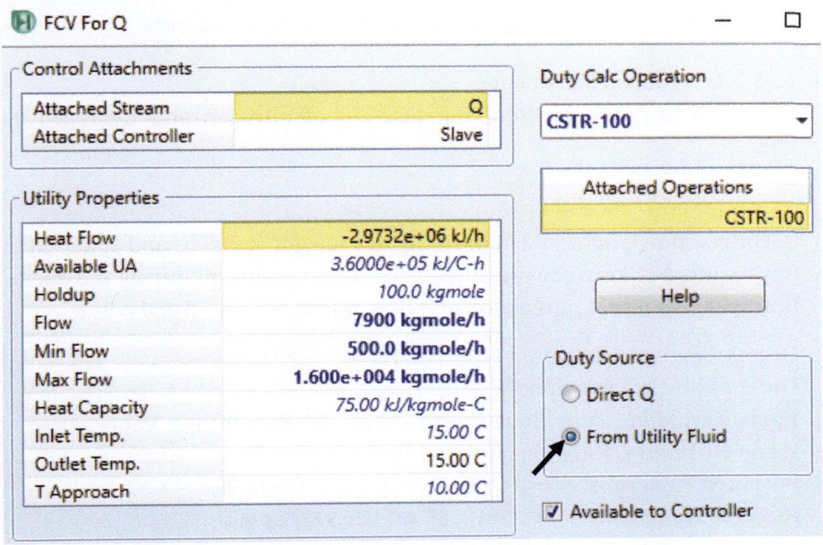

Figure 13.46 Specifying the minimum and maximum of Q-utility. See the comment in figure 13.43.

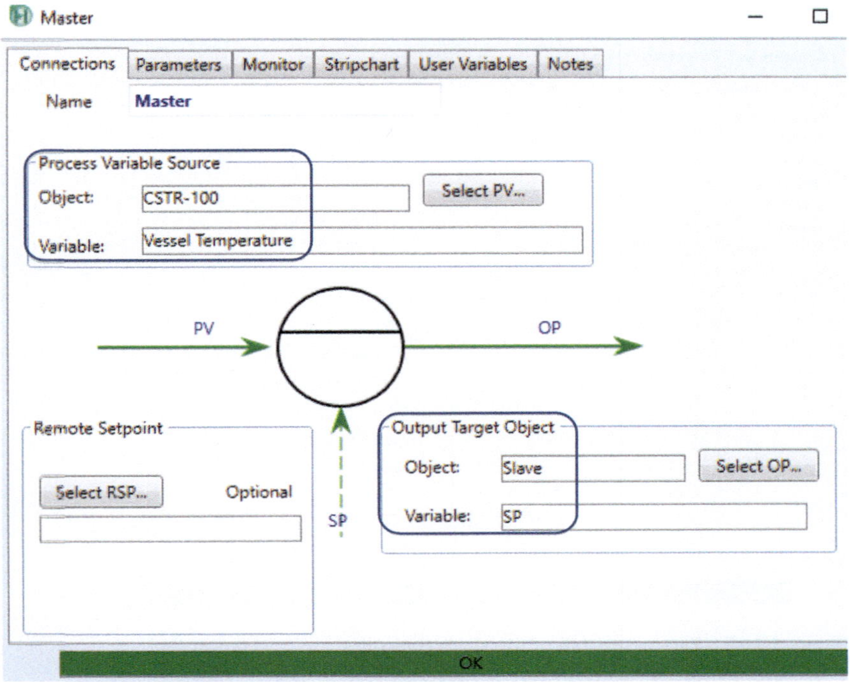

Figure 13.47 Connection page of master controller with its signals.

13.3.3.1 Problem Statement

Ammonia is produced from the reaction of nitrogen and hydrogen as follows:

$$0.5N_2 + 1.5H_2 = NH_3 \tag{13.5}$$

The reaction is achieved in the plug flow reactor at 200 °C and 15,000 kPa; it is required to keep the pressure of the reactor constant during the operation. The conditions of reaction and flow rate of feed are listed below.

Order of Forward Reaction: $N_2 = 0.5$, $H_2 = 1.5$, $NH_3 = 0$
Order of Reverse reaction: $N_2 = 0$, $H_2 = 0$, $NH_3 = 1$
Basis: Partial Pressure, **Base Component**: N_2, **Rxn Phase**: Vapor phase,
Basis Unit: atm, **Rate Units**: kmol/m^{3}s
Forward Reaction: $A = 10{,}000$, $E = 91{,}000$ kJ/kg mol,
Reverse Reaction: $A' = 1.3e10$, $E' = 1.41e5$ kJ/kg mol,
Feed: $T = 200$ °C, $P = 15{,}500$ kPa

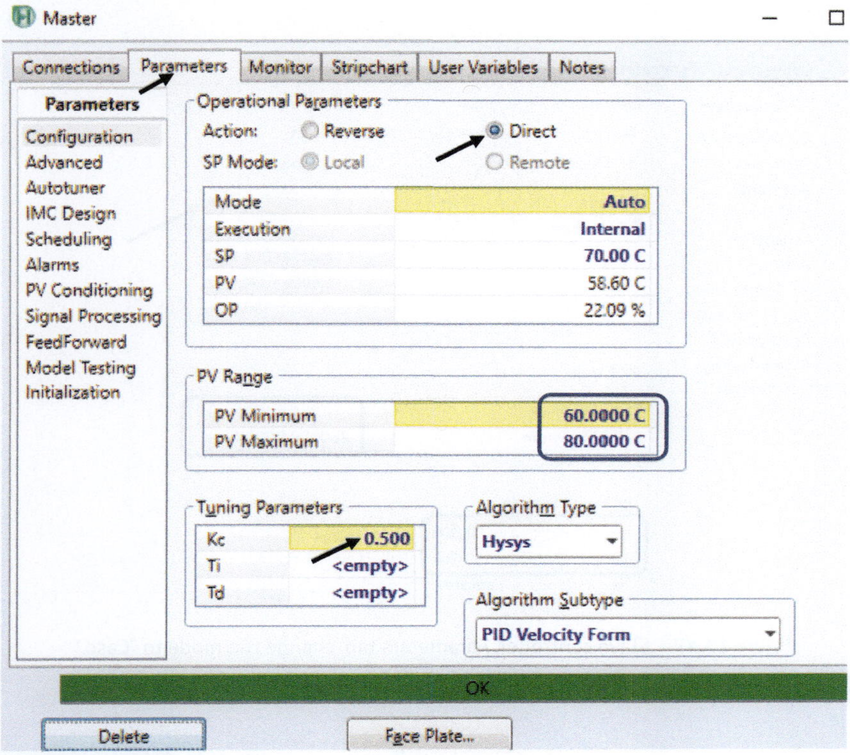

Figure 13.48 Parameters tab of the master controller.

The composition in Molar Flow is shown below:

Component Name	Comp Mole Flow, kmol/h
Methane	294.44
Nitrogen	3417.5
Hydrogen	10,253
Ammonia	0.0
Argon	134

Use a Split range control system to control the pressures of the inlet and outlet valve streams.

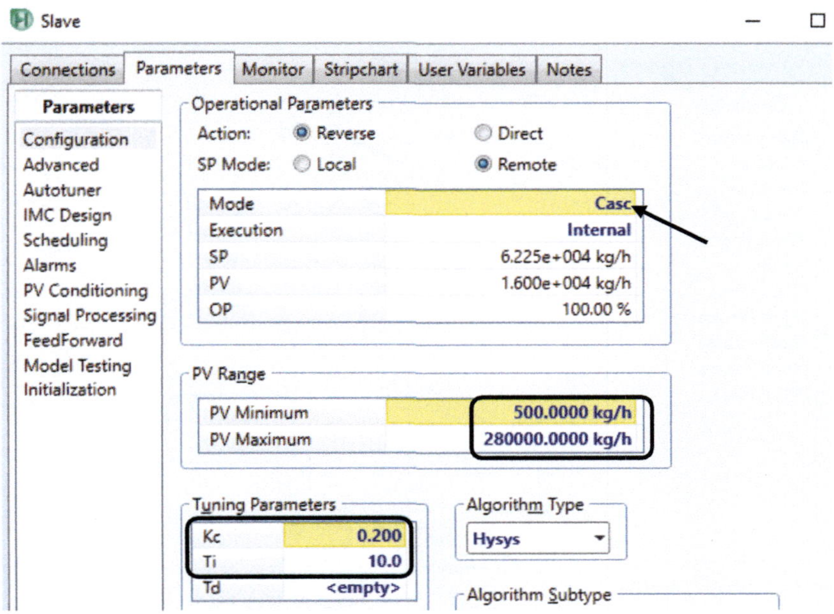

Figure 13.49 Slave controller, Parameters tab, change the mode to "Casc."

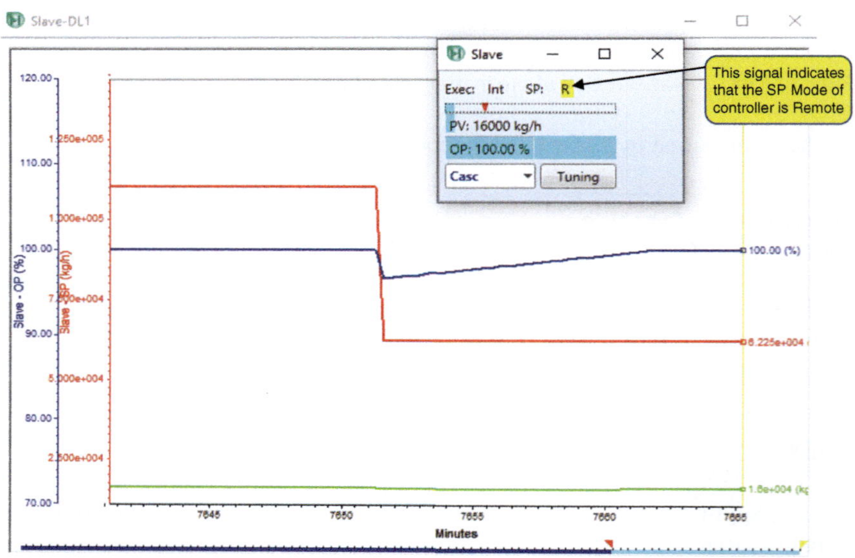

Figure 13.50 Curves of the slave controller.

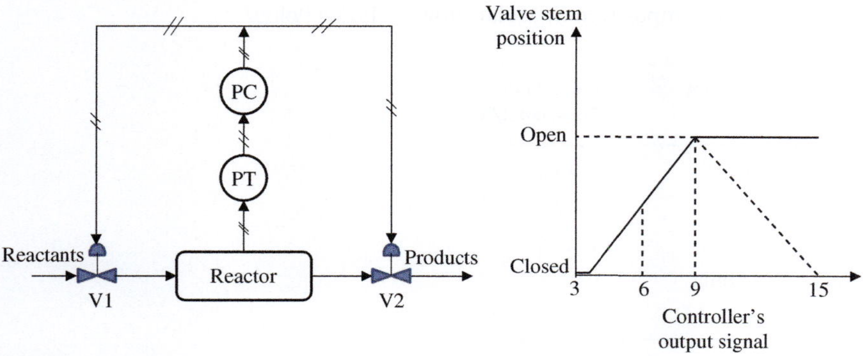

Figure 13.51 Mechanism of split range control.

13.3.3.2 Simulation

1) Open a new case and add the components: CH_4, H_2, N_2, NH_3, and Ar.
2) Property package: Peng Robinson
3) Add a Reaction-Set: **Set-1, Reaction type**: Kinetics,
   Enter the stoichiometry of the components according to the reaction (13.5); other data will be as given above and as shown in Figure 13.52.
4) Add the Reaction Set to the fluid package (**FP**).
5) Enter the Simulation environment, and create the feed stream as follows:
   Name = Feed, $T = 200\,°\mathrm{C}$, $P = 15{,}500\,\mathrm{kPa}$,

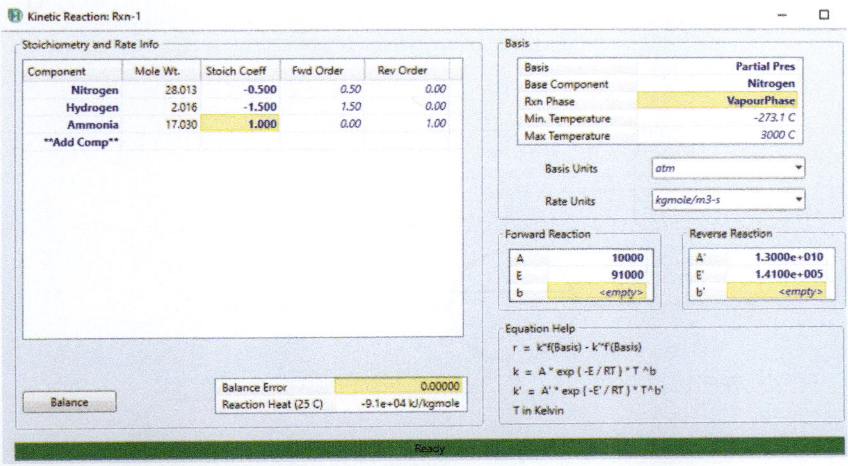

Figure 13.52 Window of kinetics reaction data of ammonia production.

The composition in Molar Flow is shown below:

Component Name	Comp Mole Flow, kmol/h
Methane	294.44
Nitrogen	3417.5
Hydrogen	10,253
Ammonia	0.0
Argon	134

6) Insert a valve (Inlet = Feed, Outlet = F), $\Delta P = 500\,\text{kPa}$.
7) Add a plug flow reactor, Inlet = F, Outlet = P, $\Delta P = 200\,\text{kPa}$, Length = 0.9692 m, Diameter = 3.0 m, Void fraction = 0.5. Add the reaction set-1 to the reactor.
8) Insert a valve (Inlet = P, Outlet = Product), $\Delta P = 500\,\text{kPa}$. The process flow diagram will appear as in Figure 13.53.

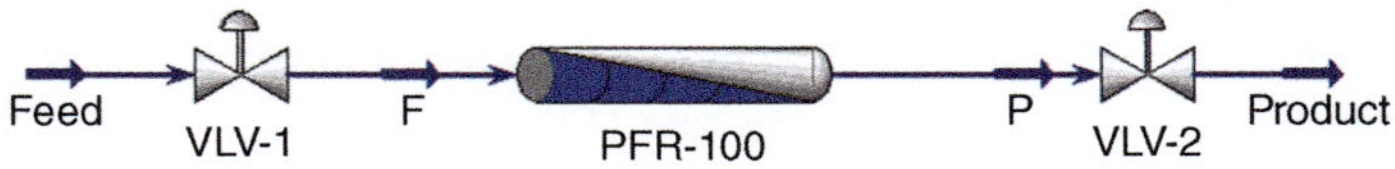

Figure 13.53 Process flow diagram of plug flow reactor of ammonia production in steady state.

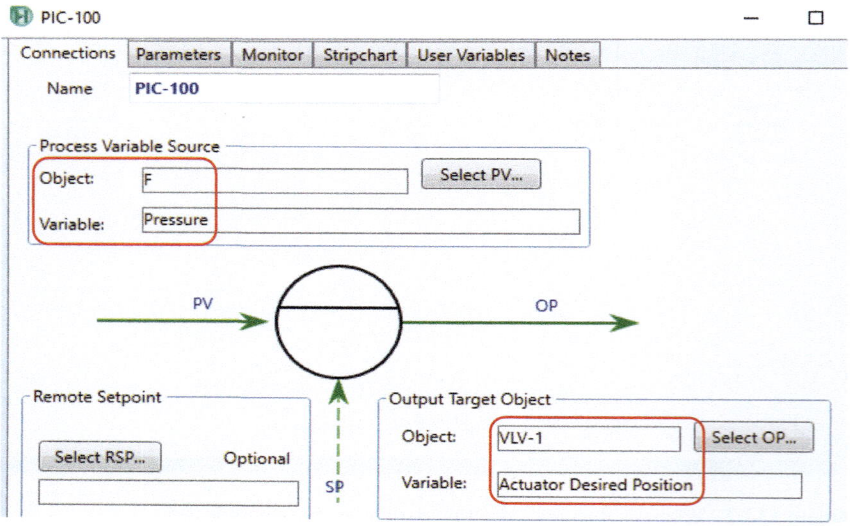

Figure 13.54 Window of the PIC-100 controller.

9) Open each valve, Rating tab, and click the **Size Valve** button.
10) From the object palette, Dynamics group, add two PID controllers.
11) Open the first PID controller, IC-100, click the **Select PV** button, Object = F, Variable = Pressure, Figure 13.54. Click the **Select OP** button, Object = VLV-1, Variable = Actuator Desired Position.
12) Fill the Parameters tab as follows: Action = Reverse, PV Minimum = 0.0, PV Maximum = 25,000 kPa.
$K_c = 2$, $Ti = 2$.
13) Open the second PID controller, IC-101, click the **Select PV** button, Object = P, Variable = Pressure. Click the **Select OP** button, Object = VLV-2, Variable = Actuator Desired Position.
14) Fill the Parameters tab as follows: Action = Direct, PV Minimum = 0.0, PV Maximum = 25,000 kPa.
$K_c = 2$, $Ti = 2$.
15) Add a split range controller, SPLT-100, OP # 1: click the **Select PV** button, Object = PFR-100, Variable = Pressure drop. Click the **Select OP** button, Object = PIC-100, Variable = Sp, Figure 13.55.
16) Repeat the step 14 for OP # 2: Click **Select OP** button, Object = PIC-101, Variable = Sp, Figure 13.56.
17) Switch to the **Parameters** tab, **Configuration** page; Set the min and max values of PV and SP as in Figure 13.57.

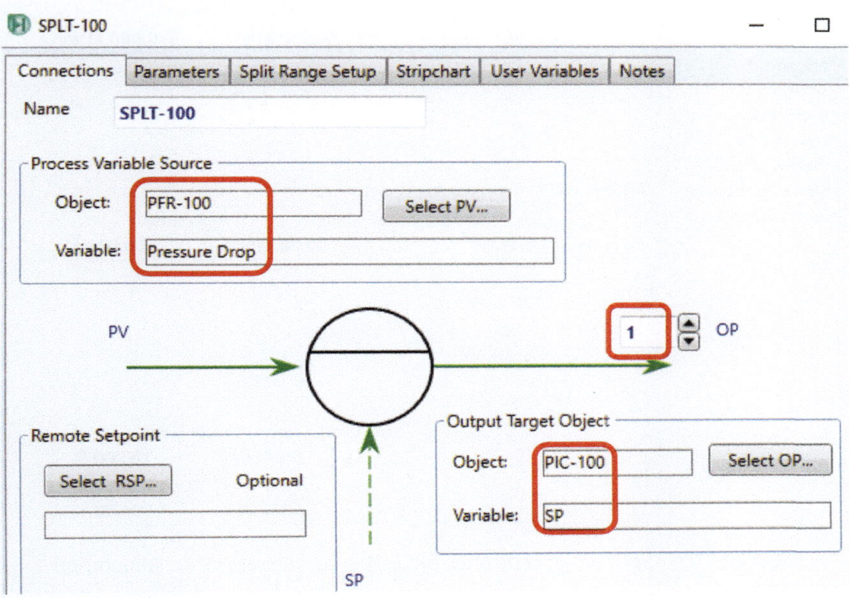

Figure 13.55 Split range controller, SPLT-100, connection tab, OP # 1.

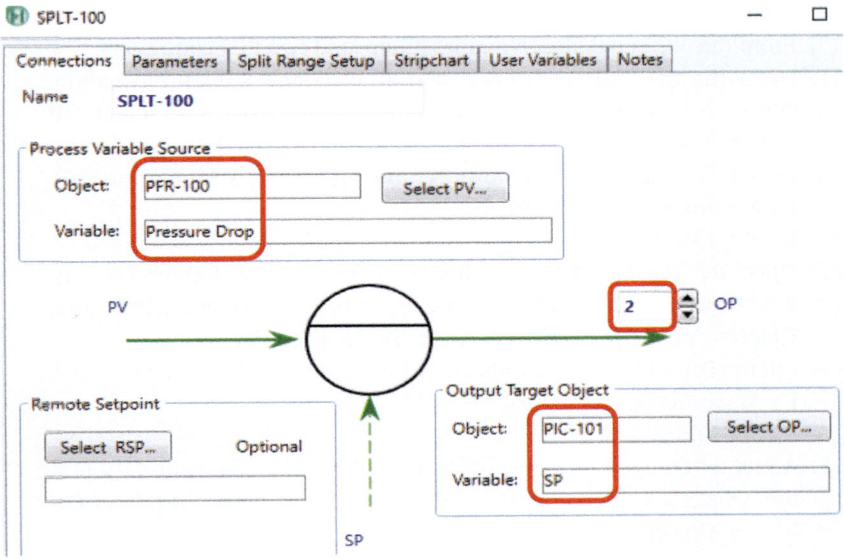

Figure 13.56 Split range controller, SPLT-100, connection tab, OP # 2.

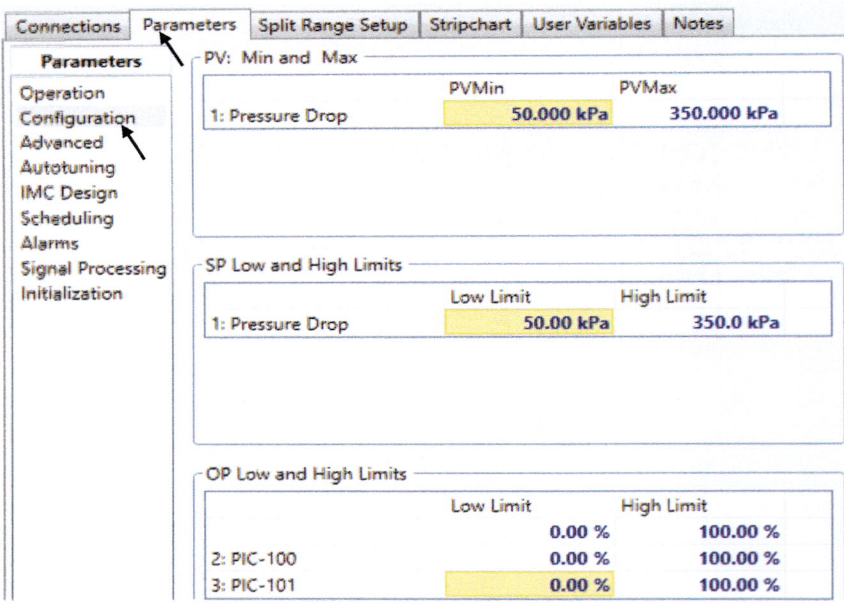

Figure 13.57 Split range controller, SPLT-100, parameters tab, configuration page.

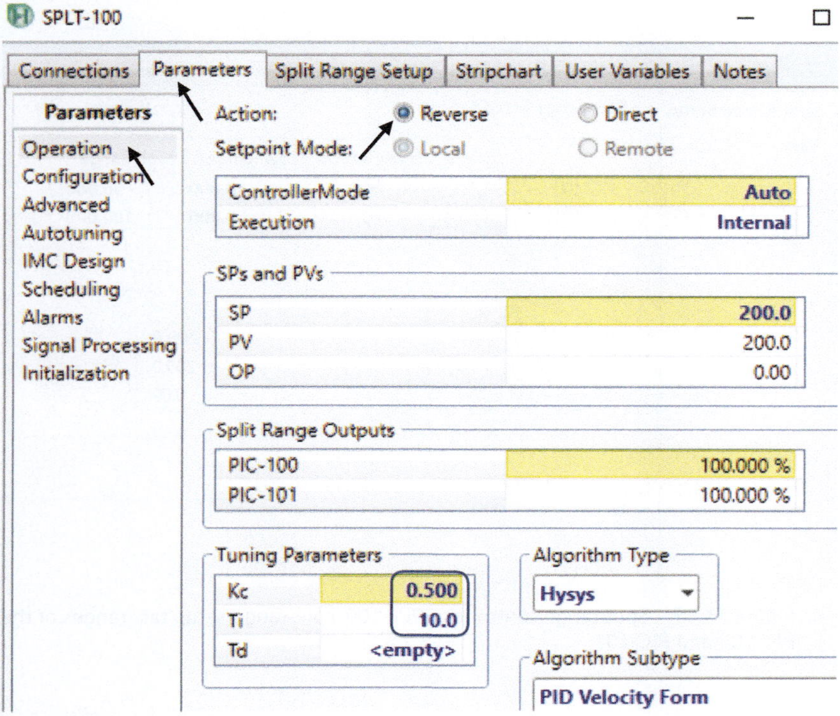

Figure 13.58 Split range controller, SPLT-100, parameters tab, operation page.

18) **Parameters** tab, **Operation** page; Action = Reverse, Controller Mode = Auto, SP = 200, $K_c = 2$, $T_i = 2$, Figure 13.58.
19) Switch to the **Split Range Setup** tab, Set the ranges of the PIC-100, and PIC-101 as in Figure 13.59.
20) Open the PIC-100 controller, Parameters tab, set the SP Mode to * Remote, and Mode to "Casc," Figure 13.60. Repeat the same procedure for PIC-101 controller, Figure 13.61.
21) The final PFD of the Split Range control of plug flow reactor of ammonia production is shown in Figure 13.62.
22) Click the **StripChart** tab of SPLT-100, Select (SP, PV, OP Only), Deactivate the OP_3, then click Display, Figure 13.63. Close the StripChart table
23) From the **Dynamics** tab of the Ribbon, click **Integrator** tab Integrator, and reduce the Acceleration to 0.02.

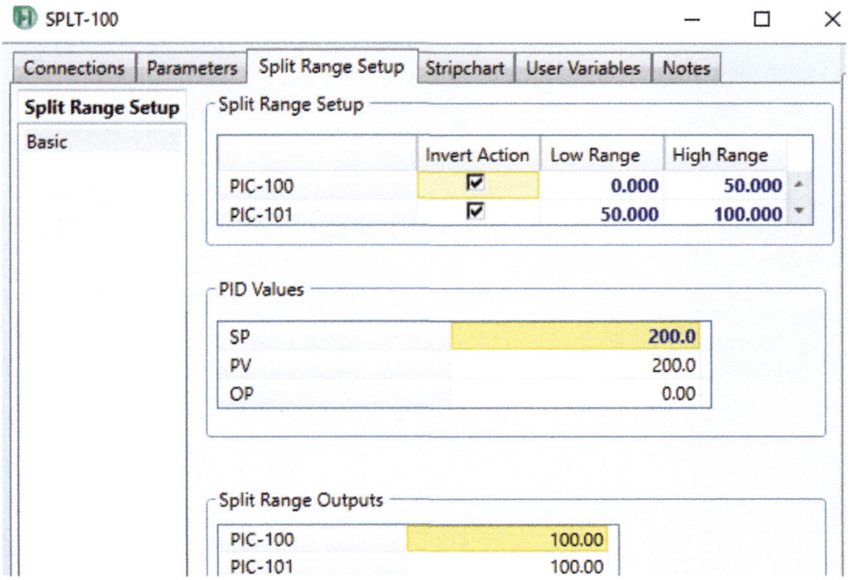

Figure 13.59 Split range controller, SPLT-100, split range setup tab, ranges of the PIC-100 and PIC-101.

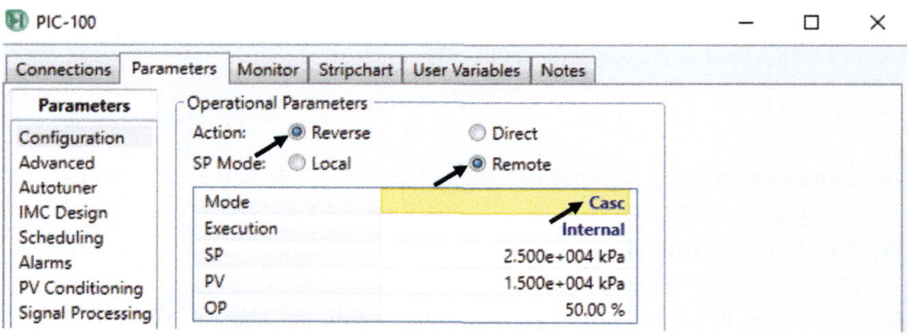

Figure 13.60 PIC-100 controller, parameters tab, setting the SP mode to * remote, and mode to "Casc."

Click the **Dynamic Mode** tab 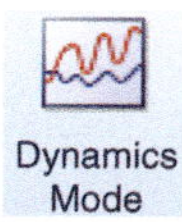

, from the new message, click Yes, then of the new window, click **Make Change** button. Click **Finish**.

24) Click **Dynamics Mode** again, from the new message, click Yes.

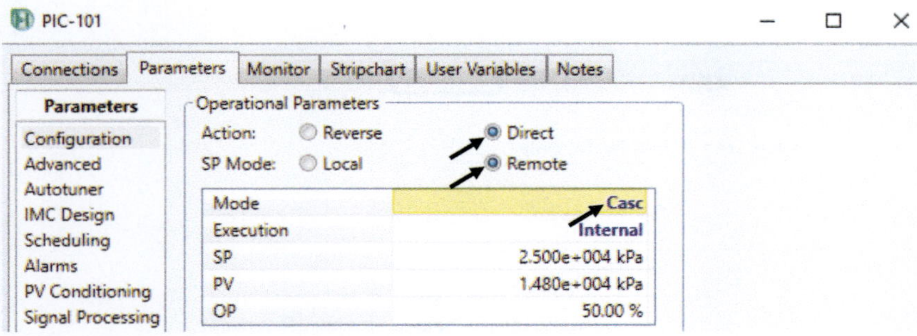

Figure 13.61 PIC-101 controller, parameters tab, setting the SP mode to remote, and mode to "Casc."

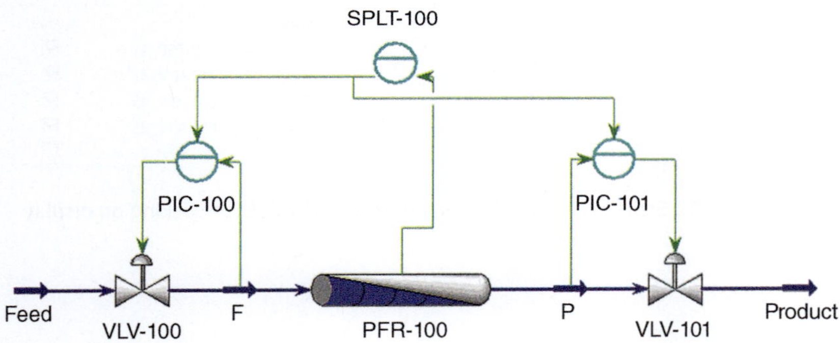

Figure 13.62 Final PFD of split range control of plug flow reactor of ammonia production.

25) Now we are in the dynamics mode, click **RUN** to display variable signals as in Figure 13.64.

It can be seen from Figure 13.64, that the initial positions of the outputs of controllers PIC-100 (1) and PIC-101 (2) are 0.0% and 100%, respectively, but when change the set point of the SPLT-100 (pressure drop of the reactor) from 200 to 204.5 kPa, the OP (output) of the PIC-100 (1) increases and the PIC-101 (2) decreases in order to achieve the requirement of set point.

13.4 Summary

As mentioned above, we can say that the controller may be simple, such as PID controller, or more complex, such as Ratio, Cascade, and Split Range controller. The first one can be easily created and operated and may serve

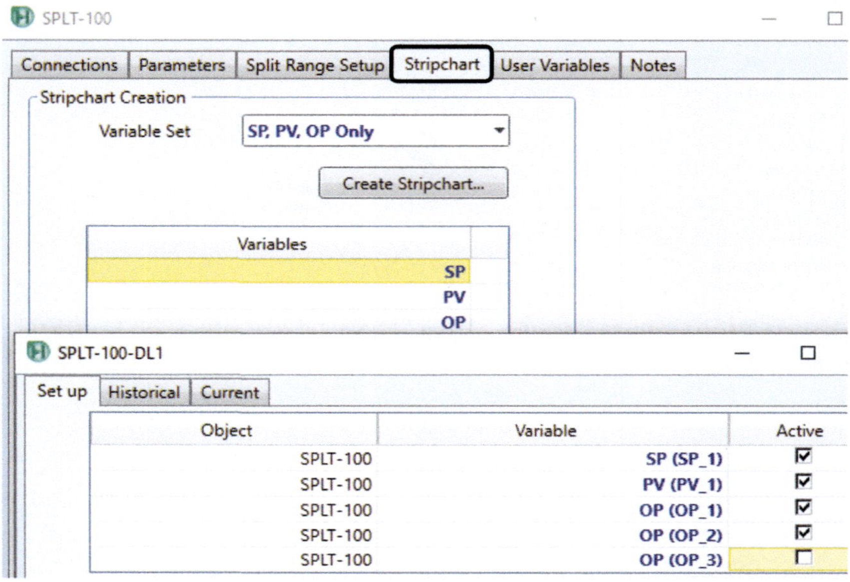

Figure 13.63 Stripchart tab, selecting (SP, PV, OP Only), then clicking on display tab.

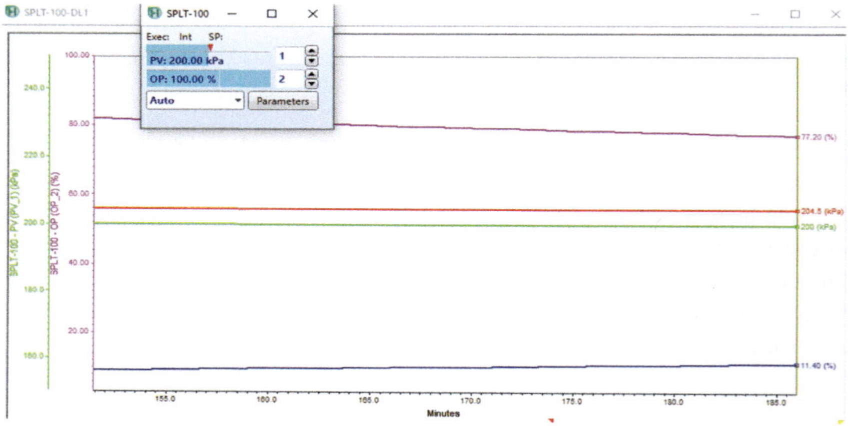

Figure 13.64 Screen of the stripchart of the split range controller after changing the set point from 200 to 204.5 kPa.

the required purpose, but it may be less accurate for more complex industrial operations and result in some delay in the response. The second type is more accurate and faster, but it is more expensive and complex than the first type.

Accurate representation of the behavior of systems in common dynamic conditions, like dead time, disturbances, valve failure, and feed characteristic changes, is made possible by the dynamic analysis capabilities found in commercial process simulators. For a process designer or operator, these kinds of scenarios are interesting since they enable modifying the control systems and anticipating emergency scenarios in a plant.

References

1 Seborg, D.E., Edgar, T.F., Mellichamp, D.A., and Doyle III, F.J. (2011). *Process Dynamics and Control*, 3e. John Wiley & Sons, Inc.

2 Stephanopoulos, G. (1985). *Chemical Process Control, An Introduction to Theory and Practice*, 1e. Prentice-Hall.

3 Dissinger, G., Levin, J., and Peers, Z. (2015). *Jump Start: Aspen HYSYS ® Dynamics V7.3, A Brief Tutorial (and Supplement to Training and Online Documentation)*. Aspen Technology, Inc. AspenTech®, aspenONE®.

4 (2006). *Dynamic Modeling Manual, Aspen Hysys*. Aspen Technology, Inc.

5 Coughanowr, D.R. and LeBlanc, S.E. (2009). *Process Systems Analysis and Control*, 3e. McGraw-Hill.

6 Levin, J. (2013). *Jump Start: Aspen HYSYS ® Dynamics V8, A Brief Tutorial (and Supplement to Training and Online Documentation)*. Aspen Technology, Inc. AspenTech®, aspenONE®.

7 (2005). *HYSYS® 2004.2, Tutorials and Applications*. Aspen Technology, Inc.

8 Luyben, W.L. (2002). *Plantwide Dynamic Simulator in Chemical Processing and Control*. Marcel Dekker, Inc.

9 Jenkins, S. (2018). Cyclohexane Production from Benzene and Hydrogen. *Chemical Engineering*; New York 125 (2): 28.

10 İnce, E. (2011). Kinetic of esterification of ethyl alcohol by acetic acid on a catalytic resin. *Pamukkale University Journal of Engineering Sciences* 8 (1).

11 Mariano Martín Martín (2020). *Introduction to Software for Chemical Engineers*, 2e. CRC Press, Inc.

14

Economic Evaluation

In the Aspen HYSYS the economic evaluation will be estimated through multi-steps; the user must enter the prices of the material streams, choose the type of the utility, and choose the type of the equipment used [1]. After that the software will determine the sizing of equipment, calculate the capital cost, operating cost, utility cost, and installing cost [2].

14.1 Problem Statement

One can take the gas processing plant (Section 13.2) as an example to estimate the economic evaluation.

14.2 Simulation

Open the case of gas processing (steady state) that was constructed in the previous chapter, Section 13.2. Delete the Balance, Adjust operations, and Sale DP stream. The process flow diagram will be as shown in Figure 14.1.

A description of the process was mentioned in Section 13.2.1.

The procedure for the Economic Evaluation can be listed in the following steps [1–3]:

1) Entering the buying and selling prices of the feed and product streams: Double-click the Feed1 stream and go to the **Cost Parameters** form under the **Worksheet** tab, select **Mass Flow** for **Stream Price Basis,** and enter **0.2 Cost/kg** for **Stream Price**, Figure 14.2. By the same way, enter the prices of the following streams:
   Feed2 = 0.2 Cost/kg,
   Sale gas = 0.8 Cost/kg,

Chemical Process Simulations using Aspen Hysys, First Edition. Khalid W. Hameed.

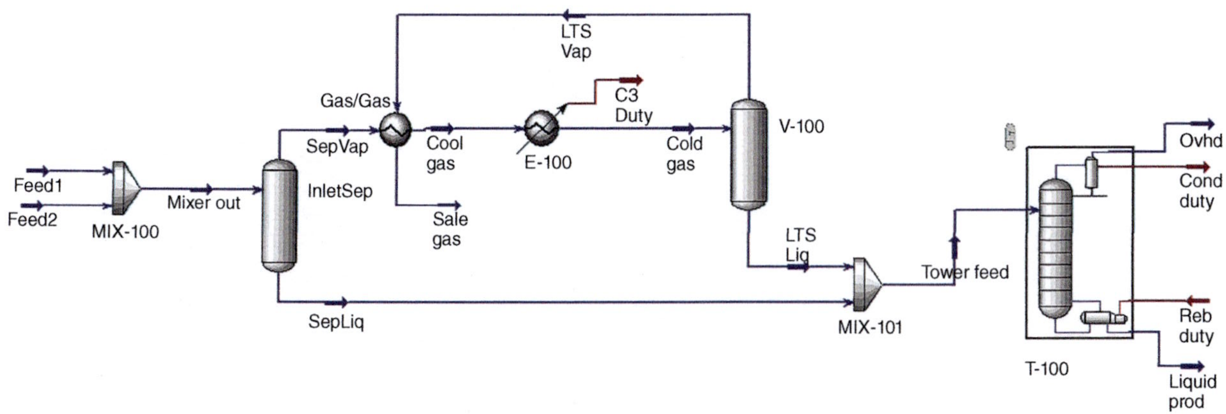

Figure 14.1 PFD of the gas processing where the economic evaluation will be estimated.

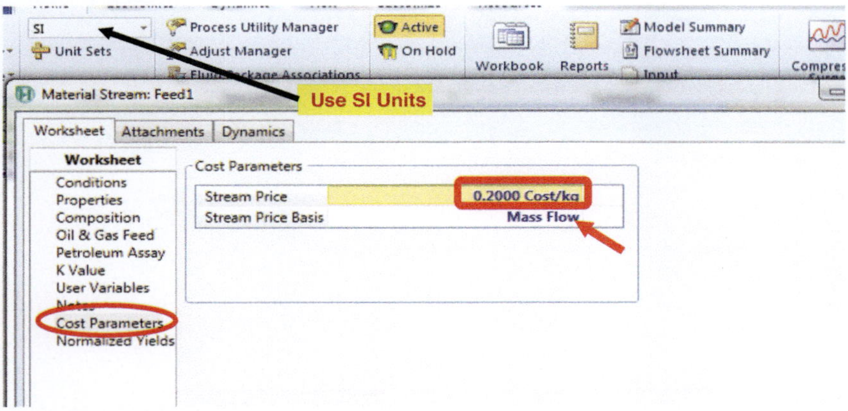

Figure 14.2 Entering the price of the Feed1 stream.

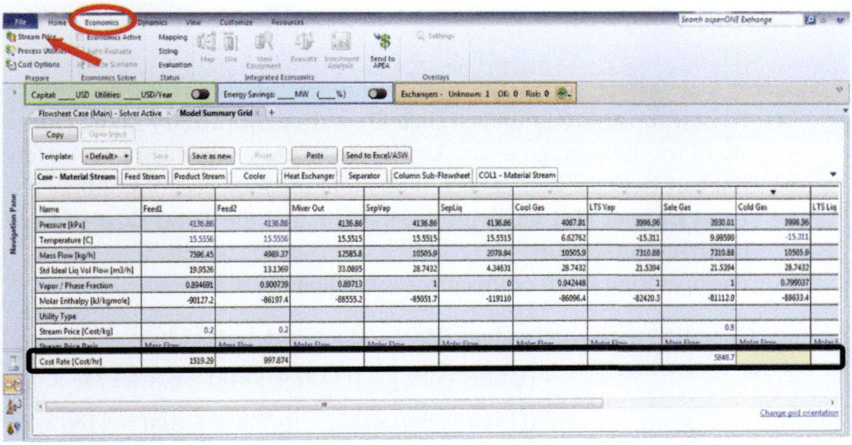

Figure 14.3 Presenting the prices of the material streams.

Liq-Prod = 0.8 Cost/kg,
Ovhd = 0.5 Cost/kg.

2) To view the total stream costs, go to the **Economics** tab in the ribbon and select **Stream Price**, Figure 14.3.
3) Estimating the cost of utilities. Double-click the energy stream **C3 Duty**, from the **Utility Type**, select **Refrigerant 1** (Freon 12).
   By the same way above:
   Cond-Q = Refrigerant 1,
   Reb-Duty = LP Steam (Low pressure steam).

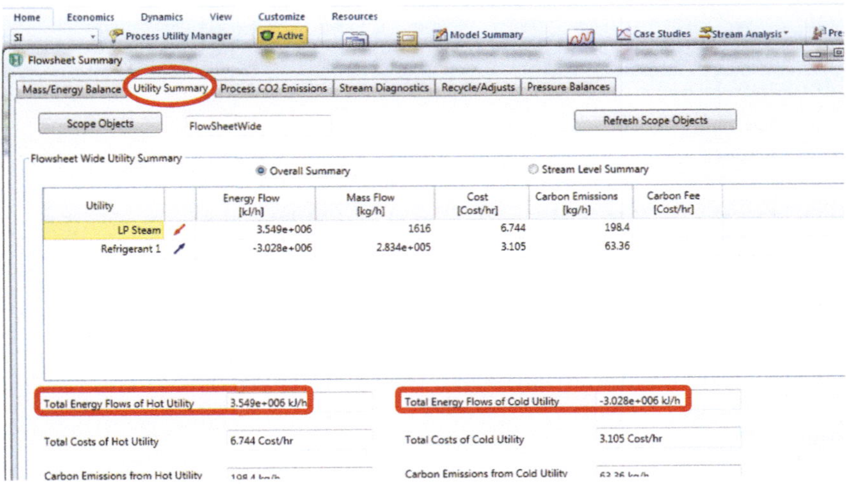

Figure 14.4 Utility summary/cost of each utility and the total costs of utilities.

4) To view the utility summary, click **Flowsheet Summary** in the **Home** tab of the Ribbon, go to the **Utility Summary** tab, Figure 14.4. Here you can view the cost of each utility and the total cost of utilities. The **Total Cost of Hot Utilities** is **$6.744/h**, and the **Total Cost of Cold Utilities** is **$3.105/h**.
5) The operating profit of this process is equal to:

$$\begin{aligned}\text{Operating profit} &= [\text{Product sales}] - [\text{Raw materials costs}] \\ &\quad - [\text{Utility costs}] = [(1427 + 5848) \times 0.8 + 1746 \times 0.5] \\ &\quad - [(1519 + 998) \times 0.2] - [6.744 + 3.105] = 6180\ \$/\text{h}.\end{aligned} \tag{14.1}$$

6) From the **Economic** tab, click **Cost Options**, from the new window, change the Operating Life of Plant to 20 years, and Operational Year to 7920 h/yr, Figure 14.5. When finishing, close the **Cost Options** window.
7) The next step is to evaluate the capital costs of the process.
Go to the **Economics** tab, Figure 14.6, activate the **Economics Active** radio, and wait some minutes. This will enable the **Economic Analysis** functionality in Aspen HYSYS.
8) Click the **Map** button. From the new window, press **OK** to continue. The **Map Preview** window will appear that allows you to change the mapping for certain unit operations, Figure 14.7. The default selection

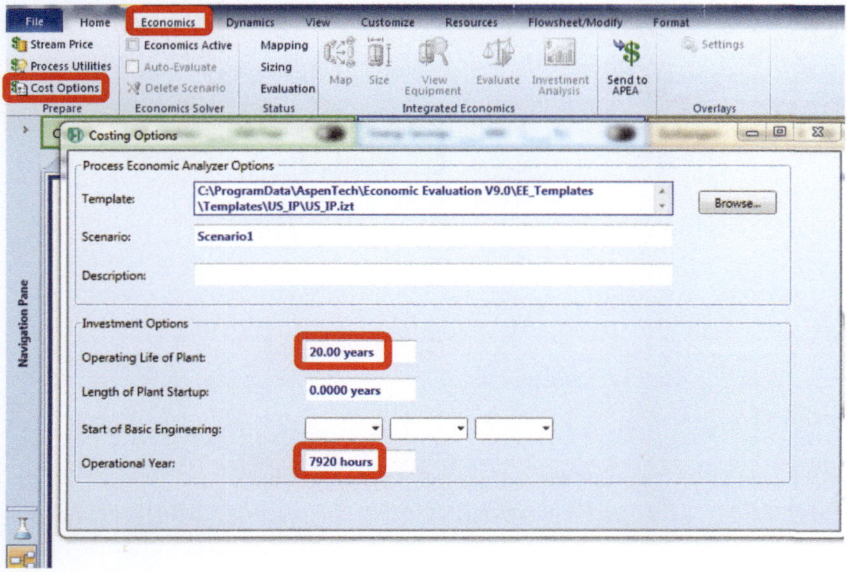

Figure 14.5 Setting the operating life of plant and operational year.

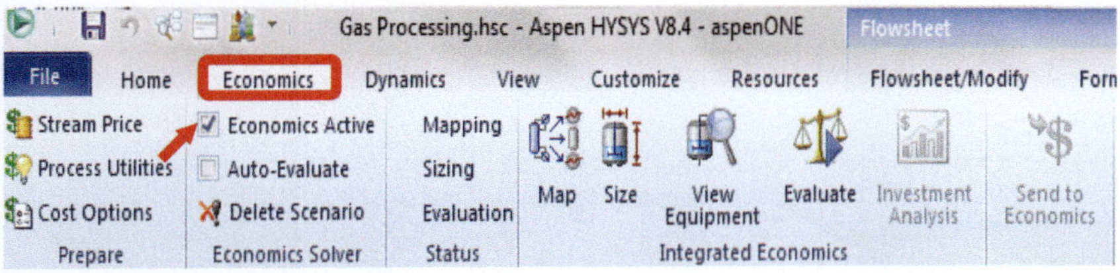

Figure 14.6 Activating the economics active radio.

of the type of equipment is presented under **Description**, and it can be adopted or select another type as follows:

The map function is a key step in determining project scope and cost. This function enables unit operations from the simulation model to be mapped to "real-world" equipment so that preliminary equipment sizing can be performed.

9) Select **Gas/Gas (HEATX)** and then select **Heat exchanger**, heaters. From drop-list menu under the **Equipment Type**, click OK, Figure 14.8. From the new group, select **Heat Exchanger**, OK. From the new menu, select **U-Tube shell and Tube heat exchanger**, Figure 14.9, OK.

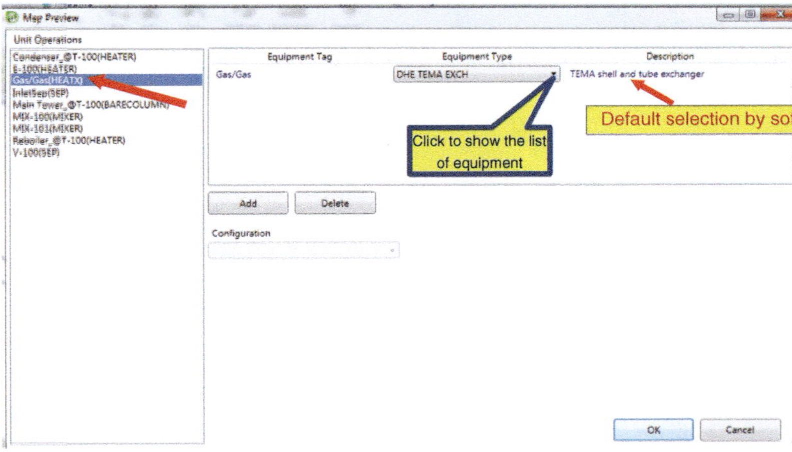

Figure 14.7 Selecting the gas/gas (HEATX) from the map.

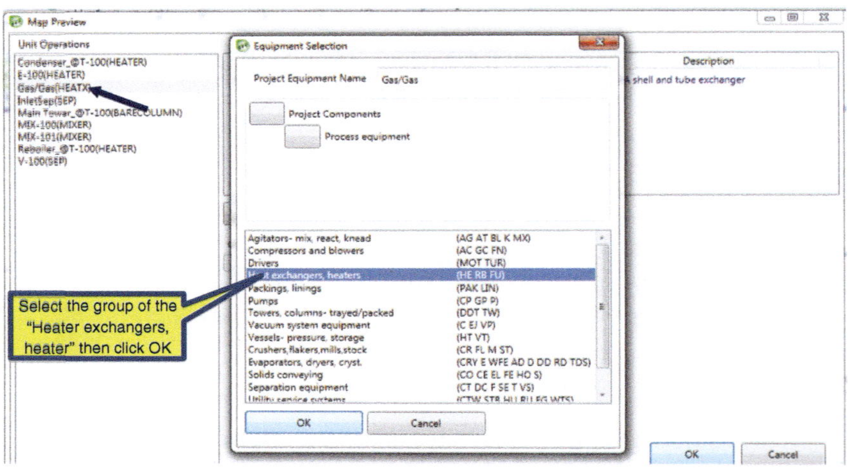

Figure 14.8 Procedure of selecting the type of unit operation in the map.

10) Again, as above: for the main tower section of the distillation column, **Main_TS@T-100(BARECOLUMN)** → Towers, Columns-Trayed/Packed → Tower-Single diameter → Packed Tower.
For reboiler of the distillation column,
Reboiler_@T-100(HEATER) → Heat exchangers, heaters → Reboiler → Kettle type reboiler with floating head.
Leave the other equipment as default.

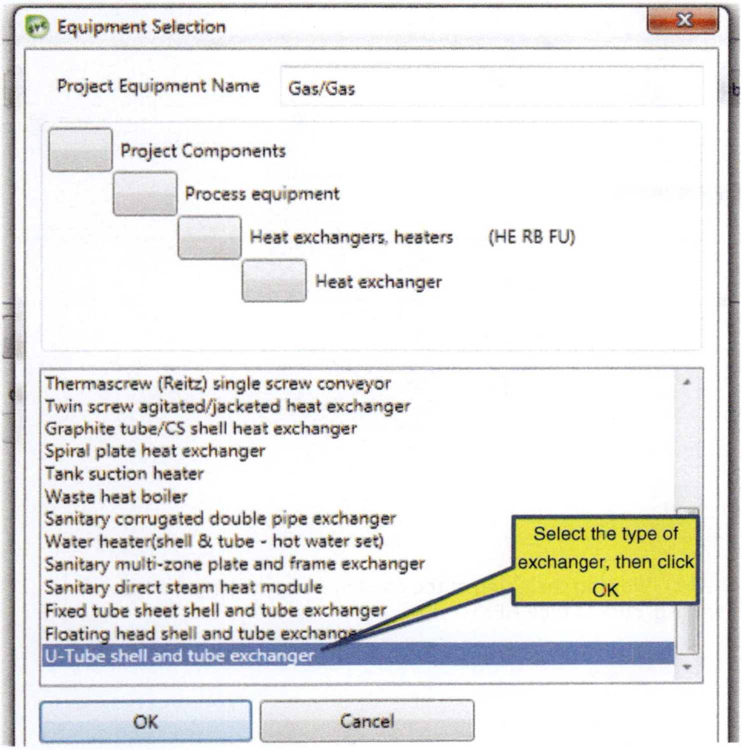

Figure 14.9 Procedure of selecting the type of unit operation in the map, gas/gas exchanger (continued).

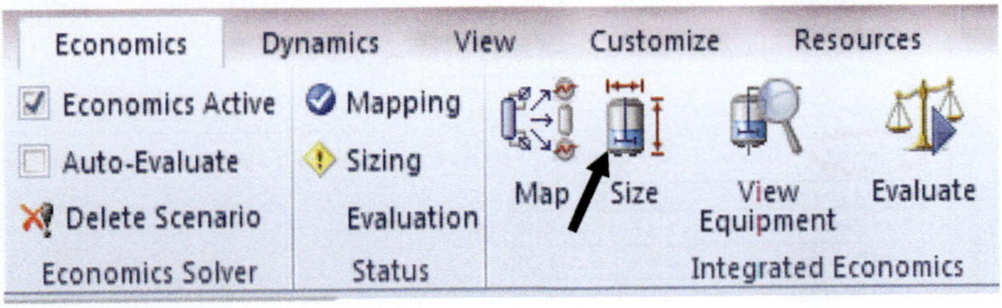

Figure 14.10 Selecting the sizing icon for executing the sizing of the equipment.

11) Next, click the **Size** icon, Figure 14.10, wait some minutes. The sizing of equipment is complete.
12) Select **View Equipment** to view the results of the sizing.

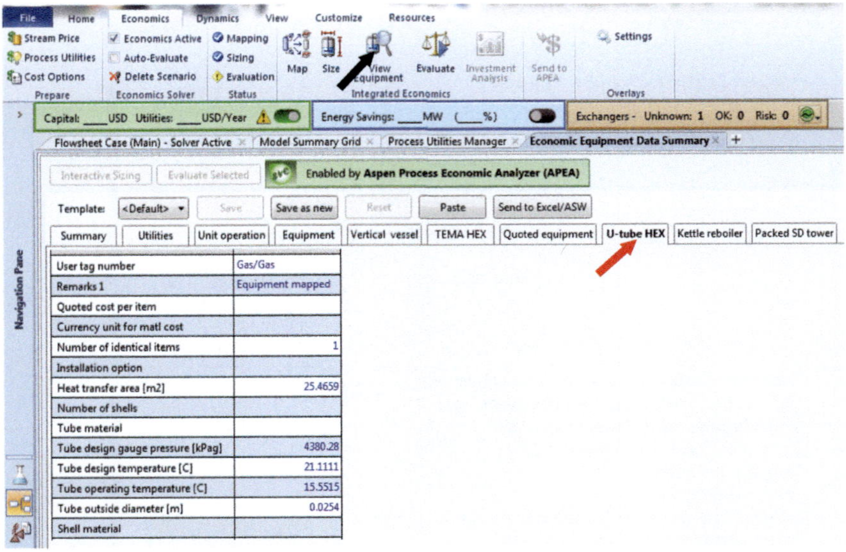

Figure 14.11 Viewing the sizing of the equipment by clicking the type of equipment (e.g. click U-tube HEX).

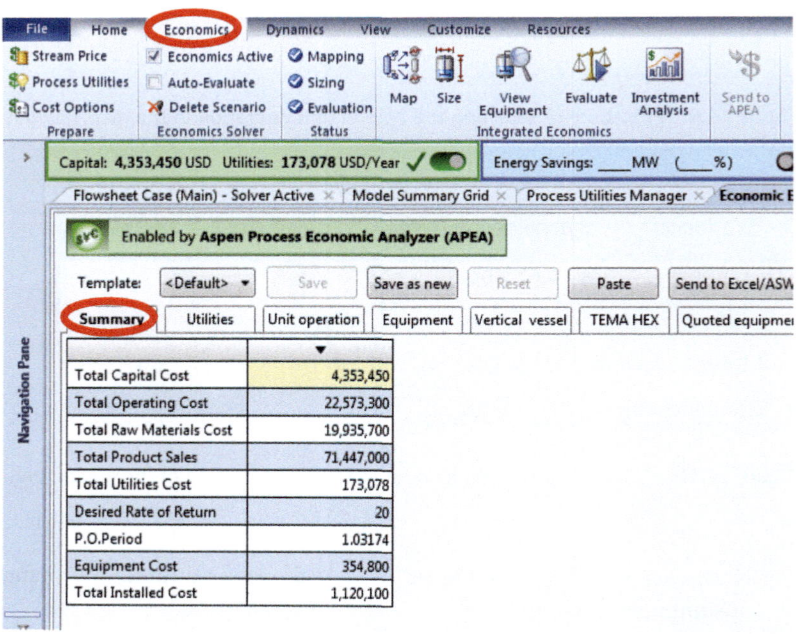

Figure 14.12 Summary results of economic evaluation.

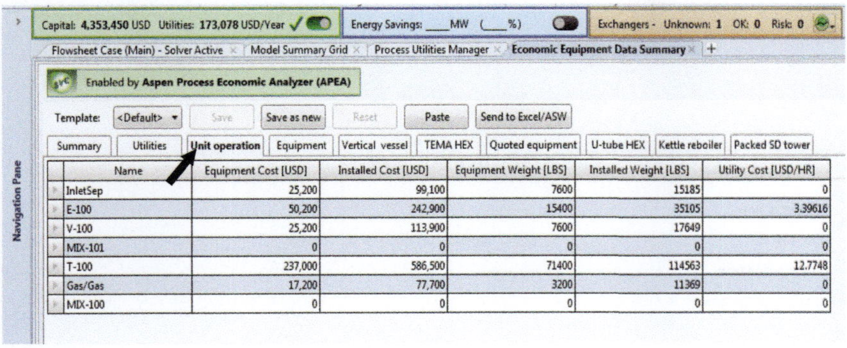

Figure 14.13 Details of cost of unit operation.

The **Economic Equipment Data Summary** will open. The sizing of each equipment can be seen by clicking its icon, e.g. click the **U-tube HEX** to view the details of Gas/Gas Exchanger, Figure 14.11.

13) The software is now ready to be evaluated. Click the **Evaluate** button in the Ribbon, and the economic engine will perform the analysis. It may take a few moments.
14) Click **View Equipment**, and go to the **Equipment** tab to view any errors that occurred during the evaluation. These errors will tell you what inputs or changes are required in order to cost the simulation more realistically. The summary of the economic evaluation is ready now, Figure 14.12. For more details, one can see from the other tabs, such as the **Unit Operation** tab, Figure 14.13.
15) Click the **Investment Analysis** button, wait few moments. This will open up a Microsoft Excel spreadsheet that summarizes the results. In the Excel spreadsheet there will be the following sheets [1, 2]:
Run Summary, Executive Summary, Cash Flow, Project Summary, Equipment, Utility Summary, Utility Resource Summary, Raw Material Summary, and Product Summary.

The **Executive Summary** sheet is a very useful sheet which displays the project name, capacity, plant location, description, scheduling, and investment information. This is shown in Figure 14.14.

The **Cash Flow** sheet is also useful and displays various costs and assumptions that went into making the economic estimations, Figure 14.15.

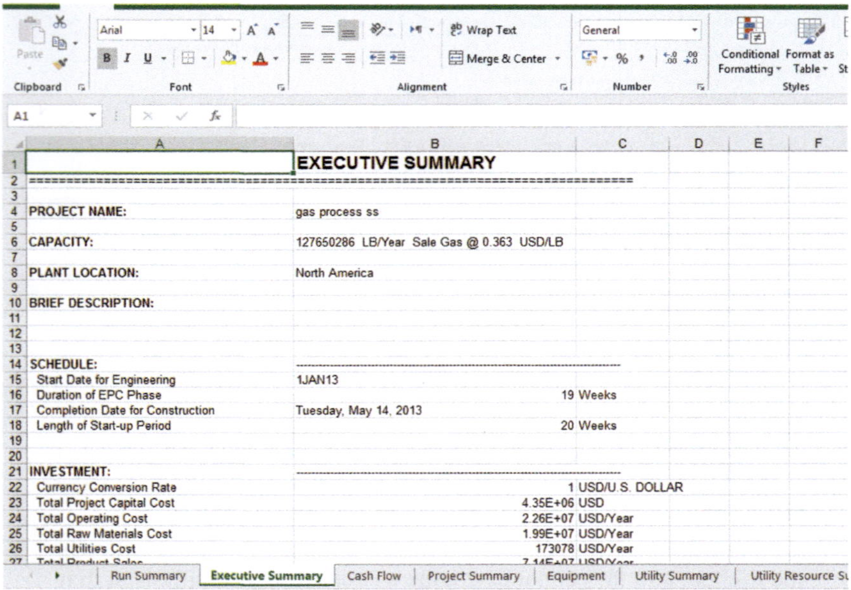

	A	B	C
1		**EXECUTIVE SUMMARY**	
2	================================	================================	
3			
4	**PROJECT NAME:**	gas process ss	
5			
6	**CAPACITY:**	127650286 LB/Year Sale Gas @ 0.363 USD/LB	
7			
8	**PLANT LOCATION:**	North America	
9			
10	**BRIEF DESCRIPTION:**		
11			
12			
13			
14	**SCHEDULE:**	------------------------------	
15	Start Date for Engineering	1JAN13	
16	Duration of EPC Phase	19	Weeks
17	Completion Date for Construction	Tuesday, May 14, 2013	
18	Length of Start-up Period	20	Weeks
19			
20			
21	**INVESTMENT:**	------------------------------	
22	Currency Conversion Rate	1	USD/U.S. DOLLAR
23	Total Project Capital Cost	4.35E+06	USD
24	Total Operating Cost	2.26E+07	USD/Year
25	Total Raw Materials Cost	1.99E+07	USD/Year
26	Total Utilities Cost	173078	USD/Year
27	Total Product Sales	7.14E+07	USD/Year

Figure 14.14 Executive summary sheet obtained from investment analysis.

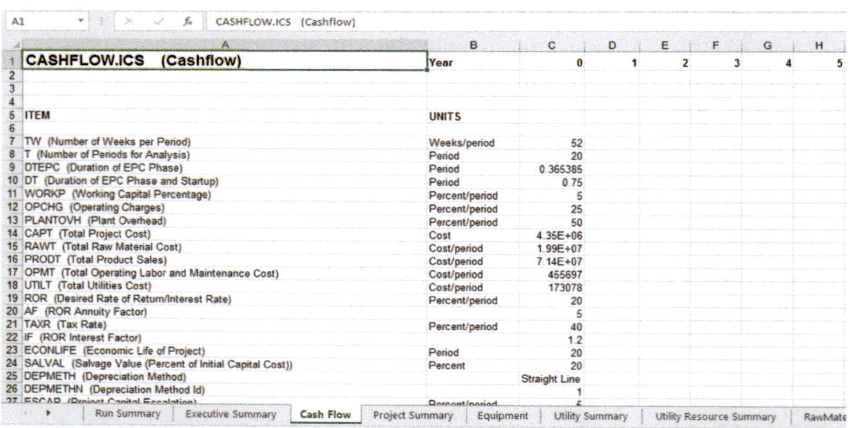

	A	B	C	D	E	F	G	H
1	**CASHFLOW.ICS (Cashflow)**	**Year**	0	1	2	3	4	5
2								
3								
4								
5	**ITEM**	**UNITS**						
6								
7	TW (Number of Weeks per Period)	Weeks/period	52					
8	T (Number of Periods for Analysis)	Period	20					
9	DTEPC (Duration of EPC Phase)	Period	0.365385					
10	DT (Duration of EPC Phase and Startup)	Period	0.75					
11	WORKP (Working Capital Percentage)	Percent/period	5					
12	OPCHG (Operating Charges)	Percent/period	25					
13	PLANTOVH (Plant Overhead)	Percent/period	50					
14	CAPT (Total Project Cost)	Cost	4.35E+06					
15	RAWT (Total Raw Material Cost)	Cost/period	1.99E+07					
16	PRODT (Total Product Sales)	Cost/period	7.14E+07					
17	OPMT (Total Operating Labor and Maintenance Cost)	Cost/period	455697					
18	UTILT (Total Utilities Cost)	Cost/period	173078					
19	ROR (Desired Rate of Return/Interest Rate)	Percent/period	20					
20	AF (ROR Annuity Factor)		5					
21	TAXR (Tax Rate)	Percent/period	40					
22	IF (ROR Interest Factor)		1.2					
23	ECONLIFE (Economic Life of Project)	Period	20					
24	SALVAL (Salvage Value (Percent of Initial Capital Cost))	Percent	20					
25	DEPMETH (Depreciation Method)		Straight Line					
26	DEPMETHN (Depreciation Method Id)		1					
27	ESCAP (Project Capital Escalation)	Percent/period	5					

Figure 14.15 Cash flow sheet.

14.3 Summary

With the use of the economic analysis tool, Aspen HYSYS may swiftly generate initial, rough estimations of the size and expense of an activity. When comparing various process designs to determine which one has the best likelihood of being successful, this is quite helpful. If an activity appears profitable at this level of research, costing engineers will take the basic design and fine-tune it into a more sophisticated costing program, such as the Aspen Capital Cost Estimator. When employing a conceptual design produced by a process simulator, the ability to properly estimate associated costs is vitally important and could mean the difference between a firm's success or failure.

References

1 (2012). *Ammonia Synthesis with Aspen HYSYS® V8.0, Part 3 Process Economic Analysis.* Aspen Technology, Inc.

2 (2008). *Aspen Process Economic Analyzer V7.0 User Guide.* Aspen Technology Industry.

3 McCarthy, D., Beck, R., and Dissinger, G. (2011). *Getting Started with Relative Economics in Aspen HYSYS.* Aspen Technology Inc.

15

Plant-Wide Projects

15.1 Cyclohexane Production Plant

The hydrogenation of benzene to produce cyclohexane is a well-known reaction wherein one mole of benzene reacts with three moles of hydrogen in the presence of a suitable hydrogenation catalyst to form one mole of cyclohexane. The following paragraphs describe a process for the production of cyclohexane from benzene and hydrogen that involves liquid-phase hydrogenation of benzene in the presence of a nickel-based catalyst. Figure 15.1 presents a simplified flow diagram of the process. Initially, benzene is fed to the primary reactor along with fresh and recycled hydrogen. The hydrogenation reaction is carried out in a bubble column reactor in the presence of a nickel catalyst (reactor is simulated as conversion reactor where the conversion is 98% of benzene), [1].

The operating conditions of benzene feed and hydrogen feed are 38°C, 2310 kPa, 100 kmol/h, and 100% benzene, and 50°C, 2310 kPa, 310 kmol/h, and (0.98 H_2, 0.005 N_2, and 0.015 CH_4), respectively. These two feed streams are mixed together and heated to 150°C in a Heat-Exchanger (H.E.1), then sent to the reactor, where the reaction is adiabatic. The vapor from the reactor is used to heat the feed F1 in H.E.1, then cooled by cold water (20°C, 140 kPa, 100% water) in the Heat Exchanger H.E.2 to 50°C. The output of the cold water temperature is 40°C. The pressure drops of the reactor and VLV-100 are 100 kPa each [2].

Distillation T-100: Number of stages = 11, Condenser type: Full reflux, the pressures of the reboiler and condenser are 1450 and 1380 kPa, respectively, reflux ratio = 0.9 by mole, and overhead rate = 3.4 kmol/h.

The outlet pressure of compressor = 2310 kPa.

Chemical Process Simulations using Aspen Hysys, First Edition. Khalid W. Hameed.

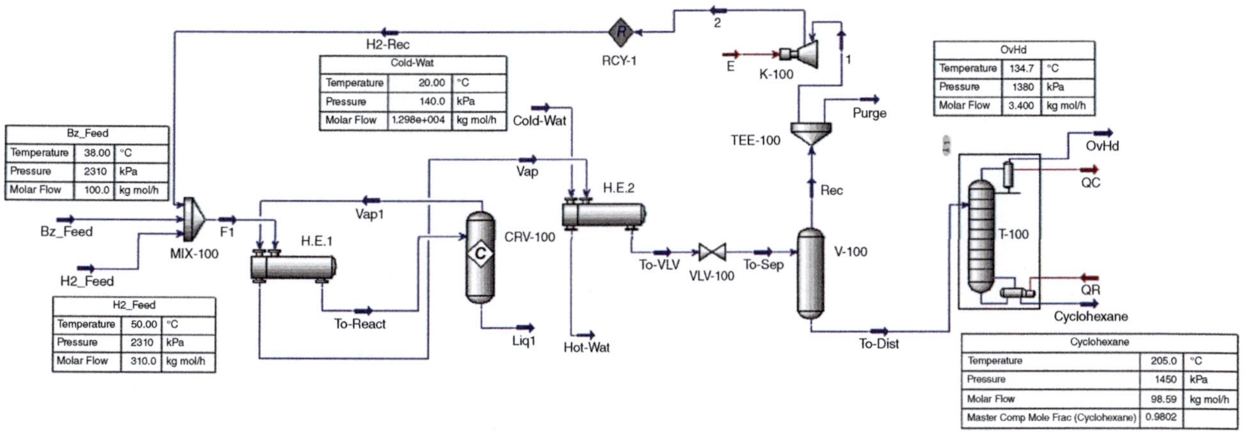

Figure 15.1 PFD of the cyclohexane production plant.

The split ratio of TEE-100: stream-1 = 0.7.

The pressure drops of H.E.1 in the shell and tube are 30 and 50 kPa, respectively.

The pressure drops of H.E.2 in the shell and tube are 20 kPa each.

15.1.1 Simulation

1) Open a new case of Aspen HYSYS and enter the following components: Benzene, Cyclohexane, Hydrogen, Nitrogen, Methane, and Water.
   Fluid Package: Peng–Robinson.
2) Construct the reaction set, type Conversion, where the reaction is:

$$\text{Bz} + 3\text{H}_3 \rightarrow \text{C}_6\text{H}_{12}\ (\text{conversion} = 98\%\ \text{with respect to Bz}) \qquad (15.1)$$

3) Enter the **Simulation** environment and Construct the two feed streams:
   Bz-Feed with the conditions (38°C, 2310 kPa, 100 kmol/h, and 100% Benzene).
   H_2-Feed with the conditions (38°C, 2310 kPa, 310 kmol/h, and mole fractions: 0.98 H_2, 0.005 N_2, and 0.015 CH_4).
4) Mix the two feed streams by Mixer (MIX-100), where the outlet stream is F1.
5) Insert a Heat Exchanger, H.E.1, and enter the streams as follows:
   Tube side: Inlet = Vap1, Outlet = Vap,
   Shell side: Input = F1, Output = To-React.
   Pressure drops of tube side and shell side are 50 and 30 kPa, respectively.
   Temperature of To-React stream is 150°C.
6) Insert a conversion reactor and connect streams as follows:
   Inlet = To-React, Vapor Outlet = Vap1, and Liquid Outlet = Liq1.
   Pressure drop = 100 kPa. For the Reaction tab, Select **Set-1**.
7) Insert another Heat Exchanger, H.E.2 and enter the streams as follows:
   Tube side: Inlet = Cold wat (20°C, 140 kPa, 100% water), Outlet = Hot wat (set its temperature = 40°C).
   Shell side: Inlet = Vap, Outlet = To-VLV (set its temperature = 50°C)
   Set the pressure drops of tube side and shell side as 20 kPa each.
8) Insert a valve (VLV-100), Input = To-Valve, and output = To-Sep.
   Pressure drop of valve = 100 kPa.
9) Insert a separator, V-100: Inlet = To-Sep, Vapor Outlet = Rec, Liquid Outlet = To-Dist.
10) Insert a Distillation column, T-100, with the following conditions:
    Number of stages = 11,
    Condenser type = Full reflux,

Condenser energy stream = QC, Reboiler energy stream = QR,
Overhead vapor stream = OVhd, Bottom stream = Cyclohexane,
Pressures of condenser and reboiler are: 1380 and 1450 kPa, respectively,
Reflux ratio = 0.9, Overhead vapor rate = 3.4 kmol/h.
Click **Done**, then click **Run.**

11) Insert TEE-100, Inlet = Rec stream, Outlet = 1, and Purge streams. Flow ratio = 0.7 for stream-1.
12) Insert a Compressor: Inlet = stream-1, Outlet = stream-2, pressure of stream-2 is setting to pressure of Bz-Feed (2310 kPa) using Set-Operation.
13) Add a Recycle operation, Inlet = stream-2, Outlet = stream H_2-Rec.
14) Connect the H_2-Rec with the mixer MIX-100

Now, the results of any stream/unit operation can be seen.

15.2 Production of Dimethyl_ether

Dimethyl_ether, DME, is produced from methanol according to the following reaction [3]:

$$2\text{Methanol} \rightarrow \text{DME} + H_2O \tag{15.2}$$

The production process is performed as shown in Figure 15.2. The raw material, stream 1 (25°C, 1 atm, 100% methanol), is pumped to 16.8 atm by Pump (P-100). Stream 3 is combined with stream 16 to the vessel (V-100). Stream 17 has a flow rate of 270 kmol/h. Stream 4 is sent to the heater (E-100), where it is heated to a temperature of 250°C before being sent to a packed bed reactor (conversion reactor, CRV-101) to form DME. The reaction is achieved adiabatically and the reactor effluent is cooled to 100°C in the Heat Exchanger (E-101), by water (20°C, 200 kPa, 2000 kmol/h) inlet to shell. The process fluid is then throttled to 10 atm through (VLV-101) before entering a distillation column (T-100). Here, the dimethyl ether is separated from the other components as distillate, Stream 9. The bottom product, Stream 10, is throttled to 6.9 atm through (VLV-102) and sent to a distillation column (T-101) where the methanol is separated from the top, Stream 12, which is pumped and recycled to mix with stream 3. Pressure of stream 15 is equal to the pressure of stream 3 using Set operation (SET-1).

(***Hint***: *Complete the all flowsheet, then add recycle to the MIX-100*)

Fluid package: NRTL, from Binary Coeff. click **Unknowns only**.

Data:

- VLV-100: $\Delta P = 170$ kPa,
- E-100: $\Delta P = 11$ kPa,
- Reaction (conversion 30%),

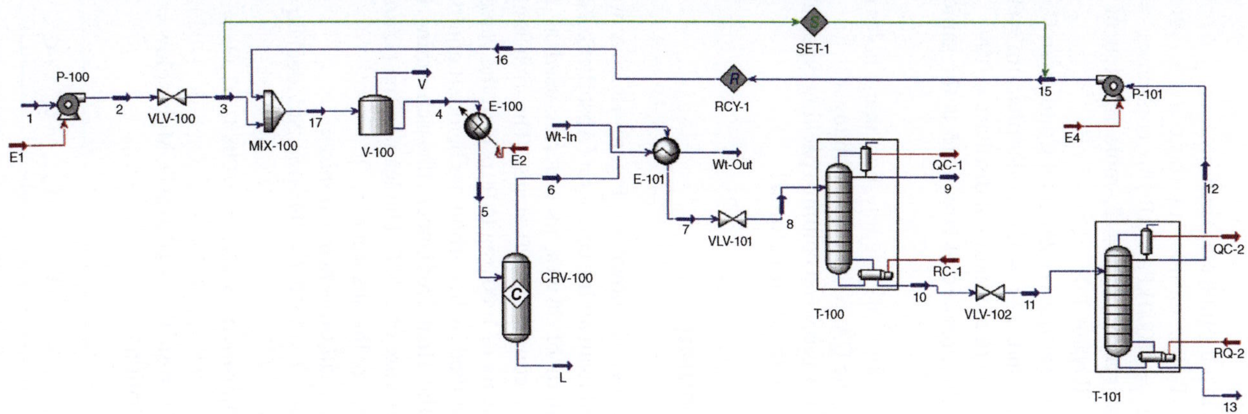

Figure 15.2 PFD of dimethyl_ether production plant.

- Rxn Phase = Vapor phase,
- CRV-100: $\Delta P = 9$ kPa,
- H.E. E-101: $\Delta P = 30$ kPa for tube and shell,
- Distillation column T-100: number of trays = 15, Total Condenser, condenser pressure = 1000 kPa, ΔP in condenser = 0, reboiler pressure = 1010 kPa, reflux ratio = 2, mole fraction of DME in distillate = 0.99999 (it will occur 1.0).
- Distillation T-101: number trays = 25, Total Condenser, Condenser pressure = 680 kPa, ΔP in condenser = 0, Reboiler pressure = 690 kPa, Reflux ratio = 2.7, mole fraction of methanol in distillate = 0.99
- SET-1: For Keeping the pressure of stream-15 is the same of stream-30 pressure.

1) If the Distillation (T-100) is of sieve tray, how many holes in each tray?
2) What is the value of the UA in the Heat Exchanger?
3) If the distillation T-101 is packed column (packing type Pall, 1-in.), what is the height and diameter of the column?

15.3 Gas Sweetening

A gas-sweetening process is shown in Figure 15.3. A water-saturated natural gas stream is fed to an amine contactor (Absorber). Diethanolamine (DEAmine) at a strength of 28 wt% in water is used as the absorbing medium. The absorber consists of 20 stages. The rich amine is flashed from the absorber pressure of 1000 psia to 90 psia in the valve (VLV-100) to release most of the absorbed hydrocarbon gas before it enters the lean/rich amine exchanger. In the Heat Exchanger, the rich amine is heated to a distillation feed temperature of 200°F. The lean amine (Lean From L/R) is cooled and recycled back to the absorber [4, 5].

Data: Fluid package: *Acid Gas–chemical solvent*

Stream of DEA to Cont: $T = 95$°F, $P = 995$ psia, Std ideal liq vol flow = 190 USGPM,

Mass fraction composition: CO_2 = 0.0018, H_2O = 0.7187, and DEA = 0.2795.

Stream of Sour Gas: $T = 86$°F, $P = 1000$ psia, Molar Flow = 25 MMSCFD, Mole fraction composition:

N_2	CO_2	H_2S	C_1	C_2	C_3	i-C_4	n-C_4	i-C_5	n-C_5	n-C_6	n-C_7	H_2O
0.0016	0.0413	0.0172	0.8692	0.0393	0.0093	0.0026	0.0029	0.0014	0.0012	0.0018	0.0072	0.005

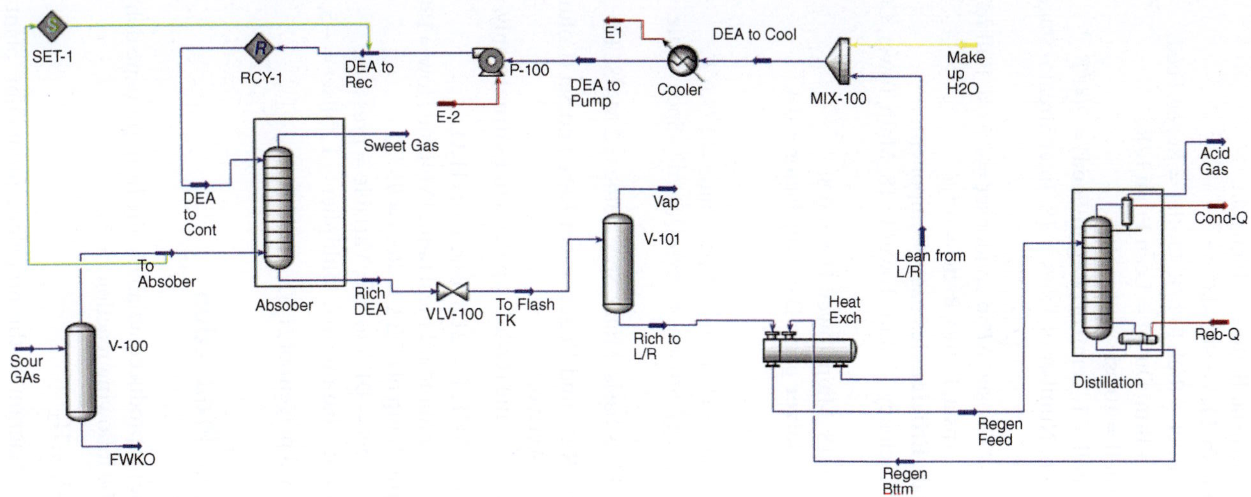

Figure 15.3 PFD of the gas sweetening.

- Absorber: Number of trays = 20,
- Pressures: Top = 995 psia, Bottom = 1000 psia,
- Pressure of (**To Flash TK**) = 90 psia,
- Heat Exch: **Tube**: Inlet = Rich to L/R, Outlet = Regen Feed,
- **Shell**: Inlet = Regen Btm, Outlet = Lean From L/R,
- ΔP for tube and shell = 10 psi,
- Tube Passes per shell = 1, Temp of (**Regen Feed**) = 200°F
- Distillation Column: Number of Trays = 18, Inlet Stream/Stage = Regen Feed/4,
- Condenser Press = 27.5 psia, ΔP in condenser = 2.5 psi, Full reflux,
- Reboiler Press = 31.5 psia, Reflux Ratio = 0.5,
  - **Parameters** tab; **Efficiencies** page: Component:
    Click Specify Efficiency 1_Main Tower to 18_Main Tower, CO_2 = 0.15, and
    1_Main Tower to 18_Main Tower, H_2S = 0.8,
    Parameters tab, **Solver** page: Damping factor = 0.4,

Column Specifications:

1) Column Temp: Stage = Condenser, Spec Value = 179.6°F,
2) Column Duty: Energy Stream = Reb-Q@col2, Spec value = 1.356e7 Btu/h,
3) Ovhd Vap Rate: Flow basis: Molar, Spec Value = 2 MMSCFD

Delete the **Reflux Rate** and **Regen Bttm Rate** and Set (Column Temp) and Reboiler Duty to **Active,**

* Set the (Reflux ratio) and (Ovhd Vap Rate) to **Estimate** only,

- **Make up H_2O:** T = 70 °F, P = 21.5 psia, 100% H_2O,
- Stream (DEA to Cool): Standard Ideal Liquid Volume Flow = 190 USGPM,
- **Cooler:** ΔP = 5 psi, Temp of (DEA to Rec) = 95 °F,
- **Set Operation:** Target = DEA to Rec, Variable = pressure,
- Source = To Absorber, Parameters: Multiplier = 1, Offset = −5.

What is the removal percent of H_2S?

15.4 Propylene Production

Propylene (propene) is produced adiabatically from propane dehydrogenation according to the following reaction [6, 7]:

$$C_3H_8 \rightarrow C_3H_6 + H_2 \tag{15.3}$$

The process flow diagram of the propylene production plant is shown in Figure 15.4. The conversion of propane is 40% in the vapor phase.

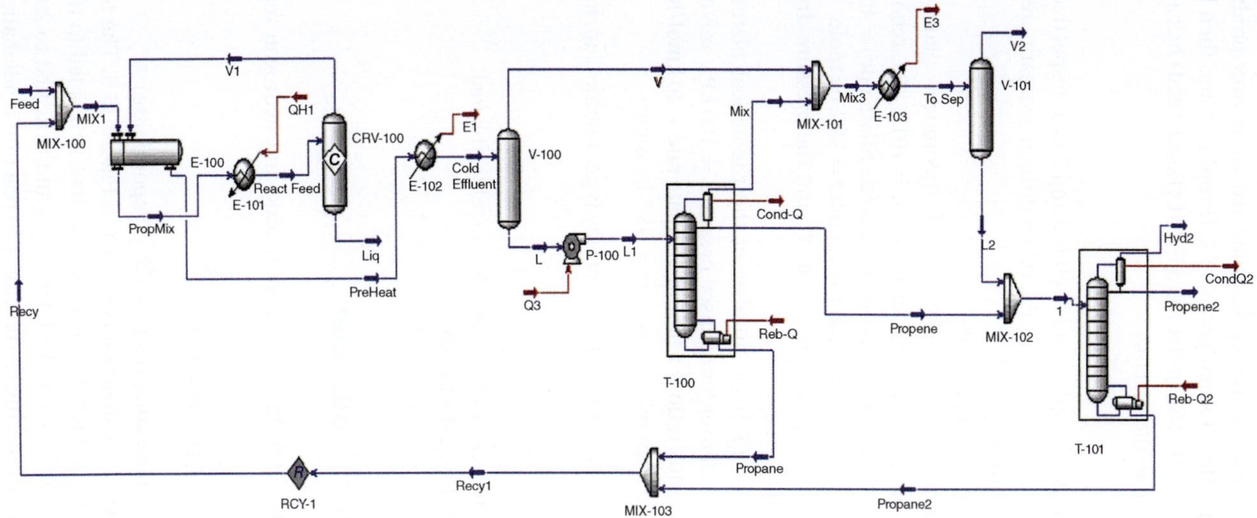

Figure 15.4 PFD of the propylene production plant.

The propane is supplied in the feed stream at (−45°C and 110 kPa). The flow of the Mix1 stream is 5000 kg/h, preheated using over-products (V1) of the reactor in the Heat Exchanger to 60°C (PropMix), and then heated to 600°C by a heater E-101. The outlet from E-101 (React-Feed) has a pressure of 105 kPa and is fed to the reactor.

Data:

- Heat Exchanger: ΔP for tubes and shell is 5 and 2 kPa, respectively,
- The temperature and pressure of the (Cold effluent) stream are (−75°C) and 1 atm, respectively,
- Pressure of Stream (L1) is 120 kPa,
- Distillation column (T-100): Number of theoretical plates (stages) are 20, Partial condenser, condenser pressure = 101.3 kPa, reboiler pressure = 110 kPa, reflux ratio = 3, propene flow (distillate rate) = 40 kmol/h, and the purity of the Propane from the reboiler = 98% by mole,
- Temperature of (To_Sep) stream is −80°C, and the pressure drop in the cooler2 (E-103) = 0.
- Distillation column (T-101): Number of the theoretical plates (stages) are 20, Partial condenser, condenser pressure = 100 kPa, reboiler pressure = 102 kPa, reflux ratio = 5, liquid distillate rate = 40 kmol/h, and the purity of propane from reboiler stream = 98% by mole.

* If the Distillation column (T-100) is of tray type, Bubble Cap, how many caps per active area?

Hint:

(1) Don't add the recycle until the all Flowsheet is finished.
(2) Fluid package: **Peng–Robinson.**

15.5 Methanol Production

Methanol is produced from the carbon dioxide and hydrogen reaction as follows [1, 8]:

$$CO_2 + 3H_2 \rightarrow CH_3OH + H_2O \tag{15.4}$$

(Conversion 20%, Base component = CO_2, Vapor phase)

where the process flow diagram is shown in Figure 15.5. The feed (0.25 CO_2 and 0.75 H_2 mole fractions) at 40°C and 4000 kPa is fed to the mixer, where the Mix stream has a flow of 3500 kg/h and is heated to 200°C by a heater (E-100) before entering the reactor. The reaction is adiabatic, and the product from the reactor is cooled to 40°C by a cooler (E-101), then the Vap1

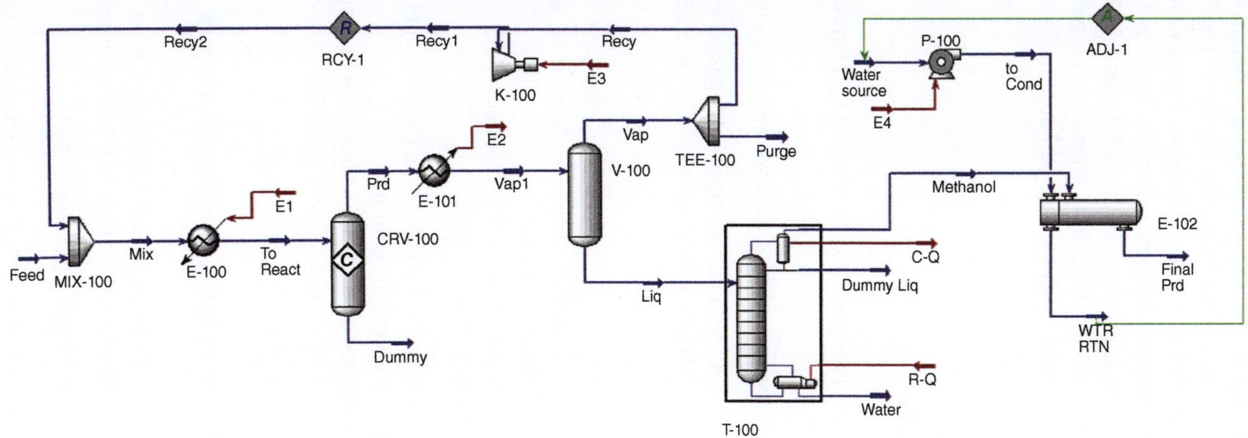

Figure 15.5 PFD of the methanol production plant.

stream is separated into vapor and liquid in the separator (V-100). The vapor is split in the TEE-100 and the Recy stream is compressed to 4000 kPa by a compressor (E-100), recycled and mixed with the feed. The liquid, which contains the most methanol and water, is sent to the distillation column (T-100) to separate the methanol from the water. The methanol is condensed and cooled to 40°C in the condenser (Heat Exchanger E-102). The water source (25°C, 101.3 kPa, and 10 000 kg/h) is raised to a pressure of 250 kPa by a pump (P-100) and used to condense the methanol in (E-102).

Data:

- Fluid Package: SRK,
- E-100: $\Delta P = 50$ kPa,
- Reactor: $\Delta P = 100$ kPa,
- E-101: $\Delta P = 50$ kPa,
- Parameters of TEE-100: Flow ratio: Purge = 0.0001 (*if it appears to be 0.000, neglect*).

Note: *Don't add the Recycle until the distillation column is converged.*

- Distillation column, T-100: Number of stages = 17, partial condensation,
- Condenser pressure = 1000 kPa, Reboiler pressure = 1015 kPa, liquid Distillate = 0 (Dummy),

Specifications:

1) Column Component Recovery: Draw = Methanol @Col1, Spec Value = 0.96, Component = Methanol
2) Component mole fraction of methanol in reboiler = 0.05
   - E-102: $\Delta P = 50$ kPa for tubes and shell.

(1) If the Distillation column is of packed type (Intalox saddles [ceramic] 0.5 in.), find the dimension of the column.
(2) What is the mass flow rate of the water source so as to keep the temperature of (WTR RTN) at 45°C? Use Adjust Operation.

15.6 Production of Vinyl_Chloride_Monomer

Vinyl_chloride_monomer, VCM, (C_2H_3Cl) is produced from the pyrolysis of 1,2-dichloroethane ($C_2H_4Cl_2$) according to the following reaction [8–11]:

$$C_2H_4Cl_2 \rightarrow HCl + C_2H_3Cl \text{ conversion} = 55\% \quad (15.5)$$

The process flow diagram is shown in Figure 15.6. Stream 1 at (20°C, and 30 bar) of pure $C_2H_4Cl_2$ is mixed with the Rec2 stream to produce stream 2,

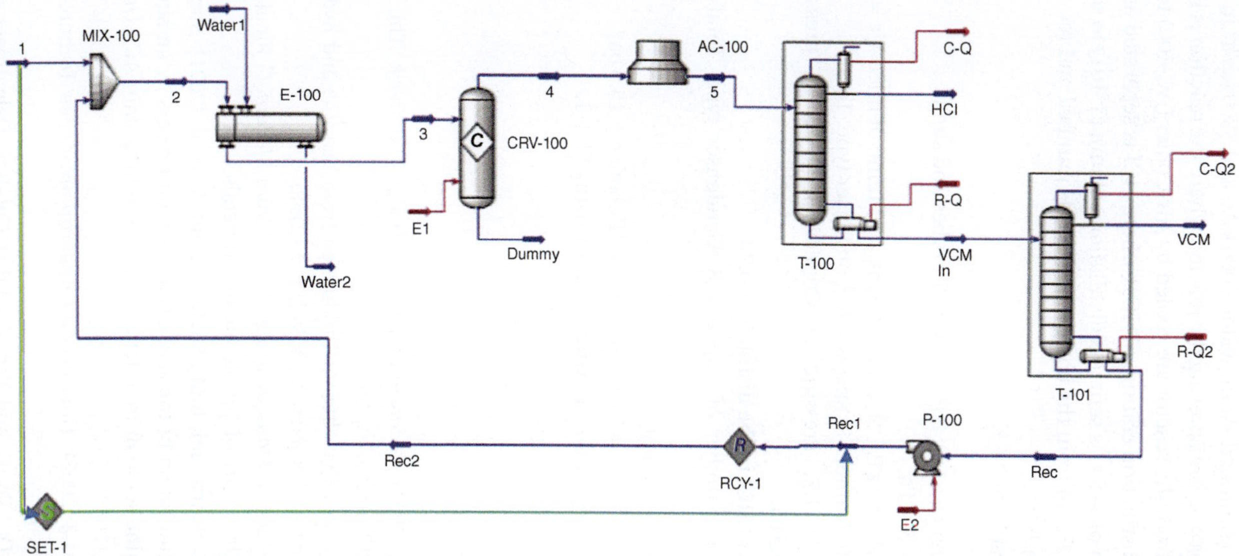

Figure 15.6 PFD of the vinyl_chloride_monomer production plant.

which has a mole flow rate of 500 kmol/h. Stream 2 is heated to 170°C using a stream water1 (saturated steam water, 1000 kPa, and 3800 kg/h) in a Heat Exchanger (E-100) before entering the reactor, where the reaction is isothermal. The gases from the reactor are cooled by Air Cooler (AC-100) to 70°C, then enter the distillation column (T-100) where HCl is separated as distillate. The bottom of T-100 is sent to a distillation column (T-101) to separate VCM. The Rec stream from the bottom of T-101 is pumped and recycled to mix with stream 1.

Data: Fluid Package: SRK

- Heat Exchanger (E-100): $\Delta P = 50$ kPa for tubes and shell,
- Reactor: $\Delta P = 50$ kPa,
- Air Cooler: $\Delta P = 50$ kPa, Number of Fans = 3, Temp of inlet Air = 20°C,
- Column T-100: Number of stages = 8, Condenser type: Total,
  Reflux ratio = 1.5, Pressure of Condenser = 25 bar, pressure of Reboiler = 25.5 bar,
  Mole fraction of HCl in the distillate = 0.99,
- Column T-101: Number of stages = 8, Condenser type: Total Reflux ratio = 1.5
  Distillate rate = 272 kmol/h,
  Pressure of Condenser = 15 bar, pressure of Reboiler = 15.5 bar
- Pressure of Rec1 stream is set with stream 1 using SET-1.

References

1 Speight, J.G. (2002). *Chemical and Process Design Handbook*. The McGraw-Hill Companies, Inc.

2 Jenkins, S. (2018). Cyclohexane production from benzene and hydrogen. *Chemical Engineering*, New York Vol. 125 (2): 8.

3 Ateah, M.A. (2015). *Manufacturing of DME from Methanol*" Final report. King Fahd University of Petroleum and Minerals.

4 Darani, N.S., Behbahani, R.M., Shahebrahimi, Y. et al. (2021). Simulation and optimization of the acid gas absorption process by an aqueous diethanolamine solution in a natural gas sweetening unit. *ACS Omega* 6 (18): 12072–12080.

5 HYSYS® 2004.2 (2005). *Tutorials and Applications*. Aspen Technology, Inc.

6 Bai, P., Liu, D., Wu, P., and Yan, Z. (2014). Catalytic Dehydrogenation of Propane to Propene: Catalyst Development, Mechanistic Aspects and

Reactor Design. *Reviews in Advanced Sciences and Engineering* 3 (2): 180–195.

7 Kaneko, S., Arakawa, T., Ohshima, M.A. et al. (2009). Dehydrogenation of propane combined with selective hydrogen combustion over Pt—Sn bimetallic catalysts. *Applied Catalysis A: General* 356 (1): 80–87.

8 Borisut, P. and Nuchitprasittichai, A. (2019). Methanol production via CO_2 hydrogenation: sensitivity analysis and simulation—based optimization. *Frontiers in Energy Research* 7: 81.

9 Dimian, A.C. and Bildea, C.S. (2008). *Chemical Process Design: Computer-Aided Case Studies*. Weinheim: Wiley-VCH Verlag GmbH & Co. KGaA.

10 Andliyani, L., Alfandi, I., Inda, N., Putri, K., and Muchtar, Z. (2023). Production of vinyl chloride from ethylene: technology review and process selection, *International Engineering Student Conference*, Palembang (September 2023).

11 Naqvi, M.K. and Kulshreshtha, A.K. (1995). Vinyl chloride manufacture: Technology Trends and an Energy Economic Perspective. *Polymer-Plastics Technology and Engineering* 34 (2): 213–226.

Index

Chemical Process Simulations using Aspen Hysys, First Edition. Khalid W. Hameed.

Printed and bound by CPI Group (UK) Ltd, Croydon, CR0 4YY
17/09/2026
14964330-0004